Thermodynamics and Statistical Mechanics

Equilibrium by Entropy Maximisation

Humpty Dumpty sat on a wall
Humpty Dumpty had a great fall
All the King's horses and all the King's men
Couldn't put Humpty together again.

Thermodynamics and Statistical Mechanics
Equilibrium by Entropy Maximisation

Phil Attard

Australian Research Council
University of South Australia

ACADEMIC PRESS

An imprint of Elsevier Science

Amsterdam Boston London New York Oxford Paris
San Diego San Francisco Singapore Sydney Tokyo

This book is printed on acid–free paper.

Academic Press
An Imprint of Elsevier Science
84 Theobald's Road, London WC1X 8RR, UK
http://www.academicpress.com

Academic Press
An Imprint of Elsevier Science
525 B Street, Suite 1900, San Diego, California 92101–4495, USA
http://www.academicpress.com

ISBN 0–12–066321–X

Library of Congress Control Number: 2002103392

A catalogue record for this book is available from the British Library

Printed and bound in Great Britain by MPG Books Ltd, Bodmin, Cornwall
02 03 04 05 06 07 MP 9 8 7 6 5 4 3 2 1

Thermodynamics and Statistical Mechanics
Equilibrium by Entropy Maximisation

Short Contents

Contents

Preface

Thermodynamics deals with the general principles and laws that govern the behaviour of matter and with the relationships between material properties. The origins of these laws and quantitative values for the properties are provided by statistical mechanics, which analyses the interaction of molecules and provides a detailed description of their behaviour. This book presents a unified account of equilibrium thermodynamics and statistical mechanics using entropy and its maximisation.

A physical explanation of entropy based upon the laws of probability is introduced. The equivalence of entropy and probability that results represents a return to the original viewpoint of Boltzmann, and it serves to demonstrate the fundamental unity of thermodynamics and statistical mechanics, a point that has become obscured over the years. The fact that entropy and probability are objective consequences of the mechanics of molecular motion provides a physical basis and a coherent conceptual framework for the two disciplines. The free energy and the other thermodynamic potentials of thermodynamics are shown simply to be the total entropy of a subsystem and reservoir; their minimisation at equilibrium is nothing but the maximum of the entropy mandated by the second law of thermodynamics and is manifest in the peaked probability distributions of statistical mechanics. A straightforward extension to nonequilibrium states by the introduction of appropriate constraints allows the description of fluctuations and the approach to equilibrium, and clarifies the physical basis of the equilibrium state.

Although this book takes a different route to other texts, it shares with them the common destination of explaining material properties in terms of molecular motion. The final formulae and interrelationships are the same, although new interpretations and derivations are offered in places. The reasons for taking a detour on some of the less-travelled paths of thermodynamics and statistical mechanics are to view the vista from a different perspective, and to seek a fresh interpretation and a renewed appreciation of well-tried and familiar results. In some cases this reveals a shorter path to known solutions, and in others the journey leads to the frontiers of the disciplines.

The book is basic in the sense that it begins at the beginning and is entirely self-contained. It is also comprehensive and contains an account of all of the modern techniques that have proven useful in modern equilibrium, classical statistical mechanics. The aim has been to make the subject matter broadly accessible to advanced students, whilst at the same time providing a reference text for graduate scholars and research scientists active in the field. The later chapters deal with more advanced applications, and while their details may be followed step-by-step, it may require a certain experience and sophistication to appreciate their point and utility. The emphasis throughout is on fundamental principles and upon the relationship between various approaches. Despite this, a deal of space is devoted to applications, approximations, and computational algorithms; thermodynamics and statistical mechanics were in the final analysis developed to describe the real world, and while their generality and universality are intellectually satisfying, it is their practical application that is their ultimate justification. For this reason a certain pragmatism that seeks to convince by physical explanation rather than to convict by mathematical sophistry pervades the text; after all, one person's rigor is another's mortis.

The first four chapters of the book comprise statistical thermodynamics. This takes the existence of weighted states as axiomatic, and from certain physically motivated definitions, it deduces the familiar thermodynamic relationships, free energies, and probability distributions. It is in this section that the formalism that relates each of these to entropy is introduced. The remainder of the book comprises statistical mechanics, which in the first place identifies the states as molecular configurations, and shows the common case in which these have equal weight, and then goes on to derive the material thermodynamic properties in terms of the molecular ones. In successive chapters the partition function, particle distribution functions, and system averages, as well as a number of applications, approximation schemes, computational approaches, and simulation methodologies, are discussed. Appended is a discussion of the nature of probability.

The paths of thermodynamics and statistical mechanics are well-travelled and there is an extensive primary and secondary literature on various aspects of the subject. Whilst very many of the results presented in this book may be found elsewhere, the presentation and interpretation offered here represent a sufficiently distinctive exposition to warrant publication. The debt to the existing literature is only partially reflected in the list of references; these in general were selected to suggest alternative presentations, or further, more detailed, reading material, or as the original source of more specialised results. The bibliography is not intended to be a historical survey of the field, and, as mentioned above, an effort has been made to make the book self-contained.

At a more personal level, I acknowledge a deep debt to my teachers, collaborators, and students over the years. Their influence and stimulation are impossible to quantify or detail in full. Three people, however, may be fondly acknowledged: Pat Kelly, Elmo Lavis, and John Mitchell, who in childhood, school, and PhD taught me well.

Chapter 1

Prologue

1.1 Entropy in Thermodynamics and Statistical Mechanics

All systems move in the direction of increasing entropy. Thus Clausius introduced the concept of entropy in the middle of the 19th century. In this – the second law of thermodynamics – the general utility and scope of entropy is apparent, with the implication being that entropy maximisation is the ultimate goal of the physical universe. Thermodynamics is based upon entropy and its maximisation. The fact that the direction of motion of thermal systems is determined by the increase in entropy differentiates thermodynamics from classical mechanics, where, as Newton showed in the 17th century, it is energy and its minimisation that plays the primary role.

It was quickly clear that entropy in some sense measured the disorder of a system, but it was not until the 1870's that Boltzmann articulated its physical basis as a measure of the number of possible molecular configurations of the system. According to Boltzmann, systems move in the direction of increasing entropy because such states have a greater number of configurations, and the equilibrium state of highest entropy is the state with the greatest number of molecular configurations. Although there had been earlier work on the kinetic theory of gases, Boltzmann's enunciation of the physical basis of entropy marks the proper beginning of statistical mechanics.

Despite the fact that thermodynamics and statistical mechanics have entropy as a common basis and core, they are today regarded as separate disciplines. Thermodynamics is concerned with the behavior of bulk matter, with its measurable macroscopic properties and the relationships between them, and with the empirical laws that bind the physical universe. These laws form a set of fundamental principles that have been abstracted from long experience. A relatively minor branch of thermodynamics deals with the epistemological consequences of the few axioms. For the most part pragmatism pervades the discipline, the phenomenological nature of the laws is recognised, and the main concern lies with the practical application of the results to specific systems.

Statistical mechanics is concerned with calculating the macroscopic properties of matter from the behaviour of the microscopic constituents. 'Mechanics' refers to the fact that the particles' interactions are described by an energy function or Hamiltonian, and that their trajectories behave according to the usual laws of motion, either classical or quantum. 'Statistical' denotes the fact that a measurement yields the value of an observable quantity averaged over these trajectories in time.

A typical macroscopic system has on the order of 10^{23} molecules. It is not feasible to follow the motion of the individual particles, and in any case the number of measurable thermodynamic parameters of the system is very much less than the number of possible configurations of the particles. In theory and in practice the microscopic states of the system are inaccessible. One instead seeks the probability distribution of the microstates, and the consequent macroscopic quantities follow as averages over these distributions. Despite its probabilistic nature, statistical mechanics is able to make precise predicative statements because the number of microstates is so huge that the relative fluctuations about the average are completely negligible.

Entropy has always remained central to thermodynamics, but nowadays it has been relegated to a secondary role in statistical mechanics, where it generally emerges as just another thermodynamic property to be calculated from the appropriate probability distribution. This is unfortunate because what has become obscured over the years is the fundamental basis of these probability distributions. In fact these distributions are properly derived by entropy maximisation; indeed from this basis can be derived all of equilibrium statistical mechanics. Likewise obscured in both disciplines is the nature of the various thermodynamic potentials and free energies. Despite their frequent recurrence throughout thermodynamics and statistical mechanics, it is not commonly appreciated that they represent nothing more than the maximum entropy of the system. The physical interpretation of entropy as the number or weight of the microscopic states of the system removes much of the mystery that traditionally shrouds it in thermodynamics.

The theme of this book is entropy maximisation. One aim is to show the unity of thermodynamics and statistical mechanics and to derive both from a common basis. A second goal is to derive both disciplines precisely and in their entirety, so that all of the various quantities that occur are defined and fully interpreted. A benefit of using the present self-contained maximum entropy approach consistently is the transparency of the results, which makes for relatively straightforward generalisations and extensions of the conventional results.

To begin, a simple example that clarifies the nature of entropy and the rationale for entropy maximisation will be explored. The basic notions of states, weights, and probability are then formally introduced and defined, and *the* fundamental equation for entropy that is the foundation of thermodynamics and statistical mechanics is given. The particular reservoir and constraint methodology used in the following thermodynamic chapters is presented in a general context and illustrated with reference to the initial example.

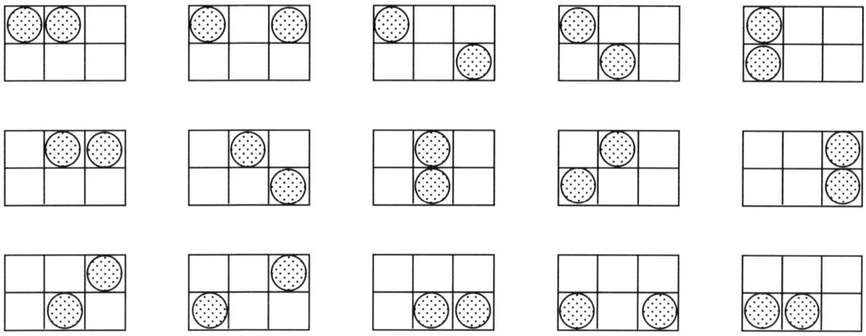

Figure 1.1: The $6!/2!4! = 15$ configurations of 2 eggs in a 6 egg carton.

1.2 An Example: Scrambled Eggs

The direction toward equilibrium is that of increasing entropy, and systems remain at equilibrium once they attain it. Withstanding even royal command, the ultimate state initiated by Humpty Dumpty's fall was irreversible, as the anonymous author clearly noted. Nowadays the fractured, final fate of the unfortunate ovoid is recognised as a state of high entropy. Here the nature of entropy and the connection between entropy and equilibrium is illustrated with a simple example.

Suppose that n eggs are placed in a carton able to hold N eggs. The number of distinct arrangements, $W_N(n)$, is the number of ways of choosing n objects from N in any order. This is $W_N(n) = {}^NC_n \equiv N!/n!(N-n)!$ (cf. Fig. 1.1). This arises because there are N possible cells in which the first egg can be placed, which leaves $N-1$ for the second, ..., and finally $N-n+1$ possible cells for the last egg. That is, there are $N!/(N-n)!$ ways of placing eggs in the carton in order. However, these eggs are indistinguishable, and the $n!$ permutations of the eggs cannot be counted as distinct configurations. This gives the remaining term in the denominator.

Suppose that the eggs are scrambled by choosing two cells at random and swapping their contents.[1] Obviously if both are empty or both are occupied nothing happens, but if only one is occupied a new configuration is generated. It will prove convenient to deal with the logarithm of the number of configurations, rather than the number of configurations itself. Accordingly one defines

$$S_N(n) = k_B \ln W_N(n). \tag{1.1}$$

This quantity is in fact the entropy (k_B is Boltzmann's constant), but for the present one need only note that because the logarithm is a monotonic function of its argument, maximizing the number of configurations is the same as maximizing the entropy.

[1] In general the transition rule determines the microstate weights and as such it is central to equilibrium statistical mechanics.

Now introduce a second carton of size M containing m eggs. The number of configurations for this carton is $W_M(m) = M!/m!(M-m)!$, and its entropy is $S_M(m) = k_B \ln W_M(m)$. The total number of distinct configurations if the two cartons are kept separate is the product $W_{N,M}(n,m) = W_N(n)W_M(m)$, since for each of the $W_M(m)$ configurations of the second system there are $W_N(n)$ configurations of the first. The total entropy is the sum of the two individual entropies, $S_{N,M}(n,m) = k_B \ln W_{N,M}(n,m) = S_N(n) + S_M(m)$. This shows the advantage of working with entropy rather than the number of configurations. Like the number of eggs, or the number of cells, it is an additive quantity, and such quantities in general are easier to deal with than products.

1.2.1 Equilibrium Allocation

What happens if interchange of eggs between the cartons is allowed via the scrambling procedure described above? Intuitively one expects that the carton with the greatest concentration of eggs will lose them to the less concentrated carton. Eventually a steady state will be reached where the eggs are as likely to be transferred in one direction as another. Here one expects the two cartons to have an equal concentration of eggs, and this steady state is called the equilibrium state. The reason that concentration is the determining quantity rather than simply the number of eggs is because at equilibrium a large carton will have proportionally more eggs than a small carton. In fact, each time an egg goes from one carton to the other an unoccupied cell goes in the opposite direction, which suggests that the steady state will treat occupied and unoccupied cells in an equivalent fashion.

To be more precise one needs to calculate the probability of moving an egg between the cartons. The probability of a cell chosen at random in the first carton being occupied is just the ratio of the number of eggs to the number of cells, namely n/N, and similarly the chance of choosing a free cell is $(N-n)/N$. For an interchange between cartons to occur the two cells must be in different cartons. The probability of choosing any cell in the first carton is $N/(N+M)$, and the probability of choosing any cell in the second carton is $M/(N+M)$. Hence the probability of the two chosen cells being in different cartons is $2NM/(N+M)^2$, the factor of 2 arising because it doesn't matter which carton is chosen first. For an egg to leave the first carton, one must choose different cartons *and* both an occupied cell in the first carton and an unoccupied cell in the second. The chance of this is just the product of the probabilities, $[2NM/(N+M)^2](n/N)[(M-m)/M] = 2n(M-m)/(N+M)^2$. Conversely, the probability of an egg going from the second carton to the first is $2m(N-n)/(N+M)^2$. For the equilibrium or steady state situation the net flux must be 0, so these two must balance. The equilibrium number of eggs in each carton is denoted by $\overline{n}$ and $\overline{m}$. Equating the fluxes, the equilibrium condition is

$$\frac{\overline{n}}{N-\overline{n}} = \frac{\overline{m}}{M-\overline{m}}, \text{ or } \frac{\overline{n}}{N} = \frac{\overline{m}}{M}. \tag{1.2}$$

The concentration of eggs in each carton is indeed equal at equilibrium, as is the ratio of occupied cells to free cells. (For simplicity, one may assume that the numbers have been chosen so that this equilibrium condition possesses an integral solution.)

1.2.2 Maximum Entropy

The number of configurations, and hence the entropy, is a maximum at the equilibrium allocation. This may be proven by showing that the number of configurations decreases monotonically moving away from equilibrium. For the case that there are too many eggs in the first carton, define $l = n - \overline{n} > 0$. If the number of configurations corresponding to n is greater than the number of configurations for $n+1$, then the entropy is increasing moving toward equilibrium. This may be proven by taking the ratio of the respective number of configurations,

$$\frac{W_N(n)W_M(m)}{W_N(n+1)W_M(m-1)} \tag{1.3}$$

$$\begin{aligned}
&= \frac{W_N(\overline{n}+l)W_M(\overline{m}-l)}{W_N(\overline{n}+l+1)W_M(\overline{m}-l-1)} \\
&= \frac{(\overline{n}+l+1)!(N-\overline{n}-l-1)!}{(\overline{n}+l)!(N-\overline{n}-l)!}\frac{(\overline{m}-l-1)!(M-\overline{m}+l+1)!}{(\overline{m}-l)!(M-\overline{m}+l)!} \\
&= \frac{\overline{n}+l+1}{N-\overline{n}-l}\frac{M-\overline{m}+l}{\overline{m}-l} \\
&> \frac{\overline{n}}{N-\overline{n}}\frac{M-\overline{m}}{\overline{m}} \\
&= 1.
\end{aligned} \tag{1.4}$$

Hence $S_N(\overline{n}+l) + S_M(\overline{m}-l) > S_N(\overline{n}+l+1) + S_M(\overline{m}-l-1)$, and an analogous argument with l changed to $-l$ gives $S_N(\overline{n}-l) + S_M(\overline{m}+l) > S_N(\overline{n}-l+1) + S_M(\overline{m}+l-1)$. That is, the entropy decreases moving away from equilibrium, which is to say that it is a concave function of the allocation of eggs that attains its maximum at the equilibrium allocation.

The total number of distinct configurations in the case that eggs are transferable between the two cartons is $W_{N+M}(n+m) = {}^{N+M}C_{n+m}$, since all cells are now accessible to all eggs, and the entropy is $S_{N+M}(n+m) = k_{\mathrm{B}} \ln W_{N+M}(n+m)$. The total number of configurations of the combined cartons must be greater than any of those of the isolated cartons with a fixed allocation of the eggs because it includes all the configurations of the latter, plus all the configurations with a different allocation of the eggs. That is, $W_{N+M}(n+m) \geq W_N(n)W_M(m)$ or in terms of the entropy, $S_{N+M}(n+m) \geq S_N(n)+S_M(m)$, for any n, m. (This is a strict inequality unless N or M is 0, or unless there are no eggs, $n+m=0$, or no empty cells, $n+m = N+M$.) One concludes that the entropy of the two cartons able to exchange eggs is greater than the entropy of the two isolated cartons with the equilibrium allocation of eggs, which is greater than the

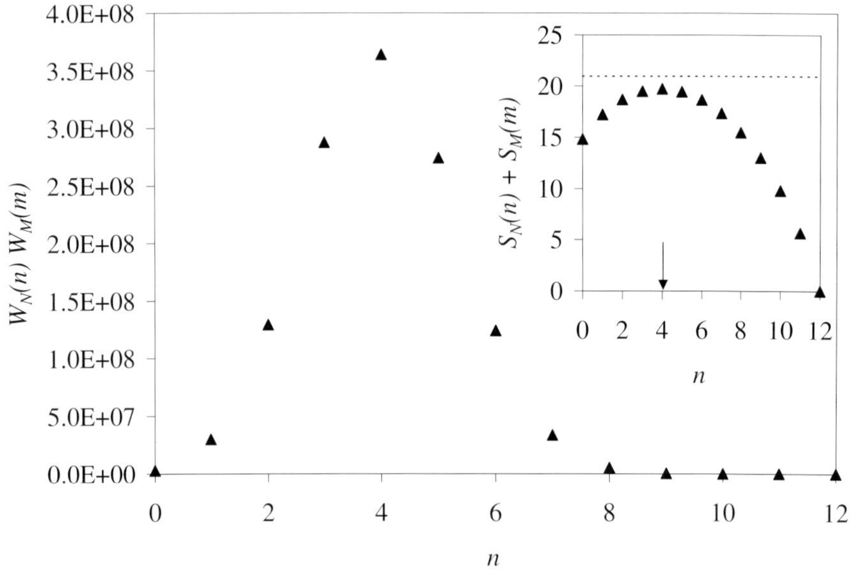

Figure 1.2: The number of configurations as a function of the number of eggs in the first carton, for a dozen eggs in total, $n + m = 12$, allocated to an ordinary carton, $N = 12$, and a twin carton, $M = 24$. The inset shows the constrained total entropy, in units of k_{B}, with the arrow indicating the equilibrium occupancy, $\overline{n} = 4$, and the dotted line giving the unconstrained entropy, $\ln[36!/12!\,24!]$.

entropy of any other allocation of eggs to the isolated cartons,

$$S_{N+M}(n+m) \geq S_N(\overline{n}) + S_M(\overline{m}) \geq S_N(n) + S_M(m), \tag{1.5}$$

where $\overline{n} + \overline{m} = n + m$, and $\overline{n}/N = \overline{m}/M$. This behaviour of the constrained entropy is shown in Fig. 1.2.

1.2.3 Direction of Motion

It has been shown that the equilibrium or steady state is the one with the maximum entropy, and now it is shown that the system out of equilibrium is most likely to move in the direction of increasing entropy. The probability that the N carton with n eggs will gain an egg, $\wp(n+1|n)$, is required. (Obviously the other carton will simultaneously lose an egg, $m \rightarrow m - 1$.) This was given above in equating the fluxes to find the equilibrium state,

$$\wp(n+1|n) = \frac{2m(N-n)}{(N+M)^2}. \tag{1.6}$$

Similarly,

$$\wp(n-1|n) = \frac{2n(M-m)}{(N+M)^2}. \tag{1.7}$$

The probability that the number of eggs in each carton is unchanged is obviously $\wp(n|n) = 1 - \wp(n+1|n) - \wp(n-1|n)$. Now if there are too few eggs in the first carton compared to the equilibrium allocation, $n < \overline{n}$ and $m > \overline{m}$, then the odds of increasing the number of eggs in this carton are

$$\begin{aligned} \frac{\wp(n+1|n)}{\wp(n-1|n)} &= \frac{m(N-n)}{n(M-m)} \\ &> \frac{N-\overline{n}}{\overline{n}} \frac{\overline{m}}{M-\overline{m}} \\ &= 1. \end{aligned} \tag{1.8}$$

Hence it is more likely that the transition $n \rightarrow n+1$ will be made than $n \rightarrow n-1$ if $n < \overline{n}$ and $m > \overline{m}$. On the other side of equilibrium one similarly sees that the transition tends in the opposite direction, so that in both cases the most likely transition is toward the equilibrium allocation of eggs. Given the monotonic decrease of entropy away from equilibrium, one concludes that the most likely flux of eggs is in the direction of increasing entropy.

1.2.4 Physical Interpretation

How are these results to be interpreted? First one distinguishes between a microstate and a macrostate. A microstate is one of the configurations of the eggs, (i.e., a specification of the occupied cells). A macrostate is a specification of the number of eggs in each carton, but not the cells that they occupy. Hence there are many microstates for each macrostate. Specifying a macrostate corresponds to isolating the cartons and to disallowing the swapping of eggs between them. This acts as a constraint on the configurations that the system can assume, and hence the number of configurations for a specified macrostate is less than the number of configurations if no macrostate is specified. In other words, the entropy of a system constrained to be in a given macrostate is less than the entropy of an unconstrained system.

The flux of eggs from the initial allocation was interpreted as the approach to equilibrium. This did not occur because the total number of configurations or the total entropy was increasing; once the cartons were allowed to exchange eggs the number of possible configurations was fixed. Rather, the approach to equilibrium was a macroscopic flux, and isolating the cartons at any instant would have given a macrostate with a number of configurations larger than before. Hence it is the entropy constrained by the current value of the quantity in flux that increases approaching equilibrium, and that reaches its maximum value for the equilibrium macrostate. Obviously one can only make such statements on average, since there is nothing to prevent a temporary reversal in the flux direction due to the transitions between microstates as one approaches equilibrium. Likewise one expects fluctuations about the equilibrium macrostate once it is attained.

The preceding paragraph describes the increase in the constrained entropy during the approach to equilibrium, but it does not explain why such an increase occurs. As mentioned, once the cartons are allowed to exchange eggs the total number of possible configurations and the unconstrained entropy is fixed, and these do not drive the approach to equilibrium. As far as the scrambling of the eggs is concerned, given a microstate there is nothing to choose between the transitions to any of the adjacent microstates, and each of these is as likely as the reverse transition. At the level of microstates there is no flux and no equilibration. The flux is occurring in the macrostates, and the key observation is that the number of microstates differs between macrostates. Hence it is the transition between macrostates that is asymmetric: if macrostates 1 and 3 are on either side of macrostate 2, and the number of corresponding microstates increases from 1 to 2 to 3, then a system in a microstate corresponding to 2 is more likely to make a transition to a microstate corresponding to 3 than to one corresponding to 1 simply because there are more of the former than the latter.[2] The transition to a macrostate with more microstates is more probable than the reverse transition (if the macrostates are ordered in terms of increasing number of microstates). Although nothing prevents the opposite transition at any instant, in the long run it is the most probable transitions between macrostates that will occur most frequently. It is the increase in configurations corresponding to the macrostates that gives an irreversible character to a macroscopic flux, and the consequent increase in the constrained entropy of the system. The equilibrium macrostate is that with the highest number of microstates. While fluctuations to nearby macrostates occur, usually the peak in the curve of the number of configurations is so sharp that the effects of such fluctuations are negligible.

1.3 Basic Notions

In this section the fundamental ideas that provide a basis for thermodynamics and statistical mechanics are set out. The concepts of states, probability, and entropy are introduced in turn.

1.3.1 States

A system possesses a fundamental set of states called *microstates* that are distinct and indivisible. Distinct means that each microstate bears a unique label, and indivisible means that no finer subdivision of the system is possible. These discrete states are ultimately quantum in nature, but one may pass to the classical continuum limit, in which case the archetypal microstate could be a position-momentum cell of fixed volume in phase space. Here the theory will initially be developed in a general and abstract way for the discrete case, as a precursor to the continuum results of classical statistical mechanics.

The *macrostates* of the system are disjoint, distinct sets of microstates. In general they correspond to the value of some physical observable, such as the energy or density of some part of the system, and they are labelled by this observable. Disjoint means that different macrostates have no microstates in

[2]This assumes that each target microstate has approximately the same number of possible source microstates (cf. the preceding footnote).

common, and distinct means that no two macrostates have the same value of the observable. In addition to macrostates, there may exist states that are sets of microstates but which are not disjoint or are not distinct.

The microstate that the system is in varies over time due to either deterministic or stochastic transitions. In consequence, transitions also occur between the macrostates of the system. The set of all states that may be reached by a finite sequence of transitions defines the possible states of the system. Hence in time the system will follow a trajectory that eventually passes through all the possible states of the system. A transition rule for the microstates may be reversible (i.e., symmetric between the forward and the reverse transitions), but will yield statistically irreversible behaviour in the macrostates over finite times.

1.3.2 Probability

A system may be considered in isolation from its surroundings. Its state is specified by the values of certain fixed quantities such as size, composition, energy, and momentum. Each of the microstates of a system has a nonnegative weight that may depend upon the state of the system. It is conventional to take the microstates of an isolated system to have equal weight, as in the ergodic hypothesis or the hypothesis of equal *a priori* probability of classical statistical mechanics, but it will be shown here that the formalism of statistical mechanics holds even for nonuniformly weighted microstates. Ultimately the weight of the microstates is determined by the transitions between them. It is the task of statistical mechanics to identify explicitly the transition rule and to construct the weights, whereas the formalism of statistical thermodynamics only requires that the microstates and their weight exist.

The probability that the system is in a particular microstate is proportional to its weight. That is, if the weight of the microstate labelled i is ω_i, and the total weight of the system is $W = \sum_i \omega_i$, then the microstate probability is

$$\wp_i = \frac{\omega_i}{W}. \tag{1.9}$$

Obviously the probability is normalised, $\sum_i \wp_i = 1$. In the event that the microstates are equally weighted, then up to an immaterial scale factor one can take the weight of the accessible microstates to be unity. (The inaccessible microstates may either be excluded from consideration, or else be given 0 weight.) In this case W is just the total number of accessible microstates, and $\wp_i = 1/W$. That is to say, uniform weight is the same as equal probability, and the probability of any one microstate is one over the total number of microstates. In the general case, the microstate weights and the total weight are dependent on the state of the system.

The theory of probability is ultimately founded upon set theory, which is the reason that macrostates were introduced as sets of microstates. The weight of a macrostate α is the sum of the weights of the corresponding microstates, $\omega_\alpha = \sum_{i\in\alpha} \omega_i$.[3] Consequently, its probability is the sum of the probabilities of

[3] Although mathematical symbolism provides a concise, precise way of communicating,

the corresponding microstates, $\wp_\alpha = \sum_{i\in\alpha} \wp_i = \omega_\alpha/W$. In the event that the microstates are equally likely this is just $\wp_\alpha = n_\alpha/W$, where $n_\alpha = \sum_{i\in\alpha}$ is the number of corresponding microstates. It is emphasised that α indexes distinct macrostates.

As just mentioned, the macrostates represent sets of microstates, and the rules for combining probabilities are derived from set theory. The nature of probability is discussed in detail in Appendix A. Briefly, an arbitrary state (set of microstates) may be denoted by a, and its probability by $\wp(a)$. The complement of a set a is $\tilde{a}$, and the probability of the system *not* being in a particular state is $\wp(\tilde{a}) = 1 - \wp(a)$. The probability of the system being in either a or b or both, which is their union $a + b$, is the sum of their probabilities less that of the common state, $\wp(a+b) = \wp(a) + \wp(b) - \wp(ab)$. The probability of the system being in both a and b (their intersection) is $\wp(ab) = \wp(a|b)\wp(b)$, where the conditional probability $\wp(a|b)$ is read as the probability of a given b.

Explicitly, given that a system is in the macrostate α, the probability that it is in a particular microstate $i \in \alpha$ is $\wp(i|\alpha) = \wp(i\alpha)/\wp(\alpha)$. Because the microstate i is a set entirely contained in the set represented by the macrostate α, their conjunction is $i\alpha = i$, and the numerator is just the probability of finding the system in the state i, $\wp(i\alpha) = \wp(i) = \omega_i/W$. The probability of the macrostate is proportional to the total weight of corresponding microstates, $\wp(\alpha) = \omega_\alpha/W$, so that the conditional probability in this case is $\wp(i|\alpha) = \omega_i/\omega_\alpha$. That is, the probability of a microstate given that the system is in a particular macrostate is the microstate's relative proportion of the macrostate's weight. In the event of equally likely microstates this reduces to $\wp(i|\alpha) = 1/n_\alpha$, which is the expected result: the probability that the system is in a particular one of the equally likely microstates given that it is in a particular macrostate is just one over the number of microstates composing the macrostate. (Note that a more precise notation would append the condition $i \in \alpha$ to the right of a vertical bar everywhere above.)

1.3.3 Entropy

The entropy of a system is defined to be the logarithm of the total weight,

$$S = k_{\mathrm{B}} \ln W. \tag{1.10}$$

This equation may be regarded as the basis of thermodynamics and statistical mechanics. If the microstates are equally likely, then this reduces to Boltzmann's original definition: the entropy is the logarithm of the total number of microstates. The prefactor is arbitrary and on aesthetic grounds would be best set to unity. However for historical reasons it is given the value $k_{\mathrm{B}} =$

in general its meaning is not independent of the context. In this book a symbol is often used simultaneously as a variable and as a distinguishing mark. In this case i and α are variables that also serve to distinguish the microstate weight ω_i from the macrostate weight ω_α. Whether $f(a)$ and $f(b)$ are the same function with arguments of different values or different functions symbolically distinguished by the appearance of their arguments depends upon the context.

1.38×10^{-23} J K^{-1}, and is called Boltzmann's constant. (It will be seen that there is a close connection between entropy, energy, and temperature, which accounts for the dimensions of entropy.)

By analogy, the entropy of a macrostate labelled α is defined as $S_\alpha = k_B \ln \omega_\alpha$. Again for equally likely microstates this is just the logarithm of the number of encapsulated microstates. One can also define the entropy of a microstate as $S_i = k_B \ln \omega_i$, which may be set to 0 if the microstates are equally likely.

The important, indeed essential, thing is the weights; it was shown above that these determine the probability of the microstates and macrostates. All of thermodynamics and statistical mechanics could be derived without ever introducing entropy, since by the above formulae the two are equivalent. However entropy is more convenient than the number of states because by definition it is a linear additive property: the total weight of two independent subsystems is $W_{\text{total}} = W_1 W_2$, whereas the total entropy is $S_{\text{total}} = S_1 + S_2$. Other linear additive quantities include the energy, volume, and particle number. It will be demonstrated in the following chapters that additivity is central to the formal development of thermodynamics and statistical mechanics.

The relationship between entropy and disorder or unpredictability is revealed by the interpretation of entropy as the number of equally likely microstates. If the system is only ever found in a few states, then those states occur with high probability: the system is said to be ordered or predictable, and it consequently has low entropy. Conversely, if a system is equally likely found in any of a large number of states, then it is quite disordered and unpredictable, and it has a high entropy. The entropy of a system with only one state is defined to be 0, and systems with more than one accessible state have an entropy that is strictly positive.

The entropy of the system may be rewritten in terms of the probability of the disjoint macrostates. (Note that the present treatment considers only *distinct* macrostates, so that all the microstates corresponding to the value α of the observable are collected together, which contrasts with many conventional treatments in which different values of the index do not necessarily correspond to different macrostates.) If macrostate α has weight ω_α, probability $\wp_\alpha = \omega_\alpha / W$, and entropy $S_\alpha = k_B \ln \omega_\alpha$, then the expression for the total entropy may be rearranged as

$$\begin{aligned} S[\wp] &= k_B \ln W \\ &= k_B \sum_\alpha \frac{\omega_\alpha}{W} \ln W \\ &= k_B \sum_\alpha \frac{\omega_\alpha}{W} \left[\ln \omega_\alpha + \ln \frac{W}{\omega_\alpha} \right] \\ &= \sum_\alpha \wp_\alpha \left[S_\alpha - k_B \ln \wp_\alpha \right]. \end{aligned} \tag{1.11}$$

This is the most general expression for the entropy of a system in terms of the probabilities of disjoint macrostates. The first term in the brackets represents the internal entropy of the macrostate, and the second term accounts for the disorder due to the system having a number of macrostates available. The internal entropy of the state α is $S_\alpha = k_B \ln \omega_\alpha$, and it arises from the conditional disorder or unpredictability associated with being in a macrostate, since this is insufficient to delineate precisely the actual microstate of the system. If the probability distribution is sharply peaked about one macrostate, then the total entropy equals the internal entropy of this macrostate, since the second term vanishes because $\ln 1 = 0$. One can replace the sum over macrostates by a sum over microstates, $S[\wp] = \sum_i \wp_i [S_i - k_B \ln \wp_i]$. For equally likely microstates, $S_i = 0$ and $\wp_i = 1/W$, it is clear that this immediately reduces to Boltzmann's original result, $S = k_B \ln W$. However, the derivation shows that in all circumstances it reduces to Boltzmann's result anyway. This point must be emphasised: in the general case of nonuniform microstates and macrostates, the entropy is a physical property of the system (it is the logarithm of the total weight) and its value is not changed by writing it as a functional of the macrostate probability, or by how the microstates are grouped into macrostates.

In the literature one often sees the expression

$$S_{GS}[\wp] = -k_B \sum_\alpha \wp_\alpha \ln \wp_\alpha. \tag{1.12}$$

This equation was originally given by Gibbs, who called it the average of the index of probability.[4] It was also derived by Shannon in his mathematical theory of communication, in which context it is called the information entropy.[5] Comparison with Eq. (1.11) shows that the internal entropy of the macrostate is missing from this formula. Hence the Gibbs–Shannon entropy must be regarded as the 'external' part of the total entropy. It should be used with caution, and depending upon the context, one may need to explicitly add the internal contribution to obtain the full entropy of the system.

In the present approach to thermodynamics and statistical mechanics, the only expressions for the entropy of the system that will be used are Eq. (1.10) and Eq. (1.11). The Gibbs–Shannon expression, Eq. (1.12), is often invoked in the principle of maximum entropy,[6] where it is used to obtain the macrostate probability (by maximisation of the entropy functional). In the present formulation, the macrostate probability may be trivially expressed in terms of the macrostate entropy, namely it is proportional to the exponential of the macrostate entropy,

$$\wp_\alpha = \frac{\omega_\alpha}{W} = \frac{1}{Z} e^{S_\alpha / k_B}, \tag{1.13}$$

[4] J. W. Gibbs, *Elementary Principles in Statistical Mechanics Developed with Special Reference to the Rational Foundation of Thermodynamics*, Yale Univ. Press, New Haven, CT, 1902; Dover, New York, 1960.

[5] C. E. Shannon and W. Weaver, *The Mathematical Theory of Communication*, Univ. of Illinois Press, Urbana, 1949.

[6] E. T. Jaynes, Information theory and statistical mechanics, *Phys. Rev.* **106** (1957), 620; **108** (1957), 171. R. D. Rosenkrantz (Ed.), *E. T. Jaynes: Papers on Probability, Statistics, and Statistical Physics*, D. Reidel, Dordrecht, 1983.

where the normalising factor is

$$Z = \sum_\alpha e^{S_\alpha/k_\mathrm{B}} = \sum_\alpha \omega_\alpha = W. \tag{1.14}$$

This expression for the macrostate probability will be used throughout, although it ought be clear that it is entirely equivalent to the original definition that the probability is proportional to the number or weight of corresponding microstates.

The Continuum

It is often the case that the states are relatively close together so that it is desirable to transform from the discrete to the continuum, as is necessary in the case of classical statistical mechanics. It is only meaningful to perform such a transformation when the state functions vary slowly between nearby states. One could simply take the continuum limit of the discrete formulae given above, transforming sums to integrals in the usual fashion, or one can derive the results for the continuum directly, as is done here.

Represent the state of the system by $\mathbf{x}$, a point in a multidimensional space, and let $\omega(\mathbf{x})$ be the nonnegative weight density measured with respect to the volume element $\mathrm{d}\mathbf{x}$. The total weight of the system is

$$W = \int \mathrm{d}\mathbf{x}\, \omega(\mathbf{x}), \tag{1.15}$$

and the total entropy is $S = k_\mathrm{B} \ln W$, as usual. The probability density is

$$\wp(\mathbf{x}) = \frac{\omega(\mathbf{x})}{W}, \tag{1.16}$$

with the interpretation that $\wp(\mathbf{x})\mathrm{d}\mathbf{x}$ is the probability of the system being within $\mathrm{d}\mathbf{x}$ of $\mathbf{x}$. Accordingly the average of a function of the state of the system is $\langle f \rangle = \int \mathrm{d}\mathbf{x}\, \omega(\mathbf{x}) f(\mathbf{x})$.

One can define the entropy of the state of the system as

$$S(\mathbf{x}) = k_\mathrm{B} \ln \left[\omega(\mathbf{x})\Delta(\mathbf{x})\right], \tag{1.17}$$

in terms of which the probability density is

$$\wp(\mathbf{x}) = \frac{e^{S(\mathbf{x})/k_\mathrm{B}}}{\Delta(\mathbf{x})W}. \tag{1.18}$$

Here $\Delta(\mathbf{x})$ is an arbitrary volume element introduced solely for convenience. (It makes the argument of the logarithm dimensionless and gives the probability density the correct dimensions.) Obviously the probability density is independent of the choice of $\Delta(\mathbf{x})$, since the one that appears explicitly cancels with the one implicit in the definition of the state entropy. The volume element has no physical consequences and is generally taken to be a constant; equally, it could be ignored altogether.

The system entropy may be written as an average of the state entropy,

$$\begin{aligned} S &= k_{\mathrm{B}} \ln W \\ &= \int \mathrm{d}\mathbf{x}\, \wp(\mathbf{x}) k_{\mathrm{B}} \ln W \\ &= \int \mathrm{d}\mathbf{x}\, \wp(\mathbf{x}) \left[S(\mathbf{x}) - k_{\mathrm{B}} \ln \frac{e^{S(\mathbf{x})/k_{\mathrm{B}}}}{W} \right] \\ &= \int \mathrm{d}\mathbf{x}\, \wp(\mathbf{x}) \left[S(\mathbf{x}) - k_{\mathrm{B}} \ln \{ \wp(\mathbf{x}) \Delta(\mathbf{x}) \} \right]. \end{aligned} \tag{1.19}$$

This is the continuum analogue of Eq. (1.11). Although the arbitrary volume element appears explicitly here, it cancels with the corresponding term in $S(\mathbf{x})$, and the total entropy is independent of the choice of $\Delta(\mathbf{x})$.

The macrostates in the case of the continuum are generally represented by hypersurfaces of the space. On the hypersurfaces particular observables that depend upon the state of the system have constant values. Although conceptually similar to the discrete case, the treatment of continuum macrostates can be more complicated in detail (see Ch. 5).

The formulae given above for entropy and probability are surprisingly powerful, despite their evident simplicity and that of their derivation. All of thermodynamics and statistical mechanics is based upon these results and the notions that underlie them.

1.4 Reservoirs

1.4.1 An Example: Egg Distribution

The example explored above, which may seem paltry, is in fact quite rich, and here it is used to introduce the reservoir formalism invoked throughout this book. In the example, the microstates, which are the distinct arrangements of eggs in the carton cells, are all equally likely, and the macrostates, which are the number of eggs in each carton, have a probability in proportion to the corresponding number of microstates. Hence the probability of there being n eggs in the N carton, given that the other carton can hold M eggs and that there is a total of $n+m$ eggs, is

$$\wp(n|N, M, n+m) = {}^N C_n \, {}^M C_m \Big/ \sum_{n=0}^{N} {}^N C_n \, {}^M C_m \, . \tag{1.20}$$

It was shown that the equilibrium allocation, $\overline{n} = N(n+m)/(N+M)$, maximises the number of configurations, and this gives the peak of the distribution. Taking the logarithm of the above probability distribution, and making Stirling's approximation for the factorials, an expansion to second order about $\overline{n}$ gives a quadratic form that when reexponentiated approximates the distribution by a Gaussian,

$$\wp(n|N, M, n+m) \approx Z^{-1} \exp \left[\frac{-N(M+N)}{2\overline{n}(N-\overline{n})M} (n-\overline{n})^2 \right], \tag{1.21}$$

where Z is the normalisation factor.

These two expressions are, respectively, the exact probability distribution and a Gaussian approximation to it. A further approximation to the distribution may be made by considering the poultry farm limit, $m \rightarrow \infty$, $M \rightarrow \infty$, $m/M = \text{const.}$ That is, the second carton is an infinite reservoir of eggs. In this limit Stirling's approximation may be used for the factorials of the entropy of the poultry farm, and a Taylor expansion may be made using the facts that $\overline{m} \gg n$ and $M - \overline{m} \gg n$, for all possible n. With $m = \overline{n} - n + \overline{m}$, and expanding to linear order, the n-dependent part of the entropy of the poultry farm is

$$\begin{aligned}
&S_M(m)/k_\text{B} \\
&= \ln \frac{M!}{m!(M-m)!} \\
&= \text{const.} - m \ln m - (M-m)\ln(M-m) \\
&= \text{const.} - (\overline{n} - n)\ln \overline{m} - (\overline{n} - n) - (n - \overline{n})\ln(M - \overline{m}) - (n - \overline{n}) \\
&= \text{const.} + n \ln \frac{\overline{m}}{M - \overline{m}}. \qquad (1.22)
\end{aligned}$$

This is linear in n, $S_M(m)/k_\text{B} = \text{const.} + \alpha n$, with the coefficient being $\alpha = -S'_M(\overline{m})/k_\text{B}$, where the prime denotes the derivative with respect to m. The poultry farm only enters the problem via this coefficient evaluated at equilibrium.

The terms neglected in the expansion are of order $n/\overline{m}$ and $n/(M - \overline{m})$, and higher, which obviously vanish in the reservoir limit. This is an important point, which emphasises the utility of dealing with entropy. If instead of entropy one used the number of configurations directly, then one would find that all terms in the expansion were of the same order, and these would have to be resummed. One would eventually obtain the same result as the single term entropy expansion, but in a more complicated and much less transparent fashion.

The probability distribution for the number of eggs in the first carton is proportional to the exponential of the total constrained (or macrostate) entropy, $S(n|N, M, n+m) = S_N(n) + S_M(m)$. In view of the reservoir expansion one has

$$\begin{aligned}
\wp(n|N,\alpha) &= \frac{1}{Z(\alpha)} \exp[S_N(n)/k_\text{B} + \alpha n] \\
&= \frac{1}{Z(\alpha)} \frac{N!}{n!(N-n)!} \left(\frac{\overline{n}}{N - \overline{n}} \right)^n, \qquad (1.23)
\end{aligned}$$

since $\overline{m}/M = \overline{n}/N$.

The egg reservoir or poultry farm estimate is compared with the exact probability distribution and with the Gaussian approximation in Fig. 1.3. It can be seen that the Gaussian approximation agrees very well with the exact result in this case. In general it is most important to get correctly the behaviour of the distribution near the peak, and experience shows that Gaussians derived as above approximate the full distribution very well. The reservoir estimate is not exact for this finite-sized system, but it is certainly a reasonable approximation.

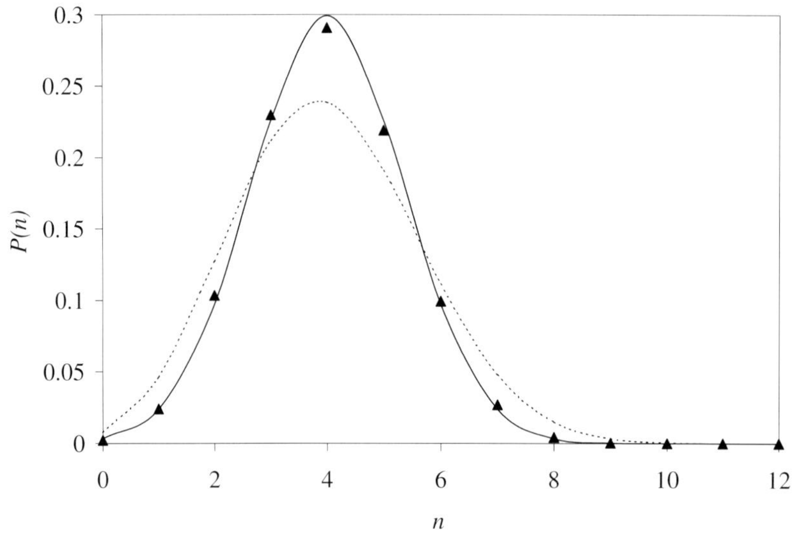

Figure 1.3: The occupancy probability, $\wp(n|N, M, n+m)$, of an ordinary carton for a dozen eggs allocated between it and a twin carton, $N = 12$, $M = 24$ and $n + m = 12$. The symbols are the exact enumeration, the solid line is the Gaussian approximation with $\overline{n} = 4$, and the dotted line results if the exchange occurs with a poultry farm with $\alpha = \ln[8/(24 - 8)]$.

When one considers that one had no information about the size of the second carton (or equivalently about the total number of eggs), it is quite a revelation to see how well it performs. The reservoir estimate must allow for all possible M and $m+n$, which is why it is broader than the exact result for this particular case ($M = 24$, $m + n = 12$). As the size of the second carton is increased (at constant $\overline{m}/M = \overline{n}/N$), the exact probability distribution becomes broader and approaches the reservoir estimate.

1.4.2 The Reservoir Formalism

An overview of the reservoir approach in thermodynamics and statistical mechanics may be summarised as follows. The more detailed analysis will be given in later chapters. Consider two systems in contact and able to exchange an extensive quantity (such as energy or particles) that is in total conserved, $x_1 + x_2 = x_{\text{total}}$. The probability distribution of the parameter for the first system is proportional to the total weight of corresponding microstates (total number if these are equally likely). Equivalently, it is proportional to the exponential of the constrained total entropy,

$$\wp(x_1|x_{\text{total}}) = \frac{1}{Z'} \exp[S(x_1)/k_{\text{B}} + S(x_2)/k_{\text{B}}]. \tag{1.24}$$

This expression is true even when the systems are of finite size. It does however assume that the interactions between the systems are negligible to the extent that the total entropy is the sum of their individual entropies, $S_{\text{total}} = S(x_1) + S(x_2)$. The denominator Z' is just the normalising factor, which at this stage is the unconstrained total entropy.

In the event that the second system is a reservoir, which means that $x_1 \ll x_2$, a Taylor expansion of its entropy may be made about x_{total} and truncated at linear order,

$$S(x_2) = S(x_{\text{total}}) - x_1 \frac{\partial S(x_{\text{total}})}{\partial x_{\text{total}}}. \tag{1.25}$$

The second term is of order x_1, whereas the first neglected term goes like $x_1^2 S_2'' \sim \mathcal{O}(x_1^2/x_2)$ (in general the entropy is extensive $S_2/x_2 \sim \mathcal{O}(1)$). This is negligible in the reservoir limit $x_1/x_2 \to 0$. The remaining neglected terms are analogously of higher order.

The constant term, $S(x_{\text{total}})$, which is independent of the subsystem 1, may be neglected (or, more precisely, cancelled with the identical factor in the denominator). One may define *field variables*, like temperature and pressure, as the derivatives of the entropy, $\lambda \equiv k_{\text{B}}^{-1} \partial S(x_{\text{total}})/\partial x_{\text{total}}$. The subscripts may now be dropped, because the reservoir only enters via λ, which has the physical interpretation as the rate of change of its entropy with x. Hence the probability distribution for the subsystem is now

$$\wp(x|\lambda) = \frac{1}{Z(\lambda)} \exp[S(x)/k_{\text{B}} - \lambda x]. \tag{1.26}$$

The equilibrium value of the exchangeable parameter, $\overline{x}$, is, by definition, the most likely macrostate, and this is given by the peak of the probability distribution. One has the implicit equation

$$\left. \frac{\partial S(x)/k_{\text{B}}}{\partial x} \right|_{x=\overline{x}} = \lambda, \tag{1.27}$$

or in view of the definition of the field variable as the derivative of the entropy, $\lambda(\overline{x}) = \lambda$. On the left-hand side is the field variable of the subsystem, and on the right-hand side is the field variable of the reservoir, and this says that equilibrium corresponds to equality of the two.

The normalisation factor for the probability distribution is called the *partition function*, and it is of course

$$Z(\lambda) = \sum_x e^{S(x)/k_{\text{B}}} e^{-\lambda x}. \tag{1.28}$$

The exponent is that part of the constrained total entropy that depends upon the subsystem, $S_{\text{total}}(x|\lambda) = S(x)/k_{\text{B}} - \lambda x$. The unconstrained, subsystem-dependent, total entropy is

$$\begin{aligned} S_{\text{total}}(\lambda) &= \sum_x \wp(x|\lambda)[S_{\text{total}}(x|\lambda) - k_{\text{B}} \ln \wp(x|\lambda)] \\ &= k_{\text{B}} \ln Z(\lambda), \end{aligned} \tag{1.29}$$

and the average value of the exchangeable quantity is given by

$$\langle x \rangle \equiv \sum_x \wp(x|\lambda) x = -\frac{\partial \ln Z(\lambda)}{\partial \lambda}. \tag{1.30}$$

One should note that in the reservoir formalism three distinct entropies appear: $S(x)$ is the entropy of the isolated subsystem in the macrostate x, $S_{\text{total}}(x|\lambda)$ is the total entropy of the subsystem and reservoir when the subsystem is constrained to be in the macrostate x, and $S_{\text{total}}(\lambda)$ is the unconstrained total entropy of the subsystem and reservoir. The latter two entropies do not include the constant contribution to the reservoir entropy that is independent of the presence of the subsystem.

One can of course generalise the formalism to include multiple extensive parameters, some exchangeable and some fixed. One can extend it to finite-sized reservoirs, where one must expand beyond linear order, and to systems in which the region of interaction is comparable to the size of the subsystem, in which case boundary terms enter. A linear additive conservative quantity is common, but one can generalise the formalism, at least in principle, to the case that it is not x itself that is conserved but some function of x, $f(x_1)dx_1 + f(x_2)dx_2 = 0$. In this case the reservoir field variable that enters the probability distribution becomes $\lambda = [k_B f(x)]^{-1} \partial S(x)/\partial x$.

Summary

- Thermodynamics deals empirically with the macroscopic behaviour of bulk matter, whereas statistical mechanics seeks to predict quantitatively that behaviour from the interactions of atoms.

- The equilibrium macrostate is that with the most microstates, and this is the state of greatest entropy. A macroscopic flux is most likely in the direction of increasing entropy.

- All systems have a set of fundamental weighted microstates. Statistical thermodynamics merely assumes the existence of the microstates and their weights, whereas statistical mechanics constructs them from the transition probabilities.

- The entropy of a state is the logarithm of the total weight of corresponding microstates. It may be expressed as a functional of the macrostate probabilities.

- The probability distribution of a parameter able to be exchanged with a second system is proportional to the exponential of the total entropy constrained by the value of the parameter. For a reservoir the constrained total entropy equals the subsystem entropy minus the parameter times the derivative of the reservoir entropy.

Chapter 2

Isolated Systems and Thermal Equilibrium

2.1 Definitions of Thermodynamic Quantities

The fundamental object treated by thermodynamics is the *isolated system*, which is one that is closed and insulated from its surroundings so that it does not interact with them. An isolated system may comprise two or more subsystems. Even though these subsystems interact with each other, macroscopic thermodynamics proceeds by treating them as quasi-isolated, which means that at any instant each subsystem is in a well-defined state and that its properties are the same as if it were in isolation in that state.[1]

The state of an isolated system is traditionally specified by the values of its energy E, volume V, and number of particles N. (For an incompressible solid, either V or N is redundant.) These variables have the important property that they do not change with time (i.e., they are conserved), so that they serve as the independent variables that specify the state of an isolated system. These particular variables represent linear additive quantities. That is, the total energy of an isolated system comprising a number of isolated subsystems is the sum of the energies of the subsystems, and similarly for the volume and particle number. Linear additivity is essential for the development of the formalism of thermodynamics.

There are a number of other linear additive conserved quantities that could be used in addition to specify the state. If the Hamiltonian that characterises the intermolecular interactions of the system is translationally invariant, as occurs when the system is free of any external force fields, then the linear momentum is conserved. Similarly a rotationally invariant Hamiltonian implies that angular momentum is conserved. These momenta are also of course linearly additive

[1]State here means macrostate. There is no need to be more precise at this stage because the formalism of thermodynamics can be derived from a series of postulates that do not invoke the microscopic interpretation.

quantities. Since most systems are enclosed in containers fixed in space, the momenta of the system itself are usually not conserved, and so these generally are not used in the macrostate specification.

Thermodynamics proceeds by asserting the existence of a function of the state of the isolated system that contains all thermodynamic information about the system. This is of course the *entropy*, $S(E, V, N)$. In the previous chapter the existence and properties of the entropy were demonstrated on the basis of the deeper postulate, namely that an isolated system possesses a set of weighted microstates, and that the entropy of a macrostate was the logarithm of the weight of the corresponding microstates. One can derive the formalism of thermodynamics without such a microscopic basis for entropy, but obviously at the cost of certain additional assumptions. The justification of the physical basis of statistical thermodynamics must be deferred until the treatment of statistical mechanics in a later chapter.

An isolated system has a well-defined temperature, T, pressure, p, and chemical potential, μ. Like all the free energies and thermodynamic potentials that are defined in what follows, the entropy is a generator of these familiar thermodynamic quantities via its partial derivatives. As a matter of logic the following expressions are taken as the definitions of these quantities, and it will be necessary to show that their behaviour reflects the behaviour of the familiar physical quantities that bear the same name. One has the inverse temperature

$$T^{-1} = \left(\frac{\partial S}{\partial E} \right)_{V,N}, \tag{2.1}$$

the pressure,

$$p = T \left(\frac{\partial S}{\partial V} \right)_{E,N}, \tag{2.2}$$

and the chemical potential,

$$\mu = -T \left(\frac{\partial S}{\partial N} \right)_{E,V}. \tag{2.3}$$

This last quantity is probably the least familiar of the three because traditional thermodynamics deals with systems with a fixed number of particles. Consequently N is usually not shown explicitly as a variable, and the chemical potential is seldom required. However the treatment of systems with variable particle number is entirely analogous to those with variable energy or volume, and so here the chemical potential is treated on an equal footing with temperature and pressure.

It is important to keep in mind the distinction between the independent variables E, V, and N, and the dependent quantities $T(E, V, N)$, $p(E, V, N)$, and $\mu(E, V, N)$, as given by the above equations. In traditional treatments this distinction is not always clear, whereas in the present development it will prove important to the correct interpretation of the formalism. On occasion when it is necessary to emphasise this distinction, dependent quantities will be overlined.

In view of the above definitions, the total differential of the entropy is

$$\begin{aligned} dS &= \left(\frac{\partial S}{\partial E}\right)_{V,N} dE + \left(\frac{\partial S}{\partial V}\right)_{E,N} dV + \left(\frac{\partial S}{\partial N}\right)_{E,V} dN \\ &= \frac{1}{T}dE + \frac{p}{T}dV - \frac{\mu}{T}dN. \end{aligned} \tag{2.4}$$

This differential form gives the above partial derivatives at a glance.

2.2 Heat Reservoir

2.2.1 Temperature Equality

The second law of thermodynamics implies that equilibrium corresponds to the maximum total entropy, and that a system prepared in a nonequilibrium macrostate will move in the direction of increasing entropy. The statistical interpretation, as exemplified in the first chapter, is that the equilibrium macrostate is the most probable macrostate (i.e., the one with the largest weight), and hence it is the state of maximum entropy. A system moves toward equilibrium because there is a greater weight of states in that direction than in the opposite direction. Here these facts are used to derive what is essentially the zeroth law of thermodynamics, namely that two systems in thermal equilibrium have the same temperature.

In what follows it will be important to distinguish between dependent and independent variables. When the energy is independently specified it will be denoted simply E, and when it is a dependent variable it will be denoted $E(N, V, T)$, or $\overline{E}(N, V, T)$, or simply $\overline{E}$. The first quantity is a well-defined property of an isolated system and $E(N, V, T)$ is given implicitly by

$$\frac{\partial S(E, N, V)}{\partial E} = \frac{1}{T}. \tag{2.5}$$

It will be shown below that the thermodynamic state is unique, and hence there is a one-to-one relationship between the thermodynamic variables. In particular this means that one may write $E_1 = E(N, V, T_1) \Leftrightarrow T_1 = T(E_1, N, V)$. An overline is used to denote the equilibrium state; $\overline{E}(N, V, T)$ is the equilibrium energy of a subsystem with N particles in a volume V in contact with a thermal reservoir of temperature T.

Consider an isolated system comprising a subsystem 1 in thermal contact with a heat reservoir 2 (Fig. 2.1). A *reservoir* in general is defined by two characteristics: it is infinitely larger than the subsystem of interest, and the region of mutual contact is infinitely smaller than the subsystem. The first property means in this case that the amount of energy exchangeable with the subsystem is negligible compared to the total energy of the reservoir. The second property ensures that interactions between the subsystem and the reservoir, while necessary to establish thermal equilibrium, are relatively negligible. That is, the total entropy of the system is equal to the sum of the entropies of the

Figure 2.1: A subsystem 1 in thermal contact with a heat reservoir 2.

subsystem and the reservoir, $S_{\text{total}} = S_1 + S_2$, and the correction due to the interactions between the subsystem and the reservoir may be neglected.

Energy partitions between the subsystem and the reservoir while the total energy of the system is conserved, $E_{\text{total}} = E_1 + E_2$. Using the properties of the reservoir, the total entropy for a given partitioning is

$$\begin{aligned}
&S_{\text{total}}(E_1|E, N_1, V_1, N_2, V_2) \\
&\quad = S_1(E_1, V_1, N_1) + S_2(E_{\text{total}} - E_1, V_2, N_2) \\
&\quad = S_1(E_1, V_1, N_1) + S_2(E_{\text{total}}, V_2, N_2) - E_1 \frac{\partial S_2(E_{\text{total}}, V_2, N_2)}{\partial E_{\text{total}}} + \ldots \\
&\quad = S_1(E_1, V_1, N_1) + \text{const.} - \frac{E_1}{T_2}. \qquad (2.6)
\end{aligned}$$

The second equality represents a Taylor expansion of the reservoir about $E_2 = E_{\text{total}}$. It is permissible to truncate this at the linear term because $E_1 \ll E_{\text{total}}$. That is, the first neglected term goes like $E_1^2 S_{EE} \sim E_1^2/E_{\text{total}}$, since both energy and entropy are extensive variables. Higher terms have higher powers of E_1/E_{total}, and these likewise vanish in the reservoir limit. In the third equality the definition of the temperature of the reservoir has been used, namely the energy derivative of the entropy. The leading term is an immaterial constant that will be neglected throughout because it is independent of E_1. Henceforth the temperature of the reservoir will be denoted by T, and the reservoir variables V_2 and N_2 upon which it depends will be suppressed, as will the consequently redundant subscript 1 for the subsystem. Hence the left-hand side will be written $S_{\text{total}}(E|N, V, T)$. The entropies on the right-hand side are the entropies of the reservoir and of the subsystem, each considered in isolation and as a function of the independent variables that are their arguments. In contrast the entropy on the left-hand side is the entropy of the total system for a particular partitioning of the energy; it will often be referred to as the *constrained* total entropy. In such cases the quantity in flux is shown to the left of a vertical bar, and the fixed variables to the right. This notation is identical to that used for the probability distributions that are derived shortly. All of the arguments of the constrained total entropy are independent.

The energy derivative of the constrained total entropy yields

$$\frac{\partial S_{\text{total}}(E|N,V,T)}{\partial E} = \frac{\partial S(E,V,N)}{\partial E} - \frac{1}{T}, \tag{2.7}$$

where the first term on the right-hand side is the reciprocal temperature of the subsystem, $1/T(E,N,V)$. The equilibrium energy $\overline{E}$ is the one that maximizes the total entropy. This corresponds to the vanishing of its derivative, or

$$T(\overline{E},N,V) = T. \tag{2.8}$$

One concludes that equilibrium corresponds to temperature equality between the subsystem and the reservoir, which is essentially the zeroth law of thermodynamics. This is an implicit equation for the equilibrium energy of the subsystem, $\overline{E} = E(N,V,T)$.

Generalised Reservoirs

One may pause here to consider the consequences of the reservoir being of finite size, so that it is no longer permissible to truncate the Taylor expansion at linear order. Assuming that the size of the region of interaction is still negligible compared to the size of both systems, the total entropy is still $S_{\text{total}} = S_1 + S_2$, and this is maximised when $T_1(\overline{E}_1, N_1, V_1) = T_2(E_{\text{total}} - \overline{E}_1, N_2, V_2)$. In this case the temperature of the second system is not constant but depends upon how much energy is accorded it at equilibrium. Obviously of less convenience than a reservoir of constant temperature, nevertheless equilibrium may still be determined. All of the results obtained here for reservoirs may be pushed through for finite-sized systems in analogous fashion.

The above analysis relied upon the conservation of energy between the subsystem and the reservoir. Such a conservation law for a linear additive property also applies to particle number and to volume, and shortly systems open with respect to these will be analysed in an analogous fashion. Some variables of interest are not conserved in this sense. In these cases one can mathematically construct a generalised reservoir that may be analysed as here, even if its physical realisation is not feasible. This is discussed in more detail in dealing with certain advanced topics in statistical mechanics in later chapters.

2.2.2 Thermodynamic Potential

The constrained total entropy (subsystem plus reservoir) determines the direction of energy flow and the equilibrium condition. An alternative to maximising the entropy is to minimise an appropriate potential. To this end a *constrained thermodynamic potential* is introduced, which may also be called a *constrained free energy*. Like the constrained total entropy, this quantity characterises the behaviour of a subsystem that is out of equilibrium.[2] The relationship with the equilibrium thermodynamic potential or free energy is discussed below.

In general the constrained thermodynamic potential is defined as the negative of the reservoir temperature times the constrained total entropy. In the present case the constrained thermodynamic potential for a subsystem of energy E in contact with a heat reservoir of temperature T is

$$\begin{aligned} F(E|N,V,T) &\equiv -TS_{\rm total}(E|N,V,T) \\ &= E - TS(E,N,V). \end{aligned} \tag{2.9}$$

By definition this is a minimum at equilibrium, and the energy flux of the subsystem is down its gradient. By virtue of its close relationship to the constrained total entropy, the constrained thermodynamic potential inherits many features of the latter, such as the appropriate concavity and bijectivity. In particular, the constrained thermodynamic potential is a convex function of energy, (and also of number and volume), which follows from the concavity of the entropy ($F'' = -TS'' > 0$). It is once more emphasised that the four arguments of the constrained thermodynamic potential are independent.

The *equilibrium thermodynamic potential* or *equilibrium free energy* is defined as the minimum value of the constrained thermodynamic potential, which in this case obviously occurs at $\overline{E} = E(N,V,T)$,

$$\overline{F}(N,V,T) \equiv F(\overline{E}|N,V,T) = \overline{E} - TS(\overline{E},N,V). \tag{2.10}$$

For the present case of constant temperature, this is called the *Helmholtz free energy*, sometimes denoted by $A(N,V,T)$. In this equation $\overline{E}$ is a dependent variable, $\overline{E}(N,V,T)$, and consequently the Helmholtz free energy is a function of just three independent variables. The entropy that appears on the right of the definition of the Helmholtz free energy is that of the isolated subsystem with the equilibrium energy $\overline{E}$. The overline on the Helmholtz free energy emphasises that it is an equilibrium property.

The constrained thermodynamic potential clearly contains more information than the equilibrium thermodynamic potential, since it is a function of four independent variables, whereas the equilibrium thermodynamic potential is only a function of three. The constrained thermodynamic potential describes the approach to energy equilibrium and energy fluctuations about the equilibrium state, whereas the equilibrium thermodynamic potential only describes the equilibrium state. The distinction made here between the two quantities is not widespread in thermodynamics, and the present nomenclature is not necessarily conventional. In most texts, the words 'thermodynamic potential' or 'free energy' in practice refer to equilibrium quantities. In the present work the strictly

[2]The properties of the constrained thermodynamic potential introduced here derive from its intimate connection with the constrained total entropy. The basis of other nonequilibrium potentials, such as the rather similar generalised Mathieu function and the Landau potential used in expansions that describe criticality and phase transitions, is less clear. For an account of these other potentials see H. B. Callan, *Thermodynamics*, Wiley, New York, 1960; L. D. Landau and E. M. Lifshitz, *Statistical Physics*, 3rd ed., Pergammon Press, London, 1980; and C. Kittel and H. Kroemer, *Thermal Physics*, 2nd ed., W. H. Freeman, New York, 1980.

equilibrium thermodynamic potential or free energy is never confused with the nonequilibrium constrained thermodynamic potential or free energy, and the separation between these conceptually different quantities will be consistently maintained. Because the constrained thermodynamic potential describes the excursions of the system to nonequilibrium states (see below), it may also be called the *fluctuation potential*.

2.2.3 Legendre Transforms

The total entropy, Eq. (2.6), may be written $S_{\text{total}}(E|T) = S(E) - E/T$, and it is extremised at $E = \overline{E}(T)$ (suppressing N and V). The function $S_{\text{total}}(T) \equiv S_{\text{total}}(E(T)|T)$ is the transformation of the entropy as a function of energy, $S(E)$, to the entropy as a function of temperature, $S_{\text{total}}(T)$. This can be regarded as a purely mathematical transformation that does not require the existence of a reservoir. Such mathematical procedures are called *Legendre transforms*. Given $f(x)$ and $\overline{y}(x) = f'(x)$, and bijectivity between x and y, such Legendre transforms take the general form $F(x|y) = f(x) - xy$, which is extremised at $x = \overline{x}(y)$. Hence $F(x|y)$ is the analogue of the constrained thermodynamic potential, and $\overline{F}(y) = F(\overline{x}(y)|y)$ is the analogue of the equilibrium thermodynamic potential. It will be seen that all of the nonequilibrium and equilibrium thermodynamic potentials derived in the following chapter via reservoir arguments can also be derived by Legendre transforms.

In mathematical content the present approach is equivalent to a Legendre transform. Physically however, the present procedure has the advantage of being based on entropy maximisation, and consequently there is a transparent interpretation available for the constrained and equilibrium thermodynamic potentials and for the distinction between them. This will prove especially valuable in the derivation and treatment of statistical mechanics below.

The Legendre transform works by adding a strictly linear function of the nonequilibrated variable. This is also the case in the present method, due to the truncated Taylor expansion of the reservoir. Since the second derivative of such a linear function vanishes, the consequence is that the convexity of the constrained thermodynamic potential is preserved by such a transformation. That is, the extremum of the constrained thermodynamic potential is guaranteed to be a minimum. More generally such a conservation law and reservoir may not physically exist for the variable of interest. Nevertheless one can imagine a generalised reservoir and still formulate the Legendre transform by addition of the appropriate linear function, thus being assured of the convexity of the resultant constrained thermodynamic potential.

2.2.4 Derivatives of the Free Energy

The utility of the constrained thermodynamic potential lies in its variational nature at equilibrium, which makes differentiation particularly simple. The temperature derivative of the Helmholtz free energy is

$$\begin{aligned}
&\frac{\partial \overline{F}(N,V,T)}{\partial T} \\
&= \left(\frac{\partial F(\overline{E}(N,V,T)|N,V,T)}{\partial T} \right)_{N,V} \\
&= \left. \frac{\partial F(E|N,V,T)}{\partial T} \right|_{E=\overline{E}} + \left. \frac{\partial F(E|N,V,T)}{\partial E} \right|_{E=\overline{E}} \frac{\partial \overline{E}(N,V,T)}{\partial T} \\
&= \left. \frac{\partial F(E|N,V,T)}{\partial T} \right|_{E=\overline{E}} \\
&= -S(\overline{E},N,V). \qquad (2.11)
\end{aligned}$$

The important point of this derivation is the passage from the second to the third equality, where the final term vanishes, $\partial F(E|N,V,T)/\partial E|_{E=\overline{E}} = 0$, because of the optimal nature of the constrained thermodynamic potential at the equilibrium state. This is a general feature that may be widely exploited. The entropy in the final equality is that of the isolated subsystem with the equilibrium energy; it may be written $\overline{S}(N,V,T)$.

One sees that because the constrained thermodynamic potential is optimised at equilibrium, differentiating the Helmholtz free energy is the same as differentiating the constrained thermodynamic potential *holding $\overline{E}$ fixed*. It is for this reason that one often sees the expression $F = E - TS$, and $\partial F/\partial T = -S$, it being implicit that although the energy that appears here is really $\overline{E}(N,V,T)$, it may be held fixed during any differentiations. The entropy may similarly be held fixed as far as temperature differentiations goes because it is really $\overline{S}(N,V,T) \equiv S(\overline{E}(N,V,T),N,V)$, and $\overline{E}$ may be held fixed.

The other derivatives of the Helmholtz free energy follow directly from the corresponding derivatives of the entropy for an isolated system, holding $\overline{E}$ fixed. They give the pressure

$$\begin{aligned}
\left(\frac{\partial \overline{F}(N,V,T)}{\partial V} \right)_{T,N} &= \left(\frac{\partial F(\overline{E}|N,V,T)}{\partial V} \right)_{\overline{E},T,N} \\
&= -T \left(\frac{\partial S(\overline{E},N,V)}{\partial V} \right)_{\overline{E},N} \\
&= -\overline{p}, \qquad (2.12)
\end{aligned}$$

and the chemical potential

$$\begin{aligned}
\left(\frac{\partial \overline{F}(N,V,T)}{\partial N} \right)_{T,V} &= \left(\frac{\partial F(\overline{E}|N,V,T)}{\partial N} \right)_{\overline{E},T,V} \\
&= -T \left(\frac{\partial S(\overline{E},N,V)}{\partial N} \right)_{\overline{E},V} \\
&= \overline{\mu}. \qquad (2.13)
\end{aligned}$$

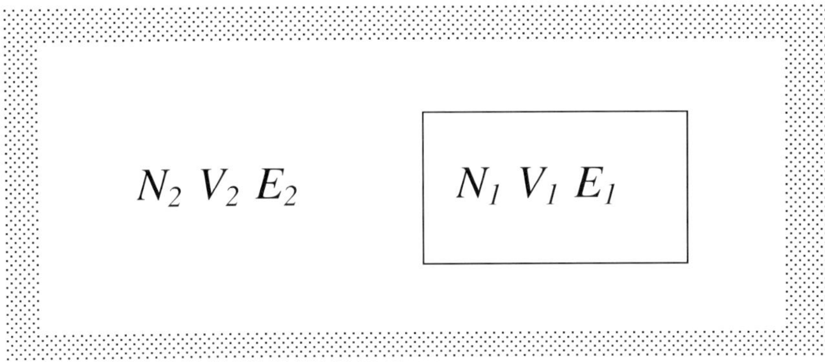

Figure 2.2: Two subsystems, identical except for energy, separated by a removable insulating partition.

From the uniqueness of the equilibrium state (see below) the pressure that appears here $\overline{p}(N, V, T)$ equals $p(\overline{E}(N, V, T), N, V)$. This demonstrates the internal consistency of thermodynamics: the volume derivative of the Helmholtz free energy for a subsystem in contact with a heat reservoir gives a pressure that is identical to that of an isolated system (with the equilibrium energy). Similar comments apply to the chemical potential, $\overline{\mu}(N, V, T) = \mu(\overline{E}(N, V, T), N, V)$ and to the subsystem entropy $\overline{S}(N, V, T) = S(\overline{E}(N, V, T), N, V)$. These three quantities, $\overline{p}$, $\overline{\mu}$, and $\overline{S}$, are properties of the subsystem in thermal equilibrium, as is emphasised by the fact that they are overlined.

In view of these results, the total differential of the Helmholtz free energy is

$$d\overline{F}(N, V, T) = -\overline{S}(N, V, T)dT - \overline{p}(N, V, T)dV + \overline{\mu}(N, V, T)dN. \tag{2.14}$$

This summarises the above results for the partial derivatives.

Finally, if one divides both sides of the definition of the Helmholtz free energy by T, $F/T = E/T - S(N, V, T)$, and differentiates with respect T^{-1} one obtains

$$\left(\frac{\partial(\overline{F}(N, V, T)/T)}{\partial(1/T)} \right)_{N,V} = \overline{E}, \tag{2.15}$$

since one can hold $\overline{E}$ (and hence $S(\overline{E}(N, V, T), N, V)$) constant as discussed above.

2.3 Properties of Thermodynamic Systems

2.3.1 Concavity of the Entropy

In the egg sample of Chapter 1, it was shown by direct calculation that the constrained total entropy was a concave function that attained its maximum at the equilibrium macrostate. Using the notions of the extensivity of the entropy and its maximisation at equilibrium, it is now demonstrated in general that the entropy is a concave function of its arguments.

Consider two systems, identical except for energy, isolated from each other by an insulating partition (Fig. 2.2). They have volume $V_1 = V_2 = V$, number of particles $N_1 = N_2 = N$, and energies E_1 and E_2. Their entropies are $S_1 = S(E_1, V, N)$ and $S_2 = S(E_2, V, N)$, and the total entropy with the partition in place is $S_{\text{total}} = S_1 + S_2$. This follows because the systems are isolated from each other, and in such cases the entropy is a linear additive quantity. If the insulating partition is removed, energy irreversibly flows from the more to the less energetic system. The total entropy at the end of this process must be greater than at the beginning,

$$S_{\text{total}}(E_1 + E_2, 2V, 2N) \geq S(E_1, V, N) + S(E_2, V, N). \tag{2.16}$$

This is simply a statement of the second law of thermodynamics, namely that energy flows in the direction of entropy increase. The microscopic interpretation is of course that relaxing the constraint imposed by the partition increases the number of microstates available to the system. From symmetry equality is expected if, and only if, $E_1 = E_2$. After equilibration the insulating partition can be reinserted, and the energy of each subsystem is $(E_1 + E_2)/2$, fluctuations about this value being relatively negligible. This equilibrium division of the energy between the two systems corresponds to the maximum entropy. Since there is no further energy flow, the removal and insertion of the partition is now a reversible process, and one must have

$$S_{\text{total}}(E_1 + E_2, 2V, 2N) = 2S\left(\frac{E_1 + E_2}{2}, V, N\right). \tag{2.17}$$

Combining these two equations one obtains

$$S\left(\frac{E_1 + E_2}{2}, V, N\right) > \frac{1}{2}[S(E_1, V, N) + S(E_2, V, N)]. \tag{2.18}$$

The case $E_1 = E_2$ is here excluded, so that the left side is strictly greater than the right. This says that any chord to the entropy curve lies below the curve, which defines a concave function. By performing a Taylor expansion as $E_1 \to E_2$, it may be seen that this is equivalent to

$$\left(\frac{\partial^2 S(E, V, N)}{\partial E^2}\right)_{V,N} < 0. \tag{2.19}$$

The above explicitly shows that the entropy of a system is a strictly concave function of energy. It is clear that analogous steps can be carried out individually for volume and for particle number, so that in general the entropy is a concave function of its arguments.

Simultaneous variations are hardly more complicated. Suppose that two isolated systems are identical apart from their energies and volumes. Allowing simultaneous exchange of both as above, one concludes that

$$S_{\text{total}}\left(\frac{E_1 + E_2}{2}, \frac{V_1 + V_2}{2}, N\right) > \frac{1}{2}[S(E_1, V_1, N) + S(E_2, V_2, N)]. \tag{2.20}$$

Expanding each term on the right hand side about the equilibrium energy and volume, the zeroth and linear terms cancel and one is left with

$$S_{EE}dE^2 + 2S_{EV}dEdV + S_{VV}dV^2 < 0, \tag{2.21}$$

where the subscripts denote partial derivatives. This result, which holds for arbitrary dE and dV, will be used several times below.

2.3.2 Uniqueness of the Thermodynamic State

It will often be convenient to interchange the dependent and the independent variables. For example, instead of dealing with $S(E, N, V)$ it may be desirable to use $E(S, N, V)$. In general an interchange such as

$$x_1 = x(y_1) \Leftrightarrow y_1 = y(x_1) \tag{2.22}$$

is valid if there is a one-to-one relationship between the conjugate variables. It will now be shown that this follows from the fact that the entropy is a strictly concave function of its arguments. Use will be made of the condition for a continuous function to be invertible, namely that its derivative must be finite and nonzero on the interior of its domain.

Energy–Entropy

Because the entropy is continuous and strictly concave, $S'' < 0$, it can have at most one turning point. Denote the point of this putative maximum by E_1, $S'(E_1) = 0$. Entropy increases with energy for $E < E_1$ ($T(E) > 0$), and decreases for $E > E_1$ ($T(E) < 0$). Only systems with positive temperatures will be treated here (because systems with negative temperatures are metastable). In view of its concavity, the gradient of the entropy may possibly be infinite at the terminal energy, but nowhere else. This is the ground state E_0, and the infinite gradient corresponds to the absolute 0 of temperature. Hence on the domain (E_0, E_1) there is a one-to-one relationship between the entropy and the energy, and one can equally well write $S(E, N, V)$ as $E(S, N, V)$.

Energy–Temperature

Now on the domain (E_0, E_1), the concavity of the entropy corresponds to $\partial(1/T)/\partial E < 0$, or $\partial T/\partial E > 0$. Furthermore, since the temperature is defined via the energy derivative of the entropy, one has $T(E)$, which means that the temperature must be a single-valued function of the energy, and $\partial T/\partial E < \infty$. Therefore, there is a one-to-one relationship between energy and temperature, and hence $T_1 = T(E_1, N, V) \Leftrightarrow E_1 = E(T_1, N, V)$. Similar arguments can be made for the volume and for the number of particles, and for their conjugate variables the pressure and the chemical potential, respectively. It will be assumed throughout that there always exists a domain upon which it is possible to interchange dependent and independent variables.

2.3.3 Thermal Stability

The constrained thermodynamic potential was constructed to be a convex function of energy. It similarly follows from the concavity of the entropy that it is also a convex function of the volume and of the particle number, $F'' = -TS'' > 0$. Because of the variational principle, the constrained thermodynamic potential is minimised by the equilibrium energy, at which point it equals the Helmholtz free energy.

It should be stressed that the equilibrium thermodynamic potential, which in this case is the Helmholtz free energy, does not obey a variational principle; it is not a minimum with respect to any of its independent variables. It is the constrained thermodynamic potential that is minimised at equilibrium.

A consequence of the convexity of the constrained thermodynamic potential is that matter is thermally stable. A temporary increase in energy in a local region of the system (of fixed volume and particle number) increases the nonequilibrium thermodynamic potential (considering the local region as a quasi-isolated subsystem). The minimisation principle for the constrained thermodynamic potential, which is equivalent to total entropy maximisation, shows that this is unfavourable (i.e., improbable), and to counteract it energy is likely to flow back out of the region, restoring it to equilibrium. Conversely, and for the same reason, energy flows back into a region following a temporary decrease. It is the minimisation of the constrained thermodynamic potential that provides the thermal stability of matter by damping local fluctuations in the energy. It ought to be clear that thermal stability does not come from any variational principle of the equilibrium Helmholtz free energy itself, but rather from the variational nature of the underlying constrained thermodynamic potential. Analogous optimisation principles will be derived in the following chapter for systems undergoing volume and particle number fluctuations, and it will similarly be shown that matter is mechanically stable.

2.3.4 Extensive and Intensive Variables

Extensive Variables

The independent variables that specified the state of the isolated system were all linear additive quantities. This implies that they are *extensive* variables, which means that they scale with the size of the system (i.e., if the number of identical isolated systems is doubled, then the total value of each of these is also doubled). The entropy itself is also extensive, as is clear in the microscopic interpretation of it as the logarithm of the total weight (i.e., the total weight of two isolated systems is the product of the individual weights, and hence the total entropy is the sum of their individual entropies). Hence for a system comprising λ identical isolated subsystems, the total entropy is $S(\lambda E, \lambda V, \lambda N) = \lambda S(E, V, N)$. By the definitions, the inverse temperature of the total system is

$$T_{\text{total}}^{-1} = \left(\frac{\partial S_{\text{total}}}{\partial E_{\text{total}}}\right)_{V_{\text{total}}, N_{\text{total}}} = \left(\frac{\partial \lambda S}{\partial \lambda E}\right)_{V,N,\lambda} = T^{-1}. \tag{2.23}$$

That is, the total temperature is the same as that of each of the independent subsystems, and similarly for the pressure and chemical potential. Such variables that are independent of the size of the system are called *field* or *intensive* variables.

Intensive Variables

For most of the present analysis the thermodynamic limit will be invoked. In this limit the system size goes to infinity while the ratios of the extensive variables are kept fixed. Consequently the intensive variables also maintain constant values. New intensive variables may be defined, such as the energy density $\epsilon \equiv E/V$ and the number density $\rho \equiv N/V$. In this thermodynamic limit the entropy density is a function of just two variables, $\sigma(\epsilon, \rho) \equiv S(E, N, V)/V$, as are the field variables,

$$\frac{1}{T} = \left(\frac{\partial V \sigma(\epsilon, \rho)}{\partial V \epsilon} \right)_{N,V} = \left(\frac{\partial \sigma(\epsilon, \rho)}{\partial \epsilon} \right)_{\rho}, \tag{2.24}$$

$$\frac{-\mu}{T} = \left(\frac{\partial V \sigma(\epsilon, \rho)}{\partial V \rho} \right)_{E,V} = \left(\frac{\partial \sigma(\epsilon, \rho)}{\partial \rho} \right)_{\epsilon}, \tag{2.25}$$

and

$$\begin{aligned} \frac{p}{T} &= \left(\frac{\partial V \sigma(\epsilon, \rho)}{\partial V} \right)_{E,N} \\ &= \left(\frac{\partial V \sigma(\epsilon, \rho)}{\partial V} \right)_{\epsilon,\rho} + \left(\frac{\partial V \sigma(\epsilon, \rho)}{\partial \epsilon} \right)_{\rho,V} \left(\frac{\partial \epsilon}{\partial V} \right)_{E} \\ &\quad + \left(\frac{\partial V \sigma(\epsilon, \rho)}{\partial \rho} \right)_{\epsilon,V} \left(\frac{\partial \rho}{\partial V} \right)_{N} \\ &= \sigma(\epsilon, \rho) - \frac{\epsilon}{T} + \frac{\rho \mu}{T}. \end{aligned} \tag{2.26}$$

That is, in the thermodynamic limit one has $T(\epsilon, \rho)$, $p(\epsilon, \rho)$, and $\mu(\epsilon, \rho)$.

Helmholtz Free Energy Density

It has just been shown that although three variables appear necessary to specify the state of the system, in the thermodynamic limit there is a nontrivial dependence on just two intensive variables. A similar result holds for the Helmholtz free energy. In terms of the number density $\rho = N/V$ and equilibrium energy density $\overline{\epsilon}(\rho, T) = \overline{E}(N, V, T)/V$, the Helmholtz free energy density is

$$\overline{f}(\rho, T) \equiv \overline{F}(N, V, T)/V = \overline{\epsilon}(\rho, T) - T\overline{\sigma}(\rho, T), \tag{2.27}$$

where the entropy density of the isolated subsystem at the equilibrium energy is $\overline{\sigma}(\rho, T) \equiv \overline{\sigma}(\overline{\epsilon}, \rho) = S(\overline{E}(N, V, T), N, V)/V$. It is straightforward to show that

$$-\overline{\sigma}(\rho, T) = \frac{\partial \overline{f}(\rho, T)}{\partial T}, \tag{2.28}$$

$$\overline{\mu}(\rho, T) = \frac{\partial \overline{f}(\rho, T)}{\partial \rho}, \tag{2.29}$$

$$-\overline{p}(\rho, T) = \overline{f}(\rho, T) - \rho\overline{\mu}(\rho, T), \tag{2.30}$$

and

$$\overline{\epsilon}(\rho, T) = \frac{\partial(\overline{f}(\rho, T)/T)}{\partial(1/T)}. \tag{2.31}$$

Thermodynamic Limit

The existence of the thermodynamic limit and the consequent properties just listed involves certain subtleties. Whereas the concept of extensivity was introduced via a system comprising independent, isolated subsystems, the thermodynamic limit applies to a single system. The two are related by imagining that this infinite system can be divided into quasi-isolated subsystems. That is, the subsystems are large enough so that the region affected by the interactions with the neighbouring subsystems is negligible compared to the size of the subsystem itself. In practice this requirement appears to hold for all physical systems, and taking the thermodynamic limit is both well defined and routine.

In the thermodynamic limit the boundaries of the system have negligible influence on its thermodynamic properties, and their nature, shape, etc. can be ignored. In practice most interest has focused on such bulk systems, and the thermodynamic limit is almost always implicitly assumed. However on occasion microscopic systems may be studied, and in these cases the effects of boundaries and of finite size can be important. Such systems exhibit additional features beyond those of bulk systems, and the above reduction to the nontrivial dependence on just two independent variables no longer holds. The shape of the container and other surface effects cannot be neglected in such small systems, and additional quantities such as the surface tension need to be defined.

Summary

- The state of an isolated system is specified by the three extensive variables: energy, volume, and number. The entropy is a function of state that is similarly extensive and that generates the thermodynamic properties of the system such as its temperature, pressure, and chemical potential. These conjugate field variables are intensive. In the thermodynamic limit, the extensive variables may be replaced by their respective densities, and the entropy density and the field variables reduce to functions of just two independent variables, namely the energy and number densities.

- The entropy of an isolated system is a strictly concave function of its arguments. Consequently the mapping between dependent and independent variables is one-to-one, which means that they may be interchanged as appropriate. The choice of independent variables commonly reflects the specific flux between a subsystem and a particular reservoir.

- The constrained total entropy of a subsystem in contact with a reservoir is a maximum at equilibrium, and a subsystem in a nonequilibrium macrostate (i.e., with the nonoptimal value of the parameter in flux) moves in the direction of increasing constrained total entropy. In the case of a constant temperature heat reservoir it is the energy that is in flux, and the equilibrium energy corresponds to temperature equality between the subsystem and the reservoir.

- The constrained thermodynamic potential is $-T$ times the constrained total entropy. It characterises the approach to equilibrium, and is minimised at equilibrium. The equilibrium thermodynamic potential is this minimum value of the constrained thermodynamic potential. It is exclusively an equilibrium quantity whose derivatives generate the equilibrium thermodynamic properties of the subsystem in contact with the reservoir. It does not obey any variational principle. In the case of a heat reservoir, the equilibrium thermodynamic potential is called the Helmholtz free energy.

Chapter 3

Various Reservoirs

In the preceding chapter it was shown how a constant temperature system has the physical realisation of a subsystem able to exchange energy with a heat reservoir, and how temperature equality represented the equilibrium state of maximum total entropy. As a corollary it follows that since a closed system may be regarded as a heat reservoir for its parts, at equilibrium the temperature is uniform throughout (at least on large enough length scales). Two essential features of the analysis were that energy is a linear additive quantity and that it is conserved. The thermodynamic behaviour of any subsystem that can exchange with a reservoir a linear additive conserved quantity may be similarly analysed, which is the content of this chapter.

Physical thermodynamic systems differ in their boundaries. Some systems may be completely insulated from their surroundings, whereas other boundaries conduct heat and allow energy exchange between the system and the surrounds. Still other systems may be open, so that both energy and particles can enter and leave. Flexible walls allow the system's volume change in response to external stimuli. These are some examples of the common thermodynamic systems that will be treated in this chapter with the reservoir formalism. Effectively this consists in essence of interchanging an extensive conserved variable with its intensive conjugate to give the equilibrium thermodynamic potential appropriate for the particular boundaries.

3.1 Constant Linear and Angular Velocity

There are seven linear additive constants of the motion of a closed system: the energy, the three components of linear momentum, and the three components of angular momentum. Hence for an isolated system comprising two isolated subsystems able to exchange all these quantities one must have for the energy $dE_1 + dE_2 = 0$, for the linear momentum $d\underline{P}_1 + d\underline{P}_2 = 0$, and for the angular momentum $d\underline{J}_1 + d\underline{J}_2 = 0$. If the second system is a reservoir, the interactions between the two systems are negligible, $S_{\text{total}} = S_1 + S_2$, and the Taylor expansion of the second system's entropy may be truncated at linear order,

$$\begin{aligned} & S_2(E_2, V_2, N_2, \underline{P}_2, \underline{J}_2) \\ & \quad = S_2(E_{\text{total}}, V_2, N_2, \underline{P}_{\text{total}}, \underline{J}_{\text{total}}) - \frac{E_1}{T_2} + \frac{\underline{P}_1 \cdot \underline{P}_2}{M_2 T_2} + \frac{\underline{J}_1 \cdot \underline{J}_2}{I_2 T_2}. \end{aligned} \tag{3.1}$$

In the appendix, Section B.2.1, it is shown that velocity and momentum are conjugate variables, and that the momentum derivative of the entropy is the velocity, Eqs. (B.22) and (B.25). Neglecting the part independent of subsystem 1, dropping the subscripts, and invoking the velocity of the reservoir, $\underline{v} = \underline{P}_2/M_2$, and its angular velocity, $\underline{\omega} = \underline{J}_2/I_2$, the total entropy is

$$S_{\text{total}}(E, \underline{P}, \underline{J}|N, V, T, \underline{v}, \underline{\omega}) = S(E, V, N, \underline{P}, \underline{J}) - \frac{E}{T} + \frac{\underline{P} \cdot \underline{v}}{T} + \frac{\underline{J} \cdot \underline{\omega}}{T}. \tag{3.2}$$

The total entropy is maximised when the derivatives with respect to energy, momentum, and angular momentum vanish. One sees that the equilibrium values are $\overline{T} = T$, $\overline{\underline{P}}/M = \underline{v}$, and $\overline{\underline{J}}/I = \underline{\omega}$. That is, at equilibrium the subsystem moves with the same velocity as the reservoir, and it rotates with the same angular velocity. A corollary is that at equilibrium the linear and angular velocities must be uniform throughout a system, as may be seen by considering the system to be composed of parts small compared to the total, but large compared to their boundary regions. That is, the system can undergo uniform translation or rotation, but equilibrium precludes internal macroscopic motions.

One could go on to derive the constrained thermodynamic potential for this composite system, and call its minimum value the equilibrium thermodynamic potential. Little of significance results, since it is clear that the momentum derivatives of this free energy simply yield the six velocities of the subsystem, which are directly related to the momenta themselves. Since macroscopic momentum contributes only trivially (it only determines that the system translates or rotates uniformly as a whole), one normally deals with systems with 0 linear and angular momenta. An exception is the case of a rotating fluid system, where the centrifugal force effectively contributes to the local chemical potential.

In the case of an external field or one-body potential, the momenta of the system are not conserved. Such a field would be provided by the walls containing the system. Henceforth momenta will not be considered explicitly.

3.2 Constant Chemical Potential

Now an isolated system comprising two subsystems able to exchange not only energy but also particles is considered (Fig. 3.1). Such a system is called an open system, and the reservoir in this case fixes the temperature and the chemical potential. The total energy $E_{\text{total}} = E_1 + E_2$ and the total particle number $N = N_1 + N_2$ are fixed.

The second system is large compared to the first, but the region of their interaction is relatively small, so that it can be considered a reservoir and

Figure 3.1: A subsystem 1 able to exchange energy and particles with a reservoir 2.

$S_{\rm total} = S_1 + S_2$. The equilibrium energy and particle number partitioned between the two systems are determined by the temperature and chemical potential, as will now be shown. Because the reservoir is so large, the Taylor expansion of its entropy may be truncated at the linear order,

$$S_2(E_2, V_2, N_2) = S_2(E_{\rm total}, V_2, N) - \frac{E_1}{T_2} + \frac{\mu_2}{T_2} N_1. \tag{3.3}$$

Neglecting the part independent of subsystem 1, and dropping the subscripts, the constrained total entropy is

$$S_{\rm total}(N, E|\mu, V, T) = S(E, V, N) - \frac{E}{T} + \frac{\mu}{T} N. \tag{3.4}$$

This gives the total entropy for a subsystem in a nonequilibrium state of energy E and particle number N, when in contact with a reservoir of temperature T and chemical potential μ.

The derivatives of the constrained total entropy are

$$\frac{\partial S_{\rm total}(N, E|\mu, V, T)}{\partial N} = \frac{-\mu(E, V, N)}{T(E, V, N)} + \frac{\mu}{T}, \tag{3.5}$$

and

$$\frac{\partial S_{\rm total}(N, E|\mu, V, T)}{\partial E} = \frac{1}{T(E, V, N)} - \frac{1}{T}. \tag{3.6}$$

The first term on the right-hand side is the derivative of the subsystem entropy, which is the subsystem chemical potential and temperature, respectively. The constrained total entropy is of course a maximum at equilibrium, which corresponds to the vanishing of these derivatives. This condition gives implicit equations for the equilibrium energy $\overline{E}(\mu, V, T)$ and the equilibrium particle number $\overline{N}(\mu, V, T)$, namely $T(\overline{E}, V, \overline{N}) = T$ and $\mu(\overline{E}, V, \overline{N}) = \mu$. That is, at equilibrium a subsystem that can exchange energy and particles with a reservoir has the same temperature and chemical potential as the reservoir.

The direction of particle flow in the nonequilibrium case depends upon the change in constrained total entropy with particle number, $dS_{\text{total}} = [-\mu(E,V,N)/T(E,V,N) + \mu/T]dN$ (at constant volume and energy). Since the total entropy must increase, $dS_{\text{total}} > 0$, particles flow to the region of lower chemical potential (for a system at the same temperature as the reservoir). In other words, an increase in the chemical potential of the reservoir will force more particles into the system.

As for the isothermal system, one defines a constrained thermodynamic potential that is the negative of the total entropy times the temperature,

$$\begin{aligned}
\Omega(E,N|\mu,V,T) &\equiv -TS_{\text{total}}(E,N|\mu,V,T) \\
&= E - TS(E,V,N) - \mu N \\
&= F(E|N,V,T) - \mu N. \qquad (3.7)
\end{aligned}$$

From the concavity of the subsystem entropy, the constrained thermodynamic potential is obviously a convex function of N and of E, and it is minimised by their equilibrium values. Thus it represents the appropriate variational principle for a nonequilibrium system able to exchange energy and particles with a reservoir characterised by chemical potential μ and temperature T. As foreshadowed above, it has the appearance of a Legendre transform with respect to N of the constrained thermodynamic potential of a closed system in thermal contact with a reservoir.

In the case of rapid thermal equilibration, one could take $\omega(N|\mu,V,T) \equiv \Omega(E(N,V,T),N|\mu,V,T) = \overline{F}(N,V,T) - \mu N$. This has the appearance of a Legendre transform of the equilibrium free energy of a constant temperature closed system (the Helmholtz free energy), and represents a variational principle for a system at the same temperature as the reservoir, but not in diffusive equilibrium with it.

The constrained thermodynamic potential is minimised by the equilibrium values of the energy $\overline{E}(\mu,V,T)$ and the number $\overline{N}(\mu,V,T)$. In the first place $\partial\Omega(E,N|\mu,V,T)/\partial E = 1 - T\partial S(E,N,V)/\partial E = 0$ when $T(E,V,N) = T$ or at the energy given by $\overline{E} = E(N,V,T)$, as found previously. Further,

$$\frac{\partial\Omega(E,N|\mu,V,T)}{\partial N} = -T\frac{\partial S(E,N,V)}{\partial N} - \mu, \qquad (3.8)$$

which vanishes at $\mu(E,V,N) = \mu$, which again is an implicit equation for $\overline{N} = N(E,V,\mu)$. These are the two equations that determine the equilibrium quantities $\overline{E}(\mu,V,T)$ and $\overline{N}(\mu,V,T)$.

From the concavity of the entropy it follows that the constrained thermodynamic potential is a convex function of particle number, $\Omega'' = -TS'' > 0$. The equilibrium thermodynamic potential or free energy of an open system, which is called the *grand potential*, equals the minimum value of the constrained thermodynamic potential,

$$\begin{aligned}
\overline{\Omega}(\mu,V,T) &\equiv \Omega(\overline{E},\overline{N}|\mu,V,T) \\
&= \overline{F}(\overline{N},V,T) - \mu\overline{N}. \qquad (3.9)
\end{aligned}$$

Again because of the variational nature of the problem, differentiating the grand potential is the same as differentiating the constrained thermodynamic potential holding $\overline{E}$ and $\overline{N}$ fixed. Accordingly,

$$\left(\frac{\partial \overline{\Omega}(\mu, V, T)}{\partial \mu} \right)_{V,T} = -\overline{N}, \tag{3.10}$$

$$\left(\frac{\partial \overline{\Omega}(\mu, V, T)}{\partial V} \right)_{\mu,T} = \left(\frac{\partial \overline{F}(\overline{N}, V, T)}{\partial V} \right)_{\overline{N},T} = -\overline{p}, \tag{3.11}$$

and

$$\left(\frac{\partial \overline{\Omega}(\mu, V, T)}{\partial T} \right)_{\mu,V} = \left(\frac{\partial \overline{F}(\overline{N}, V, T)}{\partial T} \right)_{\overline{N},V} = -\overline{S}. \tag{3.12}$$

Note that this last quantity is not the total entropy but rather $\overline{S}(\mu, V, T) \equiv S(\overline{E}(\mu, V, T), \overline{N}(\mu, V, T), V)$, which is the entropy of the (isolated) subsystem with the equilibrium energy and particle number. These results may be summarised by the total differential

$$d\overline{\Omega} = -\overline{N} d\mu - \overline{p} dV - \overline{S} dT. \tag{3.13}$$

Finally, dividing both sides of the definition by T, one readily sees that

$$\left(\frac{\partial (\overline{\Omega}/T)}{\partial (1/T)} \right)_{\mu,V} = \overline{E} - \mu \overline{N}. \tag{3.14}$$

Multicomponent System

In many cases more than one species of particle is present, so that one has N_α particles of type α in the isolated system. The corresponding chemical potential is

$$\mu_\alpha \equiv -T \left(\frac{\partial S}{\partial N_\alpha} \right)_{E,V,N_{\gamma \neq \alpha}}. \tag{3.15}$$

The total differential of the entropy now includes a sum over species, and for m different types of particles it is

$$dS(E, V, \underline{N}) = \frac{1}{T} dE + \frac{p}{T} dV - \frac{1}{T} \sum_{\alpha=1}^{m} \mu_\alpha dN_\alpha. \tag{3.16}$$

The equilibrium condition for an open system is obviously $\overline{\mu}_\alpha = \mu_\alpha(\overline{E}, V, \overline{\underline{N}})$. The constrained thermodynamic potential may be written as $\Omega(E, \underline{N} | \underline{\mu}, V, T) = E - TS(E, V, \underline{N}) - \underline{\mu} \cdot \underline{N}$.

3.3 Constant Pressure

Now the effects of variable volume are analysed for a subsystem in contact with a reservoir at constant pressure (an *isobaric* system). Consider an isolated total system consisting of two isolated subsystems separated by a moveable, diathermal partition (i.e., the partition is heat conducting). Thus the total energy $E_{\mathrm{total}} = E_1 + E_2$ and volume $V_{\mathrm{total}} = V_1 + V_2$ are fixed, but not their partitioning between the two systems. The second system is an energy and volume reservoir, and so a Taylor expansion is made for its entropy. This yields for the subsystem-dependent part of the total entropy $S_{\mathrm{total}} = S_1 + S_2$

$$S_{\mathrm{total}}(E, V|N, p, T) = S(E, N, V) - \frac{E}{T} - \frac{pV}{T}, \tag{3.17}$$

where p and T are the pressure and temperature of the reservoir. This is the entropy of a subsystem out of equilibrium with respect to energy and volume. Since the entropy must increase, the volume changes such that $0 < dS_{\mathrm{total}} = (p(E, V, N)/T(E, V, N) - p/T)dV$. That is, if the system is in thermal but not mechanical equilibrium, the volume of the system increases if its internal pressure is larger than the external pressure of the reservoir.

The total entropy is maximised at equilibrium, which determines the energy $\overline{E}(N, p, T)$ and the volume $\overline{V}(N, p, T)$ of the system. As previously one develops a minimisation procedure for equilibrium by defining a constrained thermodynamic potential that is the negative of the temperature times the total entropy,

$$\begin{aligned} G(E, V|N, p, T) &\equiv E - TS(E, N, V) + pV \\ &= F(E|N, V, T) + pV. \end{aligned} \tag{3.18}$$

This is a convex function of the system volume and energy, and is the appropriate nonequilibrium potential for a system with moveable diathermal walls in contact with a reservoir of fixed temperature and pressure. In the case of thermal equilibrium having been established, one would take $g(V|N, p, T) = \overline{F}(N, V, T) + pV$ as the appropriate constrained thermodynamic potential that controls volume equilibration.

Minimising $G(E, V|N, p, T)$ with respect to energy gives $T(E, V, N) = T$, which gives the equilibrium energy for a given volume, $\overline{E}(N, V, T)$. Minimising with respect to volume gives $p(E, V, N)/T(E, V, N) = p/T$, which gives the volume for a given energy of the system and temperature and pressure of the reservoir, $\overline{V}(E|N, p, T)$. Simultaneous solution of the two gives the full equilibrium quantities, $\overline{E}(N, p, T)$ and $\overline{V}(N, p, T)$. The latter quantity also follows directly from the minimisation of the thermally equilibrated constrained thermodynamic potential, $\partial g(V|N, p, T)/\partial V = 0$, or

$$\left. \frac{\partial \overline{F}(N, V, T)}{\partial V} \right|_{V=\overline{V}} = -p. \tag{3.19}$$

In view of the convexity of G, the *Gibbs free energy* is defined as the minimum value of the constrained thermodynamic potential,

$$\overline{G}(N, p, T) \equiv G(\overline{V}, \overline{E}|N, T, p)$$

$$= F(N,\overline{V},T) + p\overline{V}. \tag{3.20}$$

(One also has $G(N,p,T) = g(\overline{V}|N,T,p)$.) Again because of the variational nature of the problem, differentiating the Gibbs free energy is the same as differentiating the constrained thermodynamic potential holding $V = \overline{V}$ and $E = \overline{E}$ fixed. Accordingly,

$$\left(\frac{\partial \overline{G}(N,p,T)}{\partial p}\right)_{N,T} = \overline{V}, \tag{3.21}$$

which gives the equilibrium volume of the system,

$$\left(\frac{\partial \overline{G}(N,p,T)}{\partial N}\right)_{p,T} = \left(\frac{\partial \overline{F}(N,\overline{V},T)}{\partial N}\right)_{\overline{V},T} = \overline{\mu}, \tag{3.22}$$

and

$$\left(\frac{\partial \overline{G}(N,p,T)}{\partial T}\right)_{N,p} = \left(\frac{\partial \overline{F}(N,\overline{V},T)}{\partial T}\right)_{N,V} = -\overline{S}. \tag{3.23}$$

These may be summarised by the total differential,

$$d\overline{G} = \overline{V}dp + \overline{\mu}dN - \overline{S}dT. \tag{3.24}$$

Finally, dividing both sides of the definition by T, one readily sees that

$$\left(\frac{\partial (\overline{G}/T)}{\partial (1/T)}\right)_{N,p} = \overline{E} + p\overline{V}. \tag{3.25}$$

3.4 Constant Enthalpy

The quantity on the right side of the last equation is called the *enthalpy*, $H = E + pV$, and it turns out to be constant during volume changes of an isolated system, as will be shown. From the first law, for pV work at constant pressure, the change in heat of the system is $dQ = dE + pdV = d(E + pV)$, which is just the change in enthalpy.

A system isolated by adiathermal (insulated, adiabatic) walls (Fig. 3.2) has insulation that prevents heat flow across the boundaries of the system, $dQ_1 = 0$. One wall is moveable (a piston) and is acted upon by an external potential of the form $dE_2 = p_2 dV_1$. (One could imagine that the piston has area A_1 and mass M, and if the acceleration due to gravity is g, then the constant external pressure is $p_2 = Mg/A_1$.) The total energy $E_{\text{total}} = E_1 + E_2$ is fixed, which means that $dE_1 = -dE_2$. Hence changes in the volume and energy of the subsystem at constant particle number are related by

$$\frac{dE_1}{dV_1} = -p_2, \text{ or } E_1 + p_2 V_1 = \text{const}. \tag{3.26}$$

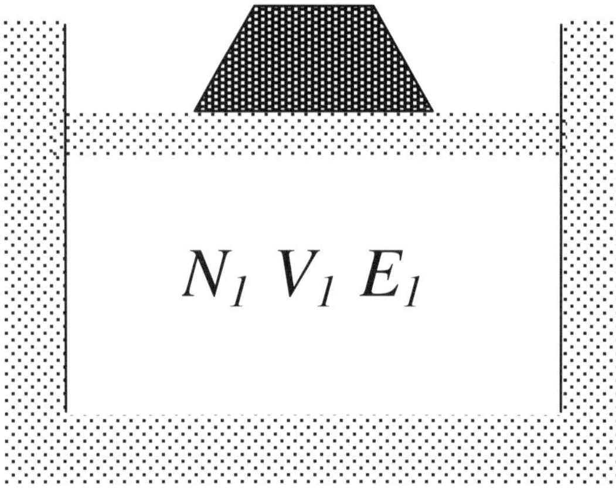

Figure 3.2: An insulated subsystem 1 with moveable piston.

The integrated form of this equation assumes that the external pressure is constant. Hence in a reversible change, when the internal pressure balances the external pressure, $p_1(E_1, V_1, N_1) = p_2$, the enthalpy of the subsystem is constant, $dH_1 = dE_1 + p_1 dV_1 = 0$. More generally, $p_1 \neq p_2$, and it is the enthalpy-like quantity $E_1 + p_2 V_1$ that is constant, where the externally applied pressure is used, not the internal pressure of the subsystem.

The total entropy is that of the subsystem alone, since in this case the external potential is mechanical, not thermodynamic, in nature. Writing the external pressure as p, the total entropy is just $S_{\text{total}}(E, V|N, p) = S_1(E, V, N)$, with $E + pV = \text{const}$. Hence (at constant particle number),

$$\begin{aligned} dS_{\text{total}} &= \frac{\partial S_1}{\partial E}\frac{dE}{dV}dV + \frac{\partial S_1}{\partial V}dV \\ &= \frac{p(E,V,N) - p}{T(E,V,N)}dV. \end{aligned} \tag{3.27}$$

The direction of entropy increase determines the approach to equilibrium. When $p > p(E, V, N)$, dS_{total} is positive when $dV < 0$ (i.e., the subsystem shrinks if the external pressure is greater than the internal pressure of the subsystem). Equilibrium corresponds to pressure equality.

In view of the constancy of the enthalpy, one regards $H = E + pV$ as one of the independent variables for this insulated, variable volume system, and one defines the constrained thermodynamic potential

$$\Phi(V|H, N, p; T) = -TS(E, V, N), \quad E \equiv H - pV. \tag{3.28}$$

Minimising this is the same as maximising the entropy. In order to give this the dimensions of energy an arbitrary temperature has been introduced. It is not essential to do this; since T only enters as a trivial multiplicative factor its value has no physical consequences. By construction Φ is a convex function of volume minimised at the equilibrium volume $\overline{V}(H, N, p)$. Hence,

$$\begin{aligned}\frac{1}{T}\frac{\partial \Phi(V|H,N,p;T)}{\partial V} &= -\frac{\partial S(E,V,N)}{\partial E}\frac{\partial (H-pV)}{\partial V} - \frac{\partial S(E,V,N)}{\partial V} \\ &= \frac{p-p(E,V,N)}{T(E,V,N)}, \quad E \equiv H - pV. \end{aligned} \tag{3.29}$$

Again, this vanishes at equilibrium, when the internal pressure of the subsystem equals the external pressure of the reservoir, $p(\overline{E}, \overline{V}, N) = p$, where $\overline{E} = H - p\overline{V}$.

One defines the thermodynamic potential of this system to be $\overline{\Phi}(H, N, p; T) = \Phi(\overline{V}|H, N, p; T)$. As usual, because of the variational nature of Φ, differentiation of $\overline{\Phi}$ can be carried out at fixed $\overline{V}$. Hence one has

$$\frac{1}{T}\left(\frac{\partial \overline{\Phi}}{\partial N}\right)_{H,p,T} = \frac{1}{T}\left(\frac{\partial \Phi}{\partial N}\right)_{H,p,T,\overline{V}} = \frac{\overline{\mu}}{\overline{T}}, \tag{3.30}$$

$$\frac{1}{T}\left(\frac{\partial \overline{\Phi}}{\partial H}\right)_{N,p,T} = \frac{1}{T}\left(\frac{\partial \Phi}{\partial H}\right)_{N,p,T,\overline{V}} = \frac{-1}{\overline{T}}, \tag{3.31}$$

and

$$\frac{1}{T}\left(\frac{\partial \overline{\Phi}}{\partial p}\right)_{H,N,T} = \frac{1}{T}\left(\frac{\partial \Phi}{\partial p}\right)_{H,N,T,\overline{V}} = \frac{\overline{V}}{\overline{T}}. \tag{3.32}$$

Obviously one also has the trivial result that $\partial\overline{\Phi}/\partial T = \overline{\Phi}/T$. These results may be summarised by the total differential,

$$d\overline{\Phi} = -\frac{T}{\overline{T}}dH + \frac{\overline{V}T}{\overline{T}}dp + \frac{\overline{\mu}T}{\overline{T}}dN + \frac{\overline{\Phi}}{T}dT. \tag{3.33}$$

3.4.1 The Approach to Equilibrium

It is worth analysing this isenthalpic system in a little more detail. In particular it was asserted that the change in internal energy was equal and opposite to the change in the external potential energy, $dE_1 = -dE_2 = -p_2 dV_1$. However the work done by the system on the piston is $dW_1 = p_1 dV_1$, and because no heat flows across the walls of the system, $dQ_1 = 0$, this implies that $dE_1 = -p_1 dV_1$. In the case of general interest, $p_1 \neq p_2$, there is a glaring contradiction between these two expressions.

In order to reconcile the two, one must recognise that the piston acquires kinetic energy as the initially unbalanced forces cause it to accelerate, so that the equation of energy conservation is in fact $dE_1 + dE_2 + dK = 0$. If the piston has a large enough mass, then its motion is slow enough for a quasi-static approach to equilibrium, which means that p_1 is well defined and uniform. However it also means that the kinetic energy is nonnegligible, and that one has $dK = (p_1 - p_2)dV_1$. One expects that the piston will execute simple harmonic motion about the point where the pressures are in balance, but this point is not the equilibrium volume found above. Because of the finite amount of kinetic energy of the piston, at a given point the internal energy of the system is lower than when the kinetic energy was neglected, and consequently so is its temperature and pressure. Nevertheless, given an equation of state one can in principle solve the equations of motion of this system.

Because no dissipative mechanisms have been introduced to this stage, the simple harmonic motion of the piston continues forever, and one can only speak of the equilibrium state in a statistical sense as the average position of the piston. In practice energy is lost due to friction and to the internal viscosity of the system. Assuming that the latter dominates, then when the motion of the piston has died out, $dK = 0$ and $dE_1 = -dE_2$, so that all of the previous equilibrium analysis holds. Effectively the system has been heated by internal motion even though it is enclosed in adiathermal walls.

Finally, it is worth inquiring of the effect of replacing the mechanical external potential by a thermodynamic one. If the second subsystem is a pressure reservoir, then p_2 remains a constant. If in addition the viscosity of the reservoir is negligible (e.g., it might comprise a low-viscosity gas), then all of the preceding comments and analysis goes through, $E_1 + p_2 V_1 = \text{const.}$, and the equilibrium energy, temperature, and volume are as found above. Contrariwise, if the viscous dissipation of the reservoir is comparable to that of the subsystem, one cannot say how the kinetic energy of the piston is divided between the two, and because of the adiathermal walls, the arbitrary internal heating does not equilibrate between them. Although equilibrium still corresponds to pressure equality, one does not know the final energy, temperature, or volume of the subsystem. This indeterminacy of linked isenthalpic systems has been noted by Callan.[1]

3.5 Constant Entropy

The reservoirs of energy, volume, and particles treated above yielded a constrained thermodynamic potential based upon the total entropy. For mathematical completeness, the entropy of the subsystem is now treated as an independently specified variable.

One considers a subsystem able to exchange energy and volume with a reservoir, but in such a way that the entropy of the subsystem is constant. The volume is regarded as the variable in flux, and the energy is regarded as a dependent variable, $E(S, V, N)$. As above the subsystem-dependent part of the constrained total entropy is

$$S_{\text{total}}(V|S, p, N; T) = S - \frac{E(S, V, N)}{T} - \frac{pV}{T}, \tag{3.34}$$

where T and p are the temperature and pressure of the reservoir, respectively. Accordingly the constrained thermodynamic potential is given by

$$H^*(V|S, p, N; T) = E(S, V, N) + pV - TS. \tag{3.35}$$

[1] H. B. Callan, *Thermodynamics*, Appendix C, Wiley, New York, 1960.

Only the first two terms depend upon the volume, and since the constrained thermodynamic potential describes the volume states of the system, one may just as well discard the final term and write the constrained thermodynamic potential as

$$H(V|S,p,N) = E(S,V,N) + pV. \tag{3.36}$$

Hence it is the enthalpy that describes volume fluctuations of an isentropic subsystem.

Since the volume derivative of the energy is the negative of the pressure, one has

$$\left(\frac{\partial H}{\partial V}\right)_{S,p,N} = -p(S,V,N) + p, \tag{3.37}$$

which shows that the equilibrium volume, $\overline{V}(S,p,N)$, that extremises this potential is the one that makes the internal pressure of the subsystem equal to the external pressure of the reservoir. Since the constrained thermodynamic potential is derived from the total entropy, one can be assured that it is a convex function of volume, and that the extremum is a minimum,

$$\left(\frac{\partial^2 H}{\partial V^2}\right)_{S,p,N} = \left(\frac{\partial^2 E}{\partial V^2}\right)_{S,N} = -\left(\frac{\partial p(S,V,N)}{\partial V}\right)_{S,N} > 0. \tag{3.38}$$

An explicit proof of this is deferred until the next section.

The thermodynamic potential, in this case the equilibrium enthalpy, is the minimum value of the constrained thermodynamic potential,

$$\overline{H}(S,p,N) = E(S,\overline{V},N) + p\overline{V}. \tag{3.39}$$

Since differentiating the enthalpy is the same as differentiating the constrained thermodynamic potential with $\overline{V}$ fixed, one has

$$\left(\frac{\partial \overline{H}(S,p,N)}{\partial p}\right)_{N,S} = \overline{V}, \tag{3.40}$$

$$\left(\frac{\partial \overline{H}(S,p,N)}{\partial N}\right)_{p,S} = \left(\frac{\partial E(S,\overline{V},N)}{\partial N}\right)_{\overline{V},S} = \overline{\mu}, \tag{3.41}$$

and

$$\left(\frac{\partial \overline{H}(S,p,N)}{\partial S}\right)_{N,p} = \left(\frac{\partial E(S,\overline{V},N)}{\partial S}\right)_{N,\overline{V}} = \overline{T}. \tag{3.42}$$

These may be summarised by the total differential,

$$d\overline{H} = \overline{V}dp + \overline{\mu}dN + \overline{T}dS. \tag{3.43}$$

In the case of the full thermodynamic potential, $\overline{H}^*(S,p,N;T)$, the results remain the same, except that the final one is replaced by

$$\left(\frac{\partial \overline{H}^*(S,p,N;T)}{\partial S}\right)_{N,p;T} = \overline{T} - T. \tag{3.44}$$

Note that there is no requirement for the temperature of the subsystem to equal that of the reservoir. In addition one has the entropy itself

$$\left(\frac{\partial \overline{H}^*(S, p, N; T)}{\partial T} \right)_{S,p,N} = -S, \tag{3.45}$$

and the equilibrium enthalpy

$$\left(\frac{\partial (\overline{H}^*(S, p, N; T)/T)}{\partial T^{-1}} \right)_{S,p,N} = -(\overline{E} + p\overline{V}). \tag{3.46}$$

3.6 Thermodynamic Second Derivatives

So far the main concern has been with the first derivatives of the constrained and equilibrium thermodynamic potentials, which yield the equilibrium value of the thermodynamic quantities. The second derivatives also yield measurable parameters, and it will be shown that these determine the statistical fluctuations about equilibrium. As seen above, there is a variety of thermodynamic systems, each characterised by their own set of dependent and independent variables. Because of the uniqueness of the thermodynamic state, it matters little which three variables one chooses as independent. This section will deal almost exclusively with equilibrium. To simplify the notation the equilibrium quantities will not always be overlined, since what depends upon what should be clear from the context. This section relies heavily upon the properties of partial derivatives outlined in Appendix C.

3.6.1 Concavity of the Thermodynamic Potentials

The generic treatment of the reservoirs above consisted in essence of interchanging an extensive conserved variable with its intensive conjugate. The total entropy was identified and used to form a constrained thermodynamic potential that was convex with respect to the nonequilibrium extensive variable. The equilibrium thermodynamic potential or free energy equalled the minimum value of the constrained thermodynamic potential. It is now shown in general that the equilibrium free energy is concave with respect to the conjugate intensive variable.

Let X be the conserved extensive variable and let the conjugate intensive (field) variable be

$$x = -\left(\frac{\partial E}{\partial X} \right)_S, \tag{3.47}$$

or

$$x = T\frac{\partial S}{\partial X}. \tag{3.48}$$

Here X could stand for E, N, or V, or it could be an additional parameter of the system. The subsystem-dependent part of the total entropy of a subsystem plus reservoir at x and T, for exchange of X, is

$$S_{\rm total}(X|x) = S(X) - \frac{xX}{T}. \tag{3.49}$$

As above, the total entropy is a concave function of the extensive variable,

$$\left(\frac{\partial^2 S_{\rm total}(X|x)}{\partial X^2}\right)_x < 0. \tag{3.50}$$

The equilibrium value of X is $\overline{X}(x)$ (or more fully $\overline{X}(x,T)$), and this gives the maximum total entropy, $\overline{S}_{\rm total}(x) = S_{\rm total}(\overline{X}(x)|x)$. It is now proven that this latter quantity is a convex function of the reservoir field x/T.

One has

$$\begin{aligned} \frac{d^2 S_{\rm total}(\overline{X}(x)|x)}{d(x/T)^2} &= \frac{d}{d(x/T)}\left[\frac{\partial S(X)}{\partial X}\frac{d\overline{X}}{d(x/T)} - \frac{x}{T}\frac{d\overline{X}}{d(x/T)} - \overline{X}(x)\right] \\ &= -\frac{d\overline{X}(x)}{d(x/T)}, \end{aligned} \tag{3.51}$$

where the cancellation occurs because $x(\overline{X}(x)) = x$ and $T(\overline{X}(x)) = T$. (On the left-hand side of each of these equations appears a property of the subsystem, and the quantity on the right-hand side is a property of the reservoir.) Now the equilibrium value is $\overline{X}$, and if X' is any other value, then by definition $S_{\rm total}(X'|x) < S_{\rm total}(\overline{X}|x)$, or

$$\frac{x\overline{X}}{T} - \frac{xX'}{T} < S(\overline{X}) - S(X'). \tag{3.52}$$

An expansion to second order about X' yields

$$\begin{aligned} S(\overline{X}) &= S(X') + (\overline{X} - X')S_X(X') + \frac{(\overline{X} - X')^2}{2}S_{XX}(X') \\ &= S(X') + (\overline{X} - X')\frac{\overline{x}'}{\overline{T}'} + \frac{(\overline{X} - X')^2}{2}S_{XX}(X'), \end{aligned} \tag{3.53}$$

so that

$$\left[\frac{x}{T} - \frac{\overline{x}'}{\overline{T}'}\right](\overline{X} - X') < \frac{(\overline{X} - X')^2}{2}S_{XX}(X'). \tag{3.54}$$

Since the right-hand side is negative due to the concavity of the entropy, dividing by $(\overline{X} - X')^2$ and taking the limit yields

$$\frac{d(\overline{x}/\overline{T})}{dX} < 0, \tag{3.55}$$

or equivalently

$$\frac{d\overline{X}}{d(x/T)} < 0. \tag{3.56}$$

One concludes that the total entropy maximum is a convex function of x/T,

$$\frac{d^2 S_{\text{total}}(\overline{X}(x)|x)}{d(x/T)^2} > 0. \tag{3.57}$$

This convexity of the total entropy shows that the equilibrium thermodynamic potential, $\overline{F}(x) \equiv F(\overline{X}(x)|x) = -TS_{\text{total}}(\overline{X}(x)|x)$, is concave,

$$\frac{d^2(\overline{F}(x)/T)}{d(x/T)^2} < 0. \tag{3.58}$$

To identify X and x in a given case, one notes that

$$\frac{d(\overline{F}(x)/T)}{d(x/T)} = \overline{X}(x). \tag{3.59}$$

3.6.2 Heat Capacity

The heat capacity is the rate of change of heat with temperature, and from the first law the heat change of the system is the change in energy of the system less the work done on the system, $dQ = dE - dW$. Neither heat nor work is a state function, and in the case of interest it matters whether the change is at constant volume ($dW = 0$) or at constant pressure ($dW = -pdV$). The heat capacity at constant volume is

$$C_V = \left(\frac{\partial Q}{\partial T}\right)_{V,N} = \left(\frac{\partial \overline{E}}{\partial T}\right)_{V,N}. \tag{3.60}$$

The energy that appears here is the energy of a subsystem in equilibrium with a heat reservoir of temperature T, $\overline{E}(N,V,T)$, and it was given above as a derivative of the Helmholtz free energy, so that one has

$$C_V = \frac{-1}{T^2}\left(\frac{\partial^2(\overline{F}/T)}{\partial(1/T)^2}\right)_{V,N}. \tag{3.61}$$

In terms of the general concavity result established in the preceding section, this is the case of $x \equiv 1$ (and $X \equiv E$), and so one concludes that

$$C_V > 0. \tag{3.62}$$

The heat capacity at constant pressure is

$$\begin{aligned} C_p &= \left(\frac{\partial \overline{E}}{\partial T}\right)_{p,N} + p\left(\frac{\partial \overline{V}}{\partial T}\right)_{p,N} \\ &= \frac{-1}{T^2}\left(\frac{\partial^2(\overline{G}/T)}{\partial(1/T)^2}\right)_{p,N}, \end{aligned} \tag{3.63}$$

where the fact that the enthalpy $E + pV$ can be written as a derivative of the Gibbs free energy has been used. One again identifies $x \equiv 1$ (and from the inverse temperature derivative of the Gibbs free energy one has $X = H$), so that

$$C_p > 0. \tag{3.64}$$

One can in fact show that $C_p > C_V$, which may be interpreted as signifying that extra heat is required to raise the temperature of a system at constant pressure because one does additional work expanding the volume against the external pressure.

3.6.3 Compressibility

The compressibility is the rate of change of volume with pressure, and this can be done at constant temperature or constant entropy. The isothermal compressibility is

$$\chi_T = \frac{-1}{\overline{V}} \left(\frac{\partial \overline{V}}{\partial p} \right)_{T,N}. \tag{3.65}$$

The volume here is $\overline{V}(N, p, T)$, which is the pressure derivative of the Gibbs free energy, so that this may be rewritten

$$\chi_T = \frac{-1}{\overline{V}} \left(\frac{\partial^2 \overline{G}}{\partial p^2} \right)_{T,N} = \frac{-1}{\overline{V}T} \left(\frac{\partial^2 (\overline{G}/T)}{\partial (p/T)^2} \right)_{T,N}. \tag{3.66}$$

One identifies $x \equiv p/T$ (and $X \equiv V$), and one concludes that the isothermal compressibility is positive, $\chi_T > 0$.

The treatment of constant entropy is slightly different. In this case the adiabatic compressibility,

$$\chi_S = \frac{-1}{\overline{V}} \left(\frac{\partial \overline{V}}{\partial p} \right)_{S,N}, \tag{3.67}$$

can be written as the second derivative of the enthalpy

$$\chi_S = \frac{-1}{\overline{V}} \left(\frac{\partial^2 \overline{H}}{\partial p^2} \right)_{S,N}. \tag{3.68}$$

For the system with constant entropy and variable volume treated above, it was claimed that the constrained thermodynamic potential, $H(V|S, p, N)$, was convex with respect to volume. This is now proved, since it is equivalent to the positivity of the isentropic compressibility. One has

$$\begin{aligned}
&\left(\frac{\partial^2 H}{\partial V^2} \right)_{S,p,N} \\
&= -\left(\frac{\partial p(S,V,N)}{\partial V} \right)_{S,N} \\
&= -\left(\frac{\partial p(E,V,N)}{\partial V} \right)_{E,N} - \left(\frac{\partial p(E,V,N)}{\partial E} \right)_{V,N} \left(\frac{\partial E(S,V,N)}{\partial V} \right)_{S,N}
\end{aligned}$$

$$
\begin{aligned}
&= -\left(\frac{\partial(Tp/T)}{\partial V}\right)_{E,N} + p\left(\frac{\partial(Tp/T)}{\partial E}\right)_{V,N} \\
&= -\frac{p}{T}\left(\frac{\partial T}{\partial V}\right)_{E,N} - T\left(\frac{\partial(p/T)}{\partial V}\right)_{E,N} + \frac{p^2}{T}\left(\frac{\partial T}{\partial E}\right)_{V,N} \\
&\quad + pT\left(\frac{\partial(p/T)}{\partial E}\right)_{V,N} \\
&= -T\left[p^2 S_{EE} - 2pS_{EV} + S_{VV}\right]. \qquad (3.69)
\end{aligned}
$$

The final quantity in brackets is negative (from Eq. (2.21), with $dE = -pdV$), and so this shows that the constrained thermodynamic potential $H(V|S, p, N)$ is a convex function of volume, as promised. From the first equality it also shows that $\chi_S > 0$, as required. Note that on the left-hand side the pressure is p, the independent variable of the constrained thermodynamic potential $H(V|S, p, N)$, whereas on the right-hand side it is $p(S, V, N)$ or $p(E, V, N)$, as appropriate.

3.6.4 Maxwell Relations

One can also give expressions for the mixed second derivatives. The *Maxwell relations* result from interchanging the order of differentiation. For example, from the temperature and volume derivatives of the Helmholtz free energy,

$$\frac{\partial^2 \overline{F}(N,V,T)}{\partial V \partial T} = \frac{\partial^2 \overline{F}(N,V,T)}{\partial T \partial V}, \qquad (3.70)$$

one obtains

$$\left(\frac{\partial \overline{S}(N,V,T)}{\partial V}\right)_{T,N} = \left(\frac{\partial \overline{p}(N,V,T)}{\partial T}\right)_{V,N}. \qquad (3.71)$$

Here the equilibrium entropy of the subsystem is $\overline{S}(N, V, T) \equiv S(\overline{E}, V, N)$, where $\overline{E} = \overline{E}(N, V, T)$. Obviously one can generate a large number of such relations in this fashion.

3.7 Equivalence of Systems

The temperature of an isolated or microcanonical system of energy E_1 may be written $T_1 = T(E_1, N, V)$. Now consider a subsystem in contact with a heat reservoir of this temperature. The equilibrium energy of this canonical system is $\overline{E}(N, V, T_1)$. By definition, this equilibrium energy is the one that gives temperature equality between the subsystem and the reservoir, $T(\overline{E}, N, V) = T_1$. Hence it is evident that $\overline{E} = E_1$. In words, the equilibrium energy of a canonical system of a given temperature is equal to the energy required for a microcanonical system to have the same temperature. The two systems are equivalent in this sense, and the result manifests the underlying consistency of thermodynamics.

The consistency extends to the other thermodynamic properties, as is obvious from their derivation as derivatives of the thermodynamic potential. For example, if the microcanonical pressure is $p_1 = p(E_1, N, V)$, and the equilibrium pressure of the canonical system is $\overline{p}(N, V, T_1)$, then by definition,

$$\overline{p} = -\left(\frac{\partial \overline{F}(N, V, T_1)}{\partial V}\right)_{N,T_1} = T_1 \left(\frac{\partial \overline{S}(\overline{E}, V, T_1)}{\partial V}\right)_{\overline{E},N} = p_1, \quad (3.72)$$

since $\overline{E} = E_1$. Hence if the energy and temperature of the two systems are chosen consistently, then the other equilibrium properties are also equal for the two systems. This is a general rule that holds for the first derivatives of all the thermodynamic potentials.

The equivalence, however, does not extend to the second and higher derivatives. It has already been asserted that the heat capacity at constant volume, which is essentially the second inverse temperature derivative of the Helmholtz free energy, is less than the heat capacity at constant pressure, which is the second derivative of the Gibbs free energy. Similarly one has

$$\left(\frac{\partial \overline{E}(p, T)}{\partial T}\right)_p = \left(\frac{\partial \overline{E}(V, T)}{\partial T}\right)_V + \left(\frac{\partial \overline{E}(p, T)}{\partial p}\right)_T \left(\frac{\partial \overline{p}(V, T)}{\partial T}\right)_V, \quad (3.73)$$

where $V = \overline{V}(p, T)$ and the fixed N has been suppressed. Evidently the temperature derivative of the equilibrium energy of an isobaric system does not equal that of a canonical system. Hence in general it is only the equilibrium properties that represent first derivatives that are equivalent between consistent systems.

3.8 Extensivity

As discussed previously, an *extensive* variable scales with the size of the system. Examples include the number of particles, the volume, the energy, and the constrained and equilibrium thermodynamic potentials. *Intensive* variables do not depend upon the size of the system, but only on other intensive variables or the ratio of pairs of extensive variables. Examples include the temperature, the pressure, and the chemical potential.

In the case of an isolated system, if the extensive variables are all scaled by λ, then the entropy itself must be scaled by λ,

$$S(\lambda E, \lambda V, \lambda N) = \lambda S(E, V, N). \quad (3.74)$$

This follows from the linear additivity of the entropy, namely that the total entropy of two isolated subsystems is the sum of their individual entropies. This shows that the entropy is a homogeneous function of first degree in its arguments. Differentiating both sides with respect to λ and setting $\lambda = 1$, one obtains a particular case of the general theorem due to Euler,

$$\begin{aligned} S(E, V, N) &= \frac{\partial S}{\partial E} E + \frac{\partial S}{\partial V} V + \frac{\partial S}{\partial N} N \\ &= \frac{1}{T} E + \frac{p}{T} V - \frac{\mu}{T} N. \end{aligned} \quad (3.75)$$

This may be rewritten as $E = TS - pV + \mu N$, with total derivative

$$dE = TdS - pdV + \mu dN + (SdT - Vdp + Nd\mu) \, . \tag{3.76}$$

Comparing this with the expression given previously, $dE = TdS - pdV + \mu dN$, one concludes that the parenthetical term must vanish,

$$SdT - Vdp + Nd\mu = 0. \tag{3.77}$$

This is known as the Gibbs–Duhem equation. It shows that the intensive variables cannot all be chosen independently, since the system variables can be used to specify independent changes in at most two of them. Hence any equilibrium intensive quantity can be written as a function of just two intensive parameters. The generalisation to a multicomponent system is straightforward. It can be used, for example, to show how changes in composition at fixed temperature and pressure cause related changes in chemical potentials.

The cases when the equilibrium thermodynamic potentials depend only upon one extensive variable are also of interest. For example, in order for the grand potential $\Omega(\mu, V, T)$ to scale with the system size it must be linearly proportional to V. The fact that its volume derivative gives the negative of the pressure shows that the latter is the proportionality constant, and one concludes that

$$\overline{\Omega}(\mu, V, T) = -\overline{p}V, \tag{3.78}$$

where the equilibrium pressure is $\overline{p}(\mu, T)$. Similarly the Gibbs free energy $\overline{G}(N, p, T)$ must be of the form

$$\overline{G}(N, p, T) = \overline{\mu}N, \tag{3.79}$$

where the equilibrium chemical potential is $\overline{\mu}(p, T)$.

An analogous treatment of the other thermodynamic potentials does not yield a single thermodynamic quantity as a coefficient. An example was given above in the treatment of the Helmholtz free energy, where Eq. (2.27) shows that

$$\overline{F}(N, V, T) = [\overline{\epsilon}(\rho, T) - T\overline{\sigma}(\rho, T)]V, \tag{3.80}$$

where the energy density is $\epsilon = E/V$, the number density is $\rho = N/V$, and the entropy density is $\sigma = S/V$. An alternative expression can be obtained by differentiating the Helmholtz free energy density $\overline{f}(\rho, T) \equiv \overline{F}(N, V, T)/V$ with respect to volume,

$$-\overline{p} = \overline{f}(\rho, T) - \frac{N}{V^2}\overline{f}'(\rho, T), \tag{3.81}$$

and with respect to particle number,

$$\overline{\mu} = \frac{1}{V}\overline{f}'(\rho, T), \tag{3.82}$$

where the prime denotes differentiation with respect to the density. Combining these two equations, one concludes that the Helmholtz free energy density may also be written

$$\overline{f}(\rho, T) = \rho\overline{\mu}(\rho, T) - \overline{p}(\rho, T). \tag{3.83}$$

This is just an intensive form of a Legendre transformation between an isothermal and an open system.

It must be emphasised that the results in this section only apply to a macroscopic system. The first correction to these terms that scale with V is in general $\mathcal{O}V^{2/3}$, which is only negligible in the thermodynamic limit.

Summary

- Subsystems with different boundaries are treated by analysing the total entropy of the subsystem plus reservoir with which it can exchange a linear additive conserved quantity. The constrained thermodynamic potential of such a subsystem is the negative of the temperature times the subsystem-dependent part of the total entropy, which has the appearance

System	Constrained potential Parameters	Equilibrium potential Differential
Isolated	– N, V, E	Entropy, S $TdS = dE + pdV - \mu dN$
Isothermal	$F = E - TS(E, V, N)$ $E\|N, V, T$	Helmholtz, $\overline{F}$ $d\overline{F} = -\overline{S}dT - \overline{p}dV + \overline{\mu}dN$
Open	$\Omega = E - TS(E, V, N) - \mu N$ $N, E\|\mu, V, T$	Grand, $\overline{\Omega}$ $d\overline{\Omega} = -\overline{S}dT - \overline{p}dV - \overline{N}d\mu$
Isobaric	$G = E - TS(E, V, N) + pV$ $E, V\|N, p, T$	Gibbs, $\overline{G}$ $d\overline{G} = -\overline{S}dT + \overline{V}dp + \overline{\mu}dN$
Isenthalpic	$\Phi = -TS(H - pV, V, N)$ $V\|N, H, p; T$	$\overline{\Phi}$ $\overline{T}d\overline{\Phi}/T = -dH + \overline{V}dp + \overline{\mu}dN$ $+\overline{\Phi}\,\overline{T}dT/T^2$
Isentropic	$H = E(S, V, N) + pV$ $V\|N, S, p$	Enthalpy, $\overline{H}$ $d\overline{H} = \overline{V}dp + \overline{\mu}dN + \overline{T}dS$

Table 3.1: Summary of the Common Thermodynamic Systems

of a Legendre transform (cf. Table 3.1). The equilibrium thermodynamic potential or free energy is the minimum value of the constrained thermodynamic potential.

- The free energy is concave with respect to the intensive reservoir variable. From this follows the positivity of the heat capacity and the compressibility, which are second derivatives of various thermodynamic potentials. The Maxwell relations follow from equating the cross-second derivatives. The thermodynamic potential or free energy is not a minimum with respect to any of its independent variables; it is in fact a concave function of some variables.

- Different thermodynamic systems are consistent as far as the equivalence of equilibrium quantities given by the first derivative of the thermodynamic potentials.

- An extensive variable scales with the size of a macroscopic system. At least one extensive variable must be specified independently, because the three intensive variables, μ, p, and T, cannot all be chosen independently.

Chapter 4

Probability and the General Formalism

The potentials for the various types of systems derived in the preceding two chapters represent particular cases of a generic reservoir approach. Here the general formalism is given, with the object of obtaining the probability distribution that corresponds to each reservoir.

4.1 Probability Distributions

4.1.1 Isolated System

An isolated system with its weighted microstates plays a fundamental role in thermodynamics and statistical mechanics.[1] The state of an isolated system is the set of values of the variables $\{A_i\}$ that are the linear additive constants of the motion. These include the energy, volume, number of each species, components of the linear and angular momenta, etc. The number of microstates and their weight depend upon the state of a system, and hence so does the entropy. The set of variables may be equivalently written $\underline{A}$, so that the entropy is $S(A_1, A_2, \ldots) \equiv S(\underline{A})$. By virtue of the additivity property these variables are all extensive, and one defines the conjugate intensive field variables as the derivatives of the entropy,

$$a_i = T\frac{\partial S(\underline{A})}{\partial A_i}. \tag{4.1}$$

Here the temperature of the system T is included in the definitions to match them to the conventional field variables. Hence if $A_1 \equiv E$, then $a_1 \equiv 1$, if $A_2 \equiv V$, then $a_2 \equiv p$, and if $A_3 \equiv N$, then $a_3 \equiv -\mu$. Strictly speaking, at

[1]If the microstates have equal weight, then the weight of a macrostate is just the number of corresponding microstates. Contrariwise, it is the sum of the weights of the corresponding microstates.

this stage the analysis is still restricted to discrete states, so that this should be interpreted as a difference equation.

4.1.2 Exchange with a Reservoir

Consider an isolated system, labelled t, comprising two subsystems labelled s and r, the latter eventually becoming the reservoir. The state of the total system is specified by the set of values of the extensive variables denoted by $\{A_{it}\}$ (equivalently $\underline{A}_t$). The state of each subsystem is denoted by $\{A_{is}\}$ and $\{A_{ir}\}$ (equivalently $\underline{A}_s$ and $\underline{A}_r$), and these label the macrostates of the total system.

The subsystems are in contact such that some, but not all, of these variables may be exchanged between them. The variations in these exchangeable variables are taken to occur independently.[2] Where necessary, the parameters that are exchangeable will be denoted $\underline{A}'$, and those that are fixed will be denoted by $\underline{A}''$, so that $\underline{A} \equiv (\underline{A}', \underline{A}'')$. Since the exchangeable variables represent linear additive conserved quantities, one has $A'_{is} + A'_{ir} = A'_{it}$, and $dA'_{is} + dA'_{ir} = 0$. (Obviously the A''_{is} and the A''_{ir} are individually constant, $dA''_{is} = dA''_{ir} = 0$.)

The region of interaction between the two subsystems is taken to be very much smaller than either subsystem, so that each subsystem may be considered effectively isolated. Consequently, the number of microstates of the total system corresponding to the macrostate $(\underline{A}_s, \underline{A}_r)$ is the product of the number of microstates in each subsystem considered as isolated and in its respective state $\underline{A}_s$ or $\underline{A}_r$. Consequently, the entropy of the system in a given macrostate is the sum of the entropies of the two subsystems,

$$S_{\text{total}}(\underline{A}_s, \underline{A}_r) = S_s(\underline{A}_s) + S_r(\underline{A}_r). \tag{4.2}$$

That is, the contribution to the total entropy due to the interactions between the two subsystems is negligible. It is emphasised that on the left side is the entropy of the total system in a given macrostate, and on the right side appears the entropies of the subsystems, each considered in isolation and in a fixed state. It is not necessary for the two subsystems to be identical in nature, so that the respective entropies may be different functions of their arguments. This possibility is signified by the subscripts on the entropy; actually little of import is gained by allowing for this general case, and these subscripts could just as well be dropped.

The probability of finding the total system in a particular macrostate (i.e., with a particular partitioning of the parameters between the two subsystems) is proportional to the exponential of the total entropy of that macrostate. (This follows because the probability is proportional to the macrostate weight, and the entropy is the logarithm of the weight.) Because of the conservation law one need only specify the macrostate of subsystem s and the total value of the

[2] The formalism needs to be modified for systems where the usual extensive variables are not independent. These include the isenthalpic and the isentropic reservoirs treated in Sections 3.4 and 3.5.

parameters, since this is the same as specifying the state of both subsystems, and one may write the macrostate probability as

$$\begin{aligned} \wp(\underline{A}'_s|\underline{A}''_s, \underline{A}_t) &= \frac{1}{Z} e^{S_{\text{total}}(\underline{A}_s, \underline{A}_r)/k_{\text{B}}} \\ &= \frac{1}{Z} e^{S_s(\underline{A}_s)/k_{\text{B}}} e^{S_r(\underline{A}_r)/k_{\text{B}}}. \end{aligned} \tag{4.3}$$

With care one may interpret this as the product of two independent probabilities, taking account of the conservation laws that make the subsystems interdependent, $\underline{A}'_r = \underline{A}'_t - \underline{A}'_s$. That is, the states of the two subsystems define the macrostate of the total system, and hence the weight of the given macrostate is the product of the weights of the states of each subsystem considered in isolation in its respective state. The left-hand side of this equation may be read as the probability of the subsystem s being in the state $\underline{A}'_s$ given the values of its nonexchangeable parameters $\underline{A}''_s$ and the total values of the parameters $\underline{A}_t$.

This is the general expression, but now the particular case that subsystem r is a reservoir is considered. This means that subsystem r is so much larger than subsystem s that changes in the values of the parameters of the latter have negligible effect on the properties of the former. In consequence one may expand the entropy of the reservoir about the total value of the parameters and truncate the expansion at the linear term,

$$\begin{aligned} S_r(\underline{A}_t - \underline{A}_s) &= S_r(\underline{A}_t) - A'_{1s} \frac{\partial S_r(\underline{A}_t)}{\partial A'_{1t}} - A'_{2s} \frac{\partial S_r(\underline{A}_t)}{\partial A'_{2t}} - \ldots \\ &= \text{const.} - A'_{1s} \frac{a'_{1r}}{T_r} - A'_{2s} \frac{a'_{2r}}{T_r} - \ldots . \end{aligned} \tag{4.4}$$

Here the $a'_{ir} \equiv T_r \partial S(A_{1r}, A_{2r}, \ldots)/\partial A'_{ir}$ are the field variables of the reservoir conjugate to the exchangeable parameters, and T_r is its temperature. The reason that it is permissible to truncate this at the linear term is that the entropy is an extensive property, so that its first derivative is of order unity, its second derivative is of order one over the size of the reservoir, etc. This is the reason that it is the entropy that is expanded and truncated; a direct expansion of the reservoir probability leads to powers of its logarithmic derivative that are all of order unity. These resum to the logarithm of the entropy expansion, so that the same final answer results, but the analysis is much less transparent. The constant contribution in the reservoir entropy expansion, $S_r(\underline{A}_t)$, which is independent of the subsystem, will be dropped because it has no physical consequence. It is more convenient to specify the reservoir by its field variables a'_{ir} than by the total fixed value of the conserved quantities. (This is permissible since, as mentioned above, the entropy is a strictly concave function of its arguments, and hence there is a one-to-one relationship between A_{ir} and a_{ir}.) The subscripts that distinguish the subsystem variables from those of the reservoir may be dropped where there is no ambiguity; the reservoir only enters via its temperature and other field variables conjugate to the parameter in flux, and it is only the field variables of the subsystem that may be confused with reservoir properties.

With this expansion the constrained total entropy becomes

$$S_{\text{total}}(\underline{A}'|\underline{a}', \underline{A}'', T) = S(\underline{A}) - \underline{A}' \cdot \underline{a}'/T. \tag{4.5}$$

This is of course just the subsystem-dependent part of the total entropy for the macrostate labelled $\underline{A}'$. Throughout, the entropy of the total system will mean this subsystem-dependent part only, and it consists of the internal subsystem entropy and the variable part of the reservoir entropy. The probability of a particular macrostate is just the exponential of this,

$$\wp(\underline{A}'|\underline{a}', \underline{A}'', T) = \frac{1}{Z(\underline{a}', \underline{A}''; T)} e^{S(\underline{A})/k_{\mathrm{B}}} e^{-\underline{A}' \cdot \underline{a}'/k_{\mathrm{B}}T}, \tag{4.6}$$

where $\underline{A} = (\underline{A}', \underline{A}'')$, and $\underline{A}' \cdot \underline{a}' = \sum_i A'_i a'_i$. Note that the constant part of the reservoir entropy has been cancelled with a similar factor in the normalisation denominator. This normalisation factor Z is called the *partition function*, and its properties will be explored in detail in Section 4.3.

4.1.3 Equilibrium

The above gives the probability of a subsystem having a particular value of a parameter that it can exchange with a reservoir. The most probable macrostate is called the *equilibrium state*, and it is denoted by an overline, $\overline{\underline{A}'}$. This is the state in which the system is most likely to be found. If the system is initially far from equilibrium, by virtue of the transitions between microstates it will soon be in one of the microstates corresponding to the macrostate with the largest number. (Almost all of the microstates of a realistic system are in a few macrostates clustered around the equilibrium one, so that the probability of observing a fluctuation far from equilibrium is exceedingly small.)

The most probable macrostate is given by the peak of the distribution. This corresponds to the maximum of the exponent, which is to say that the equilibrium macrostate is the one with the maximum constrained total entropy (equivalently, the one with the greatest weight of microstates). Equilibrium is given by the vanishing of

$$\begin{aligned} \frac{\partial S_{\text{total}}(\underline{A}'|\underline{a}'_r, \underline{A}'', T)}{\partial A'_i} &= \frac{\partial S(\underline{A})}{\partial A'_i} - \frac{a'_{ir}}{T_r} \\ &= \frac{a'_{is}}{T_s} - \frac{a'_{ir}}{T_r}, \end{aligned} \tag{4.7}$$

where $T_s(\underline{A})$ is the temperature of the subsystem. The reservoir quantities have here again been distinguished by the subscript r (equivalently, in the reservoir limit $a'_{ir} = a'_{it}$ and $T_r = T_t$). That is, equilibrium corresponds to equality between the field variables divided by their respective temperatures, $\overline{a}'_{is}/\overline{T}_s = a'_{ir}/T_r$, which is an implicit equation for the $\overline{A}'_i$. That is, $\overline{a}'_{is} = a'_{is}(\overline{\underline{A}}')$ and $\overline{T}_s = T_s(\overline{\underline{A}}')$, where the right-hand sides represent the appropriate derivative of the isolated subsystem entropy evaluated at the equilibrium macrostate. In the case of a heat reservoir (energy exchange occurs), the equilibrium amount of energy in the subsystem is the one that gives it the same temperature as the reservoir, $\overline{T}_s = T_r$. In this case it is the field variables themselves that are equalised at equilibrium, $\overline{a}'_{is} = a'_{ir}$.

It may be mentioned that the conclusion that equilibrium corresponds to equality of the intensive field variables holds even if the second subsystem is not a reservoir. In this case however the field variable of the reservoir is not constant, but depends upon how much of the extensive variable has been partitioned between the subsystems.

A system in a nonequilibrium macrostate will move toward the equilibrium macrostate, which is to say that its entropy increases. This determines the direction of flow of the exchangeable parameter since $dS_{\text{total}} = \sum_i [(\partial S/\partial A'_i) - (a'_{ir}/T_r)]\, dA'_i$, and for the total entropy to increase one must have $\sum_i [(a'_{is}/T_s) - (a'_{ir}/T_r)]\, dA'_i \geq 0$. Taking these one at a time (i.e., assuming that the rest, if any, are already equilibrated), the exchangeable parameter flows into the subsystem, ($dA'_i \geq 0$), if the value of the corresponding reservoir field variable divided by temperature is less than that of the subsystem. For the case of a heat reservoir, the energy of the subsystem increases if the inverse temperature of the reservoir is less than the inverse temperature of the subsystem, which is known as the third law of thermodynamics.

It ought be noted that in the present analysis the second law of thermodynamics is not so much an axiom as a conclusion. Equilibrium is the most probable macrostate, which is that with the greatest weight of corresponding microstates. The constrained total entropy, which is the logarithm of this, is consequently also maximal at equilibrium. This also shows clearly that the second law is *not* inviolate, since the analysis explicitly allows for entropy-decreasing spontaneous fluctuations away from the equilibrium macrostate. There is a nonzero probability that the system will spontaneously move from the equilibrium macrostate to one with lower entropy, and so the assertion that entropy never spontaneously decreases is, strictly speaking, false.

Concavity of the Entropy

For the equilibrium state of a subsystem able to exchange quantities with a reservoir to be stable, the entropy must be a concave function of its arguments, which is to say it must decrease in any direction away from equilibrium. A Taylor expansion to second order about the equilibrium state expresses this condition as

$$\underline{A}^{\mathrm{T}} \underline{\underline{S}}\, \underline{A} < 0, \tag{4.8}$$

where $\{\underline{\underline{S}}\}_{ij} = \partial^2 S/\partial A_i \partial A_j$ is the matrix of exchangeable quantities, and the A_i represents the departure from the equilibrium state. Since a system may be considered a reservoir for any of its parts, the necessary and sufficient condition for a given state of an isolated system to be stable is that this must be true for arbitrary vectors $\underline{A}$. Hence necessary conditions for stability of an isolated system may be obtained by considering one exchangeable parameter at a time, which gives $S_{ii} A_i^2 < 0$, or

$$S_{ii} < 0, \tag{4.9}$$

for any i. By considering $\underline{A}$ to be an eigenvector of $\underline{\underline{S}}$, the necessary and sufficient condition for a stable state is that all the eigenvalues of $\underline{\underline{S}}$ are negative. Considering the independent variables two at a time, this last condition implies that the determinant of the 2×2 matrix $\underline{\underline{S}}$ is positive (because it is the product of the eigenvalues). Hence one has another necessary stability condition, namely

$$S_{ii} S_{jj} - S_{ij}^2 > 0, \tag{4.10}$$

for any i and j.

This proof of the concavity of the entropy and the consequent conditions on its second derivatives holds only for those sets of parameters that can represent equilibrium with a reservoir, $\underline{A} = (\overline{\underline{A}}', \underline{A}'')$. Conversely, a system in a macrostate in which the entropy is convex cannot be in stable equilibrium with a reservoir. Since a system is a reservoir for any small part of itself, this requirement specifies the stable states of matter.

Gaussian Probability

The concavity of the entropy means that the probability distribution has a well-defined peak that allows the identification of the most probable macrostate, as was implicitly assumed above in the analysis of equilibrium. Hence one can expand the exponent to second order about the equilibrium state to obtain a quadratic form and a Gaussian distribution,

$$\wp(\underline{A}'|\underline{a}', \underline{A}'', T) \approx \frac{1}{Z_{\mathrm{G}}} \exp \frac{1}{2} \sum_{ij} (A_i' - \overline{A}_i')\, S_{ij}(\overline{\underline{A}})\, (A_j' - \overline{A}_j'), \tag{4.11}$$

where as above the subscripts on the subsystem entropy denote the second partial derivatives. The concavity of the entropy ensures that the exponent must be negative and that the probability distribution is peaked about the equilibrium state.

The average value $\langle A_i' \rangle$ is approximately equal to the equilibrium value $\overline{A}_i'$. Writing the average as $\left\langle \overline{A}_i' + (A_i' - \overline{A}_i') \right\rangle$, one sees from the parity of the summand that the correction term $\left\langle A_i' - \overline{A}_i' \right\rangle$ vanishes when the Gaussian expression above is used for the probability. The difference between the average value and the equilibrium value depends upon the third derivative of the entropy, which is on the order of the reciprocal of the square of the system size. Hence for macroscopic systems one may equate the equilibrium value to the average value with negligible error. It is only for finite-sized systems that there is a measurable difference between the two. This point will be revisited below in Section 4.3.4.

4.2 Constrained and Equilibrium Potentials

4.2.1 Definitions

Instead of dealing with the total entropy one may introduce a *constrained thermodynamic potential*, which has the general form

$$\begin{aligned} F(\underline{A}'|\underline{a}', \underline{A}''; T) &\equiv -TS_{\text{total}}(\underline{A}'|\underline{a}', \underline{A}'', T) \\ &= \underline{A}' \cdot \underline{a}' - TS(\underline{A}). \end{aligned} \tag{4.12}$$

The temperature T that appears here is always the reservoir temperature, whether or not energy is an exchangeable parameter.[3] With this definition the probability of the subsystem having a particular allocation of macrostates is

$$\wp(\underline{A}'|\underline{a}', \underline{A}''; T) = \frac{1}{Z(\underline{a}', \underline{A}'', T)} \exp[-F(\underline{A}'|\underline{a}', \underline{A}''; T)/k_{\text{B}}T], \tag{4.13}$$

which has the appearance of the well-known Boltzmann distribution, except that it is the constrained thermodynamic potential, not the energy, that appears in the exponent.

Since the total entropy is a maximum at equilibrium, by definition the constrained thermodynamic potential, which is its negative, is a minimum. Hence on average spontaneous transitions between macrostates occur down the gradient in the constrained thermodynamic potential. By virtue of its close relationship to the total entropy, the constrained thermodynamic potential inherits many features of the latter, such as the appropriate concavity and bijectivity. In particular, the constrained thermodynamic potential is a convex function of the A_i, which follows from the concavity of the entropy, $\partial^2 F/\partial A_i^2 = -T\partial^2 S(\underline{A})/\partial A_i^2 > 0$. The variational nature of the constrained thermodynamic potential is quite an advantage and it is exploited extensively below.

The equilibrium thermodynamic potential or free energy of the system is defined to be the minimum value of the constrained thermodynamic potential,

$$\begin{aligned} \overline{F}(\underline{a}', \underline{A}''; T) &\equiv F(\overline{\underline{A}}'|\underline{a}', \underline{A}''; T), \\ &= \overline{\underline{A}}' \cdot \underline{a}' - TS(\overline{\underline{A}}', \underline{A}''). \end{aligned} \tag{4.14}$$

The equilibrium thermodynamic potential is overlined to make it clear that it is an equilibrium property of the subsystem. The entropy that appears explicitly here is that of the subsystem, constrained to have the equilibrium value of the exchangeable parameters. It is neither the constrained total entropy nor the unconstrained total entropy of the system.

[3]Recall that the field variable conjugate to the energy is unity, which is the reason that T is shown explicitly as an extra argument in the constrained thermodynamic potential. In the event that the energy is not an exchangeable parameter, the additional variable that is the reservoir temperature effects the thermodynamic potential in only a trivial fashion.

The distinction between the constrained thermodynamic potential and the equilibrium or optimised thermodynamic potential is quite important. Unfortunately this distinction does not appear in conventional thermodynamic or statistical mechanics texts, and the present nomenclature is not standard. Here the nonequilibrium quantity will always be called the constrained thermodynamic potential. It may also be called the fluctuation potential because it characterises the fluctuations about equilibrium. The optimised thermodynamic potential is what conventionally appears in text books and is exclusively an equilibrium property. Although the same symbol is here used for both, they may be distinguished by the overline and by the number of arguments. A certain number of parameters are necessary to fix the state of an isolated system (the number of components in $\underline{A}$), and the equilibrium thermodynamic potential has exactly this number of independent variables $(\underline{a}', \underline{A}'')$. The constrained thermodynamic potential, however, has an augmented set of independent parameters $(\underline{A}'|\underline{a}', \underline{A}'')$, which exceeds the necessary number by the number of exchangeable parameters. This confirms that the optimised thermodynamic potential or free energy is purely an equilibrium property, whereas the constrained thermodynamic potential characterises the nonequilibrium state.

4.2.2 Derivatives

The utility of the constrained thermodynamic potential lies in its variational nature, which makes differentiation particularly simple. The derivative of the equilibrium thermodynamic potential with respect to any of the field variables a_i' except that conjugate to energy is

$$
\begin{aligned}
\frac{\partial \overline{F}(\underline{a}', \underline{A}''; T)}{\partial a_i'} &= \left(\frac{\partial F(\overline{\underline{A}}'|\underline{a}', \underline{A}''; T)}{\partial a_i'} \right)_{A_j'', T, a_{j \neq i}} \\
&= \left. \frac{\partial F(\underline{A}'|\underline{a}', \underline{A}''; T)}{\partial a_i'} \right|_{\underline{A}'=\overline{\underline{A}'}} \\
&\quad + \sum_i \frac{\partial \overline{A}_i'(\underline{a}', \underline{A}''; T)}{\partial a_i'} \left. \frac{\partial F(\underline{A}'|\underline{a}', \underline{A}''; T)}{\partial A_i'} \right|_{\underline{A}'=\overline{\underline{A}'}} \\
&= \left. \frac{\partial F(\underline{A}'|\underline{a}', \underline{A}''; T)}{\partial a_i'} \right|_{\underline{A}'=\overline{\underline{A}'}} . \\
&= \overline{A}_i', \quad A_i' \neq E.
\end{aligned} \tag{4.15}
$$

The terms in the sum of the second equality do not contribute because the derivative of the constrained thermodynamic potential vanishes at equilibrium. One may similarly differentiate the equilibrium thermodynamic potential with respect to the reservoir temperature,

$$
\frac{\partial \overline{F}(\underline{a}', \underline{A}''; T)}{\partial T} = \left. \frac{\partial F(\underline{A}'|\underline{a}', \underline{A}''; T)}{\partial T} \right|_{\underline{A}'=\overline{\underline{A}}'} = -S(\overline{\underline{A}}). \tag{4.16}
$$

Again the $\overline{A}'_i$ have been held fixed because of the variational nature of the constrained thermodynamic potential. Similarly, dividing the equilibrium thermodynamic potential by the reservoir temperature and differentiating with respect to the inverse temperature one obtains

$$\frac{\partial(\overline{F}(\underline{a}', \underline{A}''; T)/T)}{\partial T^{-1}} = \left. \frac{\partial(F(\underline{A}'|\underline{a}', \underline{A}''; T)/T)}{\partial T^{-1}} \right|_{\underline{A}'=\overline{\underline{A}}'} = \underline{a}' \cdot \overline{\underline{A}}'. \tag{4.17}$$

These results hold whether or not energy is an exchangeable parameter.

Differentiating with respect to one of the nonexchangeable extensive variables of the subsystem A''_i is almost identical to differentiating with respect to the a'_i. Because of the variational nature of the constrained thermodynamic potential the $\overline{A}'_i$ may again be regarded as fixed and one obtains

$$\begin{aligned} \frac{\partial \overline{F}(\underline{a}', \underline{A}''; T)}{\partial A''_i} &= \left. \frac{\partial F(\underline{A}'|\underline{a}', \underline{A}''; T)}{\partial A''_i} \right|_{\underline{A}'=\overline{\underline{A}}'} \\ &= -T \left. \frac{\partial S(\underline{A})}{\partial A''_i} \right|_{\underline{A}'=\overline{\underline{A}}'} \\ &= -T \frac{\overline{a}''_{is}}{\overline{T}_s}. \end{aligned} \tag{4.18}$$

Recall that if energy is an exchangeable parameter, $\overline{T}_s = T$.

These results give the total differential of the equilibrium thermodynamic potential,

$$d\overline{F}(\underline{a}', \underline{A}''; T) = -S(\overline{\underline{A}})dT + \sum_i{}^{*} \overline{A}'_i da'_i - T \sum_i \frac{\overline{a}''_{is}}{\overline{T}_s} dA''_i. \tag{4.19}$$

Here the asterisk on the summation indicates that the energy term should be excluded in the event that it is an exchangeable parameter.

One sees that because the constrained thermodynamic potential is optimised at equilibrium, differentiating the equilibrium thermodynamic potential or free energy is the same as differentiating the constrained thermodynamic potential *holding the* $\overline{A}'_i$ *fixed*. In conventional texts the free energy is usually expressed as $F = \underline{A}' \cdot \underline{a}' - TS(\underline{A})$, either it being implicitly understood that the exchangeable parameters that appear here are really $\overline{\underline{A}}'(\underline{a}', \underline{A}''; T)$ and that the constrained free energy is stationary with respect to their variations or, more likely, it being assumed that the A'_i are independent variables. These conventional texts typically differentiate the free energy with fixed $\overline{A}'_i$, without any explicit justification. Although the same answer results in the end, it is only the present distinction between the constrained and equilibrium thermodynamic potentials that shows why it is so. Similarly, the present approach makes it clear why the entropy that appears may similarly be held fixed as far as differentiations with respect to the a'_i go (another common and necessary procedure), because it is not $S(\underline{a}', \underline{A}'')$ that appears, but rather $S(\overline{\underline{A}}', \underline{A}'')$, and the $\overline{A}'_i$ may be held fixed.

4.2.3 Concavity of the Equilibrium Free Energy

From the concavity of the entropy, $\partial^2 S(\underline{A})/\partial A_i^2 < 0$, it follows that the constrained thermodynamic potential is a convex function, $\partial^2 F(\underline{A}'|\underline{a}', \underline{A}''; T) \ /\partial A_i^2 = -T\partial^2 S(\underline{A}) \ /\partial A_i^2 > 0$. The curvature of the equilibrium thermodynamic potential is now derived.

First the second derivative with respect to inverse temperature is treated. The a_i' and the A_i'' are held fixed, and the equilibrium quantities are functions of them and the temperature $\overline{A}_i'(\underline{a}', \underline{A}''; T)$. For a slightly different temperature $\tilde{T}$, the equilibrium values shift to $\tilde{A}_i \equiv \overline{A}_i'(\underline{a}', \underline{A}''; \tilde{T})$. Since the constrained thermodynamic potential is a minimum at the appropriate equilibrium values, one must have $F(\overline{A}'|\underline{a}', \underline{A}''; T) < F(\tilde{A}'|\underline{a}', \underline{A}''; T)$, or writing this in full, rearranging, and expanding to second order the entropy about $\underline{\tilde{A}}$,

$$\begin{aligned} \frac{\overline{\underline{A}}' \cdot \underline{a}'}{T} - \frac{\underline{\tilde{A}}' \cdot \underline{a}'}{T} &< S(\overline{\underline{A}}) - S(\underline{\tilde{A}}) \\ &= [\overline{\underline{A}} - \underline{\tilde{A}}] \cdot \frac{\underline{a}'}{\tilde{T}} + \frac{1}{2}[\overline{\underline{A}} - \underline{\tilde{A}}] \cdot S_{AA}(\underline{\tilde{A}}) \cdot [\overline{\underline{A}} - \underline{\tilde{A}}]. \end{aligned} \tag{4.20}$$

Here has been used the fact that $\partial S/\partial A_i|_{A_i = \tilde{A}_i} = a_i(\underline{\tilde{A}})/T(\underline{\tilde{A}}) = a_i/\tilde{T}$. By the concavity of the entropy the quadratic term is negative, and so $[\overline{\underline{A}}' \cdot \underline{a}' - \underline{\tilde{A}}' \cdot \underline{a}'][1/T - 1/\tilde{T}] < 0$. Dividing by the square of the difference in inverse temperatures, and taking the limit $\tilde{T} \to T$, this proves that

$$\frac{\partial^2 [\overline{F}(\underline{a}', \underline{A}''; T)/T]}{\partial(1/T)^2} = \frac{\partial(\overline{\underline{A}}' \cdot \underline{a}')}{\partial(1/T)} < 0, \tag{4.21}$$

where the first equality follows from Eq. (4.17). This derivative is at constant a_i' and A_i''.

One may similarly consider two nearby points, $\overline{\underline{A}}' \equiv \overline{\underline{A}}'(\underline{a}', \underline{A}''; T)$ and $\underline{\tilde{A}}' \equiv \overline{\underline{A}}'(\underline{\tilde{a}}', \underline{A}''; \tilde{T})$. From the variational nature of the constrained thermodynamic potential one has $F(\underline{\tilde{A}}'|\underline{a}', \underline{A}''; T) > F(\overline{\underline{A}}'|\underline{a}', \underline{A}''; T)$, or

$$\begin{aligned} \underline{\tilde{A}}' \cdot \underline{a}' - TS(\underline{\tilde{A}}', \underline{A}'') &> \overline{\underline{A}}' \cdot \underline{a}' - TS(\overline{\underline{A}}', \underline{A}'') \\ &= \overline{\underline{A}}' \cdot \underline{a}' - T\left[S(\underline{\tilde{A}}', \underline{A}'') + (\overline{\underline{A}}' - \underline{\tilde{A}}') \cdot \frac{\partial \tilde{S}}{\partial \underline{\tilde{A}}'} \right. \\ &\quad \left. + \frac{1}{2}(\overline{\underline{A}}' - \underline{\tilde{A}}') \cdot \frac{\partial^2 \tilde{S}}{\partial \underline{\tilde{A}}' \partial \underline{\tilde{A}}'} \cdot (\overline{\underline{A}}' - \underline{\tilde{A}}') \right]. \end{aligned} \tag{4.22}$$

This may be rearranged to give

$$\begin{aligned} (\overline{\underline{A}}' - \underline{\tilde{A}}') \cdot \left[\frac{\underline{a}'}{T} - \frac{\underline{\tilde{a}}'}{\tilde{T}} \right] &< \frac{1}{2}(\overline{\underline{A}}' - \underline{\tilde{A}}') \cdot \frac{\partial^2 \tilde{S}}{\partial \underline{\tilde{A}}' \partial \underline{\tilde{A}}'} \cdot (\overline{\underline{A}}' - \underline{\tilde{A}}') \\ &< 0. \end{aligned} \tag{4.23}$$

Since the $\overline{A}'_i$ and the $\tilde{A}'_i$ are the first derivatives of the thermodynamic potential, the difference between them may be written in terms of the second derivative. The elements of the second derivative matrix are defined as

$$F_{ij} = \frac{\partial^2[\overline{F}(\underline{a}', \underline{A}''; T)/T]}{\partial(a'_i/T)\partial(a'_j/T)} = \frac{\partial \overline{A}'_i(\underline{a}', \underline{A}''; T)}{\partial(a'_j/T)}. \tag{4.24}$$

The meaning of the partial derivative is that the a'_k/T, $k \neq i$ and $k \neq j$, and all the A''_k are held fixed. (If $A'_i \neq E$ and $A'_j \neq E$, this means that T is constant and the a'_k themselves are held fixed.) With this the inequality may be written as a quadratic form,

$$\left[\frac{\underline{a}'}{T} - \frac{\tilde{\underline{a}}'}{\tilde{T}}\right] \cdot \underline{\underline{F}} \cdot \left[\frac{\underline{a}'}{T} - \frac{\tilde{\underline{a}}'}{\tilde{T}}\right] < 0. \tag{4.25}$$

Following the analysis for the concavity of the entropy given above, one concludes that all the eigenvalues of $\underline{\underline{F}}$ are negative. Hence it follows that

$$F_{ii} < 0, \text{ and } F_{ii}F_{jj} - F_{ij}^2 > 0. \tag{4.26}$$

These inequalities for the second derivative of the equilibrium thermodynamic potential determine the sign of physical quantities such as the heat capacity or the compressibility.

4.3 Partition Function

4.3.1 Total Entropy

The normalising factor for the probability distribution is called the *partition function*. It is given by

$$Z(\underline{a}', \underline{A}''; T) = \sum_{A'_1, A'_2, \ldots} e^{S(\underline{A})/k_B} e^{-\underline{A}' \cdot \underline{a}'/k_B T}, \tag{4.27}$$

and it plays a key rôle in statistical mechanics because it acts as a generating function for the thermodynamic properties of the subsystem.

The partition function is closely related to the unconstrained total entropy (at least that part that depends on the presence of the subsystem). By definition, the total entropy is the average of the entropy of the macrostate less the average of the logarithm of the macrostate probability. Since the probability distribution is the exponential of the constrained total entropy normalised by the partition function, direct substitution yields

$$\begin{aligned} S_{\text{total}}(\underline{a}', \underline{A}'', T) &= \sum_{A'_1, A'_2, \ldots} \wp(\underline{A}'|\underline{a}', \underline{A}'', T) \left[S_{\text{total}}(\underline{A}'|\underline{a}', \underline{A}'', T) \right. \\ &\quad \left. - k_B \ln \wp(\underline{A}'|\underline{a}', \underline{A}'', T)\right] \\ &= k_B \ln Z(\underline{a}', \underline{A}'', T). \end{aligned} \tag{4.28}$$

By virtue of this essential equivalence with the unconstrained total entropy, the logarithm of the partition function is the generating function of statistical mechanics.

4.3.2 Averages

The average value of a function of state is the sum over the states of that function times the state probability. Hence the average value of an exchangeable parameter of the subsystem is

$$\begin{aligned}
\langle A'_i \rangle &= \sum_{A'_1, A'_2, \ldots} A'_i \, \wp(\underline{A}' | \underline{a}', \underline{A}'', T) \\
&= \sum_{A'_1, A'_2, \ldots} \frac{\partial(\underline{A}' \cdot \underline{a}'/T)}{\partial(a'_i/T)} \frac{e^{S(\underline{A})/k_\mathrm{B}} e^{-\underline{A}' \cdot \underline{a}'/k_\mathrm{B}T}}{Z(\underline{a}', \underline{A}''; T)} \\
&= -k_\mathrm{B} \frac{\partial \ln Z(\underline{a}', \underline{A}''; T)}{\partial(a'_i/T)}. \qquad (4.29)
\end{aligned}$$

In the event that $A'_i \neq E$, this is $\langle A'_i \rangle = -k_\mathrm{B} T \partial \ln Z(\underline{a}', \underline{A}''; T)/\partial a'_i$.

Similarly, since the field parameters of the subsystem $\underline{a}_s$ depend upon the entire set of extensive variables $\underline{A}_s = (\underline{A}'_s, \underline{A}''_s)$, the a''_{is} also vary with the flux of the A'_{is}. Their average value is given by

$$\begin{aligned}
\langle a''_{is}/T_s \rangle &= \sum_{A'_1, A'_2, \ldots} (a''_{is}/T_s) \, \wp(\underline{A}' | \underline{a}'_r, \underline{A}''; T) \\
&= \sum_{A'_1, A'_2, \ldots} \frac{\partial S(\underline{A})}{\partial A''_i} \frac{e^{S(\underline{A})/k_\mathrm{B}} e^{-\underline{A}' \cdot \underline{a}'_r/k_\mathrm{B}T}}{Z(\underline{a}'_r, \underline{A}''; T)} \\
&= k_\mathrm{B} \frac{\partial \ln Z(\underline{a}'_r, \underline{A}''; T)}{\partial A''_i}. \qquad (4.30)
\end{aligned}$$

(Here the labels on the field variables have been reintroduced to distinguish those of the subsystem from those of the reservoir.)

The average value of the field variables of the subsystem conjugate to those of the exchangeable parameters is given by

$$\begin{aligned}
\langle a'_{is}/T_s \rangle &= \sum_{A'_j, j \neq i} \sum_{A'_i} (a'_{is}/T_s) \, \wp(\underline{A}' | \underline{a}'_r, \underline{A}''; T) \\
&= \sum_{A'_j, j \neq i} \int_0^\infty \frac{\mathrm{d}A'_i}{\Delta} \frac{\partial S(\underline{A})}{\partial A'_i} \frac{e^{S(\underline{A})/k_\mathrm{B}} e^{-\underline{A}' \cdot \underline{a}'_r/k_\mathrm{B}T}}{Z(\underline{a}'_r, \underline{A}''; T)} \\
&= \frac{k_\mathrm{B}}{Z(\underline{a}'_r, \underline{A}''; T)} \sum_{A'_j, j \neq i} \left[e^{S(\underline{A})/k_\mathrm{B}} e^{-\underline{A}' \cdot \underline{a}'_r/k_\mathrm{B}T} \Big|_0^\infty \right. \\
&\quad \left. + \frac{a'_{ir}}{k_\mathrm{B}T} \int_0^\infty \frac{\mathrm{d}A'_i}{\Delta} e^{S(\underline{A})/k_\mathrm{B}} e^{-\underline{A}' \cdot \underline{a}'_r/k_\mathrm{B}T} \right] \\
&= \frac{a'_{ir}}{T_r}. \qquad (4.31)
\end{aligned}$$

Here the sum over A'_i has been converted to an integral and an integration by parts used, the integrated portion vanishing because the tail of the probability distribution is negligible. In this case the average value of the field parameters of the subsystem conjugate to the exchangeable quantities equals precisely those fixed by the reservoir.

4.3.3 Fluctuations

The logarithm of the partition function is the unconstrained total entropy of the subsystem and reservoir, and as such its derivatives give the average value of the exchangeable parameters and the field variables of the subsystem. Likewise the second derivatives give the fluctuations in these quantities. One has

$$\begin{aligned}
& k_{\rm B}^2 \frac{\partial^2 \ln Z(\underline{a}', \underline{A}''; T)}{\partial (a_i'/T)^2} \\
&= -k_{\rm B} \frac{\partial \left\langle A_i' \right\rangle}{\partial (a_i'/T)} \\
&= -k_{\rm B} \frac{\partial}{\partial (a_i'/T)} \frac{1}{Z(\underline{a}', \underline{A}''; T)} \sum_{A_1', A_2', \ldots} A_i' e^{S(\underline{A})/k_{\rm B}} e^{-\underline{A}' \cdot \underline{a}'/k_{\rm B} T} \\
&= \left\langle (A_i')^2 \right\rangle - \left\langle A_i' \right\rangle^2 \\
&= \left\langle \Delta (A_i')^2 \right\rangle,
\end{aligned} \tag{4.32}$$

where the departure from the average is $\Delta(x) \equiv x - \langle x \rangle$. The meaning of the partial derivatives is that the a_j'/T, $j \neq i$, are held fixed.

The final quantity is called the fluctuation (or deviation about the average), and as the average of the square of a quantity it must be positive. The fluctuation indicates the width of the probability distribution, being relatively small for a sharply peaked distribution. Since the total entropy scales with the size of the subsystem, and the denominator on the left-hand side is intensive, then the fluctuation must be extensive. Accordingly, the relative root mean square fluctuation, $\sqrt{\langle \Delta(A_i')^2 \rangle} / \langle A_i' \rangle$, is $\mathcal{O}V^{-1/2}$, since the exchangeable parameter is extensive. This indicates that the relative error in the measurement of a thermodynamic quantity will be negligibly small for a macroscopic system.

One may similarly show that the cross-second derivative is

$$\begin{aligned}
k_{\rm B}^2 \frac{\partial^2 \ln Z(\underline{a}', \underline{A}''; T)}{\partial (a_i'/T) \partial (a_j'/T)} &= \left\langle A_i' A_j' \right\rangle - \left\langle A_i' \right\rangle \left\langle A_j' \right\rangle \\
&= \left\langle \Delta(A_i') \Delta(A_j') \right\rangle.
\end{aligned} \tag{4.33}$$

This gives the cross-correlation of the exchangeable quantities, and it can be positive, negative, or 0. In general, such cross-fluctuations vanish when there is no correlation between the quantities. One can, however, combine this with the above result in such a way as to obtain a positive quantity,

$$k_{\rm B}^2 \left(\frac{\partial}{\left\langle A_i' \right\rangle \partial (a_i'/T)} - \frac{\partial}{\left\langle A_j' \right\rangle \partial (a_j'/T)} \right)^2 \ln Z(\underline{a}', \underline{A}''; T)$$

$$= \left\langle \left(\frac{A'_i}{\langle A'_i \rangle} - \frac{A'_j}{\langle A'_j \rangle} \right)^2 \right\rangle. \tag{4.34}$$

The averages that appear in the denominators on the left-hand side are here regarded as constants that are not differentiated.

If instead of holding the a'_i/T fixed one holds the a'_i fixed and differentiates with respect to $1/T$ one obtains

$$\begin{aligned} k_{\mathrm{B}}^2 \left(\frac{\partial^2 \ln Z(\underline{a}', \underline{A}''; T)}{\partial (1/T)^2} \right)_{\underline{a}', \underline{A}''} &= -k_{\mathrm{B}} \left(\frac{\partial \left\langle \underline{a}' \cdot \underline{A}' \right\rangle}{\partial (1/T)} \right)_{\underline{a}', \underline{A}''} \\ &= \left\langle \Delta(\underline{a}' \cdot \underline{A}')^2 \right\rangle. \end{aligned} \tag{4.35}$$

This gives the fluctuation of the exchangeable work terms in total.

The second derivative with respect to the fixed extensive variables of the subsystem is similarly straightforward. One has

$$\begin{aligned} \frac{\partial^2 \ln Z(\underline{a}', \underline{A}''; T)}{\partial (A''_i)^2} &= \frac{\partial \left\langle a''_{is}/k_{\mathrm{B}} T_s \right\rangle}{\partial A''_i} \\ &= \left\langle (a''_{is}/k_{\mathrm{B}} T_s)^2 \right\rangle - \left\langle a''_{is}/k_{\mathrm{B}} T_s \right\rangle^2 + \left\langle \frac{\partial^2 S(\underline{A})/k_{\mathrm{B}}}{\partial (A''_i)^2} \right\rangle \\ &= \left\langle \Delta(a''_{is}/k_{\mathrm{B}} T_s)^2 \right\rangle + \left\langle \frac{\partial^2 S(\underline{A})/k_{\mathrm{B}}}{\partial (A''_i)^2} \right\rangle. \end{aligned} \tag{4.36}$$

(Note that T_s is the subsystem temperature, and a''_{is} is the field variable of the subsystem.) Since the subsystem-dependent part of the constrained total entropy, or $\ln Z$, is extensive, the left-hand side must be $\mathcal{O}V^{-1}$, as is the final term on the right-hand side (the numerator goes like V, the denominator like V^2). One concludes that the fluctuation must be of the same order. This is one power of volume lower than the order of the averand (the quantity being averaged), which is consistent with the observation above that $\left\langle \Delta(A'_i)^2 \right\rangle = \mathcal{O}V$, since the averand in this case is $(A'_i)^2 = \mathcal{O}V^2$. The fluctuation of the field variable is positive, whereas the second derivative of the subsystem entropy is negative due to its concavity, which means that the sign of the quantity on the left is not determined in this case. However, rewriting the result,

$$\frac{\partial^2 \ln Z(\underline{a}', \underline{A}''; T)}{\partial (A''_i)^2} - \left\langle \frac{\partial^2 S(\underline{A})/k_{\mathrm{B}}}{\partial (A''_i)^2} \right\rangle = \left\langle \Delta(a''_{is}/k_{\mathrm{B}} T_s)^2 \right\rangle, \tag{4.37}$$

determines that the left-hand side is positive.

Similarly the cross-second derivative is

$$\begin{aligned} \frac{\partial^2 \ln Z(\underline{a}', \underline{A}''; T)}{\partial A''_i \partial A''_j} &= \frac{\partial \left\langle a''_{is}/k_{\mathrm{B}} T_s \right\rangle}{\partial A''_j} \\ &= \left\langle a''_{is} a''_{js}/(k_{\mathrm{B}} T_s)^2 \right\rangle - \left\langle a''_{is}/k_{\mathrm{B}} T_s \right\rangle \left\langle a''_{js}/k_{\mathrm{B}} T_s \right\rangle \\ &\quad + \left\langle \frac{\partial^2 S(\underline{A})/k_{\mathrm{B}}}{\partial A''_i \partial A''_j} \right\rangle, \end{aligned} \tag{4.38}$$

which may be combined with the above to yield

$$\left\langle \left(\frac{\partial}{\langle a''_{is}/k_{\mathrm{B}}T_s \rangle \, \partial A''_i} - \frac{\partial}{\langle a''_{js}/k_{\mathrm{B}}T_s \rangle \, \partial A''_j} \right)^2 \left[\ln Z(\underline{a}', \underline{A}''; T) - S(\underline{A})/k_{\mathrm{B}} \right] \right\rangle$$
$$= \left\langle \left(\frac{a''_{is}/k_{\mathrm{B}}T_s}{\langle a''_{is}/k_{\mathrm{B}}T_s \rangle} - \frac{a''_{js}/k_{\mathrm{B}}T_s}{\langle a''_{js}/k_{\mathrm{B}}T_s \rangle} \right)^2 \right\rangle, \tag{4.39}$$

which shows that the left-hand side must be positive. Again the averages in the denominators on the left-hand side are held fixed during the differentiations. Also the derivatives of the partition function can be removed from the average.

Finally, differentiation with respect to both a field and an extensive variable yields

$$-k_{\mathrm{B}} \frac{\partial^2 \ln Z(\underline{a}', \underline{A}''; T)}{\partial(a'_i/T)\partial A''_j} = \frac{\partial \langle A'_i \rangle}{\partial A''_j} = -\frac{\partial \langle a''_{js}/T_s \rangle}{\partial(a'_i/T)}$$
$$= \langle A'_i a''_{js}/k_{\mathrm{B}}T_s \rangle - \langle A'_i \rangle \langle a''_{js}/k_{\mathrm{B}}T_s \rangle . \tag{4.40}$$

Accordingly, the quantity

$$\left(\frac{\partial}{\langle A'_i \rangle \, \partial(a'_i/k_{\mathrm{B}}T)} - \frac{\partial}{(a''_{js}/k_{\mathrm{B}}T_s)\partial A''_j} \right)^2 \ln Z(\underline{a}', \underline{A}''; T)$$
$$= \left\langle \left(\frac{A'_i a'_i}{k_{\mathrm{B}}T} - \frac{A''_j a''_{js}}{k_{\mathrm{B}}T_s} \right)^2 \right\rangle \tag{4.41}$$

is positive. In this case both sides of the equation are of order V.

4.3.4 Partition Function and the Thermodynamic Potential

In the formalism the partition function plays an essential part. To be specific, the derivatives of its logarithm generate the average values of the thermodynamic quantities of the subsystem and their fluctuation. The reason that this is such a central quantity lies in the fact that the logarithm of the partition function is the unconstrained total entropy of the subsystem and reservoir, Eq. (4.28),

$$S_{\mathrm{total}}(\underline{a}', \underline{A}''; T) = k_{\mathrm{B}} \ln Z(\underline{a}', \underline{A}''; T). \tag{4.42}$$

This last equality appears to contradict certain long-standing notions in conventional statistical mechanics, where it is the thermodynamic potential or free energy that is equated to the logarithm of the partition function. In fact the two notions are consistent for macroscopic systems where fluctuations are relatively negligible. One has

$$-k_{\mathrm{B}}T \ln Z(\underline{a}', \underline{A}''; T)$$
$$= -k_{\mathrm{B}}T \ln \sum_{A'_1, A'_2, \ldots} e^{S(\underline{A})/k_{\mathrm{B}}} e^{-\underline{A}' \cdot \underline{a}'/k_{\mathrm{B}}T}$$

$$
\begin{aligned}
&= -k_{\mathrm{B}}T \ln \int \prod_i \frac{\mathrm{d}A'_i}{\Delta_i} e^{S(\underline{A})/k_{\mathrm{B}}} e^{-\underline{A}' \cdot \underline{a}'/k_{\mathrm{B}}T} \\
&\approx -k_{\mathrm{B}}T \ln \Bigg[e^{S(\overline{\underline{A}})/k_{\mathrm{B}}} e^{-\overline{\underline{A}}' \cdot \underline{a}'/k_{\mathrm{B}}T} \\
&\qquad \times \int \prod_i \frac{\mathrm{d}A'_i}{\Delta_i} \exp \sum_{ij} (A'_i - \overline{A}'_i) S_{ij}(\overline{\underline{A}})(A'_j - \overline{A}'_j)/2k_{\mathrm{B}} \Bigg] \\
&= \overline{\underline{A}}' \cdot \underline{a}' - TS(\overline{\underline{A}}) - \frac{k_{\mathrm{B}}T}{2} \ln \left[\mathrm{Det}\{S'_{ij}(\overline{\underline{A}})\} \prod_i \frac{-\Delta_i^2}{2\pi k_{\mathrm{B}}} \right].
\end{aligned}
\tag{4.43}
$$

Here a WKB approximation has been made (i.e., the exponent has been expanded to second order and the Gaussian integral evaluated). The Δ_i are length scales that arise in the conversion of the sum to an integral (see the next chapter) and $\{S'_{ij}\}$ is the matrix of second derivatives of the subsystem entropy (those of the exchangeable parameters only). The first two terms on the right-hand side give precisely the equilibrium thermodynamic potential or free energy. The remaining term is a logarithmic correction that for macroscopic systems may be neglected (it is only when fluctuations about equilibrium are large compared to the mean that this term may become important). To leading order the equilibrium thermodynamic potential is indeed equal to the negative of the temperature times the logarithm of the partition function,

$$
\overline{F}(\underline{a}', \underline{A}''; T) = -k_{\mathrm{B}}T \ln Z(\underline{a}', \underline{A}''; T). \tag{4.44}
$$

A consequence of this equality is that for macroscopic systems the derivatives of the equilibrium thermodynamic potential equal the derivatives of the partition function. That is, equilibrium values are the same as average values, as has been already demonstrated in Section 4.1.3. Similar statements apply to the second derivatives, so that the positivity of the fluctuation determines the sign of the corresponding derivatives of the equilibrium quantities.

4.4 Specific Reservoirs

The general formalism just derived may be applied to the different reservoirs treated in the preceding two chapters. Whereas in those chapters only the equilibrium results were obtained explicitly, here the emphasis is on the probability distribution and average values.

4.4.1 Constant Temperature

For a heat reservoir, the exchangeable parameter is energy $A'_1 = E$, and the fixed parameters are the volume $A''_1 = V$ and the particle number $A''_2 = N$. The relevant field variable of the reservoir is $a_1' = 1$. Hence the constrained total entropy of the subsystem and the reservoir, Eq. (4.5), is

$$S_{\text{total}}(E|N,V,T) = S(E,N,V) - E/T, \tag{4.45}$$

which gives for the constrained thermodynamic potential, Eq. (4.12),

$$F(E|N,V,T) = E - TS(E,N,V). \tag{4.46}$$

According to Eq. (4.6), the probability of the subsystem actually having an energy E given that it has N particles and volume V, and that the temperature of the reservoir is T, is

$$\wp(E|N,V,T) = \frac{1}{Z(N,V,T)} e^{S(E,N,V)/k_{\text{B}}} e^{-E/k_{\text{B}}T}. \tag{4.47}$$

The partition function, Eq. (4.27), is of course

$$Z(N,V,T) = \sum_E e^{S(E,N,V)/k_{\text{B}}} e^{-E/k_{\text{B}}T}, \tag{4.48}$$

and the unconstrained total entropy of the subsystem plus reservoir, Eq. (4.28), is just $S_{\text{total}}(N,V,T) = k_{\text{B}} \ln Z(N,V,T)$. The logarithmic derivatives of the partition function give the average energy

$$\langle E \rangle = -k_{\text{B}} \frac{\partial \ln Z(N,V,T)}{\partial(1/T)}, \tag{4.49}$$

from Eq. (4.29), and from Eq. (4.30), the average pressure

$$\langle p_s/T_s \rangle = k_{\text{B}} \frac{\partial \ln Z(N,V,T)}{\partial V}, \tag{4.50}$$

and the average chemical potential

$$\langle -\mu_s/T_s \rangle = k_{\text{B}} \frac{\partial \ln Z(N,V,T)}{\partial N}, \tag{4.51}$$

where the subscript s signifies a field variable of the subsystem.

The equilibrium thermodynamic potential for this constant temperature system is the Helmholtz free energy, and from Eq. (4.21) its second inverse temperature derivative is

$$\frac{\partial^2[\overline{F}(N,V,T)/T]}{\partial(1/T)^2} = \frac{\partial \overline{E}(N,V,T)}{\partial(1/T)} < 0, \tag{4.52}$$

or

$$C_V \equiv \frac{\partial \overline{E}(N,V,T)}{\partial T} > 0. \tag{4.53}$$

This agrees with the result established in Ch. 3 that the heat capacity at constant volume is positive. The same result emerges from Eq. (4.26), because in this case energy is the sole exchangeable parameter. The analogous result for the partition function, Eq. (4.32), may be written

$$\frac{k_{\mathrm{B}}}{T^2}\frac{\partial^2 \ln Z(N,V,T)}{\partial(1/T)^2} = \frac{\partial \langle E \rangle}{\partial T} = \frac{1}{k_{\mathrm{B}}T^2}\left\langle \Delta(E)^2 \right\rangle > 0. \tag{4.54}$$

In so far as equilibrium values equal average values for macroscopic systems, this is in agreement with the bound for the heat capacity. More precisely, since $C_V = -(1/k_{\mathrm{B}}T^2)\partial^2\beta\overline{F}/\partial\beta^2$ (where $\beta \equiv 1/k_{\mathrm{B}}T$), and since $\left\langle \Delta(E)^2 \right\rangle = \partial^2 \ln Z(N,V,T)/\partial\beta^2$, and since in the thermodynamic limit $\beta\overline{F}(N,V,T) = -\ln Z(N,V,T)$, the relationship between energy fluctuations and the heat capacity is given by

$$\left\langle \Delta(E)^2 \right\rangle = k_{\mathrm{B}}T^2 C_V. \tag{4.55}$$

The isothermal compressibility will be shown to be positive in the pressure reservoir case below. In the present temperature reservoir case one has

$$\begin{aligned}
\chi_T^{-1} &\equiv -V\frac{\partial p(N,V,T)}{\partial V} \\
&= V\frac{\partial^2 \overline{F}(N,V,T)}{\partial V^2} \\
&= -k_{\mathrm{B}}TV\frac{\partial^2 \ln Z(N,V,T)}{\partial V^2} \\
&= -k_{\mathrm{B}}TV\left[\left\langle \Delta(p/k_{\mathrm{B}}T_s)^2 \right\rangle + \left\langle \frac{\partial^2 S(E,N,V)/k_{\mathrm{B}}}{\partial V^2} \right\rangle\right].
\end{aligned} \tag{4.56}$$

Since $\chi_T > 0$ this shows that

$$\left\langle \frac{\partial^2 S(E,N,V)/k_{\mathrm{B}}}{\partial V^2} \right\rangle < -\left\langle \Delta(p/k_{\mathrm{B}}T_s)^2 \right\rangle. \tag{4.57}$$

4.4.2 Constant Chemical Potential

For a subsystem of fixed volume $A_1'' = V$ able to exchange with a reservoir energy $A_1' = E$ and particles $A_2' = N$, the relevant field variables of the reservoir are the temperature, and hence $a_1' = 1$, and the chemical potential, $a_2' = -\mu$. Hence the constrained total entropy of the subsystem and the reservoir, Eq. (4.5), is

$$S_{\text{total}}(E|N,V,T) = S(E,N,V) - E/T + \mu N/T, \tag{4.58}$$

which gives for the constrained thermodynamic potential,

$$\Omega(E,N|\mu,V,T) = E - \mu N - TS(E,N,V). \tag{4.59}$$

The probability of the subsystem actually having an energy E and particle number N, given that its volume is V, and that the temperature and chemical potential of the reservoir are T and μ, respectively, is the exponential of the constrained total entropy,

$$\wp(E, N|\mu, V, T) = \frac{1}{Z(\mu, V, T)} e^{S(E,N,V)/k_{\rm B}} e^{-E/k_{\rm B}T} e^{\mu N/k_{\rm B}T}. \tag{4.60}$$

The partition function normalises this

$$Z(\mu, V, T) = \sum_{E,N} e^{S(E,N,V)/k_{\rm B}} e^{-E/k_{\rm B}T} e^{\mu N/k_{\rm B}T}, \tag{4.61}$$

and the unconstrained total entropy of the subsystem plus reservoir is just $S_{\rm total}(\mu, V, T) = k_{\rm B} \ln Z(\mu, V, T)$. The logarithmic derivatives of the partition function give, from Eq. (4.29), the average energy

$$\langle E \rangle = -k_{\rm B} \left(\frac{\partial \ln Z(\mu, V, T)}{\partial (1/T)} \right)_{\mu/T,V}, \tag{4.62}$$

and the average particle number,

$$\langle N \rangle = k_{\rm B} T \left(\frac{\partial \ln Z(\mu, V, T)}{\partial \mu} \right)_{N,T}, \tag{4.63}$$

and from Eq. (4.30), the average pressure

$$\langle p_s/T_s \rangle = k_{\rm B} \left(\frac{\partial \ln Z(\mu, V, T)}{\partial V} \right)_{\mu,T}. \tag{4.64}$$

The equilibrium thermodynamic potential for this constant temperature system is the grand potential, and from Eq. (4.21) its second inverse temperature derivative is

$$\left(\frac{\partial^2 [\Omega(\mu, V, T)/T]}{\partial (1/T)^2} \right)_{\mu,V} = \frac{\partial (\overline{E}(\mu, V, T) - \mu \overline{N}(\mu, V, T)}{\partial (1/T)} < 0, \tag{4.65}$$

or

$$\frac{\partial \overline{E}(\mu, V, T)}{\partial T} > \frac{\mu \partial \overline{N}(\mu, V, T)}{\partial T}. \tag{4.66}$$

These derivatives are of course at constant volume and chemical potential. The analogous result for average quantities follows from the derivative of the partition function, Eq. (4.35). It also follows from Eq. (4.26) that

$$\left(\frac{\partial \overline{E}(\mu, V, T)}{\partial T} \right)_{V,\mu/T} > 0 \tag{4.67}$$

and that

$$\left(\frac{\partial \overline{N}(\mu, V, T)}{\partial \mu} \right)_{V,T} > 0. \tag{4.68}$$

The same two conclusions, with the equilibrium value replaced by the average value, follows from the second derivative of the logarithm of the partition function, Eq. (4.32).

The second volume derivative, which follows from Eq. (4.37), is

$$\left(\frac{\partial^2 \ln Z(\mu, V, T)}{\partial V^2}\right)_{\mu,T} - \left\langle \left(\frac{\partial^2 S(E, N, V)/k_{\mathrm{B}}}{\partial V^2}\right)_{E,N} \right\rangle = \left\langle \Delta(p_s/k_{\mathrm{B}}T_s)^2 \right\rangle . \tag{4.69}$$

Now the extensivity arguments used in Section 3.8 apply as well to the partition function, and for macroscopic systems one must have

$$S_{\mathrm{total}}(\mu, V, T) = k_{\mathrm{B}} \ln Z(\mu, V, T) = V \left\langle p_s/k_{\mathrm{B}}T_s \right\rangle + \mathcal{O}V^{2/3}, \tag{4.70}$$

where the average pressure divided by temperature is a function of the chemical potential and temperature of the reservoir but not of the volume. The coefficient of the linear term is of course the volume derivative of the partition function found above. This means that

$$k_{\mathrm{B}} \left(\frac{\partial^2 \ln Z(\mu, V, T)}{\partial V^2}\right)_{\mu,T} = \left(\frac{\partial \left\langle p_s/k_{\mathrm{B}}T_s \right\rangle}{\partial V}\right)_{\mu,T} = \mathcal{O}V^{-4/3}, \tag{4.71}$$

which may be neglected compared to terms of order V^{-1}. This is of course consistent with the above assertion that the average of the ratio of the subsystem pressure and temperature, which are themselves intensive, only depends upon intensive variables and is independent of the volume. Hence one concludes that for macroscopic systems,

$$\left\langle \left(\frac{\partial^2 S(E, N, V)}{k_{\mathrm{B}}\partial V^2}\right)_{E,N} \right\rangle = \left\langle \left(\frac{\partial [p(E, N, V)/k_{\mathrm{B}}T(E, N, V)]}{\partial V}\right)_{E,N} \right\rangle = -\left\langle \Delta(p_s/k_{\mathrm{B}}T_s)^2 \right\rangle + \mathcal{O}V^{-4/3}. \tag{4.72}$$

Both sides of this equation are clearly negative. Note that in Eq. (4.57), the averages are $\langle \ldots \rangle_{N,V,T}$, whereas here they are $\langle \ldots \rangle_{\mu,V,T}$. In general the reservoirs of subsystems with different boundaries may be chosen to be consistent in the values of average quantities, but the fluctuations in these quantities differ.

4.4.3 Constant Pressure

For a subsystem with a fixed number of particles $A_1'' = N$ and able to exchange energy $A_1' = E$ and volume $A_2' = V$ with a reservoir, the relevant field variables of the latter are the temperature, so that $a_1' = 1$, and the pressure, $a_2' = p$. Hence the constrained total entropy of the subsystem and the reservoir, Eq. (4.5), is

$$S_{\mathrm{total}}(E, V|N, p, T) = S(E, N, V) - E/T - pV/T, \tag{4.73}$$

which gives for the constrained thermodynamic potential

$$G(E, V|N, p, T) = E + pV - TS(E, N, V). \tag{4.74}$$

According to either Eq. (4.6) or Eq. (4.12), the probability of the subsystem actually having an energy E and volume V, given that it has N particles and that the temperature and pressure of the reservoir are T and p, respectively, is

$$\wp(E, V|N, p, T) = \frac{1}{Z(N, p, T)} e^{S(E,N,V)/k_B} e^{-E/k_B T} e^{-pV/k_B T}. \tag{4.75}$$

The normalising partition function is

$$Z(N, p, T) = \sum_{E,V} e^{S(E,N,V)/k_B} e^{-E/k_B T} e^{-pV/k_B T}, \tag{4.76}$$

and the unconstrained total entropy of the subsystem plus reservoir is just $S_{\text{total}}(N, p, T) = k_B \ln Z_p(N, p, T)$. From Eq. (4.29) one obtains the average energy

$$\langle E \rangle = -k_B \left(\frac{\partial \ln Z(N, p, T)}{\partial(1/T)} \right)_{p/T,N}, \tag{4.77}$$

and the average volume,

$$\langle V \rangle = -k_B T \frac{\partial \ln Z(N, p, T)}{\partial p}, \tag{4.78}$$

and from Eq. (4.30) one obtains the average chemical potential

$$\langle \mu_s / T_s \rangle = k_B \frac{\partial \ln Z(N, p, T)}{\partial N}. \tag{4.79}$$

The equilibrium thermodynamic potential for this constant temperature system is the Gibbs free energy, and its second inverse temperature derivative was analysed earlier, Eq. (3.63). The same result also follows from the generic analysis, Eq. (4.21),

$$\left(\frac{\partial^2 [\overline{G}(N, p, T)/T]}{\partial(1/T)^2} \right)_{N,p} = \frac{\partial[\overline{E}(N, p, T) + p\overline{V}(N, p, T)]}{\partial(1/T)} < 0, \tag{4.80}$$

or

$$\frac{\partial \overline{E}(N, p, T)}{\partial T} > \frac{-p \partial \overline{V}(N, p, T)}{\partial T}. \tag{4.81}$$

These derivatives are of course at constant number and pressure. The analogous result for average quantities follows from the derivative of the partition function, Eq. (4.35). It also follows from Eq. (4.26) that

$$\left(\frac{\partial \overline{E}(N, p, T)}{\partial T} \right)_{N,p/T} > 0, \tag{4.82}$$

or

$$\left(\frac{\partial \overline{E}(N, p, T)}{\partial T} \right)_{N,p} + \frac{p}{T} \left(\frac{\partial \overline{E}(N, p, T)}{\partial p} \right)_{N,T} > 0, \tag{4.83}$$

and that

$$\left(\frac{\partial \overline{V}(N,p,T)}{\partial p} \right)_{N,T} < 0, \tag{4.84}$$

which was found in the preceding chapter, Eq. (3.65). The same conclusions for the average values follow from the analogous second derivatives of the logarithm of the partition function, Eq. (4.32). In particular, the isothermal compressibility in the thermodynamic limit is

$$\begin{aligned} \chi_T &\equiv \frac{-1}{\overline{V}} \frac{\partial \overline{V}(N,p,T)}{\partial p} \\ &= \frac{-1}{\overline{V}} \frac{\partial^2 \overline{G}(N,p,T)}{\partial p^2} \\ &= \frac{k_{\mathrm{B}}T}{\overline{V}} \frac{\partial^2 \ln Z(N,p,T)}{\partial p^2} \\ &= \frac{k_{\mathrm{B}}T}{\overline{V}} \left\langle \Delta(V/k_{\mathrm{B}}T_s)^2 \right\rangle, \end{aligned} \tag{4.85}$$

which is positive.

The second number derivative that follows from Eq. (4.37) is

$$\begin{aligned} &\left(\frac{\partial^2 \ln Z(N,p,T)}{\partial N^2} \right)_{p,T} - \left\langle \left(\frac{\partial^2 S(E,N,V)/k_{\mathrm{B}}}{\partial N^2} \right)_{E,V} \right\rangle \\ &= \left\langle \Delta(\mu_s/k_{\mathrm{B}}T_s)^2 \right\rangle. \end{aligned} \tag{4.86}$$

Again the extensivity arguments used in Section 3.8 apply, and for macroscopic systems one must have

$$S_{\mathrm{total}}(N,p,T) = k_{\mathrm{B}} \ln Z(N,p,T) = N \left\langle \mu_s/k_{\mathrm{B}}T_s \right\rangle + \mathcal{O}N^{2/3}. \tag{4.87}$$

The coefficient of the linear term is of course the number derivative of the partition function found above, and is independent of N. Hence to order N^{-1},

$$k_{\mathrm{B}} \left(\frac{\partial^2 \ln Z(N,p,T)}{\partial N^2} \right)_{p,T} = \left(\frac{\partial \left\langle \mu_s/k_{\mathrm{B}}T_s \right\rangle}{\partial N} \right)_{p,T} = 0, \tag{4.88}$$

the first neglected term being of order $\mathcal{O}N^{-4/3}$. Hence one concludes that for macroscopic systems,

$$\begin{aligned} -\left\langle \left(\frac{\partial^2 S(E,N,V)}{\partial N^2} \right)_{E,V} \right\rangle &= \left\langle \left(\frac{\partial [\mu(E,N,V)/T(E,N,V)]}{\partial N} \right)_{E,V} \right\rangle \\ &= \left\langle \Delta(\mu_s/k_{\mathrm{B}}T_s)^2 \right\rangle + \mathcal{O}N^{-4/3}. \end{aligned} \tag{4.89}$$

It is straightforward to show that the fluctuations in the enthalpy, $H = E + pV$, are given by

$$\left\langle \Delta(H)^2 \right\rangle = \frac{\partial^2 \ln Z(N,p,T)}{\partial \beta^2}, \tag{4.90}$$

where $\beta = 1/k_{\rm B}T$. One can also define a type of heat capacity,

$$\tilde{C}_p \equiv \left(\frac{\partial \overline{H}(N,p,T)}{\partial T} \right)_{N,p} = \frac{-\beta}{T} \frac{\partial^2 \beta \overline{G}(N,p,T)}{\partial \beta^2}, \tag{4.91}$$

and in the thermodynamic limit one has

$$\left\langle \Delta(H)^2 \right\rangle = \frac{T}{\beta} \tilde{C}_p, \tag{4.92}$$

which shows that $\tilde{C}_p$ is positive. Note that the usual heat capacity at constant pressure is related to the heat capacity defined above by

$$\begin{aligned} C_p \equiv \left(\frac{\partial \overline{E}(N,p,T)}{\partial T} \right)_{N,p} &= \frac{-\beta}{T} \frac{\partial^2 \beta \overline{G}(N,p,T)}{\partial \beta^2} - p \frac{\partial^2 \overline{G}(N,p,T)}{\partial T \partial p} \\ &= \tilde{C}_p - \alpha p \overline{V}(N,p,T), \end{aligned} \tag{4.93}$$

where the coefficient of thermal expansion is

$$\alpha \equiv \frac{1}{\overline{V}(N,p,T)} \left(\frac{\partial \overline{V}(N,p,T)}{\partial T} \right)_{N,p}. \tag{4.94}$$

4.4.4 Constant Enthalpy

The case of an insulated subsystem undergoing volume fluctuations was treated in Section 3.4. In this case the enthalpy, $H \equiv E + pV$, is constant. Here E and V are the energy and volume of the subsystem, and p is the pressure of the reservoir. (Here the reservoir is regarded as mechanical in nature, with negligible viscous dissipation.) The work done by the reservoir is isentropic, $dS_2 = -dE_1 - p_2 dV_1 = 0$, so that the volume-dependent part of the constrained total entropy is just the entropy of the subsystem itself, $S_{\rm total}(V|H,N,p) = S(E,N,V)$, $E \equiv H - pV$. Consequently the constrained thermodynamic potential is just $\Phi(V|H,N,p;T) = -TS(E,N,V)$, $E \equiv H - pV$, and so Eq. (4.6) gives the probability of the subsystem having a volume V, given that it has N particles and enthalpy H and that the pressure of the reservoir is p,

$$\wp(V|H,N,p) = \frac{1}{Z(H,N,p)} e^{S(E,N,V)/k_{\rm B}}, \quad E \equiv H - pV. \tag{4.95}$$

Note that the arbitrary temperature T does not enter this expression.

The partition function, Eq. (4.27), is of course

$$Z(H,N,p) = \sum_V e^{S(E,N,V)/k_{\rm B}}, \quad E \equiv H - pV, \tag{4.96}$$

and since the part of the total entropy that depends upon the presence of the subsystem is the subsystem entropy itself, Eq. (4.28) gives the unconstrained entropy of the subsystem, $S(H, N, p) = k_{\mathrm{B}} \ln Z(H, N, p)$. The general formulae for the logarithmic derivatives of the partition function, Eqs. (4.29) and (4.30), cannot be used in this case because they were derived for independent extensive parameters. However direct differentiation yields the average inverse temperature,

$$\left\langle T_s^{-1} \right\rangle = k_{\mathrm{B}} \frac{\partial \ln Z(H, N, p)}{\partial H}, \tag{4.97}$$

the average chemical potential,

$$\langle -\mu_s / T_s \rangle = k_{\mathrm{B}} \frac{\partial \ln Z(H, N, p)}{\partial N}, \tag{4.98}$$

and the average volume,

$$\langle -V / T_s \rangle = k_{\mathrm{B}} \frac{\partial \ln Z(H, N, p)}{\partial p}. \tag{4.99}$$

The subscripts denote the field parameters of the subsystem, which of course vary with its state.

Taking the second derivatives yields for number

$$\frac{\partial^2 \ln Z(H, N, p)}{\partial N^2} = \left\langle \Delta(\mu_s / k_{\mathrm{B}} T_s)^2 \right\rangle + \left\langle \frac{\partial^2 S(H - pV, N, V)/k_{\mathrm{B}}}{\partial N^2} \right\rangle, \tag{4.100}$$

for pressure

$$\frac{\partial^2 \ln Z(H, N, p)}{\partial p^2} = \left\langle \Delta(V_s / k_{\mathrm{B}} T_s)^2 \right\rangle + \left\langle \frac{\partial^2 S(H - pV, N, V)/k_{\mathrm{B}}}{\partial p^2} \right\rangle, \tag{4.101}$$

and for enthalpy,

$$\frac{\partial^2 \ln Z(H, N, p)}{\partial H^2} = \left\langle \Delta(1 / k_{\mathrm{B}} T_s)^2 \right\rangle + \left\langle \frac{\partial^2 S(H - pV, N, V)/k_{\mathrm{B}}}{\partial H^2} \right\rangle. \tag{4.102}$$

These are not particularly illuminating.

For there to be an equilibrium state, the probability distribution must have a peak, which means that

$$\frac{\partial^2 S(H - pV, V, N)}{\partial V^2} < 0, \tag{4.103}$$

or

$$p \left. \frac{\partial p(E, V, N)}{\partial E} \right|_{\overline{E}, \overline{V}} > \left. \frac{\partial p(E, V, N)}{\partial V} \right|_{\overline{E}, \overline{V}}. \tag{4.104}$$

Here of course $p(E, V, N)$ is the pressure of an isolated system, as given by the volume derivative of its entropy, and one also has $p(\overline{E}, \overline{V}, N) = p$, and $\overline{E} = H - p\overline{V}$.

4.4.5 Constant Entropy

In Section 3.5 a constant entropy subsystem with fluctuating volume was considered. In this case the energy was a dependent variable, and so again it is not possible to use the generic analysis given in the first half of this chapter, which was confined to independent fluctuating quantities. Nevertheless, the derivation of the volume probability distribution and average values remains straightforward. The constrained thermodynamic potential is just $H^*(V|S,p,N;T) = -TS_{\rm total}(V|S,p,N;T)$, where the total entropy is given by Eq. (3.34),

$$S_{\rm total}(V|S,p,N;T) = S - \frac{E(S,V,N)}{T} - \frac{pV}{T}. \tag{4.105}$$

Here T and p are the temperature and pressure of the reservoir, and S is the fixed entropy of the subsystem. The volume probability distribution is just the exponential of the total entropy,

$$\begin{aligned} \wp(V|S,p,N;T) &= \frac{1}{Z'} e^{S/k_{\rm B}} e^{-E(S,V,N)/k_{\rm B}T} e^{-pV/k_{\rm B}T} \\ &= \frac{1}{Z(S,p,N;T)} e^{-E(S,V,N)/k_{\rm B}T} e^{-pV/k_{\rm B}T}. \end{aligned} \tag{4.106}$$

The constant factor $e^{S/k_{\rm B}}$ has been dropped because it is independent of the volume. The exponent can also be written $-H(V|S,p,N)/k_{\rm B}T$, where in Section 3.5 this enthalpy was just the volume-dependent part of the constrained thermodynamic potential. Whereas the reservoir temperature did not enter in dealing with the equilibrium state, it does reappear in dealing with the probability distribution.

The partition function is just

$$Z(S,p,N;T) = \sum_V e^{-E(S,V,N)/k_{\rm B}T} e^{-pV/k_{\rm B}T}, \tag{4.107}$$

and the unconstrained total entropy of the subsystem plus reservoir, Eq. (4.28), is just $S_{\rm total}(S,p,N;T) = k_{\rm B} \ln Z(S,p,N;T)$. Direct differentiation yields the average volume,

$$\langle V \rangle = -k_{\rm B}T \frac{\partial \ln Z(S,p,N;T)}{\partial p}, \tag{4.108}$$

the average chemical potential,

$$\langle \mu_s \rangle = -k_{\rm B}T \frac{\partial \ln Z(S,p,N;T)}{\partial N}, \tag{4.109}$$

and the average temperature of the subsystem,

$$\langle T_s \rangle = -k_{\rm B}T \frac{\partial \ln Z(S,p,N;T)}{\partial S}. \tag{4.110}$$

These may be compared with their equilibrium counterparts in Section 3.5. In addition one has the average enthalpy

$$\langle E + pV \rangle = -k_{\rm B} \frac{\partial \ln Z(S,p,N;T)}{\partial T^{-1}}, \tag{4.111}$$

which in conjunction with the average volume gives the average energy.

It is now shown that the enthalpy is a concave function of the pressure. Denote the equilibrium volume by $\overline{V} = V(S,p,N)$, and the equilibrium volume due to an almost identical pressure by $\tilde{V} = V(S,\tilde{p},N)$. Due to the variational nature of the constrained thermodynamic potential one has

$$\begin{aligned} & H(\tilde{V}|S,p,N) \\ & \quad > \; H(\overline{V}|S,p,N) \\ & \quad = \; p\overline{V} + E(S,\tilde{V},N) + (\overline{V}-\tilde{V})\tilde{p} + \frac{1}{2}(\overline{V}-\tilde{V})^2 \tilde{E}_{VV}. \end{aligned} \tag{4.112}$$

The left-hand side is just $E(S,\tilde{V},N) + p\tilde{V}$, and so one has

$$(\overline{V}-\tilde{V})(p-\tilde{p}) - \frac{1}{2}(\overline{V}-\tilde{V})^2 \frac{\partial p(S,\tilde{V},N)}{\partial \tilde{V}} < 0. \tag{4.113}$$

Dividing by $(p-\tilde{p})^2$ and taking the limit $\tilde{p} \to p$, one sees that the first term is just $\partial\overline{V}/\partial p$, and that it is partially cancelled by the second term, which is just $(\partial\overline{V}/\partial p)^2(\partial\overline{p}/\partial V)/2$, so that one concludes that the isentropic compressibility is positive,

$$\chi_s \equiv \frac{-1}{\overline{V}} \frac{\partial \overline{V}(S,p,N)}{\partial p} > 0. \tag{4.114}$$

Here the uniqueness of the thermodynamic state has been used, $V_1 = V(S,p_1,N) \Leftrightarrow p_1 = p(S,V_1,N)$.

There is an analogous result for the logarithm of the partition function. One has

$$(k_{\mathrm{B}}T)^2 \frac{\partial^2 \ln Z(S,p,N;T)}{\partial p^2} = -k_{\mathrm{B}}T \left(\frac{\partial \langle V \rangle}{\partial p} \right)_{S,N,T} = \langle \Delta(V)^2 \rangle > 0. \tag{4.115}$$

4.4.6 Equivalence of Systems

The different systems, each specified by its fixed variables and the field variables of the reservoir, were shown in Section 3.7 to be consistent in their equilibrium values. By this it is meant that if the set of extensive variables of each (fixed values and equilibrium values) were the same, then so are their field variables, which are the first derivatives of the respective thermodynamic potentials. This consistency did not extend to the second derivatives.

It was also shown, in Section 4.3.4, that for macroscopic systems, the logarithm of the partition function equalled the thermodynamic potential for that particular system. It follows that average values, which are the derivative of the former, equal equilibrium values, with the consequence that different systems are also consistent as far as average values go. Again this does not hold for the second derivatives, so that fluctuations differ for different systems.

This may be illustrated explicitly for the case of energy fluctuations in canonical and grand canonical systems. For a system at constant temperature one has

$$\begin{aligned}
\left\langle \Delta_T(E)^2 \right\rangle &\equiv \left\langle (E - \langle E \rangle_T)^2 \right\rangle_T \\
&= \left(\frac{\partial^2 Z(N,V,T)}{\partial (1/k_\mathrm{B}T)^2} \right)_{N,V} \\
&= -\left(\frac{\partial \langle E \rangle_T}{\partial (1/k_\mathrm{B}T)} \right)_{N,V} .
\end{aligned} \tag{4.116}$$

Here the subscript T is used explicitly to indicate a canonical quantity. For a system able to exchange both energy and particles one has

$$\begin{aligned}
&\left\langle \Delta_\mu(E)^2 \right\rangle \\
&\equiv \left\langle (E - \langle E \rangle_\mu)^2 \right\rangle_\mu \\
&= \left(\frac{\partial^2 Z(\mu,V,T)}{\partial (1/k_\mathrm{B}T)^2} \right)_{\mu/T,V} \\
&= -\left(\frac{\partial \langle E \rangle_\mu}{\partial (1/k_\mathrm{B}T)} \right)_{\mu/T,V} \\
&= -\left(\frac{\partial \langle E \rangle_\mu}{\partial (1/k_\mathrm{B}T)} \right)_{N,V} - \left(\frac{\partial \langle E \rangle_\mu}{\partial \langle N \rangle_\mu} \right)_{T,V} \left(\frac{\partial \langle N \rangle_\mu}{\partial (1/k_\mathrm{B}T)} \right)_{\mu/T,V} \\
&= -\left(\frac{\partial \langle E \rangle_T}{\partial (1/k_\mathrm{B}T)} \right)_{N,V} - \left(\frac{\partial \langle E \rangle_T}{\partial N} \right)_{T,V} \left(\frac{\partial \langle E \rangle_\mu)}{\partial (\mu/k_\mathrm{B}T)} \right)_{T,V} \\
&= \left\langle \Delta_T(E)^2 \right\rangle - \left(\frac{\partial \langle E \rangle_T}{\partial N} \right)_{T,V} \left(\frac{\partial \langle E \rangle_T}{\partial \langle N \rangle_\mu} \right)_{T,V} \left(\frac{\partial \langle N \rangle_\mu}{\partial (\mu/k_\mathrm{B}T)} \right)_{T,V} \\
&= \left\langle \Delta_T(E)^2 \right\rangle + \left(\frac{\partial \langle E \rangle_T}{\partial N} \right)_{T,V}^2 \left\langle \Delta_\mu(N)^2 \right\rangle .
\end{aligned} \tag{4.117}$$

Here the consistency of the systems, $\langle E \rangle_T = \langle E \rangle_\mu$, and $\langle N \rangle_\mu = N$, a Maxwell relation for the cross-second derivative, $\partial^2 Z / \partial (1/T) \partial (\mu/T)$, and also certain rules for the manipulation of partial derivatives have been used. The result says that the energy fluctuations of a grand canonical system equal those of a canonical system, plus those that result from the fluctuations in number. This is a typical example of the relationship between the fluctuations of the various systems.

Summary

- It is the constrained total entropy whose exponential gives the probability distribution for parameters exchangeable with a reservoir. The former in general is the entropy of the isolated subsystem less the sum of the product of the parameters and their conjugate reservoir field variables. The constrained thermodynamic potential is the negative of the reservoir temperature times the constrained total entropy, and hence it appears like a Boltzmann factor in determining the probability distribution.

- The partition function normalises the probability distribution, and its logarithm gives the subsystem-dependent part of the unconstrained total entropy. Its logarithmic derivative with respect to a reservoir field variable gives the average of the conjugate exchangeable variable, and that with respect to a fixed extensive variable gives the average of the conjugate field variable. The second logarithmic derivatives give various fluctuations about the averages, which show that the unconstrained total entropy is a convex function of the reservoir field variables.

- If the state of an isolated subsystem can represent a state of equilibrium with a reservoir, then the entropy of the isolated subsystem in that state must be a concave function of its arguments. Since a system can be considered a reservoir for its parts, in general the stable states of matter are characterised by a concave entropy.

- The thermodynamic potential is the minimum value of the constrained thermodynamic potential and is a property of the equilibrium state. Its derivatives yield the equilibrium values of the exchangeable parameters and the field variables. These derivatives may be evaluated using the variational nature of the constrained thermodynamic potential.

- The equilibrium thermodynamic potential is a concave function of the inverse reservoir temperature, and also of the reservoir field variables.

- In the thermodynamic limit (macroscopic systems), equilibrium values equal average values, and the equilibrium thermodynamic potential equals the logarithm of the partition function, which is equivalent to the unconstrained total entropy. Different systems are equivalent as far as their first derivatives (average values and equilibrium values), but differ in their second derivatives (fluctuations).

Chapter 5

Classical Statistical Mechanics

The thermodynamic analysis of the preceding chapters has been of an abstract and general character. Such is the nature of statistical thermodynamics, which is based on asserting the existence of weighted microstates rather than on explicitly constructing them. This suffices to deduce the laws of thermodynamics, and the interrelationships between the thermodynamic parameters. Likewise a great deal of statistical mechanics can be carried out on rather idealised models, particularly systems that exist in denumerable states. In these cases nothing beyond what has been already derived is required. However in order to use statistical mechanics on realistic systems, it remains to make an explicit and quantitative connection with molecular properties, to put egg in the shell, as it were.

The further development of the theory raises several fundamental issues. One must identify the set of states that will serve as the microstates. In the case of classical statistical mechanics, which is the main concern in what follows, these will turn out to have a continuous distribution, and so one must develop the formalism for a continuum. The expression for the continuum entropy was briefly discussed in Section 1.3.3, and a summary of the general laws for probability and entropy in the discrete and continuous cases is given in the Appendix, Section A.3. Here a detailed derivation for the particular case of classical statistical mechanics is given, and from the equations of motion an explicit expression for the weight attached to the points in the continuum is obtained.

The chapter begins with an analysis of an isolated system that evolves with Hamilton's equations of motion, and which consequently has constant energy. The probability distribution of such a system is generally taken to be constant, the rationale being that *states of equal energy are equally likely*. Whilst such a statement, once accepted, is straightforward to implement for discrete states, some detailed analysis is required to justify and to interpret it for the continuum that is classical statistical mechanics. Usually this is done on the basis of the ergodic hypothesis, but here a different approach is taken based upon a comparison of time, phase space, and hypersurface probability densities. The sense in which the consequent phase space probability density is uniform is discussed and shown to agree with the conventional result deduced from Liouville's theorem and the ergodic hypothesis. The chapter concludes with specific results for the particular reservoirs analysed in the preceding chapters.

5.1 Constant Energy Hamiltonian System

5.1.1 Phase Space and Hamilton's Equations

The basis of thermodynamics is that all systems possess a set of weighted microstates that are distinct and indivisible, and that entropy is the logarithm of the sum of their weights. The first nontrivial task of classical statistical mechanics is to identify the microstates.

Whereas the analysis up till now has invoked discrete states, classical mechanics deals with continuous variables. For an isolated system comprising N particles, an obvious choice for the space governed by classical mechanics is *phase space*, which is the space of canonical positions and momenta of all the particles. In a three-dimensional universe, phase space has $6N$ dimensions, with a point in phase space being denoted by $\mathbf{\Gamma}$, or by $\mathbf{p}^N\mathbf{q}^N$, where $\mathbf{q}_i$ is the position of the particle labelled i, and $\mathbf{p}_i$ is its momentum. It is clear that specifying a point in phase space specifies the state of the system, since this reflects exactly what all the particles are doing at any instant.[1] Instead of the momenta one could specify the particles' velocities. The particles' accelerations cannot be specified independently since these are determined by their mutual forces, which in turn are determined by their positions (and in some cases their velocities). Hence whilst one requires the positions and the velocities or the momenta to be specified to pin down the state of the system, one does not need to specify the accelerations or higher time derivatives, since these are already determined. This means that a point in phase space can be used to represent the state of the system, and that phase space itself can be used as the continuum analogue of the set of microstates of the system. At this stage there is nothing to prefer phase space over other possible representations (e.g., Lagrangian mechanics specifies instead the velocities and the positions of the particles).

The word 'canonical' in the above is important because it signifies the fact that the time development of the system is governed by Hamilton's equations of motion. That is, there exists a *Hamiltonian*, $\mathcal{H}(\mathbf{\Gamma})$, which is the energy of the system and whose derivatives give the time rate of change of the positions and momenta of the particles,

$$\dot{\mathbf{q}}_i = \frac{\partial \mathcal{H}(\mathbf{\Gamma})}{\partial \mathbf{p}_i}, \text{ and } \dot{\mathbf{p}}_i = -\frac{\partial \mathcal{H}(\mathbf{\Gamma})}{\partial \mathbf{q}_i}. \tag{5.1}$$

[1]This level of description obviously ignores any internal coordinates of the particles, or orientations in the case of nonspherical particles, or configurations of any particles external to the system. It is straightforward to generalise the formalism for these cases.

Here the vector derivative means that each of these represents three equations, one for each vector component.

An isolated system follows a trajectory through phase space that may be denoted $\mathbf{\Gamma}(t|\mathbf{\Gamma}_0)$, where the notation signifies that the trajectory was at $\mathbf{\Gamma}_0$ at time $t = 0$, $\mathbf{\Gamma}(0|\mathbf{\Gamma}_0) = \mathbf{\Gamma}_0$. This trajectory is fully determined by Hamilton's equations. Energy is conserved on such a trajectory, since

$$\begin{aligned} \frac{d\mathcal{H}(\mathbf{\Gamma})}{dt} &= \dot{\mathbf{\Gamma}} \cdot \frac{\partial \mathcal{H}}{\partial \mathbf{\Gamma}} \\ &= \sum_{i=1}^{N} \left[\dot{\mathbf{p}}_i \cdot \frac{\partial \mathcal{H}}{\partial \mathbf{p}_i} + \dot{\mathbf{q}}_i \cdot \frac{\partial \mathcal{H}}{\partial \mathbf{q}_i} \right] \\ &= \sum_{i=1}^{N} [\dot{\mathbf{p}}_i \cdot \dot{\mathbf{q}}_i - \dot{\mathbf{q}}_i \cdot \dot{\mathbf{p}}_i] \\ &= 0. \end{aligned} \tag{5.2}$$

Recall that it is an essential requirement of the formalism of thermodynamics that the energy of an isolated system be conserved.

5.1.2 Averages and Probability Densities

Amongst other things statistical mechanics yields the average properties of the system. Many such properties can be expressed as a molecular function of the phase state of the system, $f(\mathbf{\Gamma})$. Examples include the pressure, where f is essentially the intermolecular forces manifest in the so-called virial of Claussius, the density, where f counts the number of molecules in a particular region, and the kinetic energy. To obtain these averages one requires a probability density, of which there are several possibilities.

There are three different ways to obtain the average, and each has its own probability density. The most well known is the temporal average, which is the average of the phase space function along a trajectory in time,

$$\begin{aligned} \langle f \rangle_\tau &= \int_0^\tau \mathrm{d}t\, \wp(t) f(\mathbf{\Gamma}(t|\mathbf{\Gamma}_0)) \\ &= \frac{1}{\tau} \int_0^\tau \mathrm{d}t\, f(\mathbf{\Gamma}(t|\mathbf{\Gamma}_0)). \end{aligned} \tag{5.3}$$

Here τ is an experimental time scale that is long compared to molecular time scales, and in consequence the average is independent of the starting position of the trajectory. The second line assumes that the temporal probability density is uniform, $\wp(t) = 1/\tau$. In other words, the weight of a state is linearly proportional to the time the system spends in the state, and a time average is a simple average.

Alternatively, the phase space average of f may also be defined. Taking phase space itself as the set of possible microstates, the conditional probability density for an isolated system that has energy within $\mathrm{d}E$ of E may be denoted

$\wp(\mathbf{\Gamma}|E)$. This density gives the probability that the system will be found within $\mathrm{d}\mathbf{\Gamma}$ of $\mathbf{\Gamma}$, and is defined in terms of the temporal average as

$$\wp(\mathbf{\Gamma}_1|E) = \frac{1}{\tau}\int_0^\tau \mathrm{d}t\, \delta(\mathbf{\Gamma}(t|\mathbf{\Gamma}_0) - \mathbf{\Gamma}_1). \tag{5.4}$$

With it the average of any function of the state of the system is

$$\langle f \rangle_{\mathbf{\Gamma}} = \int \mathrm{d}\mathbf{\Gamma}\, \wp(\mathbf{\Gamma}|E) f(\mathbf{\Gamma}). \tag{5.5}$$

Direct substitution and interchange of the order of integration shows that phase and time averages are equivalent, $\langle f \rangle_{\mathbf{\Gamma}} = \langle f \rangle_\tau$.

One can also define a surface average. In this case one invokes a coordinate system for the surface, $\boldsymbol{\gamma}(\mathbf{\Gamma})$. One can write equally well $\mathbf{\Gamma}(\boldsymbol{\gamma}, E)$, since the hypersurface has dimension $6N - 1$. The surface probability density may be written

$$\wp^{(E)}(\boldsymbol{\gamma}_1) = \frac{1}{\tau}\int_0^\tau \mathrm{d}t\, \delta(\boldsymbol{\gamma}(t|\boldsymbol{\gamma}_0) - \boldsymbol{\gamma}_1). \tag{5.6}$$

In terms of it the surface average is

$$\langle f \rangle_{\boldsymbol{\gamma}} = \oint_E \mathrm{d}\boldsymbol{\gamma}\, \wp^{(E)}(\boldsymbol{\gamma}) f(\mathbf{\Gamma}(\boldsymbol{\gamma}, E)), \tag{5.7}$$

and again one may confirm the equivalence of surface and time averages, $\langle f \rangle_{\boldsymbol{\gamma}} = \langle f \rangle_\tau$. Here the integral is over the energy hypersurface of thickness $\mathrm{d}E$ to which the isolated system is constrained, and $\mathrm{d}\boldsymbol{\gamma}$ is the element of area.

The element of volume is related to the element of area by

$$\mathrm{d}\mathbf{\Gamma} = \mathrm{d}\boldsymbol{\gamma}\, \mathrm{d}n = \mathrm{d}\boldsymbol{\gamma}\, \mathrm{d}E/|\nabla\mathcal{H}(\mathbf{\Gamma})|, \tag{5.8}$$

where $\mathrm{d}n$ is the differential length of the normal to the energy hypersurface. The gradient of the energy hypersurface converts this length, which is measured in units of $\mathbf{\Gamma}$, to the differential energy in energy units. From this equation one sees that the distance between two energy hypersurfaces is not constant in phase space.

The phase space probability density is given in terms of the surface probability density by

$$\begin{aligned}\wp(\mathbf{\Gamma}|E) &= \wp^{(E)}(\boldsymbol{\gamma}) \left|\frac{\mathrm{d}\boldsymbol{\gamma}}{\mathrm{d}\mathbf{\Gamma}}\right|, \ |\mathcal{H}(\mathbf{\Gamma}) - E| < \mathrm{d}E \\ &= \wp^{(E)}(\boldsymbol{\gamma}) \frac{|\nabla\mathcal{H}(\mathbf{\Gamma})|}{\mathrm{d}E}, \ |\mathcal{H}(\mathbf{\Gamma}) - E| < \mathrm{d}E \\ &= \wp^{(E)}(\boldsymbol{\gamma})\, |\nabla\mathcal{H}(\mathbf{\Gamma})|\, \delta(\mathcal{H}(\mathbf{\Gamma}) - E),\end{aligned} \tag{5.9}$$

where $\boldsymbol{\gamma} = \boldsymbol{\gamma}(\mathbf{\Gamma})$. The Dirac δ follows as the limit that $\mathrm{d}E \to 0$ in the passage to the final line. The distinction between the two probability densities is that $\wp(\mathbf{\Gamma}|E)$ is defined throughout phase space, whereas the surface probability density $\wp^{(E)}(\gamma)$ is only defined on the energy hypersurface. The former is used when phase space itself is considered to be the set of all possible states of the system.

With this expression one can readily verify that surface and the phase space averages are equal,

$$\begin{aligned} \langle f \rangle_{\mathbf{\Gamma}} &= \int \mathrm{d}\mathbf{\Gamma}\, \wp(\mathbf{\Gamma}|E) f(\mathbf{\Gamma}) \\ &= \int \frac{\mathrm{d}\gamma\, \mathrm{d}E}{|\nabla \mathcal{H}(\mathbf{\Gamma})|} |\nabla \mathcal{H}(\mathbf{\Gamma})|\, \delta(\mathcal{H}(\mathbf{\Gamma}) - E) \wp^{(E)}(\gamma) f(\mathbf{\Gamma}) \\ &= \oint_E \mathrm{d}\gamma\, \wp^{(E)}(\gamma) f(\mathbf{\Gamma}(\gamma, E)) \\ &= \langle f \rangle_{\gamma}\,. \end{aligned} \tag{5.10}$$

This is of course necessary for the internal consistency of statistical mechanics.

The factor relating the two probability densities may be called the characteristic function of the energy hypersurface,

$$\begin{aligned} \chi(\mathcal{H}, \mathbf{\Gamma}) &= |\nabla \mathcal{H}(\mathbf{\Gamma})|\, \delta(\mathcal{H}(\mathbf{\Gamma}) - E) \\ &= |\nabla g_H(\mathbf{\Gamma})|\, \delta(g_H(\mathbf{\Gamma}) - G). \end{aligned} \tag{5.11}$$

The final line, in which $g_H(\mathbf{\Gamma}) \equiv g(\mathcal{H}(\mathbf{\Gamma}))$ and $G \equiv g(E)$, demonstrates that the characteristic function is invariant with respect to the representation of the hypersurface. On physical grounds this is an essential requirement, and all of the following results, which are given in the 'natural' representation of the first line, are unchanged by any other representation. The characteristic function gives the area of the energy hypersurface,

$$\begin{aligned} A(E) &= \oint_E \mathrm{d}\gamma \\ &= \int \mathrm{d}\mathbf{\Gamma}\, |\nabla \mathcal{H}(\mathbf{\Gamma})|\, \delta(\mathcal{H}(\mathbf{\Gamma}) - E), \end{aligned} \tag{5.12}$$

as may be verified by direct substitution. In view of this, the characteristic function may also be called the area density of the hypersurface in phase space.

Multiple Constraints

Although the focus here is on an isolated system with only energy conserved, one can envisage more general cases with multiple conserved quantities. In general, m constraints in n dimensions give an $(n - m)$-dimensional hypersurface with area

$$S_m^n = \int \mathrm{d}x^n\, \mathcal{D}(F^m, x^n) \prod_{i=1}^{m} \delta(F_i(x^n) - f_i). \tag{5.13}$$

Here the $F_i(x^n)$ are the constrained functions of the hyperspace and the f_i are their constant values. For the area to be invariant to different representations of the hypersurface, the area density must be of the form of a sum of products of all the F_i. Further, dimensional considerations show that it must be the derivatives of the F_i that occur. Finally, the area density must be symmetric with regard to each of the F_i and each of the dimensions of the space. The function that fulfils all of these requirements is

$$\begin{aligned}
&\mathcal{D}(F^m, x^n)^2 \\
&= \sum_{i_1=1}^{n-m+1} \sum_{i_2=i_1+1}^{n-m+2} \cdots \sum_{i_m=i_{m-1}+1}^{n} \left(\sum_{\mathcal{P}(\underline{i})} \epsilon(\underline{i}) \frac{\partial F_1}{\partial x_{i_1}} \frac{\partial F_2}{\partial x_{i_2}} \cdots \frac{\partial F_m}{\partial x_{i_m}} \right)^2 \\
&= \sum_{i_1<i_2<\ldots<i_m}^{n} \left| \nabla^{(n)} F^m \right|^2 . \qquad (5.14)
\end{aligned}$$

Here $\epsilon = \pm 1$ gives the parity of the permutation $\mathcal{P}$. The dyadic matrix whose determinant appears in the second form has elements $\partial F_i / \partial x_j$, where $i \in \{1, 2, \ldots, m\}$ and $j \in \{i_1, i_2, \ldots, i_m\}$.

5.1.3 Phase Space Probability Density

The surface probability density follows directly from the uniform temporal probability density, $\wp(t) = 1/\tau$,

$$\begin{aligned}
\wp^{(E)}(\boldsymbol{\gamma}) &= \wp(t) \left| \frac{\mathrm{d}t}{\mathrm{d}\boldsymbol{\gamma}} \right| \\
&\propto \frac{1}{|\dot{\boldsymbol{\gamma}}|}. \qquad (5.15)
\end{aligned}$$

Mathematically, this result holds if the experimental time scale τ is short compared to the system recurrence time, so that the trajectory only passes within $\mathrm{d}\boldsymbol{\gamma}$ of $\boldsymbol{\gamma}$ once (assuming smoothness of $\wp^{(E)}(\boldsymbol{\gamma})$ so that it may be continued to the neighbourhood of the trajectory), and if τ is long enough so that it covers a representative selection of points on the hypersurface. (By definition, if the experimental temporal average is representative, then so must be the points visited by this finite trajectory.) The physical argument that underpins this result follows from the temporal representation, Eq. (5.6). There the δ-function in the integrand picks out the amount of time the system spends in the vicinity of a particular phase point on the energy hypersurface. The amount of time is $\tau_{\boldsymbol{\gamma}} = \mathrm{d}\boldsymbol{\gamma}/|\dot{\boldsymbol{\gamma}}|$. The probability density is in addition inversely proportional to the area element $\mathrm{d}\boldsymbol{\gamma}$, and the combination of these two factors gives the above result. Note that the velocity of the trajectory is not a constant of the motion. Hence one may conclude from this result that the surface probability density is *not* uniform on the surface.

In practice it is difficult to give explicit expressions for the curvilinear coordinates of the hypersurface, and it is much more convenient to remain in phase space itself. Hence it is preferable to deal with the phase space probability density rather than this energy hypersurface probability density. The time that the system spends in an elementary cell $\mathrm{d}\boldsymbol{\Gamma}$ remains τ_γ, as given above (because only the area element through which the trajectory passes is relevant), and the phase space probability density is proportional to this and inversely proportional to the volume element $\mathrm{d}\boldsymbol{\Gamma} = \mathrm{d}\gamma \mathrm{d}E/\left|\nabla\mathcal{H}(\boldsymbol{\Gamma})\right|$,

$$\begin{aligned} \wp(\boldsymbol{\Gamma}|E) &\propto \frac{|\nabla\mathcal{H}(\boldsymbol{\Gamma})|}{|\dot{\gamma}|}\frac{1}{\mathrm{d}E}, \quad |\mathcal{H}(\boldsymbol{\Gamma}) - E| < \mathrm{d}E \\ &\propto \frac{|\nabla\mathcal{H}(\boldsymbol{\Gamma})|}{|\dot{\boldsymbol{\Gamma}}|}\delta(\mathcal{H}(\boldsymbol{\Gamma}) - E). \end{aligned} \tag{5.16}$$

Note that $|\dot{\boldsymbol{\Gamma}}| = |\dot{\gamma}|$ because $|\gamma_1 - \gamma_2| = |\boldsymbol{\Gamma}(\gamma_1, E) - \boldsymbol{\Gamma}(\gamma_2, E)| = |\boldsymbol{\Gamma}_1 - \boldsymbol{\Gamma}_2|$. This result can also be obtained directly from Eq. (5.15) using the conversion rule (5.9).

The interpretation of this result is straightforward. The denominator is the speed of the trajectory, and represents the fact that the probability of a region of phase space is proportional to the time that the system spends in that region. The numerator is the area of the energy hypersurface at that point, and represents the fact that the probability of a region of phase space is proportional to the amount of the hypersurface to which the system is confined that lies in the region.[2]

Hamiltonian System

From the above it follows that the phase space probability density is therefore

$$\wp(\boldsymbol{\Gamma}|E) = \frac{|\nabla\mathcal{H}(\boldsymbol{\Gamma})|}{|\dot{\boldsymbol{\Gamma}}|Z'(E)}\delta(\mathcal{H}(\boldsymbol{\Gamma}) - E). \tag{5.17}$$

The normalising partition function or total weight, $Z'(E)$, is discussed in the next section. Note that whilst this result relies upon the fact that $\mathcal{H}(\boldsymbol{\Gamma})$ is a constant of the motion (i.e., that the system is confined to a hypersurface of constant $\mathcal{H}$), Hamilton's equations have *not* been assumed in deriving this result. So this is a general expression for a system that evolves with arbitrary equations of motion and that has a single constant of the motion represented by some arbitrary function $\mathcal{H}(\boldsymbol{\Gamma})$.

However for a Hamiltonian system, it is a remarkable coincidence that in the space of canonical positions and momenta,

$$\begin{aligned} |\nabla\mathcal{H}(\boldsymbol{\Gamma})|^2 &= \sum_{i=1}^N \frac{\partial\mathcal{H}}{\partial\mathbf{q}_i}\cdot\frac{\partial\mathcal{H}}{\partial\mathbf{q}_i} + \frac{\partial\mathcal{H}}{\partial\mathbf{p}_i}\cdot\frac{\partial\mathcal{H}}{\partial\mathbf{p}_i} \\ &= \sum_{i=1}^N \dot{\mathbf{p}}_i\cdot\dot{\mathbf{p}}_i + \dot{\mathbf{q}}_i\cdot\dot{\mathbf{q}}_i \\ &= |\dot{\boldsymbol{\Gamma}}|^2. \end{aligned} \tag{5.18}$$

[2] One could take these two points as the axiomatic basis for the more general probability density for non-Hamiltonian systems and for systems with multiple constraints. Little work has been done on this, and the fact that for equilibrium or steady state systems the probability density must be constant in time places restrictions on the allowed equations of motion.

Hence the gradient in the numerator cancels with the speed in the denominator, giving a particularly simple expression for the probability density in phase space (on the energy hypersurface),

$$\wp(\mathbf{\Gamma}|E, N, V) = \frac{\delta(\mathcal{H}(\mathbf{\Gamma}) - E)}{Z'(E, N, V)}. \tag{5.19}$$

Here it has been signified that the energy, the number of particles, and the volume of the isolated system are specified.

This result shows that phase space points with the same energy are equally likely. This is the most precise way of interpreting the discrete notion that *states of equal energy are equally likely.* One cannot say that the phase space probability density is uniform in phase space, since the Dirac δ is highly nonuniform, nor that the surface probability density is uniform on the energy hypersurface, since it is proportional to the inverse of the trajectory speed, which is not a constant of the motion. Keeping in mind the distinction between the two probability densities, one could perhaps say that the phase space probability density is uniform on the energy hypersurface, which paraphrases the statement at the beginning of this paragraph. This description of Eq. (5.19) appears to be consistent with the hypothesis of equal *a priori* probabilities enunciated by Tolman: "the distribution of members of the ensemble over different states agrees with what is known to be the actual state of the system of interest but is otherwise uniform in the phase space."[3]

Note how the canonical relationship between positions and momenta embodied in Hamilton's equations of motion was central to this result, and that it is the 'natural' representation of the energy hypersurface that simplifies the final formula. It is important to understand that this is a result specific for Hamiltonian systems with only energy conserved, and using the 'natural' representation of the energy hypersurface. If any other representation of the energy hypersurface is used, it is important to convert this expression correctly (or else to start from the invariant representation, Eq. (5.17)), since ignoring the energy-dependent factor that arises gives unphysical results for the energy derivatives of the partition function.

With these results the average of a function of phase space for a Hamiltonian system of energy E is

$$\begin{aligned} \langle f(\mathbf{\Gamma})\rangle &= \frac{1}{Z'(E, N, V)} \int \mathrm{d}\mathbf{\Gamma}\, f(\mathbf{\Gamma})\delta(\mathcal{H}(\mathbf{\Gamma}) - E) \\ &= \frac{1}{Z'(E, N, V)} \oint \frac{\mathrm{d}\gamma}{|\dot{\gamma}|} f(\mathbf{\Gamma}(\gamma, E)). \end{aligned} \tag{5.20}$$

5.1.4 Continuum Entropy of a Hamiltonian System

The partition function that normalises the phase space probability density is meant to represent the total weight of the system, and its logarithm is the entropy. Two subtleties need to be dealt with before implementing this explicitly.

[3] R. C. Tolman, *The Principles of Statistical Mechanics*, Oxford Univ. Press, Oxford, 1938.

In the discrete case the microstates were distinct, and the total weight was just the sum of the weights of the microstates. In the case of phase space, because the particles are indistinguishable, the physical state of the system labelled by $\mathbf{\Gamma}$ is also labelled by the point $\mathbf{\Gamma}'$ corresponding to some permutation of the indices of the particles. For example, the microstate represented by particle 1 at $(\mathbf{p}_a, \mathbf{q}_a)$ and particle 2 at $(\mathbf{p}_b, \mathbf{q}_b)$ is exactly the same microstate as that with particle 2 at $(\mathbf{p}_a, \mathbf{q}_a)$ and particle 1 at $(\mathbf{p}_b, \mathbf{q}_b)$. Since the particles are indistinguishable one cannot count these two arrangements as distinct states. Hence an integral over phase space overcounts the number of distinct states by a factor equal to the number of permutations of identical particles. To rectify this in the present case a factor of $N!$ must be introduced into the denominator of the phase space probability density, which has the effect of reducing $Z'(E, N, V)$ by the same factor. This is important for systems with variable numbers of particles and for number derivatives of the entropy.

The second point is that it is desirable for the partition function to be dimensionless, and that the probability density genuinely have the dimensions of phase space density. This is convenient for taking the logarithm of the partition function, and also for transforming the probability density to different coordinate systems. Noting that Plank's constant has the dimensions of momentum times length, $h = 6.63 \times 10^{-34}$ J s, one can define a volume element for phase space of $\Delta = h^{3N}$. One also needs to cancel the dimensions of the energy δ-function in the natural representation by introducing another constant Δ_E with the dimensions of energy.

In view of these two points, the phase space probability density may be taken to be

$$\wp(\mathbf{\Gamma}|E, V, N) = \frac{\Delta_E \delta(\mathcal{H}(\mathbf{\Gamma}) - E)}{N! h^{3N} Z(E, N, V)}. \tag{5.21}$$

With this the dimensionless partition function is

$$\begin{aligned} Z(E, N, V) &= \frac{\Delta_E}{N! h^{3N}} \int \mathrm{d}\mathbf{\Gamma}\, \delta(\mathcal{H}(\mathbf{\Gamma}) - E) \\ &= \frac{\Delta_E}{N! h^{3N}} \oint_E \frac{\mathrm{d}\gamma}{|\dot{\gamma}|}. \end{aligned} \tag{5.22}$$

Note that it doesn't matter whether the system volume V is included explicitly as a limit on the spatial integrals or included in the Hamiltonian as a one-body potential. It is interesting to note that whereas $\mathrm{d}\gamma$ itself is not a constant of the motion, the integrand on the second equality is. (This follows from Liouville's theorem, since both $\mathrm{d}\mathbf{\Gamma}$ and $\mathrm{d}E$ are constants of the motion and the integrand is their ratio.) The denominator of the second equality may be replaced by $|\nabla \mathcal{H}(\mathbf{\Gamma}(\gamma, E))|$ for this Hamiltonian system.

The constant Δ_E that appears here has the dimensions of energy. It may be interpreted as the width of the energy shell. Accordingly the Dirac δ that appears is really a coarse-grained 'top-hat' function that is only nonzero when the Hamiltonian is within Δ_E of E, at which point it has value $1/\Delta_E$. The value of Δ_E has no physical consequence, and it is a convenience, not a necessity, to introduce it.

The appearance of Planck's constant is suggestive of the quasi-classical argument that to each quantum state there corresponds a cell in phase space of volume h^{3N}. However such an interpretation is not essential, and any constant with the above dimensions will do. The value of the constant has no physical consequence, which means that one could even proceed with no constant at all. The fact that the constant depends upon N will not even affect the exchange of particles with a second system; since $N_1 + N_2 = N$, one has a total factor of $h^{3N_1}h^{3N_2} = h^{3N}$, which is independent of the partitioning of particles between the two systems. In other words, Planck's constant will become an additive constant for the chemical potential, and only differences in chemical potential are physically meaningful. This is a general result: for an exchangeable parameter A, one can always multiply the partition function or probability density by $\Delta_A \exp \lambda_A A$ with no physical consequence.

The entropy of an isolated Hamiltonian system with energy E is

$$\begin{aligned} S(E,N,V) &= k_{\mathrm{B}} \ln Z(E,N,V) \\ &= k_{\mathrm{B}} \ln \frac{\Delta_E}{N!h^{3N}} \int \mathrm{d}\mathbf{\Gamma}\, \delta(\mathcal{H}(\mathbf{\Gamma}) - E) \\ &= k_{\mathrm{B}} \ln \frac{\Delta_E}{N!h^{3N}} \oint_E \frac{\mathrm{d}\gamma}{|\nabla\mathcal{H}(\mathbf{\Gamma}(\gamma,E))|} \\ &= k_{\mathrm{B}} \ln \frac{\Delta_E}{N!h^{3N}} \oint_E \frac{\mathrm{d}\gamma}{|\dot{\gamma}|}. \end{aligned} \tag{5.23}$$

The more general expression for the probability density that uses an arbitrary representation of the energy hypersurface is

$$\wp(\mathbf{\Gamma}|E,N,V) = \frac{\Delta_E}{N!h^{3N}} \frac{|\nabla g_H(\mathbf{\Gamma})|\delta(g_H(\mathbf{\Gamma}) - G)}{|\dot{\mathbf{\Gamma}}|Z(E,N,V)}. \tag{5.24}$$

In this case the entropy may be written

$$\begin{aligned} S(E,N,V) &= k_{\mathrm{B}} \ln \frac{\Delta_E}{N!h^{3N}} \int \mathrm{d}\mathbf{\Gamma}\, \frac{|\nabla g_H(\mathbf{\Gamma})|}{|\dot{\mathbf{\Gamma}}|} \delta(g_H(\mathbf{\Gamma}) - G) \\ &= k_{\mathrm{B}} \ln \frac{\Delta_E}{N!h^{3N}} \oint_E \frac{\mathrm{d}\gamma}{|\dot{\gamma}|}. \end{aligned} \tag{5.25}$$

Evidently the entropy of the isolated system is indeed independent of the representation of the energy hypersurface.

The partition function of an isolated system gives the total weight of the energy hypersurface, and the energy weight density may be written $\omega(E,N,V) = Z(E,N,V)/\Delta_E$. In terms of the weight density of phase space, $\omega(\mathbf{\Gamma})$, this is

$$\omega(E,N,V) = \int \mathrm{d}\mathbf{\Gamma}\, \omega(\mathbf{\Gamma})\delta(\mathcal{H}(\mathbf{\Gamma}) - E). \tag{5.26}$$

Comparing this with the expression for the partition function, one concludes that phase space has uniform weight density,

$$\omega(\mathbf{\Gamma}) = \frac{1}{N!h^{3N}}. \tag{5.27}$$

One can define the entropy of a point in phase space in terms of this weight density,

$$S(\mathbf{\Gamma}) = k_{\mathrm{B}} \ln \left[\omega(\mathbf{\Gamma}) \Delta(\mathbf{\Gamma}) \right]. \tag{5.28}$$

The consequent phase space probability density is

$$\wp(\mathbf{\Gamma}) = e^{S(\mathbf{\Gamma})/k_{\mathrm{B}}} / \Delta(\mathbf{\Gamma}) Z. \tag{5.29}$$

Here has been introduced an arbitrary volume element $\Delta(\mathbf{\Gamma})$ that has no physical consequence. The one that appears explicitly in the probability density cancels with that implicit in the entropy. It is included here so that the logarithm is taken of a dimensionless number and so that the probability density has explicitly the correct dimensions. Also, it ensures that the entropy is invariant with respect to transformations of phase space. For convenience, one could take it to be constant, or even set it equal to $N!h^{3N}$.

In view of the above, the phase space probability density conditional upon the system having energy within $\mathrm{d}E$ of E may therefore alternatively be written

$$\wp(\mathbf{\Gamma}|E, N, V) = \frac{\omega(\mathbf{\Gamma})}{\omega(E, N, V)} \delta(\mathcal{H}(\mathbf{\Gamma}) - E). \tag{5.30}$$

This is the continuum analogue of the discrete expressions for conditional probability. Hence the entropy of a point in phase space given that the system has energy within $\mathrm{d}E$ of E is

$$\begin{aligned} S(\mathbf{\Gamma}|E, N, V) &= k_{\mathrm{B}} \ln \left[\omega(\mathbf{\Gamma}) \Delta(\mathbf{\Gamma}) \Delta_E \delta(\mathcal{H}(\mathbf{\Gamma}) - E) \right] \\ &= k_{\mathrm{B}} \ln \left[\frac{\Delta(\mathbf{\Gamma})}{N!h^{3N}} \Delta_E \delta(\mathcal{H}(\mathbf{\Gamma}) - E) \right]. \end{aligned} \tag{5.31}$$

One need not be alarmed at the appearance of the Dirac δ here, since in the coarse-grained interpretation it equals $1/\Delta_E$ for any point on the energy hypersurface, and hence the logarithm of the product of the final two terms is 0 whenever the system is on the hypersurface. Accordingly, the phase space probability density for a system with energy E may be written,

$$\wp(\mathbf{\Gamma}|E, N, V) = \frac{e^{S(\mathbf{\Gamma}|E,N,V)/k_{\mathrm{B}}}}{\Delta(\mathbf{\Gamma}) Z(E, N, V)}. \tag{5.32}$$

With this result the entropy of the isolated system may be written

$$\begin{aligned} & S(E, N, V) \\ &= k_{\mathrm{B}} \ln Z(E, N, V) \\ &= \int \mathrm{d}\mathbf{\Gamma}\, \wp(\mathbf{\Gamma}|E, N, V) k_{\mathrm{B}} \ln Z(E, N, V) \end{aligned}$$

$$\begin{aligned} &= \int \mathrm{d}\mathbf{\Gamma}\, \wp(\mathbf{\Gamma}|E,N,V)\left\{S(\mathbf{\Gamma}|E,N,V) - k_\mathrm{B} \ln\left[\wp(\mathbf{\Gamma}|E,N,V)\Delta(\mathbf{\Gamma})\right]\right\} \\ &= -k_\mathrm{B} \int \mathrm{d}\mathbf{\Gamma}\, \wp(\mathbf{\Gamma}|E,N,V) \ln\left[\wp(\mathbf{\Gamma}|E,N,V) N! h^{3N}\right]. \end{aligned} \tag{5.33}$$

The last equality follows because the δ-function does not contribute to the phase space entropy on the hypersurface, as discussed above (i.e., $\Delta_E \delta(\mathcal{H}(\mathbf{\Gamma}) - E)$ is equal to either 1 or 0, with consequence that the probability times the logarithm of this product is always 0). Note that while the microstate volume affects the microstate entropy, it affects neither the entropy of the energy macrostate nor the probability density, as is evidenced by the final equality. In other words, whether $\Delta(\mathbf{\Gamma})$ is known or unknown, it has no physical consequences.

The expression for the entropy represented by the last equality was given by Gibbs, who called it the average of the index of probability.[4] The discrete version was used by Shannon, who called it the information entropy.[5] It has the property that of all normalised probability distributions of the form $g(\mathbf{\Gamma})\delta(\mathcal{H}(\mathbf{\Gamma}) - E)$, it is maximised by $g(\mathbf{\Gamma}) = \text{const}$. Whilst this is the correct probability distribution in the present case of a Hamiltonian system with only energy conserved, it is not valid more generally. The penultimate equality must be regarded as the general expression for the entropy as a functional of the probability density, and the final equality as a particular representation that holds for a uniform probability distribution. In practice there appears to be no reason to use these entropy functionals to determine the probability density since they require explicit knowledge of $S(\mathbf{\Gamma}|\mathbf{f})$, which already gives the probability density directly, Eq. (5.32).

5.1.5 Ergodic Hypothesis and Liouville's Theorem

The usual justification for the uniform probability density of an isolated system is the *ergodic hypothesis*, originally enunciated by Boltzmann. This consists of two separate assertions: (i) that for an isolated system all points in phase space with a given energy lie on a single trajectory, and (ii) that the probability density in phase space is uniform along this trajectory.[6]

Boltzmann and Gibbs imagined an ensemble of replica systems, the ith member of which was in the state $\mathbf{\Gamma}_i$. From this viewpoint, $\wp(\mathbf{\Gamma})$ represents the fraction of the members of the ensemble in the vicinity of $\mathbf{\Gamma}$. The ensemble evolves in time according to the trajectories of the individual members. One may denote the original probability density at $t = 0$ by $\wp_0(\mathbf{\Gamma})$, and the probability density at time t by $\wp(\mathbf{\Gamma}, t)$. One has $\wp(\mathbf{\Gamma}_1, t) = \wp_0(\mathbf{\Gamma}(-t|\mathbf{\Gamma}_1))$ (see below), so that the

[4] J. W. Gibbs, *Elementary Principles in Statistical Mechanics Developed with Special Reference to the Rational Foundation of Thermodynamics*, Yale Univ. Press, New Haven, CT, 1902; Dover, New York, 1960.

[5] C. E. Shannon and W. Weaver, *The Mathematical Theory of Communication*, Univ. of Illinois Press, Urbana, 1949.

[6] A. Münster, *Statistical Thermodynamics*, Vol. 1, Springer-Verlag, Berlin, 1969, discusses the role that phase space and the ergodic hypothesis has played in the axiomatic development of statistical mechanics.

ensemble picture can be applied to nonequilibrium systems. The time evolution of the probability density (or ensemble) may be obtained following a trajectory,

$$\frac{d\wp(\mathbf{\Gamma}_1, t)}{dt} = \left(\frac{\partial\wp(\mathbf{\Gamma}_1, t)}{\partial t}\right)_{\mathbf{\Gamma}_1} + \dot{\mathbf{\Gamma}}_1 \cdot \left(\frac{\partial\wp(\mathbf{\Gamma}_1, t)}{\partial\mathbf{\Gamma}_1}\right)_t, \tag{5.34}$$

where $\mathbf{\Gamma}_1 = \mathbf{\Gamma}(t|\mathbf{\Gamma}_0)$. Note that it is not necessary to assume that the members of the ensemble all have precisely the same energy.

The probability density in the neighbourhood of a fixed position in phase space changes due to the flux across the boundary of members of the ensemble, each following their individual trajectory. The time rate of change of the number of members of the ensemble in the region Ω at time t is

$$\begin{aligned} \frac{\partial N_\Omega(t)}{\partial t} &= \int_\Omega \mathrm{d}\mathbf{\Gamma}\, \frac{\partial\wp(\mathbf{\Gamma}, t)}{\partial t} \\ &= -\oint_\Omega \mathrm{d}\mathbf{\Gamma}\, \hat{\mathbf{n}} \cdot \dot{\mathbf{\Gamma}}\, \wp(\mathbf{\Gamma}, t) \\ &= -\int_\Omega \mathrm{d}\mathbf{\Gamma}\, \frac{\partial}{\partial\mathbf{\Gamma}} \cdot \left[\dot{\mathbf{\Gamma}}\, \wp(\mathbf{\Gamma}, t)\right]. \end{aligned} \tag{5.35}$$

For a small enough region this gives

$$\frac{\partial\wp(\mathbf{\Gamma}, t)}{\partial t} = -\frac{\partial}{\partial\mathbf{\Gamma}} \cdot \left[\dot{\mathbf{\Gamma}}\, \wp(\mathbf{\Gamma}, t)\right]. \tag{5.36}$$

This is the conservation law for members of the ensemble.

From Hamilton's equations one may readily verify the adiabatic incompressibility of phase space,

$$\frac{\partial}{\partial\mathbf{\Gamma}} \cdot \dot{\mathbf{\Gamma}} = 0, \tag{5.37}$$

since in full this is

$$\sum_{i=1}^N \left[\frac{\partial\dot{\mathbf{q}}_i}{\partial\mathbf{q}_i} + \frac{\partial\dot{\mathbf{p}}_i}{\partial\mathbf{p}_i}\right] = \sum_{i=1}^N \left[\frac{\partial^2\mathcal{H}}{\partial\mathbf{q}_i\partial\mathbf{p}_i} - \frac{\partial^2\mathcal{H}}{\partial\mathbf{p}_i\partial\mathbf{q}_i}\right] = 0. \tag{5.38}$$

Using this and the above partial time derivative, Liouville's theorem follows:

$$\frac{d\wp(\mathbf{\Gamma}(t|\mathbf{\Gamma}_0), t)}{dt} = 0. \tag{5.39}$$

This says that phase space density is conserved along a trajectory.

A corollary of Liouville's theorem is that the volume of a region of phase space is conserved during its time evolution. This follows because trajectories cannot cross, and so the number of members of the ensemble inside a region must stay within that region during the evolution of its boundaries. Hence the volume of the region must be constant in order for the density to remain fixed.

If the region that was Ω at time $t = 0$ evolves to Ω' at $t = t'$, then the fact that the volume is conserved is expressed by

$$\int_{\Omega} \mathrm{d}\mathbf{\Gamma} = \int_{\Omega'} \mathrm{d}\mathbf{\Gamma}. \tag{5.40}$$

However Hamilton's equations represent a transformation of phase space to itself, $\mathbf{\Gamma}' = \mathbf{\Gamma}(t'|\mathbf{\Gamma})$, and so one also has

$$\int_{\Omega} \mathrm{d}\mathbf{\Gamma} = \int_{\Omega'} \mathrm{d}\mathbf{\Gamma}' \mathcal{J}(\mathbf{\Gamma}, \mathbf{\Gamma}'). \tag{5.41}$$

Comparing these two, since the volume is arbitrary, one concludes that the Jacobean of the time evolution of phase space is unity, $\mathcal{J}(\mathbf{\Gamma}, \mathbf{\Gamma}') = 1$. This last result can of course be shown directly from Hamilton's equations, and hence reversing the argument provides and alternative proof of the conservation of phase space volumes during their evolution.

The conservation of the phase space volume element $\mathrm{d}\mathbf{\Gamma}$ means that the quantity $\mathrm{d}\gamma \mathrm{d}n = \mathrm{d}\gamma \mathrm{d}E/|\nabla \mathcal{H}(\mathbf{\Gamma})|$ is also a constant of the motion. However, since energy is conserved, one concludes that the area of a region of the energy hypersurface is *not* a constant of the motion, but rather it is the quantity $\mathrm{d}\gamma/|\nabla \mathcal{H}(\mathbf{\Gamma})|$ that is invariant. It will be recalled that it was this quantity that appeared in the expression for the entropy, Eq. (5.23).

Using the above result for the Jacobean, the probability density at time t may be written in terms of the original probability distribution,

$$\begin{aligned} \wp(\mathbf{\Gamma}, t) &= \int \mathrm{d}\mathbf{\Gamma}_0 \, \wp_0(\mathbf{\Gamma}_0) \, \delta(\mathbf{\Gamma} - \mathbf{\Gamma}(t|\mathbf{\Gamma}_0)) \\ &= \int \mathrm{d}\mathbf{\Gamma}_1 \, \wp_0(\mathbf{\Gamma}(-t|\mathbf{\Gamma}_1)) \, \delta(\mathbf{\Gamma} - \mathbf{\Gamma}_1) \\ &= \wp_0(\mathbf{\Gamma}(-t|\mathbf{\Gamma})), \end{aligned} \tag{5.42}$$

since $\mathbf{\Gamma}_1 = \mathbf{\Gamma}(t|\mathbf{\Gamma}_0)$ means that $\mathbf{\Gamma}_0 = \mathbf{\Gamma}(-t|\mathbf{\Gamma}_1)$. This describes how the current probability density evolves from the initial distribution.

Liouville's theorem holds for both equilibrium and nonequilibrium systems. In the case of equilibrium one has the stronger result that the probability density at any point must be stationary,

$$\frac{\partial \wp(\mathbf{\Gamma}, t)}{\partial t} = 0. \tag{5.43}$$

From Liouville's theorem, Boltzmann deduced that the probability density in phase space must be uniform along a trajectory. He considered two regions of phase space, the second of which evolves from the first after time t (see Fig. 5.1). The two regions have of course the same volume, but different shapes. Next he considered these two regions as fixed in phase space, and noted the times at which a trajectory in phase space entered and left the first region (t_A and t_B), and similarly for the second region (t_C and t_D). Since the boundary of the second region represents the evolution of the boundary of the first region

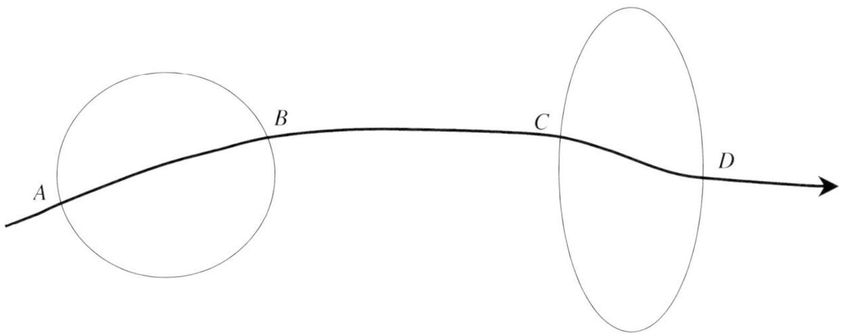

Figure 5.1: A single trajectory passing through the original and the evolved regions of phase space.

after time t, one must have $t_C - t_A = t$, and $t_D - t_{\rm B} = t$, from which it follows that the trajectory spent an equal amount of time in the two regions, $t_{\rm B} - t_A = t_D - t_C$. (That is, the 'width' of the evolved region measured at a trajectory must be proportional to the velocity of the trajectory.) Since the probability of a region of phase space is proportional to the time that a trajectory spends in it, Boltzmann deduced that evolved volumes of phase space lying on a single trajectory are equally likely.

The same conclusion may be reached directly from Liouville's theorem. As shown above, probability density is conserved on a trajectory, as are volumes of phase space. Hence the probability of an evolved region of phase space is the same as that of the original region, which is precisely what Boltzmann concluded from his equal time argument.

For a system in equilibrium, the partial temporal derivative of the probability density vanishes, Eq. (5.43), which is to say that the probability density in phase space is unchanging at any fixed point on a trajectory. However, as just argued, the probability density also moves along with the trajectory. These two statements can only be reconciled if the probability density is uniform along the trajectory. Hence one concludes that

$$\wp(\mathbf{\Gamma}(t|\mathbf{\Gamma}_0)) = \wp(\mathbf{\Gamma}_0), \text{ for all } t. \tag{5.44}$$

Now the first part of the ergodic hypothesis is required, namely that a single trajectory covers all of the energy hypersurface. Hence one may draw the stronger conclusion that the phase space probability density is uniform on the entire energy hypersurface,

$$\wp(\mathbf{\Gamma}) = \wp(\mathbf{\Gamma}_0), \text{ for all } \mathbf{\Gamma} \text{ such that } \mathcal{H}(\mathbf{\Gamma}) = \mathcal{H}(\mathbf{\Gamma}_0). \tag{5.45}$$

It is clear that the probability distribution given above,

$$\wp(\mathbf{\Gamma}|E, N, V) = \frac{\delta(\mathcal{H}(\mathbf{\Gamma}) - E)}{N! h^{3N} Z(E, N, V)}, \tag{5.46}$$

is a manifestation of this conclusion. This shows the precise sense in which the probability density of an isolated system is uniform, namely that phase space points with the same energy are equally likely, or that the phase space probability density is uniform on the energy hypersurface.

5.2 Various Reservoirs

5.2.1 Constant Temperature

An isolated system composed of two subsystems able to exchange energy is now treated. The respective volumes are implicit in the Hamiltonian and distinguish the particles confined to each subsystem. The Hamiltonian can be written as the sum of the two Hamiltonians of the systems considered in isolation, $\mathcal{H} = \mathcal{H}_1 + \mathcal{H}_2$. Provided that the region of interaction is small compared to each subsystem, the remaining contribution $\mathcal{H}_{12}$ is negligible. The weight or number of microstates attached to subsystem 1 having particular values of energy is the weight of the trajectory of the total system that corresponds to this macrostate,

$$\begin{aligned} & Z(E_1|E, V_1, N_1, V_2, N_2) \\ &= \frac{\Delta_E^2}{h^{3N}} \int \frac{\mathrm{d}\mathbf{p}^{N_1}\,\mathrm{d}\mathbf{q}^{N_1}}{N_1!} \int \frac{\mathrm{d}\mathbf{p}^{N_2}\,\mathrm{d}\mathbf{q}^{N_2}}{N_2!}\,\delta(E_1 - \mathcal{H}_1)\,\delta(E - \mathcal{H}) \\ &= \frac{\Delta_E^2}{h^{3N}} \int \frac{\mathrm{d}\mathbf{p}^{N_1}\,\mathrm{d}\mathbf{q}^{N_1}}{N_1!}\delta(E_1 - \mathcal{H}_1) \int \frac{\mathrm{d}\mathbf{p}^{N_2}\,\mathrm{d}\mathbf{q}^{N_2}}{N_2!}\delta(E - E_1 - \mathcal{H}_2) \\ &= Z(E_1, N_1, V_1)\, Z(E - E_1, N_2, V_2). \end{aligned} \tag{5.47}$$

Here $N = N_1 + N_2$, and because the particles are confined to two distinct regions, one must correct the integration by $N_1!N_2!$ rather than by $N!$ in order to count distinct microstates. The final product form relies upon the additivity and independence of the variables for each subsystem. The probability density of the partitioning is of course proportional to this weight. As in the discrete case, the weight of microstates of the total system corresponding to a macrostate is the product of the weights of the subsystems considered as isolated in the given respective state.

For the first subsystem the entropy is

$$S(E_1, N_1, V_1) = k_{\mathrm{B}} \ln Z(E_1, N_1, V_1), \tag{5.48}$$

where

$$Z(E, N, V) = \frac{\Delta_E}{N!h^{3N}} \int \mathrm{d}\mathbf{p}^N \mathrm{d}\mathbf{q}^N\, \delta(\mathcal{H}(\mathbf{p}^N, \mathbf{q}^N) - E), \tag{5.49}$$

Similarly, the entropy of the second subsystem is $S(E - E_1, N_2, V_2) = k_{\mathrm{B}} \ln Z(E - E_1, N_2, V_2)$, and assuming that it is a reservoir, it may be expanded to linear order,

$$S(E - E_1, N_2, V_2) = \mathrm{const.} - \frac{E_1}{T_2}. \tag{5.50}$$

As usual the reservoir temperature T_2 is the energy derivative of its entropy. Since the constrained total weight is the product of the weights of the subsystem and the reservoir in the energy macrostate, the constrained total entropy is the sum of the two entropies.

The energy probability density for the subsystem is proportional to the constrained total weight, or, equivalently, to the exponential of the constrained total entropy,

$$\wp(E|N,V,T) = \frac{\Delta_E^{-1}}{Z(N,V,T)} e^{S(E,N,V)/k_B} e^{-E/k_B T}. \tag{5.51}$$

As usual the subscripts are now dropped because the reservoir only enters via its temperature T. The final factor is the so-called Boltzmann factor, which says that for an isothermal system the probability of a state is proportional to the exponential of the negative of its energy divided by temperature. This exponent is just the state-dependent reservoir entropy and one must include in addition a factor for the subsystem (or internal) entropy of the state (see below). The density is of course normalised by the dimensionless partition function, which is

$$\begin{aligned} Z(N,V,T) &= \frac{1}{\Delta_E}\int \mathrm{d}E\, e^{S(E,N,V)/k_B} e^{-E/k_B T} \\ &= \frac{1}{N!h^{3N}}\int \mathrm{d}E\, e^{-E/k_B T}\int \mathrm{d}\mathbf{p}^N\, \mathrm{d}\mathbf{q}^N\, \delta(\mathcal{H}(\mathbf{p}^N,\mathbf{q}^N)-E) \\ &= \frac{1}{N!h^{3N}}\int \mathrm{d}\mathbf{p}^N\, \mathrm{d}\mathbf{q}^N\, e^{-\mathcal{H}(\mathbf{p}^N,\mathbf{q}^N)/k_B T}, \end{aligned} \tag{5.52}$$

where the final line follows upon interchanging the order of integration. This is the most usual form of the partition function that one encounters in statistical mechanics and the notation Z is rather widespread. Note that $Z(E,N,V)$ and $Z(N,V,T)$ are two completely different functions; here and throughout the form of the arguments is used to indicate the type of the function.

It is most useful to have the phase space probability density rather than the energy probability density. Denoting a point in the phase space of the subsystem by $\mathbf{\Gamma} = \mathbf{p}^N\mathbf{q}^N$, one has

$$\begin{aligned} \wp(\mathbf{\Gamma}|N,V,T) &= \int \mathrm{d}E_1\, \wp(\mathbf{\Gamma}|E_1,N,V)\wp(E_1|N,V,T) \\ &= \int \mathrm{d}E_1\, \frac{\Delta_E \delta(\mathcal{H}(\mathbf{\Gamma})-E_1)}{N!h^{3N}Z(E_1,N,V)}\, \frac{e^{S(E_1,N,V)/k_B} e^{-E_1/k_B T}}{\Delta_E Z(N,V,T)} \\ &= \frac{1}{N!h^{3N}Z(N,V,T)} e^{-\mathcal{H}(\mathbf{\Gamma})/k_B T}. \end{aligned} \tag{5.53}$$

The first equality is a standard application of the laws of probability;[7] the first quantity in the integrand is the probability that the isolated subsystem is at the given point in its phase space given that it has energy E_1 (this term is

[7] One has $\wp(a|c) = \sum_b \wp(ab|c) = \sum_b \wp(a|bc)\wp(b|c) = \sum_b \wp(a|b)\wp(b|c)$, the final equality holding if and only if $b \cap c = b$.

independent of the reservoir temperature). The cancellation of the weight or partition function in the denominator follows because its logarithm is defined to be the entropy that appears in the exponent in the numerator. Finally, the Dirac δ is used to undo the integration. This phase space probability density has the desired dimensions and is correctly normalised. The derivation of this result provides the justification for the widely used Boltzmann factor, which sets the probability proportional to the exponential of the negative of the energy divided by $k_{\rm B}T$. Compared to the isothermal distribution of the energy macrostates, no subsystem entropy appears here because points in phase space have equal weight density, Eq. (5.27).

The total entropy of the subsystem and reservoir may be written

$$\begin{aligned} S(N,V,T) &= k_{\rm B} \ln Z(N,V,T) \\ &= k_{\rm B} \int_V \mathrm{d}\mathbf{r}^N \, \mathrm{d}\mathbf{p}^N \, \wp(\mathbf{r}^N, \mathbf{p}^N | N, V, T) \ln Z(N,V,T) \\ &= k_{\rm B} \int_V \mathrm{d}\mathbf{r}^N \, \mathrm{d}\mathbf{p}^N \, \wp(\mathbf{r}^N, \mathbf{p}^N | N, V, T) \\ &\quad \times \left[\ln \frac{e^{-\beta \mathcal{H}_N(\mathbf{r}^N, \mathbf{p}^N)}}{N! h^{3N}} - \ln \wp(\mathbf{r}^N, \mathbf{p}^N | N, V, T) \right] \\ &= \frac{-1}{T} \left\langle \mathcal{H}_N(\mathbf{r}^N, \mathbf{p}^N) \right\rangle - k_{\rm B} \int_V \mathrm{d}\mathbf{r}^N \, \mathrm{d}\mathbf{p}^N \, \wp(\mathbf{r}^N, \mathbf{p}^N | N, V, T) \\ &\quad \times \ln \left[N! h^{3N} \wp(\mathbf{r}^N, \mathbf{p}^N | N, V, T) \right]. \end{aligned} \tag{5.54}$$

Here the first term may be recognised as the average reservoir entropy (at least the subsystem-dependent part). Conversely, the second term may be described as the entropy of the subsystem.

In the first chapter an analogous expression for the entropy was given as the average of the internal entropy of the microstates less the average of the logarithm of the probability. In the present case the position-momentum microstates of the isolated subsystem are all equally likely, and hence they have no internal entropy as such. However, when the subsystem is in contact with the heat reservoir there is an entropy associated with each microstate due to the loss of energy from the reservoir. This is the first term, which plays the role of the internal entropy of the microstate.

As also mentioned in the first chapter, this first term is often neglected and the entropy is said to be simply the average of the logarithm of the microstate or macrostate probability, (cf. Eq. (1.12)). In the context of the principle of maximum entropy, which seeks to obtain the equilibrium probability distribution by maximising the second contribution only (see Section A.6), the reservoir contribution, the first term, is belatedly added by carrying out the maximisation with a Lagrange multiplier representing a constraint on the average energy. This is precisely the first term above, and so one sees that the artificial constraint introduced into the maximum entropy formalism effectively adds the reservoir contribution that is inadvertently neglected.

The thermodynamic properties of the constant temperature system carry over directly from the general analysis carried out for the discrete case. The constrained thermodynamic potential is the temperature times the constrained total entropy, Eq. (4.12),

$$F(E|N, V, T) = E - TS(E, N, V), \tag{5.55}$$

which of course describes the probability of fluctuations about equilibrium. The equilibrium thermodynamic potential, which in this case is called the Helmholtz free energy, is the minimum value of the constrained thermodynamic potential, Eq. (4.14),

$$\overline{F}(N, V, T) = F(\overline{E}|N, V, T) = \overline{E} - TS(\overline{E}, N, V), \tag{5.56}$$

where the equilibrium energy $\overline{E}(N, V, T)$ is the most likely energy of the subsystem. The Helmholtz free energy is as often denoted by $A(N, V, T)$ as by $F(N, V, T)$. The partial derivatives of the Helmholtz free energy give the equilibrium thermodynamic properties of the subsystem. According to Eq. (4.19), the total differential of the Helmholtz free energy is

$$d\overline{F}(N, V, T) = -S(\overline{E}, N, V)dT - \overline{p}dV + \overline{\mu}dN, \tag{5.57}$$

which gives its partial derivatives at a glance.

The formal analysis showed that the unconstrained total entropy, Eq. (4.28), is $S_{\text{total}}(N, V, T) = k_{\text{B}} \ln Z(N, V, T)$. Likewise, the average energy is the inverse temperature derivative of this, Eq. (4.29), and the average pressure and the average chemical potential of the subsystem are its volume and number derivatives, Eq. (4.30).

As discussed in Section 4.3.4, statistical mechanics usually equates the free energy, not the entropy, to the logarithm of the partition function. As was shown there for macroscopic systems the two are equivalent. Explicitly, in the present constant temperature case,

$$\begin{aligned}
&-k_{\text{B}}T \ln Z(N, V, T) \\
&= -k_{\text{B}}T \ln \int \mathrm{d}E \, \frac{1}{\Delta_E} e^{-E/k_{\text{B}}T} e^{S(E,N,V)/k_{\text{B}}} \\
&\approx -k_{\text{B}}T \ln \left[e^{-\overline{E}/k_{\text{B}}T} e^{S(\overline{E},N,V)/k_{\text{B}}} \int \frac{\mathrm{d}E}{\Delta_E} e^{S_{EE}(\overline{E},N,V)(E-\overline{E})^2/2k_{\text{B}}} \right] \\
&= \overline{E} - TS(\overline{E}, N, V) + \frac{k_{\text{B}}T}{2} \ln \frac{-2\pi k_{\text{B}}}{\Delta_E^2 S_{EE}(\overline{E}, N, V)}.
\end{aligned} \tag{5.58}$$

The right-hand side is the Helmholtz free energy with a logarithmic correction that for macroscopic systems may be neglected (it is only when fluctuations about equilibrium are large compared to the mean that this term may become important). To leading order the free energy is indeed equal to the negative of the temperature times the logarithm of the partition function,

$$\overline{F}(N, V, T) = -k_{\text{B}}T \ln Z(N, V, T). \tag{5.59}$$

Since the derivatives of the two are likewise equal, one may conclude that for macroscopic systems the equilibrium energy, pressure, and chemical potential are equal to the average energy, pressure, and chemical potential, respectively.

5.2.2 Constant Chemical Potential

Open systems do not have a fixed number of particles, and they may be analysed by considering an isolated system comprising two subsystems at rest and with fixed volumes, but allowing energy and particle exchange. At any instant there is a well-defined number of particles in each subsystem such that $N_1 + N_2 = N$. As usual, the probability of a particular macrostate factorises into the product of weights for the quasi-isolated subsystems,

$$\wp(E_1, N_1|E, N, V_1, V_2) = \frac{Z(E_1, N_1, V_1)\, Z(E - E_1, N - N_1, V_2)}{Z(E, N, V_1, V_2)}, \tag{5.60}$$

where the total weight of the trajectory is

$$\begin{aligned} &Z(E, N, V_1, V_2) \\ &= \frac{\Delta_E}{h^{3N}} \sum_{N_1=0}^{N} \int_{V_1} \frac{\mathrm{d}\mathbf{p}^{N_1}\, \mathrm{d}\mathbf{q}^{N_1}}{N_1!} \int_{V_2} \frac{\mathrm{d}\mathbf{p}^{N_2}\, \mathrm{d}\mathbf{q}^{N_2}}{N_2!} \delta(E - \mathcal{H}), \end{aligned} \tag{5.61}$$

with $N_1 + N_2 = N$. It is clear that this counts each distinct configuration of the system with equal weight.

Taking the second subsystem to be a reservoir and expanding its entropy to second order one has

$$\begin{aligned} S(E - E_1, N - N_1, V_2) &\equiv k_{\mathrm{B}} \ln Z(E - E_1, N - N_1, V_2) \\ &= \text{const.} - \frac{E_1}{T} + \frac{\mu N_1}{T}, \end{aligned} \tag{5.62}$$

where the reservoir chemical potential μ equals minus the temperature times the number difference of its entropy. The probability that subsystem 1 has a particular energy and particle number becomes

$$\wp(E, N|\mu, V, T) = \frac{e^{S(E,N,V)/k_{\mathrm{B}}} e^{-[E-\mu N]/k_{\mathrm{B}}T}}{\Delta_E Z(\mu, V, T)}. \tag{5.63}$$

Here the partition function is

$$\begin{aligned} Z(\mu, V, T) &= \sum_{N=0}^{\infty} \int \frac{\mathrm{d}E}{\Delta_E} e^{S(E,N,V)/k_{\mathrm{B}}} e^{-[E-\mu N]/k_{\mathrm{B}}T} \\ &= \sum_{N=0}^{\infty} \frac{1}{h^{3N}} \int \mathrm{d}E\, e^{-[E-\mu N]/k_{\mathrm{B}}T} \int \frac{\mathrm{d}\mathbf{p}^N\, \mathrm{d}\mathbf{q}^N}{N!} \delta(E - \mathcal{H}) \\ &= \sum_{N=0}^{\infty} \frac{e^{\mu N/k_{\mathrm{B}}T}}{h^{3N}} \int \frac{\mathrm{d}\mathbf{p}^N\, \mathrm{d}\mathbf{q}^N}{N!} e^{-\mathcal{H}(\mathbf{q}^N, \mathbf{p}^N)/k_{\mathrm{B}}T}. \end{aligned} \tag{5.64}$$

The probability density is of course uniform on a hypersurface of constant energy and number,

$$\wp(\mathbf{q}^N, \mathbf{p}^N, N|\mu, V, T) = \frac{e^{\mu N/k_{\mathrm{B}}T} e^{-\mathcal{H}(\mathbf{q}^N, \mathbf{p}^N)/k_{\mathrm{B}}T}}{N! h^{3N} Z(\mu, V, T)}. \tag{5.65}$$

The total entropy of the subsystem and reservoir may be written

$$\begin{aligned}
S(\mu, V, T) & \qquad\qquad (5.66) \\
&= k_\mathrm{B} \ln Z(\mu, V, T) \\
&= k_\mathrm{B} \sum_{N=0}^{\infty} \int_V \mathrm{d}\mathbf{r}^N \, \mathrm{d}\mathbf{p}^N \, \wp(\mathbf{r}^N, \mathbf{p}^N | \mu, V, T) \ln Z(\mu, V, T) \\
&= k_\mathrm{B} \sum_{N=0}^{\infty} \int_V \mathrm{d}\mathbf{r}^N \, \mathrm{d}\mathbf{p}^N \, \wp(\mathbf{r}^N, \mathbf{p}^N | \mu, V, T) \\
&\quad \times \left[\ln \frac{e^{\beta\mu N} e^{-\beta\mathcal{H}_N(\mathbf{r}^N, \mathbf{p}^N)}}{N! h^{3N}} - \ln \wp(\mathbf{r}^N, \mathbf{p}^N | \mu, V, T) \right] \\
&= \frac{1}{T} \langle \mu N - \mathcal{H}_N(\mathbf{r}^N, \mathbf{p}^N) \rangle - k_\mathrm{B} \sum_{N=0}^{\infty} \int_V \mathrm{d}\mathbf{r}^N \, \mathrm{d}\mathbf{p}^N \, \wp(\mathbf{r}^N, \mathbf{p}^N | \mu, V, T) \\
&\quad \times \ln \left[N! h^{3N} \wp(\mathbf{r}^N, \mathbf{p}^N | \mu, V, T) \right]. \qquad (5.67)
\end{aligned}$$

Here the first term may be recognised as the average of the subsystem-dependent part of the reservoir entropy. It represents the internal entropy of the microstates once the subsystem is in contact with the reservoir. Conversely, the second term may be described as the entropy of the subsystem. More precisely, it is in fact the difference between the total unconstrained entropy of the full system and the average entropy of the reservoir.

The constrained thermodynamic potential is

$$\Omega(E, N | \mu, V, T) = E - TS(E, N, V) - \mu N, \tag{5.68}$$

and its minimum value is the equilibrium thermodynamic potential, which is called the grand potential,

$$\overline{\Omega}(\mu, V, T) = \Omega(\overline{E}, \overline{N} | \mu, V, T) = \overline{E} - TS(\overline{E}, \overline{N}, V) - \mu\overline{N}. \tag{5.69}$$

Here the equilibrium energy $\overline{E}(\mu, V, T)$ and the equilibrium number $\overline{N}(\mu, V, T)$ of the subsystem are the most likely values. The former is the inverse temperature derivative of the grand potential divided by temperature, and the latter is its chemical potential derivative. Its temperature derivative gives the equilibrium entropy, and its volume derivative the equilibrium pressure, as summarised in the total derivative, Eq. (4.19),

$$d\overline{\Omega}(\mu, V, T) = -\overline{N} d\mu - \overline{p} dV - S(\overline{E}, \overline{N}, V) dT. \tag{5.70}$$

The average number and the average pressure of the subsystem are the chemical potential and the volume derivative of the total unconstrained entropy, respectively, which to leading order equals the grand potential.

5.2.3 Constant Pressure

Now an isolated system will be treated, such that its two subsystems can exchange energy and volume (e.g., they may be separated by a moveable, diathermal piston). This can only be done by augmenting phase space and including the changes in volume via a modified Hamiltonian, since it is only on a hypersurface in canonical phase space that the probability density is uniform. Any other method of treating volume changes cannot be guaranteed to have a uniform *a priori* distribution of volume. With P_V being the canonical momentum corresponding to the volume V, the phase space probability may be written

$$\wp(\mathbf{\Gamma}^N, V, P_V|E, N) = \frac{\Delta_E}{Z(E,N)N!h^{3N+1}}\delta(E - \mathcal{H}). \tag{5.71}$$

Hence phase space now consists of not only the positions and momenta of the particles, but also a coordinate representing the volume and another representing its conjugate momentum. The Hamiltonian contains, in addition to the usual term representing the confining effects of the volume, another term representing the kinetic energy of the volume. For simplicity the momentum can be taken to be $P_V = M_V\dot{V}$, which gives the usual quadratic form for the kinetic energy $K_V = P_V^2/2M_V$. The above probability is obviously uniform on a trajectory in this augmented phase space, and hence all of the general analysis applies.

Now explicitly treating the two subsystems of the isolated total system, their volumes are V_1 and V_2, with $V_1+V_2 = V$, their energies are E_1 and E_2, with $E_1+E_2 = E$, and their fixed particle numbers are N_1 and N_2. The subsystems are able to exchange energy and volume. The probability of a particular partitioning is

$$\begin{aligned}
&\wp(E_1, V_1, P_V|E, V, N_1, N_2) \\
&\quad = \frac{\Delta_E/h^{3N+1}}{Z(E,N,V)} \int \frac{\mathrm{d}\mathbf{p}^{N_1}\,\mathrm{d}\mathbf{q}^{N_1}}{N_1!} \int \frac{\mathrm{d}\mathbf{p}^{N_2}\,\mathrm{d}\mathbf{q}^{N_2}}{N_2!} \delta(E_1 - \mathcal{H}_1)\delta(E - \mathcal{H}) \\
&\quad = \frac{Z(E_1, N_1, V_1,)}{h\Delta_E Z(E,N,V)} Z(E - E_1 - K_V, N_2, V - V_1).
\end{aligned} \tag{5.72}$$

(The confining effects of the volume on the particles of the subsystem are included in $\mathcal{H}_1$.) One now takes the second subsystem to be a reservoir and expands its entropy to linear order as usual. One is left with the volume momentum, which by virtue of the quadratic form for the kinetic energy appears as a Gaussian multiplicative factor, $\exp[-P_V^2/2M_V k_\mathrm{B}T]$. Consequently one may integrate out the volume momentum, since this gives a constant independent of the first subsystem. Therefore the probability of the subsystem having energy between E and $E + dE$ and volume between V and $V + dV$ is

$$\wp(E, V|N, p, T) = \frac{e^{S(E,N,V)/k_\mathrm{B}} e^{-[E+pV]/k_\mathrm{B}T}}{\Delta_E \Delta_V Z(N,p,T)}, \tag{5.73}$$

where the reservoir pressure p equals the temperature times the volume derivative of its entropy. Here the partition function is

$$\begin{aligned}
Z(N,p,T) &= \frac{1}{\Delta_E\Delta_V} \int \mathrm{d}E\, dV\, e^{S(E,N,V)/k_\mathrm{B}} e^{-[E+pV]/k_\mathrm{B}T} \\
&= \frac{\Delta_V^{-1}}{h^{3N}} \int \mathrm{d}E\, dV\, e^{-[E+pV]/k_\mathrm{B}T} \int \frac{\mathrm{d}\mathbf{p}^N\,\mathrm{d}\mathbf{q}^N}{N!} \delta(E - \mathcal{H})
\end{aligned}$$

$$= \frac{\Delta_V^{-1}}{h^{3N}} \int \mathrm{d}V e^{-pV/k_{\mathrm{B}}T} \int \frac{\mathrm{d}\mathbf{p}^N \, \mathrm{d}\mathbf{q}^N}{N!} e^{-\mathcal{H}(\mathbf{p}^N, \mathbf{q})/k_{\mathrm{B}}T}. \tag{5.74}$$

In terms of phase space, the probability density is

$$\wp(\mathbf{q}^N, \mathbf{p}^N, V|N, p, T) = \frac{e^{-pV/k_{\mathrm{B}}T} e^{-\mathcal{H}(\mathbf{q}^N, \mathbf{p}^N)/k_{\mathrm{B}}T}}{\Delta_V N! h^{3N} Z(N, p, T)}. \tag{5.75}$$

This is the probability of a particular molecular configuration *and* volume of the subsystem, given the number of particles in the subsystem, and the pressure and temperature of the reservoir. Once again the value of the constant Δ_V is immaterial.

From the general analysis, the constrained thermodynamic potential is

$$G(E, V|N, p, T) = E + pV - TS(E, N, V), \tag{5.76}$$

and the equilibrium thermodynamic potential, which in this case is called the Gibbs free energy, is the minimum value of this,

$$\overline{G}(N, p, T) = G(\overline{E}, \overline{V}|N, p, T) = \overline{E} + p\overline{V} - TS(\overline{E}, N, \overline{V}). \tag{5.77}$$

Here the equilibrium energy $\overline{E}(N, p, T)$ and the equilibrium volume $\overline{V}(N, p, T)$ of the subsystem are the most likely values. The equilibrium chemical potential of the subsystem is the number derivative of the Gibbs free energy. Hence from the general expression, Eq. (4.19), the total differential of the Gibbs free energy is

$$d\overline{G}(N, p, T) = \overline{V} dp + \overline{\mu} dN - S(\overline{E}, N, \overline{V}) dT, \tag{5.78}$$

which gives its partial derivatives at a glance. The average properties of the subsystem are given by the appropriate derivative of the total unconstrained entropy, which is the logarithm of the partition function. Since to leading order the latter is essentially the Gibbs free energy, one again concludes that for macroscopic systems the equilibrium values equal the average values.

5.2.4 Constant Enthalpy

In the preceding example, the volume of the subsystem varied at constant reservoir pressure, while energy was allowed to exchange between the two. Now an insulated subsystem with variable volume is treated; the energy of the subsystem can change by the performance of mechanical work in changing its volume, but energy exchange cannot occur at the microscopic level. In this system energy and volume are no longer independent variables, and so the general formalism, which was predicated on the independence of the A_i, must be slightly modified. The problem is treated in two stages. First subsystem 1 alone is considered, with the volume variations taking place in the presence of an external potential $U(V)$. In all cases up to now a Taylor expansion of some reservoir has been made to linear order, and in view of this the external potential is taken to be $U(V) = pV$, where p at this stage is just a constant. The total energy of the subsystem is the sum of the usual internal part, $E_{\rm int} \equiv \mathcal{H}(\mathbf{p}^N, \mathbf{q}^N)$, and this external part, $H = E_{\rm int} + pV$, and during the motion of the subsystem this is constant, $dH = 0$, or $dE_{\rm int} = -pdV$. This then is the mechanical view of the problem.

Now the external potential is replaced by a second subsystem. The total system is isolated, so that the conservation laws hold: $dE_1 + dE_2 = 0$, and $dV_1 + dV_2 = 0$. Here $E_{\rm int}$ has been replaced by E_1 and V by V_1. Multiplying the second equation by the constant p, and adding them together gives $dE_1 + pdV_1 + dE_2 + pdV_2 = 0$, or $dH_1 + dH_2 = 0$. Thus the quantity H is a linear additive constant of the motion (provided of course that p is indeed a constant). In the mechanical view, $dE_1 = -pdV_1$, so that for each individual subsystem the quantity H must be constant, $dH_1 = dH_2 = 0$. Now the second subsystem is taken to be a reservoir and p may be identified with its pressure, which is permissible because this is indeed constant for any conceivable volume changes of subsystem 1. (One simply sets the state of the reservoir so that $p_2 = p$.) Since by the definition of the field variables as entropy derivatives, $T_2 dS_2 = dE_2 + p_2 dV_2 = 0$, one concludes that a reservoir performs mechanical work at constant entropy, which is not unreasonable. The fact that the energy changes of the reservoir occur independently of its entropy shows that the reservoir temperature cannot enter into the problem, which is consistent with common experience of the properties of insulation. The quantity $H = E + pV$ is called the *enthalpy*.

In view of these considerations, the probability density in phase space is

$$\begin{aligned} &\wp(\mathbf{\Gamma}^N, V_1 | H_1, N_1, H_2, N_2, V, p) \\ &= \frac{\Delta_E^2 / N! h^{3N} \Delta_V}{Z(H_1, N_1, H_2, N_2, V, p)} \delta(H_1 - pV_1 - \mathcal{H}_1)\, \delta(H_2 - pV_2 - \mathcal{H}_2), \end{aligned} \tag{5.79}$$

which is uniform on a trajectory, as specified by the individually constant enthalpies. The momentum contributions, including that of the volume, have been ignored here, since these drop out for a reservoir at rest, as above. Integrating over the phase space coordinates of the particles of the reservoir, the probability of the subsystem having a particular volume and its particles being at a particular point in phase space is

$$\begin{aligned} &\wp(\mathbf{\Gamma}^{N_1}, V_1 | H_1, N_1, H_2, N_2, V, p) \\ &= \frac{\Delta_E / N_1! h^{3N_1} \Delta_V}{Z(H_1, N_1, H_2, N_2, V, p)} Z(E_2, N_2, V_2)\, \delta(H_1 - p_1 V_1 - \mathcal{H}_1). \end{aligned} \tag{5.80}$$

The weight $Z(E_2, N_2, V_2)$ is that of the reservoir, considered as an isolated system with energy $E_2 = H_2 - pV_2$ and volume $V_2 = V - V_1$. Its logarithm yields the reservoir entropy for this particular state, and a Taylor expansion yields

$$S(E - E_1, N_2, V - V_1) = \text{const.} + \frac{pV_1}{T_2} - \frac{p_2 V_1}{T_2}. \tag{5.81}$$

Here $E_1 = H_1 - pV_1$ and $E = H_1 + H_2 - pV$, so that $E - E_1 = H_2 - p(V - V_1) = E_2$. The reservoir entropy is evidently constant with respect to the subsystem volume if the parameter p is the reservoir pressure, $p = p_2$. Accordingly the reservoir entropy as such does not affect the probability distribution. Dropping the subscripts (since everything now refers to the subsystem except for the reservoir pressure p), one has for the density in phase space

$$\wp(\mathbf{\Gamma}^N, V|H,N,p) = \frac{\Delta_E \delta(\mathcal{H}(\mathbf{\Gamma}^N) - E)}{N! h^{3N} \Delta_V Z(H,N,p)}, \quad E = H - pV. \tag{5.82}$$

Integrating over phase space, the probability of the subsystem having a particular volume is

$$\wp(V|H,N,p) = \frac{e^{S(E,N,V)/k_{\mathrm{B}}}}{\Delta_V Z(H,N,p)}, \quad E = H - pV. \tag{5.83}$$

The partition function in these distributions is

$$\begin{aligned} Z(H,N,p) &= \frac{1}{\Delta_V} \int \mathrm{d}V\, e^{S(E,N,V)/k_{\mathrm{B}}} \\ &= \int \frac{\mathrm{d}V}{\Delta_V} \int \frac{\mathrm{d}\mathbf{p}^N\, \mathrm{d}\mathbf{q}^N}{N! h^{3N}} \Delta_E \delta\big(H - pV - \mathcal{H}(\mathbf{p}^N, \mathbf{q}^N)\big). \end{aligned} \tag{5.84}$$

The subsystem-dependent part of the total entropy is the entropy of the subsystem itself, and hence the constrained thermodynamic potential is the negative of this,

$$\Phi(V|H,N,p;T) = -TS(E,N,V), \quad E = H - pV. \tag{5.85}$$

Note that because energy is not an exchangeable parameter, an arbitrary temperature T has been introduced to give the constrained thermodynamic potential the dimensions of energy; it will be seen that it actually has no physical effect. The minimum of the constrained thermodynamic potential, which gives the peak of the volume probability density, is given by the 0 of

$$\begin{aligned} \left(\frac{\partial S(E,N,V)}{\partial V}\right)_{H,N,p} &= \frac{\partial S(E,N,V)}{\partial E} \frac{\partial (H - pV)}{\partial V} + \frac{\partial S(E,N,V)}{\partial V} \\ &= \frac{-p}{T(E,N,V)} + \frac{p(E,N,V)}{T(E,N,V)}. \end{aligned} \tag{5.86}$$

The equilibrium volume $\overline{V}(H,N,p)$ is the one such that the internal pressure of the system equals the applied pressure, $\overline{p}(H,N,p) \equiv p(\overline{E},N,\overline{V}) = p$. Note that it is only in equilibrium that the parameter p that appears in the definition of the enthalpy of the subsystem is the subsystem pressure; in general it is the pressure of the reservoir.

The equilibrium thermodynamic potential is the minimum value of the constrained thermodynamic potential,

$$\begin{aligned} \overline{\Phi}(H,N,p;T) &= \Phi(\overline{V}|H,N,p,T) \\ &= -TS(\overline{E},N,\overline{V}), \quad \overline{E} = H - p\overline{V}. \end{aligned} \tag{5.87}$$

The extra parameter T in this equilibrium quantity will be seen to have no physical effect. In view of the variational nature of the constrained thermodynamic potential, differentiating the thermodynamic potential with respect to enthalpy, number, or external pressure is the same as differentiating the constrained thermodynamic potential holding the volume fixed at its equilibrium value. Hence

$$\begin{aligned}\frac{\partial \overline{\Phi}(H,N,p;T)}{\partial H} &= \left(\frac{\partial \Phi(\overline{V}|H,N,p;T)}{\partial H}\right)_{\overline{V},N,p,T} \\ &= -T\left(\frac{\partial S(E,N,\overline{V})}{\partial E}\right)_{N,\overline{V}}\left(\frac{\partial (H-p\overline{V})}{\partial H}\right)_{\overline{V},p} \\ &= \frac{-T}{\overline{T}},\end{aligned} \tag{5.88}$$

$$\begin{aligned}\frac{\partial \overline{\Phi}(H,N,p;T)}{\partial N} &= \left(\frac{\partial \Phi(\overline{V}|H,N,p;T)}{\partial N}\right)_{\overline{V},H,p,T} \\ &= -T\left(\frac{\partial S(\overline{E},N,\overline{V})}{\partial N}\right)_{\overline{E},\overline{V}} \\ &= \frac{\overline{\mu}T}{\overline{T}},\end{aligned} \tag{5.89}$$

and

$$\begin{aligned}\frac{\partial \overline{\Phi}(H,N,p;T)}{\partial p} &= \left(\frac{\partial \Phi(\overline{V}|H,N,p;T)}{\partial p}\right)_{\overline{V},H,N,T} \\ &= -T\left(\frac{\partial S(E,N,\overline{V})}{\partial E}\right)_{N,\overline{V}}\left(\frac{\partial (H-p\overline{V})}{\partial p}\right)_{\overline{V},H,N} \\ &= \frac{\overline{V}T}{\overline{T}}.\end{aligned} \tag{5.90}$$

In the above $E = H - pV$ and $\overline{E} = H - p\overline{V}$. Of course the derivative with respect to the temperature parameter T is trivial and equals the negative of the entropy. Hence the total differential of the thermodynamic potential is

$$d\overline{\Phi}(H,N,p;T) = -S(\overline{E},N,\overline{V})dT - \frac{T}{\overline{T}}dH + \frac{\overline{\mu}T}{\overline{T}}dN + \frac{\overline{V}T}{\overline{T}}dp, \tag{5.91}$$

which gives its partial derivatives at a glance.

The average properties of the subsystem are given by the appropriate derivative of the total unconstrained entropy, which is the logarithm of the partition function. The enthalpy derivative yields $\langle 1/T\rangle$, the number derivative yields $\langle -\mu/T\rangle$, and the pressure derivative yields $\langle -V/T\rangle$, where T is the temperature of the subsystem. Since to leading order the negative of the logarithm of the partition function is the free energy, one again concludes that for macroscopic systems the equilibrium values equal the average values.

Summary

- Canonical phase space, which may be taken as the set of microstates of classical statistical mechanics, has uniform weight density throughout. For an isolated Hamiltonian system, the surface probability density is inversely proportional to the local speed of the trajectory, whereas the phase space probability density is constant on the energy hypersurface.

- The entropy of a point or hypersurface in phase space is the logarithm of its weight density, and the corresponding probability density is the exponential of this entropy. In both one may introduce an arbitrary volume element of appropriate dimensions that has no physical consequence.

- The probability that a subsystem in contact with a heat reservoir is at a particular point in phase space is given by the Boltzmann factor, which is the exponential of the negative of the Hamiltonian of the point divided by $k_{\mathrm{B}}T$. For more general reservoirs and macrostates, the exponent is the constrained total entropy, which includes the subsystem's effect on the entropy of the reservoir and the entropy of the macrostate of the subsystem.

Chapter 6

Ideal Systems

In principle the thermodynamic properties of any system specified by known molecular potentials can be calculated from either the derivatives of the entropy or from system averages using the probability density, as given in the preceding chapter. In practice the explicit evaluation of these is highly nontrivial because of the huge number of dimensions involved in the integrations. To make the theory transparent with concrete examples, one needs to choose simplified Hamiltonians such that the multidimensional integrals factorise. Such is the case with noninteracting systems, which are the subject of this chapter.

Although this book is almost exclusively concerned with classical statistical mechanics, this chapter begins with quantum systems. In these the microstates are discrete, and so the integrals reduce to sums, as in the generic derivation of statistical mechanics pursued in all previous chapters but the last.

6.1 Ideal Quantum Systems

6.1.1 Fermions

Fermions, such as electrons, have the property that at most one can occupy each quantum state.[1] If the available states of the system are labelled by $j = 1, 2, \ldots,$ then the occupancy of the jth state is $n_j = 0$ or 1. Let $\mathbf{n} = \{n_j\}$ represent the microstate of the system (i.e., the occupancy of all the states). The contribution to the total energy of the jth state is taken to be $n_j\epsilon_j$ (i.e., ϵ_j is a constant that is independent of the occupancy of the other states). Accordingly, the total energy and the total number of fermions is

$$E(\mathbf{n}) = \sum_j n_j \epsilon_j, \text{ and } N(\mathbf{n}) = \sum_j n_j. \tag{6.1}$$

[1]Quantum occupancy in relation to statistical mechanics is discussed by R. K. Pathria, *Statistical Mechanics*, Pergammon, Oxford, 1972.

Assuming that all microstates with a given energy have equal weight, the probability of a particular state occupancy for an isolated system is

$$\wp(\mathbf{n}|E,N) = \frac{\delta(N(\mathbf{n}) - N)\,\delta(E(\mathbf{n}) - E)}{Z(N,E)}. \tag{6.2}$$

The entropy is of course $S(E,N) = k_{\rm B} \ln Z(E,N)$, and the temperature and chemical potential follow by differentiation.

For an open system in contact with a reservoir of temperature T and chemical potential μ, the results in Ch. 4 give the energy, number distribution as

$$\wp(E,N|\mu,T) = \frac{e^{S(E,N)/k_{\rm B}} e^{-\beta E} e^{\beta\mu N}}{\Xi(\mu,T)}. \tag{6.3}$$

Here and throughout $\beta \equiv 1/k_{\rm B}T$. From this it follows that the occupancy probability is

$$\begin{aligned} \wp(\mathbf{n}|\mu,T) &= \sum_{E,N} \wp(\mathbf{n}|E,N)\wp(E,N|\mu,T) \\ &= \frac{e^{-\beta E(\mathbf{n})} e^{\beta\mu N(\mathbf{n})}}{\Xi(\mu,T)} \\ &= \frac{1}{\Xi(\mu,T)} \prod_j e^{-\beta\epsilon_j n_j} e^{\beta\mu n_j} \\ &= \prod_j \wp_j(n_j|\mu,T). \end{aligned} \tag{6.4}$$

Because the contribution to the total energy from each state only depends upon the occupancy of that state, the system occupancy probability factorises into the product of single-state occupancy probabilities. This is a gross simplification of real fermionic systems, but it nevertheless serves to illustrate certain characteristics of their behaviour.

The occupancy probability for the jth state is given by

$$\wp_j(n|\mu,T) = \frac{e^{-\beta\epsilon_j n} e^{\beta\mu n}}{\Xi_j(\mu,T)}, \tag{6.5}$$

with the single-state fermionic partition function being

$$\Xi_j(\mu,T) = \sum_{n=0,1} e^{-\beta\epsilon_j n} e^{\beta\mu n} = 1 + e^{\beta(\mu-\epsilon_j)}. \tag{6.6}$$

The mean occupancy of the state j is

$$\langle n \rangle_j = \sum_{n=0,1} n\wp_j(n|\mu,T) = \frac{1}{(e^{\beta(\epsilon_j-\mu)}+1)} = f_{\rm FD}(\beta(\epsilon_j - \mu)). \tag{6.7}$$

The *Fermi–Dirac distribution* that appears here is $f_{\rm FD}(x) \equiv 1/(e^x + 1)$. Examples of the mean occupancy are shown in Fig. 6.1.

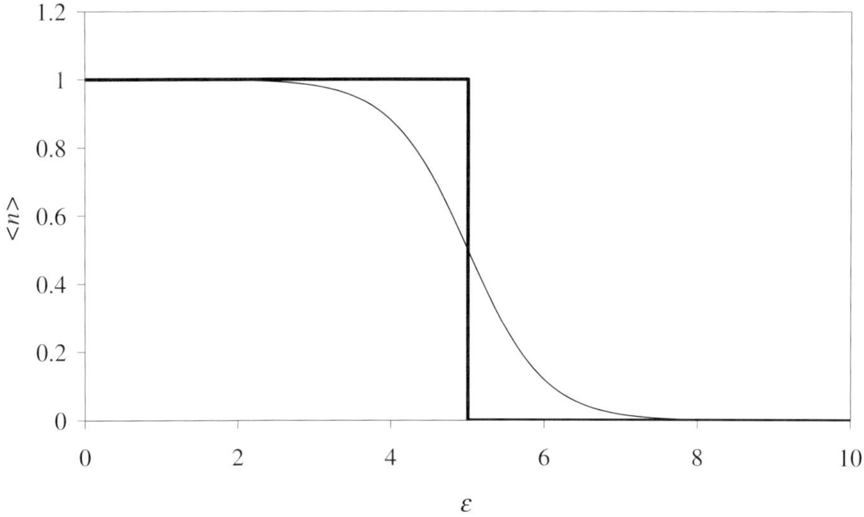

Figure 6.1: The mean state occupancy of an ideal fermion system with $\mu = 5$ for $T = 0$ (bold curve) and for $T = 0.5k_{\rm B}^{-1}$.

At $T = 0$, all states with energy less than μ are occupied, and all states with greater energy are empty,

$$\langle n \rangle_j = \begin{cases} 1, & \epsilon_j < \mu \\ 0, & \epsilon_j > \mu. \end{cases} \tag{6.8}$$

This is the ground state since the total energy is the lowest possible consistent with the exclusion principle. The energy of the highest occupied state at absolute 0 is the Fermi energy, $\epsilon_{\rm F} = \mu$. At small, nonzero temperatures, $k_{\rm B}T \ll \epsilon_{\rm F}$, fermions move from states just below the Fermi level to just above it (note that the Fermi–Dirac distribution satisfies $f_{\rm FD}(x) = 1 - f_{\rm FD}(-x)$). This has several consequences. As the reservoir temperature is increased from 0, the average number of fermions $\langle N \rangle$ remains constant if the reservoir chemical potential is fixed, since particles are simply transferred from beneath the Fermi level to above it. In other words, the chemical potential of a *closed* fermionic system in contact with a heat reservoir is approximately independent of the reservoir temperature at low temperatures, $T \ll \epsilon_{\rm F}/k_{\rm B}$. Further the heat capacity of a fermionic system is small since only those fermions in states close to the Fermi surface contribute to the change in energy with temperature.

The Fermi–Dirac distribution was derived using the grand canonical formalism, which is appropriate for a system in contact with a particle and energy reservoir. However, the result also applies to the states of an isolated system, since all of the states $i \neq j$ act as a reservoir for the state j. It is emphasised that the results are only valid for *noninteracting* fermions, which is a gross approximation to real systems, but which in some circumstances is valid.

6.1.2 Bosons

Bosons differ from fermions in that there is no restriction on the occupancy of the states, $n_j = 0, 1, 2\ldots$. However, like fermions, noninteracting bosons have ϵ_j independent of the state occupancies, and the microstate distribution also factorises,

$$\wp(\mathbf{n}|\mu, T) = \prod_j \wp_j(n_j|\mu, T), \tag{6.9}$$

with the occupancy probability for the jth state given by

$$\wp_j(n|\mu, T) = \frac{e^{-\beta\epsilon_j n} e^{\beta\mu n}}{\Xi_j(\mu, T)}. \tag{6.10}$$

For the case of bosons, however, the single-state bosonic partition function is

$$\Xi_j(\mu, T) = \sum_{n=0}^{\infty} e^{-\beta\epsilon_j n} e^{\beta\mu n} = \frac{1}{1 - e^{\beta(\mu - \epsilon_j)}}. \tag{6.11}$$

Note that in order for the sum to converge, $\mu - \epsilon_j$ must be negative. The mean occupancy of the state j is

$$\langle n \rangle_j = \sum_{n=0}^{\infty} n\wp_j(n|\mu, T) = \frac{1}{e^{\beta(\epsilon_j - \mu)} - 1} = f_{\mathrm{BE}}(\beta(\epsilon_j - \mu)). \tag{6.12}$$

(The result for the partition function may be derived from $\sum_n x^n = 1/(1 - x)$, and the mean occupancy follows by logarithmic differentiation.) Here the *Bose–Einstein distribution* is $f_{\mathrm{BE}}(x) \equiv 1/(e^x - 1)$. This says that low-energy states are favoured over high-energy ones.

The above derivation for an open system invoked an infinite sum over single-state occupancies that only converged when $\mu < \epsilon_j$. If the states are ordered such that $\epsilon_0 < \epsilon_1 < \epsilon_2 \ldots$, with ϵ_0 being the ground state energy, then the reservoir chemical potential must be less than the ground state energy for any physically realistic open bosonic subsystem. (Otherwise the number of bosons in the subsystem would be comparable to the number in the reservoir, which contradicts the notion of a reservoir.) If one focuses on an isolated system of energy E and number of bosons N, then since it acts as a reservoir for each of its states, one must have $\mu(E, N) < \epsilon_0$. Further, in the low-energy limit, $E \to N\epsilon_0$, it is clear that all the bosons must go to the ground state, $\langle n \rangle_0 = N$, and $\langle n \rangle_j = 0$, $j > 0$. The Bose–Einstein distribution shows that this can only occur if $T(E, N) \to 0$ in this limit (because the exponent in the denominator is strictly positive), and hence one must have that $\mu(E, N) \to \epsilon_0 - k_{\mathrm{B}}T \ln[1 + N^{-1}] \approx \epsilon_0 - k_{\mathrm{B}}T/N$. The same result clearly holds for a *closed* system in thermal equilibrium with a cold reservoir.

At $T = 0$ all bosons are in the ground state, $\langle n \rangle_0 = N$, and μ must approach the ground state energy (from below). As the temperature is raised, the chemical potential does not begin to substantially decrease until the thermal energy, $k_{\mathrm{B}}T$, is large enough for bosons to go from the ground to the first excited state. This is the *Bose–Einstein condensation temperature.*

High-Temperature Limit

It is now shown that the chemical potential for both bosons and fermions must decrease with increasing temperature, and that at high temperatures they become identical and classical. Since the entropy is a concave function of energy, temperature, which is its second derivative, is a strictly increasing function of energy. Hence the high-temperature limit is also the high-energy limit. As the energy of both fermionic and bosonic isolated systems is increased at constant number, then the high-energy states become occupied at the expense of low-energy states. In order for the occupancy of the low-energy states to vanish, $n_j \to 0$, the denominator of the Fermi–Dirac distribution and of the Bose–Einstein distribution must become large, which means that the exponential term dominates, $(\epsilon_j - \mu(E,N))/k_{\mathrm{B}}T(E,N) \to \infty$. This can only occur if $\mu(E,N) \to -\infty$ (and faster than linearly in temperature). These results also apply to a closed system in equilibrium with a high-temperature reservoir. One concludes that in the high-temperature limit the distinction between bosons and fermions becomes meaningless. For both types of particles, the mean occupancy becomes

$$\langle n \rangle_\epsilon = e^{-\beta(\epsilon-\mu)}, \tag{6.13}$$

which is less than unity. This is called an *uncondensed* gas, and is the sufficient and necessary condition for the application of classical statistical mechanics.

6.1.3 Density of Vibrational States

It is frequently convenient to convert the sum over states for the partition function into an integral, in which case one needs the *density of states*. A common example is the density of vibrational states, which occurs in the following elementary treatments of black-body radiation and of the heat capacity solids. A wave confined to a cubic volume $V = L^3$ and vanishing on the boundaries has the form

$$\psi(x,y,z) = A \sin k_x x \sin k_y y \sin k_z z, \tag{6.14}$$

provided that the components of the wave vector satisfy

$$k_\alpha = n_\alpha \pi/L, \quad n_\alpha = 1, 2, \ldots\ , \tag{6.15}$$

for $\alpha = x, y$, or z. One sees that the wave vector is discretised, and that the volume of k-space per wave vector is $(\pi/L)^3$. The density of modes is the volume of the spherical shell between k and $k + dk$ divided by the volume per wave,

$$\begin{aligned} f(k)\, dk &= 4\pi k^2\, dk/8(\pi/L)^3 \\ &= V k^2\, dk/2\pi^2, \end{aligned} \tag{6.16}$$

where the factor of 8 arises because only states in the first octant are allowed.

One can readily find the density of frequency states and the density of momentum states. The circular frequency is $\omega = kv$, where v is the phase velocity. For a nondispersive medium, v is independent of ω. Hence $dk = d\omega/v$ and

$$f(\omega)\, d\omega = V\omega^2\, d\omega/2\pi^2 v^3. \tag{6.17}$$

The momentum is $\mathbf{p} = \hbar\mathbf{k}$, where $\hbar$ is Planck's constant divided by 2π, and hence

$$f(p)\, dp = Vp^2\, dp/2\pi^2\hbar^3. \tag{6.18}$$

6.1.4 Photons

Historically, one of the first applications of quantum theory was to the problem of the spectral distribution of electromagnetic radiation (*black-body radiation*). More precisely, it was the attempt to describe quantitatively the radiation emitted by a hot object that lead to the quantum theory. Planck proposed that the energy of radiation of frequency ω was quantised, $\epsilon_n(\omega) = n\hbar\omega$, where the 0-point energy is omitted. (This quantisation also applies to the simple harmonic oscillator treated below.) Here n represents the number of photons of frequency ω, which is a nonconserved quantity. Assuming that there is no internal entropy associated with n, which is to say that the change in entropy due to a change in photon number comes solely from the change in energy of the reservoir, then radiation in equilibrium with a thermal reservoir of temperature T obeys a Boltzmann distribution,

$$\begin{aligned}\wp(n|\omega, T) &= Z(\omega, T)^{-1} e^{-\beta\hbar\omega n} \\ &= \frac{e^{-\beta\hbar\omega n}}{1 - e^{-\beta\hbar\omega}}.\end{aligned} \tag{6.19}$$

The average energy of frequency ω is

$$\langle\epsilon\rangle_{\omega,T} = \sum_{n=0}^{\infty} n\hbar\omega\wp(n|\omega, T) = \frac{\hbar\omega}{e^{\beta\hbar\omega} - 1}. \tag{6.20}$$

At high temperatures, $k_{\mathrm{B}}T \gg \hbar\omega$, this gives $\langle\epsilon\rangle_{\omega,T} \to k_{\mathrm{B}}T$. Conversely the amount of energy in modes beyond the thermal energy is exponentially small (i.e., when $k_{\mathrm{B}}T \ll \hbar\omega$, $\langle\epsilon\rangle_{\omega,T} \to \hbar\omega e^{-\beta\hbar\omega}$).

This result contrasts with that for the classical oscillator. For the latter, the energy is a continuous function of the (square of the) amplitude and frequency. No matter how high the resonance frequency, the classical oscillator can acquire a small amount of energy by vibrating with infinitesimal amplitude. Consequently the average energy per mode is constant and equal to $k_{\mathrm{B}}T$, which is the high-temperature limit of the quantum oscillator. The problem arises at low temperatures, $k_{\mathrm{B}}T \ll \hbar\omega$, because a quantum of energy is required to excite the quantum oscillator. When the thermal energy is less than this, the energy of the oscillator is 0, whereas it remains $k_{\mathrm{B}}T$ in the classical picture.

Now the energy per unit frequency, the *spectral distribution*, is the average energy times the density of states. Integrated over all frequencies it gives the total energy, and for a box of volume V the energy per unit volume is

$$
\begin{aligned}
E/V &= V^{-1}\int_0^\infty \langle\epsilon\rangle_{\omega,T}\, f(\omega)\,\mathrm{d}\omega \\
&= V^{-1}\int_0^\infty \frac{\hbar\omega}{e^{\beta\hbar\omega}-1}\frac{2V\omega^2}{2\pi^2c^3}\,\mathrm{d}\omega \\
&= (\pi^2/15\hbar^3c^3)(k_\mathrm{B}T)^4, \qquad (6.21)
\end{aligned}
$$

where c is the velocity of light, and the factor of 2 in the density of states arises because there are two polarisations. This shows that the energy per unit volume emitted by a hot body increases with the fourth power of the temperature and that it is independent of the nature of the body.

The history of black-body radiation is of some interest because of its connection with the development of the quantum theory. Stefan measured the total energy density and showed in 1879 that it was proportional to T^4. Shortly thereafter, in 1884, Boltzmann derived this law using purely thermodynamic arguments. In 1896 Wien showed on the basis of rather general considerations that the spectral density must be of the form $\omega^3 g(\omega/T)$. He also showed that the experimental spectral density decreased exponentially with frequency at low temperatures. Classical statistical mechanics could by this time be applied to the problem, and in 1900 Rayleigh and Jeans showed that if the allowed modes were treated as independent classical harmonic oscillators, then they would each contribute $k_\mathrm{B}T$ to the spectral density. This was in agreement with experiment at high temperatures and low frequencies, but the classical treatment predicted this constant spectral density for all temperatures, in contradiction to the measurements. Worse, because the spectral density was the same for all modes, its integral, the total energy density, diverged, and this became known as the *ultra-violet catastrophe*. Planck, also in 1900, simply postulated his spectral distribution as a way of interpolating between the high-frequency Wien's law and the low-frequency Rayleigh–Jeans result. He showed that his interpolation yielded an integrable spectral distribution, and also gave the Stefan–Boltzmann law for the energy density. He went on to discuss the physical origin of his spectral distribution, and pointed out that it would arise if the energy distribution of an oscillator of frequency ω was not continuously variable but discretised in lumps of $\hbar\omega$, $E_n = n\hbar\omega$, $n = 0, 1, \ldots$. The effect was to introduce an effective cutoff, since when $\hbar\omega \gtrsim k_\mathrm{B}T$ the oscillator was confined to the ground state $E_0 = 0$ and did not contribute the classical $k_\mathrm{B}T$ to the average energy. More precisely, in the quantum picture the average energy of a mode decreases exponentially with frequency, whereas in the classical treatment the average energy is independent of frequency. It was not until Einstein in 1905 postulated the quantisation of radiation to explain the *photo-electric effect* that Planck's discussion of the origin of his distribution could be understood in terms of the number of photons per mode.

6.1.5 Heat Capacity of Solids

The heat capacity is the energy required to change the temperature, and for a solid this mainly goes into lattice vibrations.[2] Einstein made the approximation that each atom could be regarded as an harmonic oscillator completely independent of the rest, and all with the same angular frequency ω. Regarding oscillations along each of the axes as independent, there are $3N$ such oscillators for a three-dimensional solid composed of N atoms. Each oscillator is quantised, $n = 0, 1, \ldots$, with the energy of each state being $\epsilon_n = (n + 1/2)\hbar\omega$, where Plank's constant divided by 2π is $\hbar \equiv h/2\pi = 1.06 \times 10^{-34}$.

Isolated System

Consider an isolated system comprising $3N$ oscillators of frequency ω with total energy $E = (M + 3N/2)\hbar\omega$ (i.e., M available quanta). The microstates are specified by the values of $\mathbf{n} = \{n_i\}_{i=1}^{3N}$. Given that each microstate corresponding to this energy is equally likely, their probability distribution is

$$\wp(\mathbf{n}|N, M) = \frac{\delta(M - \sum_{i=1}^{3N} n_i)}{Z(N, M)}. \tag{6.22}$$

The total weight is the number of ways of arranging M indistinguishable quanta among $3N$ distinguishable oscillators, which is just

$$Z(N, M) = \frac{(M + 3N - 1)!}{M!(3N - 1)!}. \tag{6.23}$$

(This is the number of ways of inserting $3N - 1$ partitions between M objects.) Hence the entropy is

$$\begin{aligned} S(N, M)/k_{\mathrm{B}} &= \ln \frac{(M + 3N - 1)!}{M!(3N - 1)!} \\ &\approx -M \ln \frac{M}{M + 3N} - 3N \ln \frac{3N}{M + 3N}. \end{aligned} \tag{6.24}$$

The temperature follows from the numerical difference,

$$\frac{1}{T(N, M)} = \frac{k_{\mathrm{B}}}{\hbar\omega} [S(N, M) - S(N, M - 1)] = \frac{k_{\mathrm{B}}}{\hbar\omega} \ln \frac{M + 3N - 1}{M}. \tag{6.25}$$

For $M \gg N \gg 1$ (high-energy limit), this is

$$\frac{1}{T(N, M)} = \frac{k_{\mathrm{B}}}{\hbar\omega} \frac{3N}{M}, \tag{6.26}$$

or $3Nk_{\mathrm{B}}T = M\hbar\omega \to E \to \infty$. For $N \gg M \gg 1$ (low-energy limit), one obtains

$$\frac{1}{T(N, M)} = \frac{k_{\mathrm{B}}}{\hbar\omega} \ln \frac{3N}{M}, \tag{6.27}$$

[2] Einstein's and Debye's treatments of the heat capacity are discussed and compared to experiment by F. Mandl, *Statistical Physics*, Wiley, London, 1971.

or $T = \to 0$.

The heat capacity is

$$C_V = \frac{\partial E}{\partial T(N, M)} = \frac{(\hbar\omega)^2}{k_B T^2} \frac{(3N+M)M}{3N}, \tag{6.28}$$

assuming the thermodynamic limit. At high temperatures this is $C_V \to 3Nk_B$, $M \gg N$, and at low temperatures is $C_V \to Mk_B / \ln^2(N/M)$, $M \ll N$.

Einstein Approach

For a system of oscillators in equilibrium with a thermal reservoir of temperature T, the energy probability distribution is

$$\wp(M|N,T) = \frac{e^{S(M,N)/k_B} e^{-\beta(M+N/2)\hbar\omega}}{Z(N,T)}, \tag{6.29}$$

where the entropy of the isolated system was given above. Hence the microstate probability is

$$\begin{aligned} \wp(\mathbf{n}|N,T) &= \sum_{M=0}^{\infty} \wp(\mathbf{n}, M|N,T) \\ &= \sum_{M=0}^{\infty} \wp(\mathbf{n}|M,N) \wp(M|N,T) \\ &= \sum_{M=0}^{\infty} \frac{\delta(M - \sum_{i=1}^{3N} n_i)}{Z(N,M)} \frac{e^{S(M,N)/k_B} e^{-\beta(M+N/2)\hbar\omega}}{Z(N,T)} \\ &= \frac{e^{-\beta\hbar\omega N/2}}{Z(N,T)} \prod_{i=1}^{3N} e^{-\beta h\omega n_i} \\ &= \prod_{i=1}^{3N} \wp(n_i|\omega, T). \end{aligned} \tag{6.30}$$

One sees that it factorises into the product of single-oscillator energy distributions,

$$\wp(n|\omega, T) = Z(\omega, T)^{-1} e^{-\beta\hbar\omega(n+1/2)}. \tag{6.31}$$

Here the single-oscillator canonical partition function is

$$\begin{aligned} Z(\omega, T) &= \sum_{n=0}^{\infty} \exp[-\beta\hbar\omega(n+1/2)] \\ &= \frac{e^{-\beta\hbar\omega/2}}{1 - e^{-\beta\hbar\omega}}. \end{aligned} \tag{6.32}$$

Differentiation with respect to β yields the average energy per atom per axis. The average total energy is

$$\begin{aligned} E(\omega, T) &= 3N \langle \epsilon \rangle_{\omega,T} \\ &= -3N \frac{\partial \ln Z(\omega, T)}{\partial \beta} \\ &= 3N \left[\frac{\hbar\omega}{2} + \frac{\hbar\omega}{e^{\beta\hbar\omega} - 1} \right], \end{aligned} \tag{6.33}$$

and the *heat capacity* is

$$\begin{aligned} C_V &= \frac{\partial E(\omega, T)}{\partial T} \\ &= 3Nk_{\mathrm{B}} \frac{(\beta\hbar\omega)^2 e^{\beta\hbar\omega}}{(e^{\beta\hbar\omega} - 1)^2}. \end{aligned} \tag{6.34}$$

This is essentially the heat capacity at constant volume, since taking ω fixed is equivalent to modelling the solid as incompressible. Einstein's expression for the heat capacity contains one parameter, ω, which is fitted for each solid. For $T \gg \hbar\omega/k_{\mathrm{B}}$, $C_V \sim 3Nk_{\mathrm{B}}$, which is in agreement with experiment. However at low temperatures this expression vanishes too quickly.

Debye Approach

The most drastic approximation that Einstein made was to treat the atoms as independent. Debye realised that a crystal of N atoms has $3N$ normal modes of vibration, and that each of these is independent with a frequency ω_α. Hence the total partition function is just the product of the individual oscillator partition functions, $Z(T) = \prod_{\alpha=1}^{3N} Z(\omega_\alpha, T)$, or in terms of the total entropy,

$$S(T) = \sum_{\alpha=1}^{3N} k_{\mathrm{B}} \ln Z(\omega_\alpha, T). \tag{6.35}$$

In the thermodynamic limit, the Helmholtz free energy is $F(T) = -TS(T)$. One proceeds by converting the sum over modes, which is extremely difficult to evaluate, to an integral over frequency. The density of states follows from the fact that at low frequencies the vibrations are just the elastic waves of the solid, and one has

$$f(\omega)\, d\omega = 3V\omega^2\, d\omega / 2\pi^2 \overline{v}^3. \tag{6.36}$$

Here the average velocity derives from the one longitudinal and two transverse modes, $3\overline{v}^{-3} = v_{\mathrm{L}}^{-3} + 2v_{\mathrm{T}}^{-3}$. Debye assumed that this density of frequencies held for all frequencies up to a cutoff frequency ω_{D}, whose value was determined from the total number of modes

$$3N = \int_0^{\omega_{\mathrm{D}}} f(\omega)\, d\omega. \tag{6.37}$$

This gives $\omega_{\rm D}^3 = 6\pi^2 N\overline{v}^3/V$. The total entropy becomes

$$S(T) = \int_0^{\omega_{\rm D}} k_{\rm B} \ln \left[\frac{e^{-\beta\omega\hbar/2}}{1 - e^{\beta\hbar\omega}} \right] f(\omega)\, {\rm d}\omega. \tag{6.38}$$

Differentiation yields the average energy

$$\langle E \rangle_T = \frac{9}{8} N\hbar\omega_{\rm D} + \frac{9N\hbar}{\omega_{\rm D}^3} \int_0^{\omega_{\rm D}} \frac{\omega\, {\rm d}\omega}{e^{\beta\hbar\omega} - 1}, \tag{6.39}$$

and the heat capacity is

$$C_V = 3Nk_{\rm B} \frac{3}{x_{\rm D}^3} \int_0^{x_{\rm D}} \frac{x^4 e^x\, {\rm d}x}{(e^x - 1)^2}, \tag{6.40}$$

where $x = \hbar\omega/k_{\rm B}T$. For $T \gg \hbar\omega_{\rm D}/k_{\rm B}$ ($x \to 0$), $C_V \sim 3Nk_{\rm B}$, which is the classical result. At low temperatures ($x_{\rm D} \to \infty$), one can replace the upper limit of the integral by ∞, which is then independent of temperature and equal to $4\pi^4/15$. Consequently, $C_V \sim T^3, T \to 0$, which is in agreement with experiment. Thus Debye's theory improves upon that of Einstein. Like the latter it has one parameter that can be fitted to heat capacity data for each material. However, it can also be estimated independently from measurements of the speed of sound in the solid, and the two values turn out to be in relatively good agreement.

6.2 Spin–Lattice Models

Spin–lattice models represent a rather popular application of statistical mechanics of particular interest to those with a more mathematical bent.[3] The broad aim is to obtain exact analytic expressions for specific partition functions. The models that can be so solved represent an abstraction and simplification of reality and the main use of the solutions is to illustrate explicitly the general implications of statistical mechanics. The emphasis has been on inventing soluble models and on mathematical techniques rather than on any quantitative treatment of actual physical systems.

The spin–lattice models have their origin in the quantisation of molecular spin by a magnetic field. Accordingly one has a well-defined lattice occupied by spin variables that take on discrete values. It is the coupling between neighbouring spins on the lattice that makes the evaluation of the partition function nontrivial. Typically the Hamiltonian of the system is of the form

$$\mathcal{H}(\sigma^N) = E_0 - B^{\rm ext}\mu \sum_{i=1}^{n} \sigma_i - J\mu^2 \sum_{\rm n.n.} \sigma_i\sigma_j + \ldots. \tag{6.41}$$

Here $\sigma_i \in [\pm 1, \pm 2, \ldots]$ is the ith spin variable, $B^{\rm ext}$ is the external magnetic field, μ is the spin magnetic moment, and J is the coupling between nearest

[3] R. J. Baxter, *Exactly Solved Models in Statistical Mechanics*, Academic Press, London, 1982.

neighbours on the lattice. The nonmagnetic energy, E_0, which is independent of the spin configuration, is included for completeness. The undisplayed terms can include interactions between next-nearest neighbours and other more distant pairs, and triplet and higher-order spin interactions.

6.2.1 Noninteracting Paramagnetic Spins

Consider an ideal spin–lattice model in which the spins do not interact with each other, $J = 0$. In this case the nature of the crystal lattice and its dimensionality is irrelevant. The simplest model is the so-called Ising model, in which the spin variable takes on just two values, $\sigma = \pm 1$. The remaining energy is that due to the applied magnetic field, and for each spin this is $E^{\text{ext}} = -\mu B^{\text{ext}}\sigma$, which favours 'up' spins (i.e., those aligned with the field). The number of up spins is denoted by N_+, and the number of down spins by N_-. The total number of spins is $N = N_+ + N_-$, and the magnetisation is defined as $M = \mu(N_+ - N_-)$.

In this ideal model there is no volume dependence, and the nonmagnetic energy is not dependent upon the spins. Accordingly the configurational entropy of the isolated system depends upon just two variables, which can be taken to be either the number of up and down spins or the total number of spins and the magnetisation. The entropy of the isolated system includes the spin-independent part and a purely configurational part,

$$S^{\text{id}}(M|E_0) = S_0(E_0) + k_{\text{B}} \ln \frac{N!}{N_+!N_-!}, \tag{6.42}$$

where $N_\pm = (N \pm M/\mu)/2$. One has of course $\partial S^{\text{id}}(M|E_0)/\partial E_0 = T^{-1}$. According to Eq. (B.20), the derivative of the entropy with respect to magnetisation gives the 'internal' magnetic field,

$$\left(\frac{\partial S^{\text{id}}(M|E_0)}{\partial M} \right)_{E_0} = \frac{k_{\text{B}}}{2\mu} \ln \frac{N\mu - M}{N\mu + M} \equiv \frac{-B^{\text{id}}}{T}. \tag{6.43}$$

The internal field, which is not to be confused with the local field, is equal to the external magnetic field that would produce this particular magnetisation as the equilibrium one.

This may be seen directly if the subsystem is placed in such an external magnetic field B^{ext}, in which case the external energy is $E^{\text{ext}} = -B^{\text{ext}} M$, and the total energy is $E^{\text{tot}} = E_0 + E^{\text{ext}}$. Since the magnetisation is constrained, the only effect of the external magnetic field is to determine the amount of nonmagnetic energy in the total energy. Hence the constrained total entropy of the subsystem is just

$$S^{\text{id}}(M|E^{\text{tot}}, B^{\text{ext}}) = S^{\text{id}}(M|E_0). \tag{6.44}$$

Consequently

$$\left(\frac{\partial S^{\text{id}}(M|E^{\text{tot}}, B^{\text{ext}})}{\partial M} \right)_{E^{\text{tot}}}$$

$$= \left(\frac{\partial S^{\rm id}(M|E_0)}{\partial E_0} \right)_M \left(\frac{\partial E_0}{\partial M} \right)_{E^{\rm tot}} + \left(\frac{\partial S^{\rm id}(M|E_0)}{\partial M} \right)_{E_0}$$
$$= \frac{B^{\rm ext}}{T} - \frac{B^{\rm id}}{T}, \tag{6.45}$$

where the internal field was given in Eq. (6.43). This vanishes at equilibrium, and so this is an explicit equation for the equilibrium external field, $\overline{B}^{\rm ext} = B^{\rm id}(M)$. Alternatively, one may regard this as an implicit equation for the equilibrium magnetisation, $B^{\rm id}(\overline{M}) = B^{\rm ext}$, with explicit solution

$$\overline{M} = \mu N \frac{e^{2\mu B^{\rm ext}/k_{\rm B}T} - 1}{e^{2\mu B^{\rm ext}/k_{\rm B}T} + 1} = \mu N \tanh[\mu B^{\rm ext}/k_{\rm B}T]. \tag{6.46}$$

One sees that in the high-temperature or low-field limit, $\mu B^{\rm ext}/k_{\rm B}T \to 0$, the magnetisation vanishes as $\overline{M} \sim [\mu^2 B^{\rm ext} N/k_{\rm B}T][1 + \mathcal{O}(\mu B^{\rm ext}/k_{\rm B}T)^2]$. Conversely, in the low-temperature, high-field limit, $\mu B^{\rm ext}/k_{\rm B}T \to \infty$, the magnetisation saturates at $\overline{M} \to \mu N$ as all the spins are aligned with the field.

One can rewrite this result to give the equilibrium ratio of spins,

$$\frac{\overline{N}_+}{\overline{N}_-} = \exp[2\mu B^{\rm ext}/k_{\rm B}T]. \tag{6.47}$$

That is, for this ideal spin–lattice model, the odds of a spin being parallel to the external field equals the Boltzmann factor of the energy difference between the two possible orientations.

Isothermal System

The temperature that appears in the above equations is that of the isolated system, $T(E_0)$. If the spin–lattice forms a subsystem of a thermal reservoir with which it can exchange energy, then the total entropy for a given configuration of spins and a given nonmagnetic energy is

$$S^{\rm id}(\sigma^N, E_0|B^{\rm ext}, T) = S_0(E_0) - \frac{E_0}{T} + \frac{MB^{\rm ext}}{T}, \tag{6.48}$$

where $M = \mu \sum_i \sigma_i$. The first two terms do not depend upon the spin configuration, so that the condition for thermal equilibrium is simply $\partial S_0/\partial E_0 = 1/T$. For such an equilibrated system, the probability of a particular spin configuration is the exponential of this spin-constrained total entropy, which is proportional to the Boltzmann factor of the external magnetic energy,

$$\wp^{\rm id}(\sigma^N|B^{\rm ext}, T) = \frac{1}{Z^{\rm id}(B^{\rm ext}, T)} \prod_{i=1}^N e^{\sigma_i \mu B^{\rm ext}/k_{\rm B}T}. \tag{6.49}$$

The normalising factor is

$$Z^{\rm id}(B^{\rm ext}, T) = \sum_{\sigma_1=\pm 1} \cdots \sum_{\sigma_N=\pm 1} \prod_{i=1}^N e^{\sigma_i \mu B^{\rm ext}/k_{\rm B}T}$$

$$= \left(e^{\mu B^{\rm ext}/k_{\rm B}T} + e^{-\mu B^{\rm ext}/k_{\rm B}T}\right)^N$$
$$= 2^N \cosh^N[\mu B^{\rm ext}/k_{\rm B}T]. \tag{6.50}$$

This says that in 0 field, when each spin is equally likely to point up as down, there are 2^N configurations of the system.

The probability of the macrostate M is proportional to the constrained total entropy, $\wp^{\rm id}(M|B^{\rm ext}, T) = Z^{\rm id}(B^{\rm ext}, T)^{-1} \exp S^{\rm id}(M|B^{\rm ext}, T)/k_{\rm B}$, or

$$\wp^{\rm id}(M|B^{\rm ext}, T) = \frac{N!}{N_+!N_-!} \frac{e^{MB^{\rm ext}/k_{\rm B}T}}{Z^{\rm id}(B^{\rm ext}, T)}. \tag{6.51}$$

One may readily confirm using the explicit result for the partition function that this is normalised to unity.

The average magnetic energy is the temperature derivative of the total unconstrained entropy, which is the logarithm of the partition function,

$$\left\langle E^{\rm ext}\right\rangle = \frac{-\partial \ln Z^{\rm id}(B^{\rm ext}, T)}{\partial \beta} = -(N\mu B^{\rm ext}) \tanh[\mu B^{\rm ext}/k_{\rm B}T]. \tag{6.52}$$

Hence $\langle E^{\rm ext}\rangle = -\overline{M}B^{\rm ext} = \overline{E}^{\rm ext}$. That is, the average energy and the equilibrium energy are equal.

The magnetic heat capacity is

$$\frac{\partial \overline{E}^{\rm ext}}{\partial T} = Nk_{\rm B}(\mu B^{\rm ext}/k_{\rm B}T)^2 \left(1 - \tanh^2[\mu B^{\rm ext}/k_{\rm B}T]\right). \tag{6.53}$$

This vanishes at high temperatures and low fields, because the amount of magnetic energy is low and hence so is its change with temperature. It also vanishes at low temperatures and high fields, because here the spins are nearly all aligned and changing the temperature only weakly effects their order. The maximum heat capacity is $0.44Nk_{\rm B}$, which occurs when $\mu B^{\rm ext} = 1.2k_{\rm B}T$.

Related to the heat capacity in this model are the magnetisation fluctuations. It is straightforward to show that at $M = \overline{M}(B^{\rm ext})$,

$$\frac{\partial^2 S^{\rm id}(M, E_0)/k_{\rm B}}{\partial M^2} = \frac{-1}{N\mu^2} \cosh^2[\mu B^{\rm ext}/k_{\rm B}T] = -\left\langle M^2\right\rangle^{-1}. \tag{6.54}$$

For a uniform system, the magnetisation fluctuations are the analogue of the magnetisation–magnetisation correlation function. This is a general result: the second entropy derivatives of the isolated system give the inverse of the dipole–dipole correlation tensor.

6.2.2 Ferromagnetic Spins in Mean Field Approximation

As mentioned above, it is the coupling between neighbouring spins that makes the analysis of the spin–lattice models difficult. The mean field approximation will be treated more generally in Section 6.5 below. Here it suffices to give a simple illustration of the procedure.

Suppose that in addition to the interaction with the external field there are also nearest-neighbour interactions, $J \neq 0$, but no other interactions. Ferromagnetic systems are characterised by a positive value of J, since in this case neighbouring spins tend to be aligned even in the absence of an external field. For a q-coordinated lattice in any dimension with magnetisation M, each spin is 'on average' surrounded by $q(N \pm M/\mu)/2N$ neighbouring up and down spins. Hence each spin interacts with a total surrounding magnetisation of qM/N. This expression ignores the correlation between the neighbouring spins; it would be correct if the spin orientations were distributed randomly on the lattice.

The original mean molecular field approach developed by Weiss took each atom to be coupled to every other one, $q = N$. The present approach, due to Bragg and Williams, models the finite range of molecular interactions by only including the interactions with the nearest neighbours.[4] In reality of course dipole–dipole interactions extend beyond nearest neighbours, but they do decay with distance, and so the present approach is arguably applicable when short-range, nonmagnetic interactions dominate. The coordination number of the lattice is the number of nearest neighbours of each atom. It depends upon the geometry and it increases with the dimensionality. A simple square lattice has $q = 4$ and a simple cubic lattice has $q = 6$. One could also include next-nearest neighbour and other couplings in the method.

In this mean field approximation the second term in the Hamiltonian given above is approximated by the mean field energy,

$$E^{\mathrm{mf}} = \frac{-J\mu^2}{2} \sum_{i=1}^{N} \sigma_i \frac{qM}{\mu N} = \frac{-qJM^2}{2N}. \tag{6.55}$$

The factor of one-half appears because the interaction between each pair of spins is counted twice in this expression. The external energy is $E^{\mathrm{ext}} = -B^{\mathrm{ext}} M$. It will prove useful to differentiate the entropy both at constant internal energy and at constant nonmagnetic energy. Hence the spin-independent E_0 is also included so that the Hamiltonian or total energy becomes

$$E^{\mathrm{tot}} = E_0 + E^{\mathrm{mf}} + E^{\mathrm{ext}} = E^{\mathrm{int}} + E^{\mathrm{ext}}. \tag{6.56}$$

For the isolated system with constrained magnetisation, the only effect of the external field and the coupling is on the subdivision of the total energy. The constrained entropy is equal to that of an ideal system with energy E_0 in the absence of both the external field and the coupling,

$$S(M|E^{\mathrm{tot}}, J, B^{\mathrm{ext}}) = S(M|E^{\mathrm{int}}, J) = S^{\mathrm{id}}(M|E_0). \tag{6.57}$$

Hence differentiation yields

$$\left(\frac{\partial S(M|E^{\mathrm{tot}}, J, B^{\mathrm{ext}})}{\partial M} \right)_{E^{\mathrm{tot}}}$$

[4]Mean field theory and other results for spin–lattices are discussed by C. J. Thompson, *Classical Equilibrium Statistical Mechanics*, Oxford Univ. Press, Oxford, 1988.

$$
\begin{aligned}
&= \left(\frac{\partial S(M|E^{\rm int}, J)}{\partial E^{\rm int}}\right)_M \left(\frac{\partial E^{\rm int}}{\partial M}\right)_{E^{\rm tot}} + \left(\frac{\partial S(M|E^{\rm int}, J)}{\partial M}\right)_{E^{\rm int}} \\
&= \frac{B^{\rm ext}}{T} - \frac{B}{T}.
\end{aligned} \tag{6.58}
$$

At the equilibrium magnetisation, the internal magnetic field of the coupled system equals the externally applied field, $B(\overline{M}) = B^{\rm ext}$. Explicitly one has

$$
\begin{aligned}
\frac{-B}{T} &= \left(\frac{\partial S^{\rm id}(M|E_0)}{\partial E_0}\right)_M \left(\frac{\partial E_0}{\partial M}\right)_{E^{\rm int}} + \left(\frac{\partial S^{\rm id}(M|E_0)}{\partial M}\right)_{E_0} \\
&= \frac{qJM}{NT} - \frac{B^{\rm id}}{T},
\end{aligned} \tag{6.59}
$$

where the internal field of the uncoupled system was given above, Eq. (6.43). Hence at equilibrium this and the preceding equation may be rearranged as $B^{\rm id}(\overline{M}) = B^{\rm ext} + qJ\overline{M}/N$. The right-hand side is called the local field,

$$
B^{\rm loc} = B^{\rm ext} + qJM/N; \tag{6.60}
$$

it represents the actual magnetic field seen by each atom. Accordingly, the equilibrium magnetisation gives an internal magnetic field in an ideal uncoupled system that equals the local field in the fully coupled system.

Isothermal System

For an isothermal subsystem the constrained total entropy is

$$
S(M, E^{\rm tot}|J, B^{\rm ext}, T) = S_0(E_0) + k_{\rm B} \ln \frac{N!}{N_+!N_-!} + \frac{MB^{\rm ext}}{T} + \frac{qJM^2}{2NT}. \tag{6.61}
$$

Differentiation at constant nonmagnetic energy yields

$$
\frac{\partial S}{\partial M} = \frac{k_{\rm B}}{2\mu} \ln \frac{\mu N - M}{\mu N + M} + \frac{B^{\rm ext}}{T} + \frac{qJM}{NT}. \tag{6.62}
$$

One sees that this is formally identical to the result given for the uncoupled system with the external field replaced by the local field. Amongst other things the replacement of $B^{\rm ext}$ by $B^{\rm loc}$ shows that

$$
\overline{M} = \mu N \tanh[\mu B^{\rm loc}(\overline{M})/k_{\rm B}T], \tag{6.63}
$$

which determines the equilibrium magnetisation.

One can define the dimensionless magnetisation, $x \equiv M/\mu N$, and the dimensionless coupling parameter, $\alpha = qJ\mu^2/k_{\rm B}T$. For $B^{\rm ext} = 0$ the expression for the equilibrium magnetisation becomes $\overline{x} = \tanh \alpha\overline{x}$, which has a nonzero solution only for $\alpha > 1$. The constrained total entropy in 0 field may be written in these dimensionless units as

$$
S(x|\alpha)/Nk_{\rm B} = \text{const.} - \frac{1}{2}\ln[1 - x^2] - \frac{x}{2}\ln\frac{1+x}{1-x} + \frac{1}{2}\alpha x^2. \tag{6.64}
$$

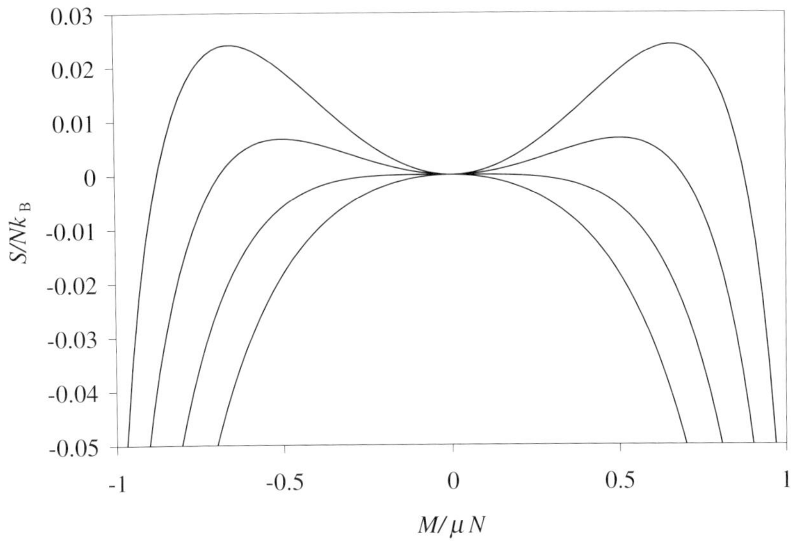

Figure 6.2: The constrained total entropy in 0 external field as a function of magnetisation for, from top to bottom, $\alpha = 1.2$, 1.1, 1, and 0.9.

This is plotted in Fig. 6.2. The location of the two maxima that appear for $\alpha > 1$ give the value of the magnetisation of the system in the absence of an applied magnetic field. That is, the system is ferromagnetic. The minimum in the entropy at $x = 0$ for $\alpha > 1$ indicates that the state of 0 magnetisation is unstable. Conversely, for $\alpha < 1$ the single maximum in the constrained entropy indicates that $\overline{M} = 0$ in the absence of an applied field. The temperature corresponding to $\alpha = 1$ is called the *critical temperature*, $T_c = qJ\mu^2/k_B$. Spontaneous magnetisation only occurs below this temperature, which may also be called the Curie temperature as it marks the transition from paramagnetic to ferromagnetic behaviour.

One can define a response function that is the rate of change of the internal field with magnetisation,

$$\begin{aligned} G^{-1} &= N\frac{\partial B}{\partial M} \\ &= -NT\frac{\partial^2 S(M|E^{\text{int}}, J)}{\partial M^2} \\ &= \frac{k_B T/\mu^2}{1-(M/\mu N)^2} - qJ. \end{aligned} \tag{6.65}$$

For a system in equilibrium, $B = \overline{B}^{\text{ext}}(M)$, and G is called the magnetic susceptibility. The fact that it is the second derivative of the entropy shows that it also gives the fluctuations in magnetisation (cf. Eq. (6.54) above). In the absence of an external magnetic field and above the critical temperature, $M = 0$.

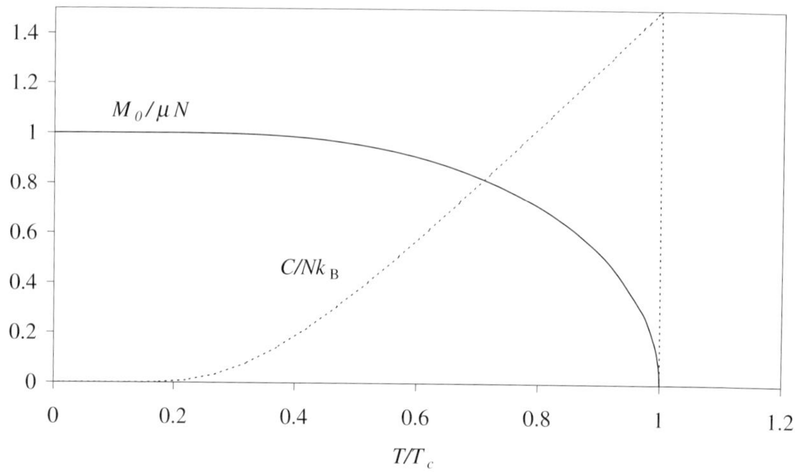

Figure 6.3: The spontaneous magnetisation (solid curve) and the heat capacity (dotted curve) in mean field approximation at subcritical temperatures.

Hence approaching the critical temperature from above, $T \to qJ\mu^2/k_B$, one sees that $G^{-1} \to 0$. In other words, the magnetic susceptibility and the magnetisation fluctuations become infinite at the critical temperature. In contrast, with the local field at equilibrium satisfying $B^{\text{id}} = \overline{B}^{\text{loc}}(M)$, the response of the magnetisation to the local field is given by

$$\begin{aligned}\chi^{-1} &= N\frac{\partial B^{\text{id}}}{\partial M}\\ &= -NT\frac{\partial^2 S^{\text{id}}(M|E_0)}{\partial M^2}\\ &= \frac{k_B T/\mu^2}{1-(M/\mu N)^2}.\end{aligned} \tag{6.66}$$

This is always nonzero, which shows that even at the critical temperature the quantity χ is well behaved. The two response functions differ by the spin interaction, $\chi^{-1} = G^{-1} + qJ$. This is also true in the more general case when the interactions are long-ranged.

The spontaneous magnetisation is nonzero below the critical temperature, $\overline{M}_0 = \mu N \tanh[T_c \overline{M}_0/\mu N T]$. This is plotted in Fig. 6.3, where it can be seen that the magnetisation saturates to μN as the temperature goes to 0, and that it vanishes as the temperature approaches the critical temperature from below. Straightforward expansions show that

$$\frac{\overline{M}_0}{\mu N} \sim \begin{cases} \sqrt{3(1-T/T_c)}, & T \to T_c^- \\ 1 - 2\exp[-2T_c/T], & T \to 0. \end{cases} \tag{6.67}$$

In 0 magnetic field the magnetic energy is $\overline{E} = -Nk_B T_c \overline{M}^2/2\mu^2 N^2$ below the critical temperature, and it is 0 above the critical temperature. Hence the

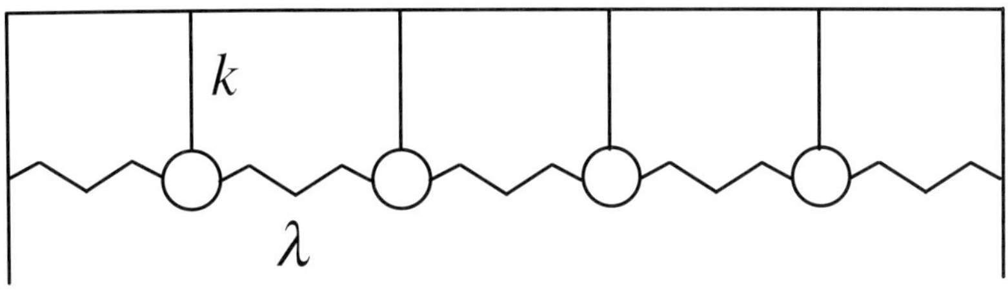

Figure 6.4: A linear crystal in which the atoms can move only in the transverse direction. The confinement of the atoms to the lattice points and their interactions with their nearest neighbours are represented by harmonic potentials of strength k and λ, respectively.

magnetic heat capacity vanishes above the critical temperature, and below the critical temperature it is given by

$$C = \frac{\partial \overline{E}}{\partial T} = Nk_{\mathrm{B}} \frac{\alpha^2 \overline{x}^2 (1 - \overline{x}^2)}{1 - \alpha(1 - \overline{x}^2)}, \tag{6.68}$$

where $\alpha = T_{\mathrm{c}}/T$ and $x = M/\mu N$. Approaching the critical temperature from below, $\alpha \sim 1 + \epsilon^2$, $\overline{x} \sim \pm\epsilon\sqrt{3}$, the heat capacity goes like $C \sim 3Nk_{\mathrm{B}}/2 + \mathcal{O}(\epsilon^2)$. This says that the critical temperature is marked by a discontinuity in the heat capacity, as is shown in Fig. 6.3.

6.3 One-Dimensional Harmonic Crystal

The simplest classical system with potential energy (as opposed to the purely kinetic ideal gas treated below) is perhaps the harmonic oscillator. Here is analysed a system of coupled oscillators, which may be considered as a model of a one-dimensional crystal (see Fig. 6.4) The system can be decoupled by normal mode analysis, and in this sense it may be considered ideal. The interesting feature of the analysis is the fact that the energy hypersurface may be mapped to a hypersphere in mode space, and it is confirmed that the general expression for the phase space probability density obtained in the preceding chapter is indeed uniform on the surface of this hypersphere, as symmetry arguments indicate it ought be.

6.3.1 Normal Modes

Let p_i be the momentum of atom i and let $q_i = r_i - r_{i0}$ be its displacement from its lattice position. The Hamiltonian for the N-atom, one-dimensional harmonic crystal is

$$H(\mathbf{p}, \mathbf{q}) = \frac{1}{2m} \sum_{i=1}^{N} p_i^2 + \frac{k}{2} \sum_{i=1}^{N} q_i^2 + \frac{\lambda}{2} \sum_{i=1}^{N-1} (q_i - q_{i+1})^2 + \frac{\lambda}{2}(q_1^2 + q_N^2), \tag{6.69}$$

where $\mathbf{p}$ and $\mathbf{q}$ are the N-dimensional vectors of the momenta and the displacements that specify a configuration of the crystal. The first term on the right is the kinetic energy (m is the mass of the atoms), the second term is the potential energy due to displacement from the lattice sites, the third term is the interaction energy between neighbouring atoms, and the final term represents the interaction of the terminal atoms with the confining walls. In order to simplify the analysis the strength of this last interaction has been taken to be the same as that between neighbouring atoms. Since $p_i = m\dot{q}_i$, differentiating the Hamiltonian yields for the acceleration

$$m\ddot{q}_i = -kq_i - \lambda(q_i - q_{i+1}) - \lambda(q_i - q_{i-1}), \;\; 2 \leq i \leq N-1. \tag{6.70}$$

This result may be extended to the terminal atoms by defining $q_0 = q_{N+1} = 0$. Alternatively, the result may be written in matrix form

$$\frac{m}{\lambda}\ddot{\mathbf{q}} = \underline{\underline{K}}\mathbf{q}, \tag{6.71}$$

where $\underline{\underline{K}}$ is a tridiagonal matrix with the nonzero off-diagonal elements equal to unity, and with diagonal elements equal to $K = -(k + 2\lambda)/\lambda$.

The eigenvalues of $\underline{\underline{K}}$ turn out to be the zeros of the Tchebyshev polynomials of the second kind,

$$\mu_n = K + 2\cos\frac{n\pi}{N+1}, \;\; 1 \leq n \leq N, \tag{6.72}$$

with corresponding eigenvector $\mathbf{u}_n$. The ith component of this eigenvector is proportional to the same polynomial of order $i-1$ evaluated at $\theta_n = n\pi/(N+1)$,

$$u_{in} = \sqrt{\frac{2}{\sqrt{N+1}}}\sin\frac{in\pi}{N+1}, \;\; 1 \leq i \leq N, \;\; 1 \leq n \leq N. \tag{6.73}$$

The eigenvectors are orthonormal

$$\mathbf{u}_n \cdot \mathbf{u}_m \equiv \sum_{i=1}^{N} u_{in}u_{im} = \delta_{nm}, \tag{6.74}$$

and the matrix formed from them is symmetric and orthogonal, and hence it is idempotent with determinant unity.

The eigenvectors represent the normal modes of the crystal, and any configuration may be expanded in terms of them,

$$\mathbf{q}(t) = \sum_{n=1}^{N} Q_n(t)\mathbf{u}_n, \;\; \mathbf{p}(t) = m\sum_{n=1}^{N} \dot{Q}_n(t)\mathbf{u}_n. \tag{6.75}$$

Substituting into Eq. (6.71) and equating coefficients yields

$$\frac{m}{\lambda}\ddot{Q}_n(t) = \mu_n Q_n(t), \tag{6.76}$$

with solution

$$Q_n(t) = \sqrt{\frac{2}{m\omega_n^2}} \left[a_n \cos\omega_n t + b_n \sin\omega_n t \right]. \tag{6.77}$$

Here the amplitudes of the normal modes are $a_n \in (-\infty, \infty)$ and $b_n \in (-\infty, \infty)$, and their frequencies are

$$\omega_n = \sqrt{-\lambda\mu_n/m}. \tag{6.78}$$

Note that the eigenvalues are negative, $\mu_n < 0$, and that the mode frequencies are positive, $\omega_n > 0$. Note also that although one must allow the amplitudes to take on all possible values, in practice larger values occur with negligible or 0 probability, according to the energy.

6.3.2 Energy Hypersurface

The energy is a constant of the motion, and direct substitution into the Hamiltonian yields after some manipulation

$$E = \sum_{n=1}^{N} \left[a_n^2 + b_n^2 \right]. \tag{6.79}$$

This equation represents the surface of a $2N$-dimensional hypersphere of radius $\sqrt{E}$. Because of the particularly simple form of the energy hypersurface in mode space, $\chi = (\mathbf{a}, \mathbf{b})$, one knows that the probability density in this space must be invariant with respect to rotations. That is, it is a function of the radius but not of the angular position.

The Hamiltonian and hence the energy hypersurface is a complicated function in phase space, but according to the results established in Ch. 5 the probability distribution of the isolated crystal is given by

$$\wp(\mathbf{p}, \mathbf{q}|E, N) = \frac{\delta(H(\mathbf{p}, \mathbf{q}) \quad E)\Delta_E}{h^N Z(E, N)}. \tag{6.80}$$

Here the constants h and Δ_E are of no consequence. No $N!$ term appears here because the oscillators are fixed to the lattice sites and are hence distinguishable. The probability distribution in mode space is

$$\wp(\mathbf{a}, \mathbf{b}|E, N) = \left| \frac{\partial(\mathbf{q}, \mathbf{p})}{\partial(\mathbf{a}, \mathbf{b})} \right| \wp(\mathbf{p}, \mathbf{q}|E, N). \tag{6.81}$$

The Jacobean of the transformation is

$$\begin{aligned}
\left| \frac{\partial(\mathbf{q}, \mathbf{p})}{\partial(\mathbf{a}, \mathbf{b})} \right| &= \left| \begin{array}{cc} \sqrt{2/m\omega_n^2}\, c_n u_{in} & \sqrt{2/m\omega_n^2}\, s_n u_{in} \\ -\sqrt{2m}\, s_n u_{in} & \sqrt{2m}\, c_n u_{in} \end{array} \right| \\
&= \left| \begin{array}{cc} \sqrt{2/m\omega_n^2} c_n u_{in} & 0 \\ 0 & \sqrt{2m}\, u_{in}/c_n \end{array} \right| \\
&= \prod_{n=1}^{N} \frac{2}{\omega_n}.
\end{aligned} \tag{6.82}$$

Here $s_n \equiv \sin\omega_n t$ and $c_n \equiv \cos\omega_n t$, and the fact that the $\mathbf{u}_n$ form an orthonormal basis has been used. Hence one has

$$\wp(\mathbf{a},\mathbf{b}|E,N) = \frac{1}{Z(E,N)}\frac{\Delta_E 2^N}{h^N\prod_n\omega_n}\delta\left(E-\sum_n[a_n^2+b_n^2]\right). \tag{6.83}$$

As anticipated, the probability density in mode space is a function of the radius of the hypersphere but not of the position on its surface.

Because the energy hypersurface is a hypersphere in mode space, it is simple to evaluate the partition function. One obtains

$$\begin{aligned} Z(E,N) &= \frac{\Delta_E 2^N}{h^N\prod_n\omega_n}\int_{-\infty}^{\infty}\mathrm{d}a^N\,\mathrm{d}b^N\,\delta\left(E-\sum_n[a_n^2+b_n^2]\right) \\ &= \frac{\Delta_E 2^N}{h^N\prod_n\omega_n}\frac{1}{2\sqrt{E}}S_{2N}(\sqrt{E}) \\ &= \frac{\Delta_E 2^N}{h^N\prod_n\omega_n}\frac{\pi^N E^{N-1}}{(N-1)!}. \end{aligned} \tag{6.84}$$

The determinant of the tridiagonal matrix satisfies the recursion relation $K_N = KK_{N-1} - K_{N-2}$, and hence it is just the Tchebyshev polynomial of the second kind, $|\underline{\underline{K}}| = U_N(K/2)$. Accordingly, the product of the mode frequencies is

$$\begin{aligned} \prod_{n=1}^{N}\omega_n &= \prod_{n=1}^{N}\sqrt{\frac{-\lambda}{m}\mu_n} \\ &= \left(\frac{\lambda}{m}\right)^{N/2}|-\underline{\underline{K}}|^{1/2} \\ &= \nu^N U_N(-K/2)^{1/2}, \end{aligned} \tag{6.85}$$

where $\nu \equiv \sqrt{\lambda/m}$. In the thermodynamic limit, $N \to \infty$, this may be rewritten more explicitly as

$$\left[\prod_{n=1}^{N}\omega_n\right]^{2/N} = \frac{2\lambda+k}{m}\prod_{j=1}^{\infty}\exp\left[\frac{-(2j-1)!}{2^{2j}(j!)^2}\left(\frac{2}{K}\right)^{2j}\right]. \tag{6.86}$$

The product on the right-hand side is obviously rapidly converging for $K \ll -2$; the case $K = -2$ is treated below. The entropy of the isolated crystal is

$$\begin{aligned} S(E,N)/k_\mathrm{B} &= \ln Z(E,N) \\ &= N\ln\frac{2\pi E}{h\nu N} + N - \frac{1}{2}\ln U_N(-K/2), \end{aligned} \tag{6.87}$$

where Stirling's approximation has been used. Differentiating with respect to energy gives the temperature, $1/T = Nk_\mathrm{B}/E$, which is the same as that of the ideal gas (see below). Accordingly the heat capacity for this one-dimensional harmonic crystal is Nk_B, which agrees with the high-temperature limit of the quantum oscillator solid treated by Einstein and by Debye. Differentiating with respect to number gives the chemical potential,

$$\begin{aligned}\mu &= -\frac{E}{N}\frac{\partial S(E,N)/k_{\mathrm{B}}}{\partial N} \\ &= -\ln\frac{2\pi E}{h\nu N}+\frac{E}{2N}\frac{U_N(-(K-1)/2)}{U_N(-K/2)}.\end{aligned} \tag{6.88}$$

This particular example models an incompressible crystal, and so the entropy does not depend upon volume. In a model in which the spring constants depended upon the volume, the pressure would follow from the volume derivative of the entropy, as usual.

Uncoupled Pendula

It is interesting to compare this result to that of independent atoms, $\lambda = 0$. In this case each atom obeys the equation of motion $m\ddot{q}_i = -kq_i$, with solution

$$q_i(t) = \sqrt{\frac{2}{m\omega^2}}\left[a_i\cos\omega t + b_i\sin\omega t\right], \tag{6.89}$$

where the resonance frequency is $\omega = \sqrt{k/m}$. As above the Hamiltonian may be written

$$\begin{aligned}H(\mathbf{q},\mathbf{p}) &= \frac{1}{2m}\mathbf{p}\cdot\mathbf{p}+\frac{k}{2}\mathbf{q}\cdot\mathbf{q} \\ &= \mathbf{a}\cdot\mathbf{a}+\mathbf{b}\cdot\mathbf{b}.\end{aligned} \tag{6.90}$$

The phase space probability is $\wp(\mathbf{q},\mathbf{p}|E,N) = \Delta_E\delta(E-H(\mathbf{q},\mathbf{p}))/h^N Z(E,N)$, where the partition function is

$$\begin{aligned}Z(E,N) &= \frac{\Delta_E}{h^N}\int_{-\infty}^{\infty}\mathrm{d}\mathbf{q}\,\mathrm{d}\mathbf{p}\,\delta(E-H(\mathbf{q},\mathbf{p})) \\ &= \frac{\Delta_E}{h^N}\left|\frac{\partial(q_i,p_i)}{\partial(a_i,b_i)}\right|^N\int_{-\infty}^{\infty}\mathrm{d}\mathbf{a}\,\mathrm{d}\mathbf{b}\,\delta\Big(E-\sum[a_i^2+b_i^2]\Big) \\ &= \frac{\Delta_E}{h^N}\frac{2^N}{\omega^N}\frac{1}{2\sqrt{E}}S_{2N}(\sqrt{E}) \\ &= \frac{\Delta_E 2^N\pi^N E^{N-1}}{h^N\omega^N(N-1)!}.\end{aligned} \tag{6.91}$$

This is clearly the limiting form of the coupled pendula result, since as $\lambda \to 0$, $\mu_n \to -k/\lambda$, and $\omega_n \to \omega$.

This is the leading term in an expansion for weakly coupled pendula. With $\lambda/k \to 0$, one has $K \to -k/\lambda$ and

$$\prod_{n=1}^{N}\omega_n^2 = \left(\frac{2\lambda+k}{m}\right)^N\prod_{n=1}^{N}\left[1+\frac{2}{K}\cos\frac{n\pi}{N+1}\right]$$

$$\begin{aligned}
&\sim \left(\frac{k}{m}\right)^N \left[1 - \frac{2\lambda}{k}\sum_{n=1}^{N}\cos\frac{n\pi}{N+1}\right. \\
&\qquad \left. + \frac{4\lambda^2}{k^2}\sum_{m>n=1}^{N}\cos\frac{n\pi}{N+1}\cos\frac{m\pi}{N+1} + \ldots\right] \\
&= \left(\frac{k}{m}\right)^N \left[1 - \frac{(N-1)\lambda^2}{k^2} + \mathcal{O}(N\lambda^4/k^4)\right]. \qquad (6.92)
\end{aligned}$$

The leading correction term to the entropy is $\mathcal{O}(\ln N\lambda/k)$. This result is valid only when the uncoupling limit is taken before the thermodynamic limit.

Fully Coupled Pendula

The opposite extreme is for fully coupled pendula, $\lambda \to \infty$, $K \to -2$. Since $U_N(1) = N + 1$, one has for the partition function in this case

$$Z(E, N) = \frac{\Delta_E 2^N \pi^N E^{N-1}}{h^N \nu^N \sqrt{N+1}(N-1)!}. \qquad (6.93)$$

In the thermodynamic limit the entropy is

$$S(E, N) = Nk_{\mathrm{B}} \ln \frac{2\pi E}{h\nu N} + Nk_{\mathrm{B}}. \qquad (6.94)$$

For both the uncoupled and fully coupled systems, the temperature–energy relationship remains that of the ideal gas. Colloquially, one says that there is a half $k_{\mathrm{B}}T$ of energy for each quadratic term in the Hamiltonian.

6.3.3 Thermal Equilibrium

In the event that the crystal is in contact with a heat reservoir, the phase space probability is the usual

$$\wp(\mathbf{q}, \mathbf{p}|N, T) = \frac{e^{-\beta H(\mathbf{q},\mathbf{p})}}{h^N Z(N, T)}, \qquad (6.95)$$

where $\beta \equiv 1/k_{\mathrm{B}}T$. The partition function is

$$\begin{aligned}
Z(N, T) &= \frac{1}{h^N}\int_{-\infty}^{\infty} \mathrm{d}\mathbf{q}\,\mathrm{d}\mathbf{p}\, e^{-\beta H(\mathbf{q},\mathbf{p})} \\
&= \frac{1}{h^N}\left|\frac{\partial(\mathbf{q}, \mathbf{p})}{\partial(\mathbf{a}, \mathbf{b})}\right| \int_{-\infty}^{\infty} \mathrm{d}\mathbf{a}\,\mathrm{d}\mathbf{b}\, \exp\left[-\beta\sum_i [a_i^2 + b_i^2]\right] \\
&= \frac{1}{h^N}\prod_{i=1}^{N}\left[\frac{2}{\omega_n}\right](\pi k_{\mathrm{B}}T)^N \\
&= \frac{(2\pi k_{\mathrm{B}}T)^N}{h^N \nu^N U_N(-K/2)^{1/2}}. \qquad (6.96)
\end{aligned}$$

6.4 Classical Ideal Gas

In general the Hamiltonian of a system consisting of classical molecules may be written as the sum of kinetic and potential energies,

$$\mathcal{H}(\mathbf{p}^N, \mathbf{q}^N) = U(\mathbf{q}^N) + \sum_{i=1}^{N} p_i^2/2m. \tag{6.97}$$

Here m is the molecular mass, and it has been assumed that the molecules are spherically symmetric so that there are only translational contributions to the kinetic energy. In an ideal gas there are no interactions between the molecules, so that

$$U(\mathbf{q}^N) = 0. \tag{6.98}$$

(The confining effects of the system walls are incorporated by limiting the configurational integrals to the volume V.)

6.4.1 Isolated System

The entropy of an isolated system is $S(E, N, V) = k_{\mathrm{B}} \ln Z(E, N, V)$, with the partition function being

$$\begin{aligned} Z^{\mathrm{id}}(E, N, V) &= \frac{\Delta_E}{N! h^{3N}} \int \mathrm{d}\mathbf{p}^N \, \mathrm{d}\mathbf{q}^N \, \delta(E - \mathcal{H}(\mathbf{p}^N, d\mathbf{q})) \\ &= \frac{\Delta_E V^N}{N! h^{3N}} \int \mathrm{d}\mathbf{p}^N \, \delta\left(E - \frac{1}{2m} \sum_{i=1}^{N} \mathbf{p}_i \cdot \mathbf{p}_i\right) \\ &= \frac{\Delta_E V^N (2m)^{3N/2}}{N! h^{3N} 2\sqrt{E}} \int \mathrm{d}x^{3N} \, \delta\left(\sqrt{E} - \sqrt{\sum_{i=1}^{3N} x_i^2}\right). \end{aligned} \tag{6.99}$$

The integral is just the surface area of a $3N$-dimensional hypersphere of radius $\sqrt{E}$, and since

$$S_n(r) \equiv \int \mathrm{d}x^n \, \delta(r - \sqrt{\mathbf{x} \cdot \mathbf{x}}) = \frac{2\pi^{n/2} r^{n-1}}{\Gamma(n/2)}, \tag{6.100}$$

one has

$$Z^{\mathrm{id}}(E, N, V) = \frac{\Delta_E V^N (2\pi m E/h^2)^{3N/2}}{N! \Gamma(3N/2) E}. \tag{6.101}$$

Using Stirling's asymptote, $\ln \Gamma(n) \sim n \ln n - n$, in the thermodynamic limit the entropy is

$$S^{\mathrm{id}}(E, N, V)/k_{\mathrm{B}} = N \ln \left[V (4\pi m E/3h^2)^{3/2}\right] - \frac{5N}{2} \ln N + \frac{5N}{2}, \tag{6.102}$$

where terms of order $\ln E$ and $\ln N$ have been neglected. Unless otherwise stated, all the following results will be restricted to the thermodynamic limit.

The temperature of the ideal gas is

$$\frac{1}{T^{\mathrm{id}}} = \frac{\partial S^{\mathrm{id}}(E,N,V)}{\partial E} = \frac{3Nk_{\mathrm{B}}}{2E}, \tag{6.103}$$

which is more commonly written $E^{\mathrm{id}} = 3Nk_{\mathrm{B}}T/2$. (For a finite-sized system this is $E^{\mathrm{id}} = (3N-2)k_{\mathrm{B}}T/2$.) Likewise the pressure is

$$\frac{P^{\mathrm{id}}}{T^{\mathrm{id}}} = \frac{\partial S^{\mathrm{id}}(E,N,V)}{\partial V} = \frac{Nk_{\mathrm{B}}}{V}, \tag{6.104}$$

which also holds for finite-sized systems, and the chemical potential is

$$\frac{-\mu^{\mathrm{id}}}{T^{\mathrm{id}}} = \frac{\partial S^{\mathrm{id}}(E,N,V)}{\partial N} = k_{\mathrm{B}} \ln\left[V(4\pi m E/3h^2)^{3/2}\right] - \frac{5k_{\mathrm{B}}}{2}\ln N. \tag{6.105}$$

In terms of the de Broglie thermal wavelength,

$$\Lambda \equiv \sqrt{\frac{2\pi\hbar^2}{mk_{\mathrm{B}}T}}, \tag{6.106}$$

where $\hbar$ is Planck's constant divided by 2π, this may be written in the more familiar form $\mu^{\mathrm{id}} = k_{\mathrm{B}}T\ln[N\Lambda^3/V]$.

The second derivatives are

$$\frac{\partial^2 S^{\mathrm{id}}(E,N,V)}{\partial E^2} = \frac{-3Nk_{\mathrm{B}}}{2E^2}, \tag{6.107}$$

$$\frac{\partial^2 S^{\mathrm{id}}(E,N,V)}{\partial V^2} = \frac{-Nk_{\mathrm{B}}}{V^2}, \tag{6.108}$$

and

$$\frac{\partial^2 S^{\mathrm{id}}(E,N,V)}{\partial N^2} = \frac{-5k_{\mathrm{B}}}{2N}. \tag{6.109}$$

These are all negative, which confirms the concavity of the entropy.

6.4.2 Isothermal System

A subsystem able to exchange energy with a reservoir of temperature T has constrained thermodynamic potential $F(E|N,V,T) = E - TS(E,N,V)$ and energy probability distribution

$$\wp(E|N,V,T) = \frac{e^{-\beta E} e^{S(E,N,V)/k_{\mathrm{B}}}}{\Delta_E Z(N,V,T)}, \tag{6.110}$$

where here and throughout $\beta \equiv 1/k_{\mathrm{B}}T$. For the present case of an ideal gas, the entropy of the subsystem $S^{\mathrm{id}}(E,N,V)$ was derived in the preceding section. The phase space probability distribution is

$$\wp^{\mathrm{id}}(\mathbf{p}^N, \mathbf{q}^N|N,V,T) = \frac{e^{-\beta\sum p_i^2/2m}}{N!h^{3N}Z^{\mathrm{id}}(N,V,T)}, \tag{6.111}$$

where the partition function represents the total weight of distinct microstates,

$$\begin{aligned} Z^{\mathrm{id}}(N,V,T) &= \frac{1}{h^{3N}N!}\int_V \mathrm{d}\mathbf{q}^N \int \mathrm{d}\mathbf{p}^N\, e^{-\beta\sum p_i^2/2m} \\ &= \frac{V^N}{h^{3N}N!}\left[\int_{-\infty}^{\infty}\mathrm{d}p\, e^{-\beta p^2/2m}\right]^{3N} \\ &= \frac{V^N\Lambda^{-3N}}{N!}. \end{aligned} \tag{6.112}$$

This result holds for finite systems as well as in the thermodynamic limit. In general the integration over the momenta is the product of $3N$ independent Gaussian integrals (or at worst the exponent is a general quadratic form), and these may be trivially evaluated as above. This is true whether or not the interaction potential is 0. When the latter does vanish, as for the present ideal gas, the configuration integral gives N products of volume; in the more general case when the interaction potential is nonzero, the configuration integral is highly nontrivial. The unconstrained total entropy is of course $S(N,V,T) = k_{\mathrm{B}}\ln Z(N,V,T)$.

The appropriate first derivatives of the partition function yield the average energy, pressure, and chemical potential. One may also explicitly confirm the temperature form of the general result established in Ch. 4, namely that the average temperature equals the reservoir temperature. One has

$$\begin{aligned} \langle 1/T_{\mathrm{s}}\rangle_{\mathrm{id}} &= \frac{\Delta_E^{-1}}{Z^{\mathrm{id}}(N,V,T)}\int_0^\infty \mathrm{d}E\,\frac{\partial S(E,N,V)}{\partial E}e^{S(E,N,V)/k_{\mathrm{B}}}e^{-E/k_{\mathrm{B}}T} \\ &= \frac{k_{\mathrm{B}}\Delta_E^{-1}}{Z^{\mathrm{id}}(N,V,T)}\left[e^{S(E,N,V)/k_{\mathrm{B}}}e^{-E/k_{\mathrm{B}}T}\Big|_0^\infty \right. \\ &\quad \left. + \frac{1}{k_{\mathrm{B}}T}\int_0^\infty \mathrm{d}E\, e^{S(E,N,V)/k_{\mathrm{B}}}e^{-E/k_{\mathrm{B}}T}\right] \\ &= \frac{1}{T}, \end{aligned} \tag{6.113}$$

where an integration by parts has been used, the integrated portion explicitly vanishing at both limits. This result holds for finite-sized systems as well as in the thermodynamic limit.

The equilibrium energy satisfies $T(\overline{E},N,V) = T$, which from the preceding section yields

$$\overline{E}^{\mathrm{id}} = 3Nk_{\mathrm{B}}T/2. \tag{6.114}$$

Inserting this into the expression for the constrained thermodynamic potential, one obtains the Helmholtz free energy for the ideal gas,

$$\overline{F}^{\mathrm{id}}(N,V,T) = k_{\mathrm{B}}T\left\{N\ln\left[N\Lambda^3/V\right] - N\right\}. \tag{6.115}$$

Differentiating with respect to number and volume yields expressions for the equilibrium chemical potential and pressure equal to those obtained by differentiating the partition function (since in the thermodynamic limit the Helmholtz free energy is essentially the logarithm of the partition function, $\overline{F}(N,V,T) = -k_{\rm B}T \ln Z(N,V,T)$, as the results in this section explicitly confirm).

The heat capacity at constant volume is

$$C_V^{\rm id} \equiv \frac{\partial \overline{E}^{\rm id}}{\partial T} = \frac{3Nk_{\rm B}}{2}, \tag{6.116}$$

which is independent of temperature. The equilibrium energy may be written as a derivative of the Helmholtz free energy, $\overline{E} = \partial \beta \overline{F}/\partial \beta$, and so the isothermal heat capacity may equivalently be written $C_V = -(\beta/T)\partial^2 \beta \overline{F}/\partial \beta^2$. The fluctuation in energy is given by

$$\left\langle \Delta(E)^2 \right\rangle_{\rm id} \equiv \left\langle (E - \langle E \rangle_{\rm id})^2 \right\rangle_{\rm id} = \frac{\partial^2 \ln Z_T^{\rm id}(N,V,T)}{\partial \beta^2} = \frac{3N}{2} k_{\rm B}^2 T^2. \tag{6.117}$$

Hence the root mean square energy fluctuation is of order one over the square root of the system size, which is negligible in the thermodynamic limit. From the equality of the Helmholtz free energy and the logarithm of the partition function, the fluctuations in energy are related to the heat capacity by

$$\left\langle \Delta(E)^2 \right\rangle_{\rm id} = -\partial^2 \beta \overline{F}/\partial \beta^2 = T^2 C_V / k_{\rm B}. \tag{6.118}$$

The inverse of the isothermal compressibility is

$$(\chi_T^{\rm id})^{-1} \equiv -V \frac{\partial \overline{p}_T^{\rm id}}{\partial V} = \frac{Nk_{\rm B}T}{V}, \tag{6.119}$$

where $\overline{p}^{\rm id} = 2\overline{E}/3V = Nk_{\rm B}T/V$. This shows that the compressibility of an ideal gas decreases with increasing temperature and density. On the other hand the fluctuations in pressure vanish,

$$\begin{aligned} \left\langle \Delta(p/k_{\rm B}T_s)^2 \right\rangle_{\rm id} &= \frac{\partial^2 \ln Z_T^{\rm id}(N,V,T)}{\partial V^2} - \left\langle \frac{\partial^2 S^{\rm id}(N,V,T)/k_{\rm B}}{\partial V^2} \right\rangle_{\rm id} \\ &= \frac{-N}{V^2} - \left\langle \frac{-N}{V^2} \right\rangle_{\rm id} = 0. \end{aligned} \tag{6.120}$$

In this case, the pressure fluctuations are not proportional solely to the second derivative of the partition function; hence the isothermal compressibility is proportional to the latter but not to the former.

The fluctuations in the chemical potential also vanish (at least to order unity in the thermodynamic limit),

$$\begin{aligned} \left\langle \Delta(\mu/k_{\rm B}T_s)^2 \right\rangle_{\rm id} &= \frac{\partial^2 \ln Z^{\rm id}(N,V,T)}{\partial N^2} - \left\langle \frac{\partial^2 S^{\rm id}(N,V,T)/k_{\rm B}}{\partial N^2} \right\rangle_{\rm id} \\ &= \frac{-1}{N+1} - \left\langle \frac{-5}{2N} \right\rangle_{\rm id} = \frac{3}{2N}. \end{aligned} \tag{6.121}$$

6.4.3 Open System

A subsystem able to exchange energy and particles with a reservoir of temperature T and chemical potential μ has constrained thermodynamic potential $\Omega(E, N|\mu, V, T) = E - TS(E, N, V) - \mu N$ and energy and number probability distribution

$$\wp(E, N|\mu, V, T) = \frac{e^{-\beta E} e^{\beta \mu N} e^{S(E,N,V)/k_{\rm B}}}{\Delta_E Z(\mu, V, T)}. \tag{6.122}$$

The phase space probability distribution for the ideal gas is

$$\wp^{\rm id}({\bf p}^N, {\bf q}^N, N|\mu, V, T) = \frac{e^{\beta\mu N} e^{-\beta \sum p_i^2/2m}}{N! h^{3N} Z^{\rm id}(\mu, V, T)}. \tag{6.123}$$

In these equations the partition function is

$$\begin{aligned} Z^{\rm id}(\mu, V, T) &= \sum_{N=0}^{\infty} \frac{e^{\beta\mu N}}{N! h^{3N}} \int_V {\rm d}{\bf q}^N \int {\rm d}{\bf p}^N \, e^{-\beta \sum p_i^2/2m} \\ &= \sum_{N=0}^{\infty} \frac{e^{\beta\mu N} V^N \Lambda^{-3N}}{N!} \\ &= \exp\left[e^{\beta\mu} V \Lambda^{-3}\right]. \end{aligned} \tag{6.124}$$

This result holds for finite systems as well as in the thermodynamic limit. The partition function for grand systems is often denoted by $\Xi(\mu, V, T)$. The unconstrained total entropy is of course $S(\mu, V, T) = k_{\rm B} \ln Z(\mu, V, T)$. The appropriate first derivatives of the partition function yield the average energy, pressure, and number.

The equilibrium energy and number satisfy temperature and chemical potential equality, $T(\overline{E}, \overline{N}, V) = T$ and $\mu(\overline{E}, \overline{N}, V) = \mu$. From the results for the isolated ideal gas these yield the average energy, $\overline{E}^{\rm id} = 3\overline{N} k_{\rm B} T/2$, and the average number of particles

$$\overline{N}^{\rm id} = V \Lambda^{-3} e^{\beta\mu}. \tag{6.125}$$

The *activity*,

$$z \equiv \Lambda^{-3} e^{\beta\mu}, \tag{6.126}$$

occurs in this equation and elsewhere. One sees from this that for the ideal gas the density equals the activity. In view of these results the grand potential for the ideal gas is

$$\begin{aligned} \overline{\Omega}^{\rm id}(\mu, V, T) &= \overline{F}^{\rm id}(\overline{N}, V, T) - \mu \overline{N}^{\rm id} \\ &= -\overline{N}^{\rm id} k_{\rm B} T. \end{aligned} \tag{6.127}$$

Differentiating with respect to volume yields the equilibrium pressure, $\overline{p}^{\rm id} = \overline{N} k_{\rm B} T/V$, and one sees that $\overline{\Omega} = -\overline{p} V$, which is the result derived from extensivity arguments.

The fluctuations in number are

$$\left\langle \Delta(N)^2 \right\rangle_{\rm id} = \left(\frac{\partial^2 \ln Z^{\rm id}}{\partial (\beta\mu)^2} \right)_{V,T} = \overline{N}^{\rm id}, \tag{6.128}$$

and those in energy are

$$\left\langle \Delta(E)^2 \right\rangle_{\rm id} = \left(\frac{\partial^2 \ln Z^{\rm id}}{\partial \beta^2} \right)_{V,\mu/T} = \frac{15\overline{N}^{\rm id}}{4\beta^2}. \tag{6.129}$$

One also has the result that

$$\begin{aligned} \left\langle \Delta(E-\mu N)^2 \right\rangle_{\rm id} &= \left(\frac{\partial^2 \ln Z^{\rm id}}{\partial \beta^2} \right)_{V,\mu} \\ &= \overline{N}^{\rm id} \beta^{-2} \left[(\beta\mu)^2 - 3\beta\mu + \frac{15}{4} \right]. \end{aligned} \tag{6.130}$$

Energy Fluctuation

While average and equilibrium values are the same for equivalent systems in contact with different reservoirs, the values of the fluctuations vary. A trivial example is the number fluctuations, which are evidently 0 in a closed system but which equally obviously do not vanish in the present open system. A less trivial example is provided by the energy fluctuations, which in this open system are clearly greater than those in the equivalent isothermal system. The present result for the ideal gas is a manifestation of the general rule,

$$\left\langle \Delta(E)^2 \right\rangle_{\mu,V,T} = \left\langle \Delta(E)^2 \right\rangle_{N,V,T} + \left\langle \Delta(N)^2 \right\rangle_{\mu,V,T} \left(\frac{\partial \left\langle E \right\rangle_{N,V,T}}{\partial N} \right)^2_{V,T}. \tag{6.131}$$

This result was derived in Section 4.4.6. It says that the energy fluctuation in an open system equals that in the equivalent isothermal system plus a term proportional to the fluctuation in particle number. The proportionality constant is the square of the rate of change of the equilibrium energy with particle number. Equivalent here means $N = \overline{N}(\mu, V, T)$, which is the same as $\mu = \overline{\mu}(N, V, T)$.

6.4.4 Isobaric System

A subsystem able to exchange energy and volume with a reservoir of temperature T and pressure p has constrained thermodynamic potential $G(E, V|N, p, T) = E - TS(E, N, V) + pV$. Given thermal equilibrium this becomes $G(V|N, p, T) = \overline{F}(N, V, T) + pV$, and the volume probability distribution for the ideal gas is

$$\wp^{\rm id}(V|N, p, T) = \frac{e^{-\beta p V} V^N \Lambda^{-3N}}{\Delta_V Z^{\rm id}(N, p, T)}. \tag{6.132}$$

The phase space probability distribution is

$$\wp^{\rm id}(\mathbf{p}^N, \mathbf{q}^N, V|N, p, T) = \frac{e^{-\beta p V} e^{-\beta \sum p_i^2/2m}}{N! h^{3N} \Delta_V Z^{\rm id}(N, p, T)}. \tag{6.133}$$

In these equations the partition function is

$$\begin{aligned} Z^{\mathrm{id}}(N,p,T) &= \int_0^\infty \mathrm{d}V \frac{e^{-\beta pV}}{\Delta_V h^{3N} N!} \int_V \mathrm{d}\mathbf{q}^N \int \mathrm{d}\mathbf{p}^N\, e^{-\beta \sum p_i^2/2m} \\ &= \int_0^\infty \mathrm{d}V \frac{e^{-\beta pV} V^N \Lambda^{-3N}}{\Delta_V N!} \\ &= \Delta_V^{-1} \Lambda^{-3N} (\beta p)^{-N-1}. \end{aligned} \tag{6.134}$$

This result holds for finite systems as well as in the thermodynamic limit.

The equilibrium energy and volume satisfy $T(\overline{E}, N, \overline{V}) = T$ and $p(\overline{E}, N, \overline{V}) = p$. From the results for the isolated ideal gas these yield

$$\overline{V}^{\mathrm{id}} = N/\beta p, \tag{6.135}$$

and $\overline{E}^{\mathrm{id}} = 3Nk_{\mathrm{B}}T/2$. Consequently the Gibbs free energy for the ideal gas is

$$\begin{aligned} \overline{G}^{\mathrm{id}}(N,p,T) &= \overline{F}^{\mathrm{id}}(N,\overline{V},T) + p\overline{V}^{\mathrm{id}} \\ &= Nk_{\mathrm{B}}T \ln[\beta p \Lambda^3]. \end{aligned} \tag{6.136}$$

The argument of the logarithm is just $N\Lambda^3/\overline{V}^{\mathrm{id}}$, and since the chemical potential of the ideal gas was shown above to be $\overline{\mu}^{\mathrm{id}} = k_{\mathrm{B}}T \ln[N\Lambda^3/\overline{V}]$ (as may be confirmed by differentiating the above), one sees that $\overline{G}^{\mathrm{id}} = \overline{\mu}^{\mathrm{id}} N$, as derived in general from extensivity arguments.

The volume fluctuation is

$$\left\langle \Delta(V)^2 \right\rangle_{\mathrm{id}} = \left(\frac{\partial^2 \ln Z^{\mathrm{id}}(N,p,T)}{\partial(\beta p)^2} \right)_{N,T} = \frac{N+1}{\beta p} = \overline{V}^{\mathrm{id}}(N,p,T), \tag{6.137}$$

the final equality holding in the thermodynamic limit. The energy fluctuation is

$$\left\langle \Delta(E)^2 \right\rangle_{\mathrm{id}} = \left(\frac{\partial^2 \ln Z^{\mathrm{id}}(N,p,T)}{\partial \beta^2} \right)_{V,\beta p} = \frac{3N}{2\beta^2}. \tag{6.138}$$

This is the same as the energy fluctuation of an isothermal system. Using the same arguments as in Section 4.4.6, one has in general that

$$\left\langle \Delta(E)^2 \right\rangle_{N,p,T} = \left\langle \Delta(E)^2 \right\rangle_{N,V,T} + \left\langle \Delta(V)^2 \right\rangle_{N,p,T} \left(\frac{\partial \langle E \rangle_{N,V,T}}{\partial V} \right)_{N,T}^2, \tag{6.139}$$

which reduces to the above since an ideal gas has $\partial \overline{E}(N,V,T)^{\mathrm{id}}/\partial V = 0$. One also has the result that in the thermodynamic limit

$$\left\langle \Delta(E+pV)^2 \right\rangle_{N,p,T;\mathrm{id}} = \left(\frac{\partial^2 \ln Z^{\mathrm{id}}(N,p,T)}{\partial \beta^2} \right)_{N,p} = \frac{5N}{2\beta^2}. \tag{6.140}$$

This may be compared to the heat capacity at constant pressure,

$$\begin{aligned} C_p &\equiv \left(\frac{\partial(\overline{E}(N,p,T)+p\overline{V}(N,p,T))}{\partial T}\right)_{N,p} \\ &= -k_{\mathrm{B}}\beta^2\left(\frac{\partial^2\beta\overline{G}(N,P,T)}{\partial\beta^2}\right)_{N,p} \\ &= \frac{5Nk_{\mathrm{B}}}{2}. \end{aligned} \tag{6.141}$$

6.4.5 Isenthalpic System

An insulated subsystem able to exchange volume with a reservoir of pressure p has constant enthalpy $H = E + pV$. The reservoir entropy is constant, and consequently the constrained thermodynamic potential is just $\Phi(V|N,H,p;T) = -TS(E,N,V)$, where $E \equiv H - pV$, and the volume probability distribution for the ideal gas is

$$\begin{aligned} \wp^{\mathrm{id}}(V|N,H,p) &= \frac{e^{S^{\mathrm{id}}(E,N,V)/k_{\mathrm{B}}}}{\Delta_V Z^{\mathrm{id}}(N,H,p)}, \quad E = H - pV \\ &= \frac{\Delta_E V^N (2\pi m(H-pV)/h^2)^{3N/2}}{N!\Gamma(3N/2)(H-pV)\Delta_V Z^{\mathrm{id}}(N,H,p)}. \end{aligned} \tag{6.142}$$

The partition function is

$$\begin{aligned} Z^{\mathrm{id}}(N,H,p) &= \frac{\Delta_E(2\pi m/h^2)^{3N/2}}{N!\Gamma(3N/2)\Delta_V}\int_0^{H/p}\mathrm{d}V\, V^N(H-pV)^{3N/2-1} \\ &= \frac{\Delta_E N(2\pi mH/h^2)^{3N/2}H^N}{\Delta_V p^{N+1}\Gamma(5N/2+1)}. \end{aligned} \tag{6.143}$$

Note that because the energy of the subsystem must be nonnegative, the maximum volume of the subsystem at constant enthalpy is H/p. In the thermodynamic limit this reduces to

$$Z^{\mathrm{id}}(N,H,p) = \frac{(2\pi mH/h^2)^{3N/2}H^N}{p^N\Gamma(5N/2)}. \tag{6.144}$$

The equilibrium volume satisfies $p(\overline{E},N,\overline{V}) = p$, where $\overline{E} = H - p\overline{V}$. This corresponds to the peak of the probability distribution, and logarithmic differentiation yields

$$\overline{V}^{\mathrm{id}}(N,H,p) = \frac{NH/p}{5N/2-1} = \frac{2H}{5p}, \tag{6.145}$$

the last equality holding in the thermodynamic limit. With this the equilibrium energy is $\overline{E}^{\mathrm{id}}(N,H,p) = 3H/5$. Consequently the thermodynamic potential for the isenthalpic ideal gas is

$$\begin{aligned}
&\overline{\Phi}^{\mathrm{id}}(N,H,p;T)\\
&= -TS^{\mathrm{id}}(\overline{E}^{\mathrm{id}},N,\overline{V}^{\mathrm{id}})\\
&= -Nk_{\mathrm{B}}T\left\{\ln\left[\frac{2H}{5p}\left(\frac{4\pi mH}{5h^2}\right)^{3/2}\right]-\frac{5}{2}\ln N+\frac{5}{2}\right\}. \qquad (6.146)
\end{aligned}$$

Because energy and volume are coupled, the averages and fluctuations of the isenthalpic system are nonstandard. For example, the general expression for the pressure derivative of the thermodynamic potential yields

$$\frac{\partial\overline{\Phi}^{\mathrm{id}}(N,H,p;T)}{\partial p}=\frac{T}{\overline{T}^{\mathrm{id}}(N,H,p)}\overline{V}^{\mathrm{id}}(N,H,p), \qquad (6.147)$$

whereas the pressure derivative of the logarithm of the partition function is

$$\begin{aligned}
-k_{\mathrm{B}}T\frac{\partial Z^{\mathrm{id}}(N,H,p)}{\partial p} &= \frac{3Nk_{\mathrm{B}}T}{2}\left\langle\frac{V}{H-pV}\right\rangle_{\mathrm{id}}\\
&= \frac{3Nk_{\mathrm{B}}T}{2}\frac{\langle V\rangle_{\mathrm{id}}}{\langle E\rangle_{\mathrm{id}}}, \qquad (6.148)
\end{aligned}$$

the final equality holding when fluctuations are negligible. This is so in the thermodynamic limit, in which case the two derivatives may be recognised as equivalent, with average values equaling equilibrium values.

An example of a fluctuation is given by the second derivative with respect to enthalpy. It is straightforward to show that

$$\frac{\partial^2\ln Z^{\mathrm{id}}(N,H,p)}{\partial H^2}=\frac{9N^2}{4}\left\langle\Delta(E^{-1})^2\right\rangle_{\mathrm{id}}. \qquad (6.149)$$

Similarly

$$\frac{\partial^2\ln Z^{\mathrm{id}}(N,H,p)}{\partial p^2}=\frac{9N^2}{4}\left\langle\Delta(V/E)^2\right\rangle_{\mathrm{id}}. \qquad (6.150)$$

In the thermodynamic limit, both sides of the first equation are of order N^{-1}, and both sides of the second equation are of order N.

Interestingly enough, integrating the partition function with respect to enthalpy gives the average energy,

$$\frac{1}{Z^{\mathrm{id}}(N,H,p)}\int_0^H \mathrm{d}H'\,Z^{\mathrm{id}}(N,H',p)=\frac{2}{3N}\langle E\rangle_{\mathrm{id}}, \qquad (6.151)$$

which gives $\langle E\rangle_{\mathrm{id}}=3NH/(5N+2)$. Integrating twice gives

$$\begin{aligned}
&\frac{1}{Z^{\mathrm{id}}(N,H,p)}\int_0^H \mathrm{d}H'\int_0^{H'}\mathrm{d}H''\,Z^{\mathrm{id}}(N,H'',p)\\
&= \frac{1}{Z^{\mathrm{id}}(N,H,p)}\int_0^H \mathrm{d}H'\,(H-H')Z^{\mathrm{id}}(N,H',p)\\
&= \frac{2}{3N}\frac{2}{3N+2}\left\langle E^2\right\rangle_{\mathrm{id}}, \qquad (6.152)
\end{aligned}$$

which gives $\langle E^2 \rangle_{\rm id} = H^2 3N(3N+2)/((5N+2)(5N+4))$. It follows that the fluctuation in energy for an isenthalpic ideal gas is in the thermodynamic limit

$$\left\langle \Delta(E)^2 \right\rangle_{\rm id} = H^2 12/125N. \tag{6.153}$$

Using the fact that the fluctuations in enthalpy vanish, $\left\langle \Delta(H)^2 \right\rangle_{\rm id} = 0$, which is the same as $\left\langle \Delta(E+pV)^2 \right\rangle_{\rm id} = 0$, one can show that the isenthalpic fluctuations in volume are in general given by $\left\langle \Delta(V)^2 \right\rangle_{\rm id} = \left\langle \Delta(E)^2 \right\rangle_{\rm id} / p^2$.

6.4.6 Isentropic System

The enthalpy $H(V|S,p,N) = E(S,V,N)+pV$ is the constrained thermodynamic potential for an isentropic system. Rearranging the expressions given above for an isolated ideal gas, one has

$$\begin{aligned} E^{\rm id}(S,V,N) &= \left[\frac{N!\Gamma(3N/2)e^{S/k_{\rm B}}}{\Delta_E V^N (2\pi m/h^2)^{3N/2}} \right]^{2/(3N-2)} \\ &= (3h^2/4\pi m) e^{-5/3} N^{5/3} V^{-2/3} e^{2S/3Nk_{\rm B}}, \end{aligned} \tag{6.154}$$

the last equality holding in the thermodynamic limit.

The volume probability distribution is given by $\wp(V|S,p,N;T) = e^{-\beta H} / \Delta_V Z(S,p,N)$. In this case it does not appear possible to evaluate exactly the partition function integral, but one may make a WKB approximation for it (below). Writing the energy as $E^{\rm id} = \alpha V^{-2/3}$ (in the thermodynamic limit), which defines the constant α, the equilibrium volume occurs at the minimum of the constrained thermodynamic potential, $\partial \overline{H}^{\rm id} / \partial V = -(2/3)\alpha \overline{V}^{\rm id}(S,p,N)^{-2/3} + p = 0$. This is the same as $\partial E^{\rm id}(S,V,N)/\partial V = -p$, and it gives explicitly

$$\overline{V}^{\rm id}(S,p,N) = \frac{N e^{-1} e^{2S/5Nk_{\rm B}}}{(2\pi m p/h^2)^{3/5}}. \tag{6.155}$$

Since $\overline{E}^{\rm id}(S,p,N) \equiv E^{\rm id}(S,\overline{V}(S,p,N),N)$, and since $\overline{T}(S,p,N) \equiv \partial \overline{E}(S,p,N) / \partial S$, one sees that this is the same as $p\overline{V}^{\rm id}(S,p,N) = Nk_{\rm B}\overline{T}^{\rm id}(S,p,N)$, which is the standard equilibrium result for the ideal gas. With this the thermodynamic potential is

$$\overline{H}^{\rm id}(S,p,N) = \frac{5}{2} p \overline{V}^{\rm id}(S,p,N) = \frac{5Ne^{-1}p^{2/5}e^{2S/5Nk_{\rm B}}}{2(2\pi m/h^2)^{3/5}}. \tag{6.156}$$

The second derivative of the enthalpy is $\overline{H}^{\rm id}_{VV} \equiv \partial^2 H(V|S,p,N)/\partial V^2 \big|_{V=\overline{V}^{\rm id}} = (10/9)\alpha \overline{V}^{\rm id}(S,p,N)^{-8/3}$. Hence the partition function may be approximated as

$$\begin{aligned} Z^{\rm id}(S,p,N;T) &= \Delta_V^{-1} \int_0^\infty {\rm d}V\, e^{-\beta H(V|S,p,N)} \\ &\approx \Delta_V^{-1} e^{-\beta \overline{H}^{\rm id}} \int_{-\infty}^\infty {\rm d}V\, e^{-\beta \overline{H}^{\rm id}_{VV}(V-\overline{V}^{\rm id}(S,p,N))^2} \\ &= \Delta_V^{-1} e^{-\beta \overline{H}^{\rm id}} \sqrt{2\pi/\beta \overline{H}^{\rm id}_{VV}}. \end{aligned} \tag{6.157}$$

One sees explicitly that in the thermodynamic limit the logarithm of the partition function equals the thermodynamic potential (times $-1/k_{\mathrm{B}}T$). The average volume is the negative of the logarithmic derivative of the partition function with respect to βp, and the fluctuations are the second derivative.

The isentropic compressibility is

$$\chi_S^{\mathrm{id}} = \frac{-1}{\overline{V}^{\mathrm{id}}(S,p,N)} \frac{\partial \overline{V}^{\mathrm{id}}(S,p,N)}{\partial p} = \frac{3}{5p}. \tag{6.158}$$

In view of the results established above, one sees that the ideal gas obeys the general law

$$\frac{C_V}{C_p} = \frac{\chi_S}{\chi_T}. \tag{6.159}$$

6.5 Mean Field Theory

The ideal systems analysed in this chapter have the property that the multidimensional partition function integral or sum could be factorised as the product of single-particle partition functions. Unfortunately in the real world such an attribute is as rare as it is desirable. Nevertheless it is possible to cast many systems in a factorised form that yields to a complete but approximate analysis. Such is the mean field approximation described in this section. First an ideal gas in an external field is treated.

6.5.1 Ideal Gas with an External One-Body Potential

The classical ideal gas treated in the preceding section has a Hamiltonian that consisted purely of the kinetic energy, $K = \sum_i p_i^2/2m$. Now to this is added a one-body potential $\psi^{\mathrm{ext}}(\mathbf{r})$, which could represent, for example, an electric or gravitational potential acting on the atoms of the gas. In this case the potential energy is

$$U(\mathbf{q}^N) = \sum_{i=1}^{N} \psi^{\mathrm{ext}}(\mathbf{q}_i). \tag{6.160}$$

Like the kinetic energy, this is the sum of single-particle terms, and so one expects that the partition integral will still factorise. In order for this to occur one needs to deal with a system in thermal equilibrium with a reservoir because this allows the energy of each individual atom to vary independently of the others, which is not the case for an isolated system of fixed total energy.

Factorisation of the Partition Function

From the Boltzmann distribution that was obtained in Ch. 2 for a constant temperature system, one obtains for the phase space probability distribution for the present ideal gas

$$
\begin{aligned}
\wp(\mathbf{\Gamma}|N,V,T) &= \frac{1}{N!h^{3N}Z(N,V,T)} e^{-\beta K(\mathbf{p}^N)} e^{-\beta U(\mathbf{q}^N)} \\
&= \frac{1}{N!h^{3N}Z(N,V,T)} \prod_{i=1}^{N} e^{-\beta p_i^2/2m} e^{-\beta \psi^{\mathrm{ext}}(\mathbf{q}_i)} \\
&= \prod_{i=1}^{N} \wp(\mathbf{p}_i, \mathbf{q}_i|V,T). \qquad (6.161)
\end{aligned}
$$

Here the one-particle probability is

$$
\wp(\mathbf{p},\mathbf{q}|V,T) = \frac{1}{h^3 Z_1(V,T)} e^{-\beta p^2/2m} e^{-\beta \psi^{\mathrm{ext}}(\mathbf{q})}, \qquad (6.162)
$$

with the single-atom partition function being

$$
\begin{aligned}
Z_1(V,T) &= \frac{1}{h^3} \int \mathrm{d}\mathbf{p}\, e^{-\beta p^2/2m} \int_V \mathrm{d}\mathbf{q}\, e^{-\beta \psi^{\mathrm{ext}}(\mathbf{q})} \\
&= \Lambda^{-3} \int_V \mathrm{d}\mathbf{q}\, e^{-\beta \psi^{\mathrm{ext}}(\mathbf{q})}. \qquad (6.163)
\end{aligned}
$$

Direct comparison with the partition function of the full system confirms the factorisation,

$$
\begin{aligned}
Z(N,V,T) &= \frac{1}{N!h^{3N}} \int \mathrm{d}\mathbf{p}^N \mathrm{d}\mathbf{q}^N e^{-\beta K(\mathbf{p}^N)} e^{-\beta U(\mathbf{q}^N)} \\
&= \frac{1}{N!} \prod_{i=1}^{N} \frac{1}{h^3} \int \mathrm{d}\mathbf{p}_i\, e^{-\beta p_i^2/2m} \int_V \mathrm{d}\mathbf{q}_i\, e^{-\beta \psi^{\mathrm{ext}}(\mathbf{q}_i)} \\
&= \frac{1}{N!} \prod_{i=1}^{N} Z_1(V,T). \qquad (6.164)
\end{aligned}
$$

Here one can see explicitly how the trivial prefactor of $N!$ arises from restricting the multidimensional integral to distinct configurations. Accordingly, it appears as an additive constant in the total entropy,

$$
S(N,V,T) = NS_1(V,T) - k_{\mathrm{B}} \ln N!, \qquad (6.165)
$$

where the entropy of each of the identical atoms is $S_1(V,T) = k_{\mathrm{B}} \ln Z_1(V,T)$.

The Density Profile

The external potential causes a density inhomogeneity; atoms are more likely to be found in regions of low potential than in regions of high potential. Probability distributions will be discussed in detail in the next chapter, but here it suffices to derive the one-particle distribution. The probability of finding a particular atom, say the first, at $\mathbf{r}$ is

$$\begin{aligned}
\wp(\mathbf{r}|N,V,T) &= \int \mathrm{d}\mathbf{p}^N\, \mathrm{d}\mathbf{q}^N\, \wp(\mathbf{\Gamma}|N,V,T)\, \delta(\mathbf{r}-\mathbf{q}_1) \\
&= \int \mathrm{d}\mathbf{p}_1\, e^{-\beta p_1^2/2m} \int_V \mathrm{d}\mathbf{q}_1\, e^{-\beta\psi^{\mathrm{ext}}(\mathbf{q}_1)} \frac{\delta(\mathbf{r}-\mathbf{q}_1)}{h^3 Z_1(V,T)} \\
&\quad \times \prod_{i=2}^{N} \left[\frac{1}{h^3 Z_1(V,T)} \int \mathrm{d}\mathbf{p}_i\, e^{-\beta p_i^2/2m} \int_V \mathrm{d}\mathbf{q}_i\, e^{-\beta\psi^{\mathrm{ext}}(\mathbf{q}_i)} \right] \\
&= \frac{e^{-\beta\psi^{\mathrm{ext}}(\mathbf{r})}}{\Lambda^3 Z_1(V,T)}. \qquad (6.166)
\end{aligned}$$

Alternatively, $\wp(\mathbf{r}|N,V,T) = \langle \delta(\mathbf{r}-\mathbf{q}_1) \rangle$. This is normalised to unity, since the atom *must* be somewhere in the volume. Notice that the external potential influences the probability distribution for an otherwise noninteracting atom in thermal equilibrium via a Boltzmann factor.

Closely related to the one-particle probability is the singlet density, which is defined as

$$\rho(\mathbf{r}) \equiv \left\langle \sum_{i=1}^{N} \delta(\mathbf{r}-\mathbf{q}_i) \right\rangle = N\wp(\mathbf{r}|N,V,T). \qquad (6.167)$$

The singlet density $\rho(\mathbf{r})$ is the average number of atoms per volume element at $\mathbf{r}$. Colloquially the singlet density is said to give the probability of finding *any* atom at the given position, but this is not strictly correct since the singlet density is normalised to N, not to 1. For a uniform system (no external potential), this is just $\rho \equiv N/V$, which is called the density. This is also the spatial average of the singlet density of a nonuniform system.

The singlet density is used to obtain one-particle averages. For example, the average energy due to the external potential is

$$\begin{aligned}
\langle U(\mathbf{q}^N) \rangle &= \left\langle \sum_{i=1}^{N} \psi^{\mathrm{ext}}(\mathbf{q}_i) \right\rangle \\
&= \left\langle \sum_{i=1}^{N} \int_V \mathrm{d}\mathbf{r}\, \psi^{\mathrm{ext}}(\mathbf{r})\, \delta(\mathbf{r}-\mathbf{q}_i) \right\rangle \\
&= \int_V \mathrm{d}\mathbf{r}\, \rho(\mathbf{r})\, \psi^{\mathrm{ext}}(\mathbf{r}). \qquad (6.168)
\end{aligned}$$

The Constrained Density Potential

The entropy of an isolated system consisting of an ideal gas of kinetic energy K was given above in Eq. (6.102),

$$S^{\mathrm{id}}(K,N,V) = N \ln \left[(V/N)(4\pi m K/3h^2 N)^{3/2} \right] + 5N/2. \qquad (6.169)$$

In terms of the kinetic energy density $\kappa \equiv K/V$ and the number density $\rho = N/V$, one may define the entropy density

$$\sigma^{\rm id}(\kappa,\rho) \equiv \frac{1}{V} S^{\rm id}(K,N,V) = \rho \ln\left[\left(\frac{4\pi m\kappa}{3h^2\rho}\right)^{3/2}\frac{1}{\rho}\right] + \frac{5\rho}{2}. \tag{6.170}$$

For the same ideal gas one has the Helmholtz free energy $F^{\rm id}(N,V,T) = \overline{K} - TS^{\rm id}(\overline{K},N,V)$ and the corresponding density $f^{\rm id}(\rho,T) = \overline{\kappa} - T\sigma^{\rm id}(\overline{\kappa},\rho)$. The corresponding constrained thermodynamic potential density is of course $f^{\rm id}(\kappa|\rho,T) = \kappa - T\sigma^{\rm id}(\kappa,\rho)$.

Now one imagines a system consisting of many isolated ideal subsystems, the ith one of which has N_i atoms, volume V_i, and kinetic energy K_i. The number of atoms in each subsystem is large enough for Stirling's approximation to be applied. The total entropy is the sum of the subsystems' entropies,

$$S^{\rm id}(\underline{K},\underline{N},\underline{V}) = \sum_i S^{\rm id}(K_i,N_i,V_i) = \sum_i V_i\sigma^{\rm id}(\kappa_i,\rho_i). \tag{6.171}$$

One may now imagine that the isolated subsystems are contiguous cells that pave a volume $V = \sum_i V_i$. Hence the ensemble of cells represents an isolated ideal system of total kinetic energy $K = \sum_i K_i$ and total number of atoms $N = \sum_i N_i$, but with the spatial distribution of each constrained. Assuming that the variation of these quantities between neighbouring cells is small, one may pass to the continuum limit so that the constrained total entropy is

$$S^{\rm id}(\kappa,\rho|V) = \int_V {\rm d}\mathbf{r}\, \sigma^{\rm id}(\kappa(\mathbf{r}),\rho(\mathbf{r})). \tag{6.172}$$

Note that no extra configurations become available when one removes the barriers between cells and allows atoms to interchange between them (provided that the kinetic energy density and number density remain constrained). Because the atoms are indistinguishable, the distinct configurations are the same in both cases, and the total entropy is just the sum of the entropies of the isolated cells.

Now adding an external potential $\psi^{\rm ext}(\mathbf{r})$, the total energy of the system may be written as $E = K + E^{\rm ext}$. The constrained density profile, $\rho(\mathbf{r})$, is arbitrary; the equilibrium density will emerge by optimisation. The corresponding external energy is

$$E^{\rm ext} = \int_V {\rm d}\mathbf{r}\, \rho(\mathbf{r})\, \psi^{\rm ext}(\mathbf{r}). \tag{6.173}$$

In terms of cells, this expression is valid provided that the external potential varies negligibly over each cell. This obviously sets an upper limit on the size of the cell, which may appear to conflict with the requirement that the cell be large enough for Stirling's approximation to be valid. However, both may be satisfied simultaneously provided that the potential is constant or slowly varying in at least one dimension.

Still constraining the profiles, the total system may now be brought into contact with a heat reservoir of temperature T such that energy exchange may occur. The constrained total entropy is now

$$\begin{aligned} S(\rho,\kappa|V,T) &= \int_V \mathrm{d}\mathbf{r}\, \sigma^{\mathrm{id}}(\kappa(\mathbf{r}),\rho(\mathbf{r})) - \frac{K+E^{\mathrm{ext}}}{T} \\ &= \int_V \mathrm{d}\mathbf{r} \left[\sigma^{\mathrm{id}}(\kappa(\mathbf{r}),\rho(\mathbf{r})) - \frac{\kappa(\mathbf{r})}{T} - \frac{1}{T}\rho(\mathbf{r})\,\psi^{\mathrm{ext}}(\mathbf{r})\right]. \end{aligned} \tag{6.174}$$

The constrained thermodynamic potential is

$$\begin{aligned} F(\rho,\kappa|V,T) &= -TS(\rho,\kappa|V,T) \\ &= \int_V \mathrm{d}\mathbf{r}\, \left[\kappa(\mathbf{r}) - T\sigma^{\mathrm{id}}(\kappa(\mathbf{r}),\rho(\mathbf{r})) + \rho(\mathbf{r})\,\psi^{\mathrm{ext}}(\mathbf{r})\right] \\ &= \int_V \mathrm{d}\mathbf{r}\, \left[f^{\mathrm{id}}(\kappa(\mathbf{r})|\rho(\mathbf{r}),T) + \rho(\mathbf{r})\,\psi^{\mathrm{ext}}(\mathbf{r})\right]. \end{aligned} \tag{6.175}$$

For thermal equilibrium of the kinetic energy this becomes

$$\begin{aligned} F(\rho|N,V,T) &= \int_V \mathrm{d}\mathbf{r}\, \left[f^{\mathrm{id}}(\rho(\mathbf{r}),T) + \rho(\mathbf{r})\,\psi^{\mathrm{ext}}(\mathbf{r})\right] \\ &= k_{\mathrm{B}}T \int_V \mathrm{d}\mathbf{r}\, \rho(\mathbf{r}) \left[-1 + \ln \Lambda^3 \rho(\mathbf{r}) + \beta\psi^{\mathrm{ext}}(\mathbf{r})\right]. \end{aligned} \tag{6.176}$$

The Helmholtz free energy density for the ideal gas that appears in the integrand derives from Eq. (6.115). Note that N has now been shown explicitly to the right of the vertical bar to indicate that only density profiles normalised to the total number of atoms are allowed.

The equilibrium density profile minimises the constrained thermodynamic potential. However, this minimisation must be carried out subject to the constraint that the total number of atoms is fixed, $\int \mathrm{d}\mathbf{r}\, \rho(\mathbf{r}) = N$. Using a Lagrange multiplier and the techniques of functional differentiation[5] one has

$$\frac{\delta\left[F(\rho|N,V,T) + \zeta \int_V \mathrm{d}\mathbf{r}\, \rho(\mathbf{r})\right]}{\delta\rho(\mathbf{r})} = k_{\mathrm{B}}T \ln \Lambda^3 \rho(\mathbf{r}) + \psi^{\mathrm{ext}}(\mathbf{r}) + \zeta, \tag{6.177}$$

which vanishes at the equilibrium profile,

$$\overline{\rho}(\mathbf{r}) = \Lambda^{-3} e^{-\zeta} e^{-\beta\psi^{\mathrm{ext}}(\mathbf{r})}. \tag{6.178}$$

The constant ζ is determined by the fact that the volume integral equals N. Since the singlet density found above is similarly normalised, and since it is likewise proportional to the Boltzmann factor of the external potential, one concludes that $e^{-\zeta} = N/Z_1(V,T)$, and that this equilibrium density profile is equal to that average density.

The thermodynamic potential for this thermally equilibrated system in an external potential is $F(\overline{\rho}|N,V,T)$. Hence direct substitution yields

$$F(N,V,T) = -Nk_{\mathrm{B}}T[\zeta + 1] = -Nk_{\mathrm{B}}T(1 - \ln[N/Z_1(V,T)]). \tag{6.179}$$

[5] Specifically, $\delta f(g(y))/\delta g(x) = [\partial f(g(y))/\partial g(y)]\delta(x-y)$. Hence if $F[g] = \int \mathrm{d}y\, f(g(y))$, then $\delta F[g]/\delta g(x) = \int \mathrm{d}y\, \delta f(g(y))/\delta g(x) = \partial f(g(x))/\partial g(x)$.

6.5.2 Mean Field Approximation

What limits the applicability of the ideal gas results to realistic systems is that it neglects the interactions between atoms. The simplest such interaction is the pair potential $u(\mathbf{r}, \mathbf{s})$. Accordingly the Hamiltonian consists of the kinetic energy, and the one- and two-body potentials, $\mathcal{H}(\mathbf{p}^N, \mathbf{q}^N) = K(\mathbf{p}^N) + U^{(1)}(\mathbf{q}^N) + U^{(2)}(\mathbf{q}^N)$, where

$$U^{(2)}(\mathbf{q}^N) = \sum_{i<j}^{N} u(\mathbf{q}_i, \mathbf{q}_j). \tag{6.180}$$

Unfortunately this interaction precludes the factorisation of the multidimensional partition function integral because each atom can no longer be treated as independent of the others.

One may introduce an approximation based upon the division of the system into cells, as used above to develop the density functional for a system in an external potential. In that case the external potential was taken to vary negligibly over the cells, so that the contribution to the energy from the ith cell was simply $N_i \psi(\mathbf{r}_i)$. Similarly one may take the contribution to the pair energy from atoms in two cells to be simply the product of the number of atoms in the cells and the pair potential evaluated between the mid-points of the cells. Passing to the continuum limit this becomes

$$\begin{aligned} E^{(2)} &= \frac{1}{2} \int_V \mathrm{d}\mathbf{r} \int_V \mathrm{d}\mathbf{s}\, \rho(\mathbf{r}) \rho(\mathbf{s}) u(\mathbf{r}, \mathbf{s}) \\ &= \frac{1}{2} \int_V \mathrm{d}\mathbf{r} \rho(\mathbf{r}) \psi^{\mathrm{mf}}(\mathbf{r}), \end{aligned} \tag{6.181}$$

where the mean field potential is

$$\psi^{\mathrm{mf}}(\mathbf{r}) = \int_V \mathrm{d}\mathbf{s}\, \rho(\mathbf{s}) u(\mathbf{r}, \mathbf{s}). \tag{6.182}$$

The factor of one-half ensures that the interaction between each pair of cells is counted once only. This approximation fails to correctly account for the interactions between atoms in the same cell, and it can also err for the interactions between atoms in neighbouring cells. It breaks down because the pair potential is usually rapidly varying at small separations, and also because the presence of an atom in one cell affects the probability of an atom being in a nearby cell, (i.e., it would be better to use the pair density $\rho^{(2)}(\mathbf{r}, \mathbf{s})$ rather than the product of singlet densities $\rho(\mathbf{r})\rho(\mathbf{s})$; the two are only equal when correlations are negligible). The mean field approximation is most appropriate for systems with a long-ranged, slowly varying pair potential, because in this case the pair energy is dominated by cells that are far-separated. The approximation is valid in this region because the pair potential is slowly varying here, and the correlations between distant cells are negligible.

The mean field approximation converts the pair potential to a density-dependent one-body potential so that it is formally identical to the external potential. Hence it contributes to the entropy in exactly the same fashion as the external potential. Using the results of the preceding section, the thermally equilibrated density constrained thermodynamic potential becomes

$$\begin{aligned} & F(\rho|N,V,T) \\ &= \int_V \mathrm{d}\mathbf{r} \left[f^{\mathrm{id}}(\rho(\mathbf{r}),T) + \rho(\mathbf{r})\,\psi^{\mathrm{ext}}(\mathbf{r}) + \frac{1}{2}\rho(\mathbf{r})\,\psi^{\mathrm{mf}}(\mathbf{r}) \right] \\ &= k_{\mathrm{B}}T \int_V \mathrm{d}\mathbf{r}\,\rho(\mathbf{r}) \left[-1 + \ln \Lambda^3 \rho(\mathbf{r}) + \beta\psi^{\mathrm{ext}}(\mathbf{r}) + \frac{\beta}{2}\psi^{\mathrm{mf}}(\mathbf{r}) \right]. \end{aligned} \tag{6.183}$$

Adding a Lagrange multiplier and differentiating one obtains

$$\frac{\delta\left[F + \zeta \int_V \mathrm{d}\mathbf{r}\,\rho(\mathbf{r})\right]}{\delta\rho(\mathbf{r})} = k_{\mathrm{B}}T \ln \Lambda^3 \rho(\mathbf{r}) + \psi^{\mathrm{ext}}(\mathbf{r}) + \psi^{\mathrm{mf}}(\mathbf{r}) + \zeta. \tag{6.184}$$

Due to the quadratic dependence of the pair energy on the density, the factor of a half is cancelled by the differentiation. The equilibrium profile follows as

$$\overline{\rho}(\mathbf{r}) = \Lambda^{-3} e^{-\zeta} e^{-\beta\overline{\psi}(\mathbf{r})} = \rho_0 e^{-\beta\overline{\psi}(\mathbf{r})}. \tag{6.185}$$

Here $\psi(\mathbf{r}) = \psi^{\mathrm{ext}}(\mathbf{r}) + \psi^{\mathrm{mf}}(\mathbf{r})$ is the total one-body potential, and $\rho_0 = \Lambda^{-3}e^{-\zeta}$ is the value of the density in any region where the total one-body potential vanishes. From the normalisation one has $\rho_0 = N / \int_V \mathrm{d}\mathbf{r}\, e^{-\beta\overline{\psi}(\mathbf{r})}$. Insertion of the equilibrium profile into the constrained thermodynamic potential yields the equilibrium thermodynamic potential in this mean field approximation. After some manipulation this is

$$F(N,V,T) = N k_{\mathrm{B}}T[\ln \Lambda^3 \rho_0 - 1] - \frac{1}{2}\int_V \mathrm{d}\mathbf{r}\,\overline{\rho}(\mathbf{r})\overline{\psi}^{\mathrm{mf}}(\mathbf{r}). \tag{6.186}$$

Many Body Interactions

It is worth mentioning that three-body and higher-order interactions also exist. These can be important but they are usually neglected because they are computationally intractable. However, they are relatively easy to incorporate into the mean field approximation. For example, the triplet energy is approximated by

$$\begin{aligned} E^{(3)} &= \frac{1}{3!}\int_V \mathrm{d}\mathbf{r} \int_V \mathrm{d}\mathbf{s} \int_V \mathrm{d}\mathbf{t}\,\rho(\mathbf{r})\rho(\mathbf{s})\rho(\mathbf{t})u^{(3)}(\mathbf{r},\mathbf{s},\mathbf{t}) \\ &= \frac{1}{3}\int_V \mathrm{d}\mathbf{r}\rho(\mathbf{r})\psi^{(3)}(\mathbf{r}), \end{aligned} \tag{6.187}$$

where

$$\psi^{(3)}(\mathbf{r}) = \frac{1}{2!}\int_V \mathrm{d}\mathbf{s} \int_V \mathrm{d}\mathbf{t}\,\rho(\mathbf{s})\rho(\mathbf{t})u(\mathbf{r},\mathbf{s},\mathbf{t}). \tag{6.188}$$

Analogous results hold for the higher many-body potentials. This gives for the equilibrium profile $\overline{\rho}(\mathbf{r}) = \rho_0 e^{-\beta\overline{\psi}(\mathbf{r})}$, where the total potential is $\psi(\mathbf{r}) = \psi^{(1)}(\mathbf{r}) + \psi^{(2)}(\mathbf{r}) + \psi^{(3)}(\mathbf{r}) + ...$ The Helmholtz free energy is

$$\begin{aligned} F(N,V,T) &= Nk_{\mathrm{B}}T[\ln\Lambda^3\rho_0 - 1] - \frac{1}{2}\int_V \mathrm{d}\mathbf{r}\,\overline{\rho}(\mathbf{r})\overline{\psi}^{(2)}(\mathbf{r}) \\ & \quad - \frac{2}{3}\int_V \mathrm{d}\mathbf{r}\,\overline{\rho}(\mathbf{r})\overline{\psi}^{(3)}(\mathbf{r}) - \ldots . \end{aligned} \tag{6.189}$$

Summary

- Ideal systems have simplified energy functions that allow the multidimensional integral (or sum) of the partition function to factorise into the product of single-particle partition functions.
- The classical ideal gas has energy $E = 3Nk_{\mathrm{B}}T/2$, pressure $p = Nk_{\mathrm{B}}T$, and chemical potential $\mu = k_{\mathrm{B}}T\ln[N\Lambda^3/V]$.
- The classical ideal heat capacity, $k_{\mathrm{B}}/2$ per harmonic degree of freedom, is correct at high temperatures but is too large at low temperatures. In general the classical approach is valid at high enough temperatures, but quantum statistics must be used when the thermal energy is small compared to the energy quantum.
- An external potential induces a density inhomogeneity that in the ideal gas is given simply by the Boltzmann factor. This may be derived by optimising the density-constrained thermodynamic potential, which is based upon the existence of a local entropy density. Pair and many-body interactions may be reduced to effective one-body interactions by the mean field approximation.

Chapter 7

Interacting Particles

7.1 Intermolecular Potentials

The preceding chapter dealt with ideal systems in which the partition function integral factorised. Such systems included the ideal gas of noninteracting molecules. In reality molecules do interact with each other, and it is these interactions that necessitate the development of special techniques for evaluating the partition function.

One may identify several types of intermolecular potentials including the Coulomb interaction due to a net charge on the molecules, dipole and multipole interactions due to permanent nonspherical charge distributions on net neutral molecules, short-range core repulsions, which ultimately arise from the Pauli exclusion principle that prevents overlap of electron clouds, and long-range dispersion attractions due to induced dipoles arising from correlated electronic fluctuations.[1] Of these only the Coulomb potential is strictly a potential; the others are effective potentials or constrained thermodynamic potentials that arise from removing various degrees of freedom from the problem. Similarly only the Coulomb potential is strictly pairwise additive; for the effective potentials it is an approximation to neglect the many-body contributions.

7.1.1 Coulomb

The potential between two atoms *in vaccuo* with point charges q_1 and q_2 is

$$u(r_{12}) = \frac{q_1 q_2}{4\pi\epsilon_0 r_{12}}, \tag{7.1}$$

where the distance between the atoms is $r_{12} = |\mathbf{r}_1 - \mathbf{r}_2|$, and $\epsilon_0 = 8.854 \times 10^{-12}$ is the permittivity of free space (SI units). The Coulomb potential is positive for similarly charged ions, which leads to a repulsive force between them. The

[1] G. C. Maitland, M. Rigby, E. B. Smith, and W. A. Wakeham, *Intermolecular Forces: Their Origin and Determination*, Clarendon Press, Oxford, 1981.

potential decays exceedingly slowly; it is in fact nonintegrable, which means that all physical systems must be overall electro-neutral.

The Coulomb potential *in vacuo* is exact, but it is frequently employed for the interaction between charges *in media*. In this case the charges are said to be dielectrically screened due to the polar or polarisable nature of the medium, and the interaction is

$$u(r_{12}) = \frac{q_1 q_2}{4\pi\epsilon_0 \epsilon r_{12}}, \tag{7.2}$$

where ϵ is the relative permittivity (dielectric constant) of the medium. It is emphasised that this expression is a continuum approximation, which avoids the explicit inclusion of the molecules of the medium, and which is exact for large separations between the charges. This is an effective potential that arises from integrating out all of the configurations of the medium. It is really a free energy for the macrostate with the charge atoms constrained at a particular separation, as may be seen by the temperature dependence of the dielectric constant.

The dielectric constant of water is about 80, which means that the charge on ions is effectively reduced by a factor of 9 in water compared to in vacuum or air. Common salt, NaCl, is a crystalline solid held together by the strong interaction of alternating positive and negative ions, but it dissolves in water because the interaction is reduced by such a large amount. This is why water is generally a better solvent than oil, which has a dielectric constant of about 5.

Dipole and Multipole Interactions

Many molecules do not bear a net charge, but do have a charge separation. For example, one end of the molecule may be positive, and the other end negative. This particular example is a dipolar molecule, and the interaction between two such molecules is just the sum of the Coulomb interactions between the four charges. In the point dipole limit one considers that the magnitude of the charge goes to infinity while the separation goes to 0 in such a way that the product remains constant and equal to the dipole moment of the molecule. This is a valid representation of a real dipole at distances not too close to the molecule, (i.e., when the intermolecular separation is greater than the distances between the intramolecular charges).

The interaction potential between two permanent point dipoles may be obtained from the dipole field tensor

$$\underline{\underline{T}} = -\nabla\nabla r^{-1} = r^{-3}[\underline{\underline{I}} - 3\hat{\mathbf{r}}\hat{\mathbf{r}}]. \tag{7.3}$$

In terms of this the electric field at $\mathbf{r} = r\hat{\mathbf{r}}$ due to a dipole at the origin is

$$\mathbf{E} = -\underline{\underline{T}} \cdot \boldsymbol{\mu} = r^{-3}[3(\boldsymbol{\mu} \cdot \hat{\mathbf{r}})\hat{\mathbf{r}} - \boldsymbol{\mu}]. \tag{7.4}$$

Here $\boldsymbol{\mu}$ signifies the strength of the dipole and its direction. Finally, the interaction potential between two dipoles is

$$u(12) = -\boldsymbol{\mu}_2 \cdot \mathbf{E}_1 = r_{12}^{-3}\left[\boldsymbol{\mu}_1 \cdot \boldsymbol{\mu}_2 - 3(\boldsymbol{\mu}_1 \cdot \hat{\mathbf{r}}_{12})(\boldsymbol{\mu}_2 \cdot \hat{\mathbf{r}}_{12})\right]. \tag{7.5}$$

Note that the dipole–dipole potential depends upon the separation and upon the mutual orientation of the dipoles and the intermolecular vector.

One can carry out an analogous procedure for more general charge distributions, and the molecule may be described in terms of a multipole expansion. For example, the linear real dipole described above has a point dipole, a point quadrupole, and higher-order point multipoles. The interaction between multipoles decays increasingly rapidly with the order of the multipole. Charge–charge interactions go like r^{-1}, charge–dipole interactions go like r^{-2}, etc. In general one therefore truncates the expansion at low order. For example, a reasonable description of the water molecule would need to include at least the dipole and quadrupole moments.

Unlike the interaction between two spherically symmetric charges, which depends only upon their separation, the interaction between dipolar and between multipolar molecules depends upon their separation and on their mutual orientation. This angular dependence makes the analysis of molecular systems a great deal more complicated than that of atomic systems. Hence here the main focus will be on atoms or spherically symmetric molecules, and the orientation dependence of the interactions will in general be neglected. The so-called simple fluids are characterised by such interactions.

7.1.2 Hard Sphere

Real molecules have a finite size and do not overlap. There is thus a strong, short-ranged repulsion between them, which has its origin in the Pauli exclusion of the electrons in the outer orbitals. It is a difficult quantum mechanical task to calculate the real interaction between molecules *ab initio*. The simplest approximation to the core repulsion is the hard-sphere potential

$$u(r) = \begin{cases} \infty & r < d \\ 0 & r > d, \end{cases} \tag{7.6}$$

where d is the hard-sphere diameter, which is typically on the order of several tenths of nanometres.

The hard-sphere fluid is the simplest, nontrivial fluid that retains some elements of reality, and is perhaps the most studied from the theoretical perspective. It does not exhibit a liquid–gas transition, because there are no attractive forces, but does show a freezing transition. Approximate solutions, analytic and numeric, exist for its equation of state and structural properties. Hard-sphere systems have also been used extensively as a reference system perturbed by more realistic intermolecular potentials.

7.1.3 Lennard–Jones

It was stated above that dipolar molecules interact due to their displaced permanent charge distributions, but even molecules such as the noble gases, which are electro-neutral and spherically symmetric in isolation, show a long-range Coulomb interaction. This is because fluctuations in the electron distribution

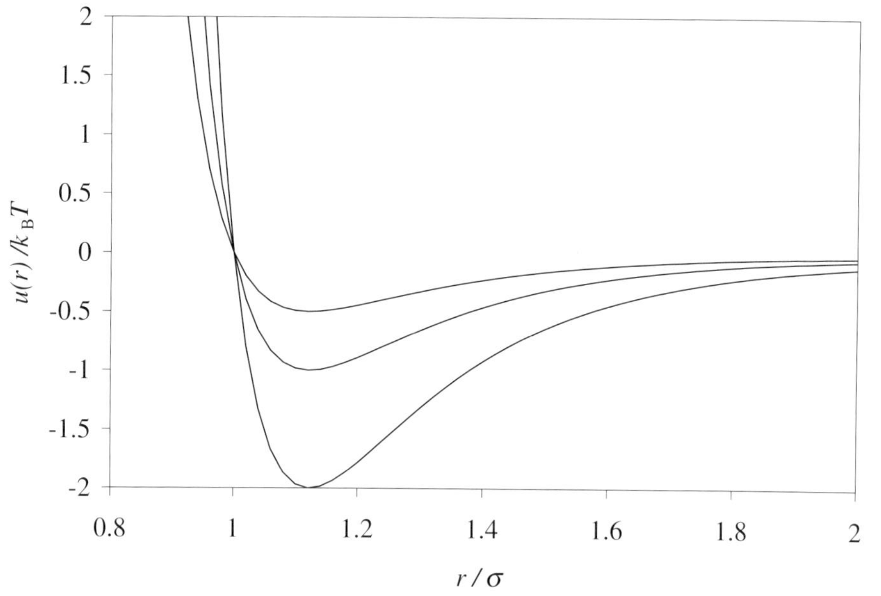

Figure 7.1: The Lennard–Jones potential for, from top to bottom, $\epsilon = 0.5$, 1, and 2 $k_\mathrm{B}T$.

about each molecule give a temporary multipole moment, and these fluctuating moments become correlated. This is the dispersion or van der Waals attraction, and it decays with the sixth power of distance. (The interaction between dipoles decays as a cubic, and the correlation between dipoles also decays as a cubic; in general, correlations lead to an attraction.) The strength of the interaction may be shown to depend upon the polarizability of the atoms.

The dispersion attraction applies at large separations; at small separations it must go over to the core repulsions discussed above, whose precise functional form is complicated. A convenient representation of the interaction potential between a pair of atoms, which shows the dispersion attraction and a core repulsion, is the Lennard–Jones 6–12 potential (see Fig. 7.1),

$$u(r) = 4\epsilon \left[\left(\frac{\sigma}{r} \right)^{12} - \left(\frac{\sigma}{r} \right)^{6} \right]. \tag{7.7}$$

This potential has a well of depth ϵ located at $r = 2^{1/6}\sigma$. The core may be taken to be at $r \lesssim \sigma$. This intermolecular potential has a long history and is a convenient analytic representation that greatly simplifies the interaction between real molecules while retaining important features of reality. A Lennard–Jones fluid shows a liquid–gas transition, and judicious choice of the two parameters allows a relatively accurate description of the thermodynamic properties of many real molecules over both the fluid and solid regions of the phase diagram.

7.1.4 Axilrod–Teller

The potentials listed above are all pair potentials: they give the energy of interaction of two atoms or molecules. Apart from the Coulomb potential *in vacuo* they are all also effective potentials ultimately derived by removing various degrees of freedom from the system. Such effective potentials also act between triplets and higher clusters of atoms. In other words, with the exception of the bare Coulomb potential, the potential energy of a cluster of atoms cannot be strictly decomposed into the sum of pair interactions.

An example of this is the dispersion interaction just treated. As mentioned, it arises from the correlated fluctuations of the electron clouds of the two atoms, and its strength is proportional to the product of their polarisabilities. In reality many-body effects act to diminish the strength of such pair-wise correlations. The dispersion interaction between three atoms is generally less than the sum of the three pair-wise interactions because the third atom may not be simultaneously correlated with the other two. The three-body excess contribution to the dispersion interaction is known as the Axilrod–Teller triple-dipole potential,

$$u(\mathbf{r}_1, \mathbf{r}_2, \mathbf{r}_3) = \nu \frac{1 + 3(\hat{\mathbf{r}}_{12} \cdot \hat{\mathbf{r}}_{13})(\hat{\mathbf{r}}_{21} \cdot \hat{\mathbf{r}}_{23})(\hat{\mathbf{r}}_{31} \cdot \hat{\mathbf{r}}_{32})}{r_{12}^3 r_{23}^3 r_{31}^3}. \tag{7.8}$$

This potential is positive for most configurations of the three atoms except for nearly linear ones. Comparing the strength of the three-body term to the Lennard–Jones pair potential, for an atom such as argon one has $\nu/(3\epsilon\sigma^9) \approx 0.04$. Hence three-body contributions to the properties of real systems can be expected to be on the order of several per cent.

7.1.5 External Potentials

As mentioned in the preceding chapter, in addition to the interactions of the atoms of the system amongst themselves, there can also be interactions that arise from external sources. Examples include the gravitational potential, electric fields due to external charges, and interactions with the walls of the container. These are usually represented as a one-body potential (i.e., they act on each atom independent of the presence of the other atoms). The chemical potential can be thought of as a constant one-body potential, although in practice it is usually treated separately for a grand canonical subsystem. As one might conclude from the existence of three-body potentials, there can also be two-body external potentials. For example, the interaction between two charges is affected by their proximity to a dielectric wall, the so-called dielectric image interaction. Such an interaction may be regarded as a two-body external potential.

External potentials in general create a density inhomogeneity so that the fluid or solid is nonuniform. Conversely, in the absence of such external potentials the system is homogeneous or uniform. Most attention has been placed on uniform systems, and these include fluid systems and disordered solids. A crystalline solid is strictly speaking an inhomogeneous system that requires an external field (possibly infinitesimal) to fix its position in space. Fluids confined to finite pores or systems near walls or inclusions are also inhomogeneous.

Although uniform systems are ubiquitous and important, in the formal development of statistical mechanics it is often convenient to retain a one-body external potential. In applications and computations one often deals with homogeneous systems by subsequently setting the external potential to 0. Conversely, three-body terms makes the system almost computationally intractable, and since they add little to the formalism they and higher many-body terms are frequently dropped from the beginning.

In the general case the potential energy of a system of N spherically symmetric atoms may be written

$$U_N(\mathbf{r}^N) = \sum_{i=1}^{N} u_1(\mathbf{r}_i) + \sum_{i<j}^{N} u_2(r_{ij}) + \sum_{i<j<k}^{N} u_3(r_{ij}, r_{jk}, r_{ki}) + \ldots . \tag{7.9}$$

The Hamiltonian is of course the sum of this and the kinetic energy. The restrictions on the sums ensure that the energy of each distinct cluster of atoms is counted only once.

7.2 Partition Function and Derivatives

As the preceding chapters have shown, the partition function plays a central role in statistical mechanics because its logarithm gives the total unconstrained entropy of the subsystem and reservoir, and its derivatives generate the average properties of the subsystem. In the case of the ideal gas the kinetic energy is the only contribution to the Hamiltonian, which means that the configurational contribution gives a factor of volume for each particle. Further, because each particle's kinetic energy depends only on its own momentum, the momentum integrals also factor. For a system in which the molecules interact, the latter property is preserved, and so one can write the partition function as the product of the ideal gas contribution and an excess.

7.2.1 Configuration Integral

For a system in contact with a thermal reservoir the canonical partition function is

$$\begin{aligned}
Z(N, V, T) &= \frac{h^{-3N}}{N!} \int \mathrm{d}\mathbf{p}^N \int_V \mathrm{d}\mathbf{r}^N \; \exp[-\beta \mathcal{H}_N(\mathbf{r}^N, \mathbf{p}^N)] \\
&= \frac{h^{-3N}}{N!} \int \mathrm{d}\mathbf{p}^N \exp[-\beta K_N(\mathbf{p}^N)] \int_V \mathrm{d}\mathbf{r}^N \; \exp[-\beta U_N(\mathbf{r}^N)] \\
&= \frac{\Lambda^{-3N}}{N!} \int_V \mathrm{d}\mathbf{r}^N \; \exp[-\beta U_N(\mathbf{r}^N)] \\
&= Z^{\mathrm{id}}(N, V, T)\, V^{-N} \int_V \mathrm{d}\mathbf{r}^N \; \exp[-\beta U_N(\mathbf{r}^N)].
\end{aligned} \tag{7.10}$$

Here and throughout the inverse temperature is $\beta = 1/k_{\mathrm{B}}T$, and the thermal de Broglie wavelength is $\Lambda = \sqrt{2\pi\hbar^2/mk_{\mathrm{B}}T}$. The nontrivial part of the partition function is called the *configuration integral* [2]

$$Q(N,V,T) \equiv \int_V \mathrm{d}\mathbf{r}^N \ \exp[-\beta U_N(\mathbf{r}^N)]. \tag{7.11}$$

The phase space probability distribution for this system in contact with a thermal reservoir is called the canonical distribution. The probability of N particles being at positions $\mathbf{r}_1, \mathbf{r}_2, \ldots, \mathbf{r}_N$ with momenta $\mathbf{p}_1, \mathbf{p}_2, \ldots, \mathbf{p}_N$ is

$$\wp(\mathbf{r}^N, \mathbf{p}^N | N, V, T) = \frac{1}{N! h^{3N} Z(N,V,T)} e^{-\beta \mathcal{H}_N(\mathbf{r}^N, \mathbf{p}^N)}. \tag{7.12}$$

The volume element of phase space for this probability density is $\mathrm{d}\mathbf{r}^N \mathrm{d}\mathbf{p}^N$. One may integrate out the momentum contributions to obtain the configuration probability distribution

$$\wp(\mathbf{r}^N | N, V, T) = \frac{1}{Q(N,V,T)} e^{-\beta U_N(\mathbf{r}^N)}. \tag{7.13}$$

Here the relevant volume element is $\mathrm{d}\mathbf{r}^N$.

Constant Chemical Potential

For a subsystem able to exchange both energy and particles with a reservoir of temperature T and chemical potential μ, the classical grand canonical distribution is

$$\wp(\mathbf{r}^N, \mathbf{p}^N, N | \mu, V, T) = \frac{e^{\beta\mu N} e^{-\beta \mathcal{H}_N(\mathbf{r}^N, \mathbf{p}^N)}}{N! h^{3N} \Xi(\mu, V, T)}. \tag{7.14}$$

Integrating over momenta gives the configuration probability

$$\wp(\mathbf{r}^N, N | \mu, V, T) = \frac{e^{\beta\mu N} e^{-\beta U_N(\mathbf{r}^N)}}{N! \Lambda^{3N} \Xi(\mu, V, T)}, \tag{7.15}$$

and integrating over configurations gives the number distribution

$$\wp(N | \mu, V, T) = \frac{e^{\beta\mu N} Z(N,V,T)}{\Xi(\mu, V, T)}. \tag{7.16}$$

The ideal contributions are number-dependent, which is of course why they are not removed here. The grand partition function is

$$\Xi(\mu, V, T) = \sum_{N=0}^{\infty} e^{\beta\mu N} Z(N,V,T) = \sum_{N=0}^{\infty} \frac{e^{\beta\mu N}}{\Lambda^{3N} N!} Q(N,V,T). \tag{7.17}$$

[2]Note that some authors use the reverse convention, denoting the partition function by Q, and the configuration integral by Z.

Constant Pressure

For a subsystem able to exchange both energy and volume with a reservoir of temperature T and pressure p, the classical grand canonical distribution is

$$\wp(\mathbf{r}^N, \mathbf{p}^N, V|N, p, T) = \frac{e^{-\beta pV} e^{-\beta \mathcal{H}_N(\mathbf{r}^N, \mathbf{p}^N)}}{\Delta_V N! h^{3N} X(N, p, T)}. \tag{7.18}$$

The configuration probability is

$$\wp(\mathbf{r}^N, V|N, p, T) = \frac{e^{-\beta pV} e^{-\beta U_N(\mathbf{r}^N)}}{\Delta_V N! \Lambda^{3N} X(N, p, T)}. \tag{7.19}$$

Integrating over configurations gives the volume distribution

$$\wp(V|N, p, T) = \frac{e^{-\beta pV} Z(N, V, T)}{\Delta_V X(N, p, T)}. \tag{7.20}$$

The ideal contributions independent of volume could be removed from this expression. It is worth mentioning that the reservoir pressure in the exponent of the first factor favours small volumes, whereas the internal pressure manifest in the partition function of the subsystem, the second factor, favours large volumes; the net effect is that the probability distribution is peaked about the equilibrium volume, and decays rapidly on either side of it. The partition function for this isothermal–isobaric system is

$$\begin{aligned} X(N, p, T) &= \int_0^\infty \frac{\mathrm{d}V}{\Delta_V} e^{-\beta pV} Z(N, V, T) \\ &= \frac{\Lambda^{-3N}}{N!} \int_0^\infty \frac{\mathrm{d}V}{\Delta_V} e^{-\beta pV} \int_V \mathrm{d}\mathbf{r}^N \, \exp[-\beta U_N(\mathbf{r}^N)]. \end{aligned} \tag{7.21}$$

7.2.2 Average Energy

It was shown in Section 4.3 that the derivatives of the logarithm of the partition function gave the average properties of the subsystem. For example, the average energy of an isothermal system is the inverse temperature derivative,

$$\begin{aligned} & \frac{-\partial \ln Z(N, V, T)}{\partial \beta} \\ &= \frac{1}{N! h^{3N} Z(N, V, T)} \int \mathrm{d}\mathbf{p}^N \int_V \mathrm{d}\mathbf{r}^N \, e^{-\beta \mathcal{H}_N(\mathbf{r}^N, \mathbf{p}^N)} \mathcal{H}_N(\mathbf{r}^N, \mathbf{p}^N) \\ &= \int \mathrm{d}\mathbf{p}^N \int_V \mathrm{d}\mathbf{r}^N \, \wp(\mathbf{r}^N, \mathbf{p}^N|N, V, T) \, \mathcal{H}_N(\mathbf{r}^N, \mathbf{p}^N) \\ &= \langle \mathcal{H}_N \rangle . \end{aligned} \tag{7.22}$$

The Hamiltonian is the sum of kinetic and potential energies. In view of the separation between the momentum and the configuration terms, the average kinetic energy may be obtained solely from the ideal partition function,

$$\langle K_N \rangle = \frac{-\partial \ln Z^{\mathrm{id}}(N, V, T)}{\partial \beta} = \frac{3N k_{\mathrm{B}} T}{2}. \tag{7.23}$$

This is often called the ideal energy. Conversely, the excess energy, which is the average potential energy, comes from the logarithmic derivative of the configuration integral,

$$\begin{aligned}\langle U_N \rangle &= \frac{-\partial \ln Q(N,V,T)}{\partial \beta} \\ &= \frac{1}{Q(N,V,T)} \int_V \mathrm{d}\mathbf{r}^N \ \exp[-\beta U_N(\mathbf{r}^N)]\, U_N(\mathbf{r}^N). \end{aligned} \tag{7.24}$$

The second logarithmic derivatives of the partition function give the fluctuation or average deviation in various quantities, as shown in Section 4.3. For the energy of the isothermal system one has

$$\begin{aligned}\frac{\partial^2 \ln Z(N,V,T)}{\partial \beta^2} &= \frac{-\partial \langle \mathcal{H}_N \rangle}{\partial \beta} \\ &= \frac{-\partial}{\partial \beta} \frac{1/N!h^{3N}}{Z(N,V,T)} \int \mathrm{d}\mathbf{p}^N \int_V \mathrm{d}\mathbf{r}^N \\ &\quad \times e^{-\beta \mathcal{H}_N(\mathbf{r}^N,\mathbf{p}^N)} \mathcal{H}_N(\mathbf{r}^N,\mathbf{p}^N) \\ &= \left\langle \mathcal{H}_N^2 \right\rangle - \langle \mathcal{H}_N \rangle^2 \\ &= \left\langle \delta(\mathcal{H}_N)^2 \right\rangle . \end{aligned} \tag{7.25}$$

It was shown in Section 4.4 that in so far as the Helmholtz free energy is equal to the logarithm of the partition function, which it is in the thermodynamic limit with consequence that the equilibrium energy is equal to the average energy, then the heat capacity at constant volume is related to the fluctuation in energy, which are the respective derivatives. The physical origin of this equivalence is that both measure the ease with which the energy of the subsystem may be changed. Since the logarithm of the partition function is an extensive quantity (i.e., it scales with the size of the system), the root mean square energy fluctuation scales with the square root of the size of the system. That is, the magnitude of the energy fluctuations increases as the system size increases However, the relative magnitude of the energy fluctuations, which is the root mean square value divided by the average energy, goes like one over the square root of the system size, which is to say that the relative fluctuations vanish in the thermodynamic limit. This result may be restated: the energy probability distribution becomes relatively sharply peaked as the system size is increased.

Other averages were expressed as derivatives of the partition function in Section 4.4. Explicit expressions for these can be obtained in a fashion similar to the above results for the energy.

7.2.3 Virial Pressure

The average pressure is the volume derivative of the logarithm of the partition function

$$\langle \beta_s p_s \rangle = \frac{\partial \ln Z(N,V,T)}{\partial V}$$

$$= \frac{\partial \ln Q(N,V,T)}{\partial V}, \tag{7.26}$$

where the second line follows because the momentum contributions do not depend on volume. When fluctuations are negligible, the subsystem temperature is equal to that of the reservoir, $T_s = T$, and β_s may be treated as a constant and taken outside of the average, $\langle \beta_s p_s \rangle = \langle \beta_s \rangle \langle p_s \rangle = \beta \langle p_s \rangle$.

This expression may be formally written as an average of the so-called virial of Clausius using a simple scaling argument. For a cubic container, $V = L^3$, one can write for the position of the ith atom $\mathbf{r}_i = (Lx_i, Ly_i, Lz_i)$, where $x_i \in [0,1]$, etc. Hence the configuration integral may be written as

$$Q(N,V,T) = L^{3N} \int_0^1 \mathrm{d}x^N\, \mathrm{d}y^N\, \mathrm{d}z^N\ \exp[-\beta U_N^{\mathrm{int}}(\mathbf{x}^N; L)]. \tag{7.27}$$

Note that the potential energy that appears here is solely that due to the intermolecular contributions. The external contributions, which provide the external forces that balance the internal pressure being obtained here, implicitly appear as the limit on the configuration integral. (See the generalised equipartition theorem below.)

The derivative of the potential energy is

$$\begin{aligned}
\frac{\partial U_N^{\mathrm{int}}(\mathbf{x}^N; L)}{\partial L} &= \sum_{i=1}^{N} \left[x_i \frac{\partial U_N^{\mathrm{int}}}{\partial (Lx_i)} + y_i \frac{\partial U_N^{\mathrm{int}}}{\partial (Ly_i)} + z_i \frac{\partial U_N^{\mathrm{int}}}{\partial (Lz_i)} \right] \\
&= L^{-1} \sum_{i=1}^{N} \mathbf{r}_i \cdot \nabla_i U_N^{\mathrm{int}}(\mathbf{r}^N) \\
&\equiv -L^{-1} \mathcal{V}^{\mathrm{int}},
\end{aligned} \tag{7.28}$$

where $\mathcal{V}$, the virial of Clausius, is essentially the sum of the force acting on each particle due to the other particles. Accordingly the *virial equation* is

$$\begin{aligned}
\langle \beta_s p_s \rangle &= \frac{1}{Q(N,V,T)} \frac{\partial Q(N,V,T)}{3L^2 \partial L} \\
&= \frac{1}{3L^2 Q(N,V,T)} \left[3NL^{3N-1} \int_0^1 \mathrm{d}\mathbf{x}^{3N}\, e^{-\beta U_N^{\mathrm{int}}(\mathbf{x}^N; L)} \right. \\
&\quad \left. - \beta L^{3N} \int_0^1 \mathrm{d}\mathbf{x}^{3N}\, e^{-\beta U_N^{\mathrm{int}}(\mathbf{x}^N; L)} \frac{\partial U_N^{\mathrm{int}}}{\partial L} \right] \\
&= \frac{N}{V} + \frac{\beta \left\langle \mathcal{V}^{\mathrm{int}} \right\rangle}{3V}.
\end{aligned} \tag{7.29}$$

The first term is the pressure of an ideal gas, and the second is the average of the intermolecular forces. For a dilute real gas, the important force between the molecules is the attractive tail, and the pressure is less than that of an ideal gas at the same density. As higher densities, the core repulsion due to the molecular size becomes more important, and the internal contribution is positive, increasing the pressure above that of the corresponding ideal gas. Accordingly for solids and liquids, the pressure usually lies above that of an ideal gas at the same density. Moreover, it also increases at a rate with density faster than that of the corresponding ideal gas, which is to say that their compressibility is lower; one often speaks of an incompressible liquid or solid.

Eular's Theorem

If the potential is a homogeneous function of order m, $U_N^{\rm int}(L\mathbf{x}_1, L\mathbf{x}_2, \ldots, L\mathbf{x}_N) = L^m U_N^{\rm int}(\mathbf{x}_1, \mathbf{x}_2, \ldots, \mathbf{x}_N)$, then $L\partial U_N^{\rm int}(\mathbf{x}^N; L)/\partial L = mU_N^{\rm int}(\mathbf{x}^N; L)$, which is a particular case of Eular's theorem. In this case the virial pressure may be expressed in terms of the average potential energy,

$$\langle \beta_s p_s \rangle = \frac{N}{V} - \frac{\beta m \left\langle U_N^{\rm int} \right\rangle}{3V}. \tag{7.30}$$

While the total potential energy itself may not be homogeneous, it may comprise several contributions that are each individually homogeneous, and the virial pressure may be expressed as the sum of the average of each energy term times its degree. Eular's theorem provides a convenient method for evaluating the pressure that avoids taking the average of the derivative of the potential.

Pair Potential

For the case of a uniform fluid in which the particles are spherical and interact only via a pair potential $u(r)$, the potential energy is $U_N^{\rm int}(\mathbf{x}^N; L) = \sum_{i<j}^N u(Lx_{ij})$. In this case the virial becomes

$$\begin{aligned} \mathcal{V}^{\rm int} &= -L\frac{\partial U_N^{\rm int}(\mathbf{x}^N; L)}{\partial L} \\ &= -L\sum_{i<j}^N \frac{\partial u(Lx_{ij})}{\partial L} \\ &= -\sum_{i<j}^N Lx_{ij}\frac{\partial u(Lx_{ij})}{\partial Lx_{ij}} \\ &= -\sum_{i<j}^N r_{ij}u'(r_{ij}), \end{aligned} \tag{7.31}$$

where the prime denotes differentiation with respect to argument. There are $N(N-1)/2$ terms in this expression, and each contributes an identical amount to the average of the virial.

Triplet Potential

The three-body interaction between three atoms in a uniform fluid depends only upon the triangle that they form, and hence the triplet potential may be written in terms of two sides of the triangle and the included angle, $u(r_{12}, r_{13}, \theta_{213})$. The triplet part of the internal potential is $U_N^{(3)}(\mathbf{x}^N; L) = \sum_{i<j<k}^N u(Lx_{ij}, Lx_{ik}, \theta_{jik})$, and its contribution to the virial is

$$\begin{aligned} \mathcal{V}^{(3)} &= -L \sum_{i<j<k}^{N} \frac{\partial u(Lx_{ij}, Lx_{ik}, \theta_{jik})}{\partial L} \\ &= -\sum_{i<j<k}^{N} \left[r_{ij} \frac{\partial u(r_{ij}, r_{ik}, \theta_{jik})}{\partial r_{ij}} + r_{ik} \frac{\partial u(r_{ij}, r_{ik}, \theta_{jik})}{\partial r_{ik}} \right] \\ &= -2 \sum_{i<j<k}^{N} r_{ij} \frac{\partial u(r_{ij}, r_{ik}, \theta_{jik})}{\partial r_{ij}}. \end{aligned} \tag{7.32}$$

The last line follows from symmetry, $u(r, s, \theta) = u(s, r, \theta)$. There are $N(N-1)(N-2)/3!$ terms in this expression. Note that it would be equally legitimate to describe the triangle formed by the three atoms by its three sides and to write $u(r_{ij}, r_{jk}, r_{ki}) = u(r_{ij}, r_{ik}, \theta_{jik})$. However, because the two sides rather than the side and the angle are being held constant, $\partial u(r_{ij}, r_{ik}, \theta_{jik})/\partial r_{ij} \neq \partial u(r_{ij}, r_{jk}, r_{ki})/\partial r_{ij}$. In this convention one has

$$\mathcal{V}^{(3)} = -3 \sum_{i<j<k}^{N} r_{ij} \frac{\partial u(r_{ij}, r_{jk}, r_{ki})}{\partial r_{ij}}. \tag{7.33}$$

7.2.4 Surface Tension

Consider a planar inhomogeneity, such as a liquid–vapour interface lying in the xy plane. In this case the coordinates may be scaled separately in each direction: $x_i = L_x \hat{x}_i$, $y_i = L_y \hat{y}_i$, and $z_i = L_z \hat{z}_i$. For simplicity consider a pair-wise additive potential, which has as derivative

$$\begin{aligned} \frac{\partial U_N(\hat{\mathbf{x}}^N; \underline{L})}{\partial L_x} &= \sum_{i<j}^{N} \frac{\partial u(r_{ij})}{\partial r_{ij}} \frac{\partial r_{ij}}{\partial L_x} \\ &= \sum_{i<j}^{N} u'(r_{ij}) \frac{L_x \hat{x}_{ij}^2}{r_{ij}}, \end{aligned} \tag{7.34}$$

since $r_{ij}^2 = L_x^2 \hat{x}_{ij}^2 + L_y^2 \hat{y}_{ij}^2 + L_z^2 \hat{z}_{ij}^2$. In this case the derivative of the Helmholtz free energy, which in the thermodynamic limit equals the logarithm of the partition function, is

$$\begin{aligned} \frac{\partial F(N,V,T)}{\partial L_x} &= \frac{-k_B T}{Q(N,V,T)} \frac{\partial}{\partial L_x} \frac{1}{L_x^N L_y^N L_z^N} \int_0^1 d\hat{\mathbf{x}}^N e^{-\beta U_N(\hat{\mathbf{x}}^N; \underline{L})} \\ &= \frac{-k_B T N}{L_x} + \left\langle \sum_{i<j}^{N} u'(r_{ij}) \frac{x_{ij}^2}{L_x r_{ij}} \right\rangle. \end{aligned} \tag{7.35}$$

Just as the pressure is the volume derivative of the Helmholtz free energy, so is surface tension conjugate to area, and one has

$$\begin{aligned} \mathrm{d}F &= -p\mathrm{d}V + \gamma\mathrm{d}A - S\mathrm{d}T + \mu\mathrm{d}N \\ &= \left[-pL_zL_y + \gamma L_y\right]\mathrm{d}L_x - S\mathrm{d}T + \mu\mathrm{d}N. \end{aligned} \tag{7.36}$$

Accordingly the above becomes

$$\beta p L_x L_y L_z - \beta\gamma L_x L_y = N - \left\langle \sum_{i<j}^{N} \beta u'(r_{ij}) \frac{x_{ij}^2}{r_{ij}} \right\rangle. \tag{7.37}$$

An analogous result holds for $\partial F/\partial L_y$. For $\partial F/\partial L_z$, since there is no interface in the z direction, the surface tension does not contribute and one has

$$\beta p L_x L_y L_z = N - \left\langle \sum_{i<j}^{N} \beta u'(r_{ij}) \frac{z_{ij}^2}{r_{ij}} \right\rangle. \tag{7.38}$$

(For a uniform system, the surface tension term vanishes in all three derivatives, and taking one-third their sum gives the virial equation derived above.) Subtracting the z expression from the x expression gives

$$\begin{aligned} \beta\gamma &= \frac{1}{L_xL_y} \left\langle \sum_{i<j}^{N} \beta u'(r_{ij}) \frac{x_{ij}^2 - z_{ij}^2}{r_{ij}} \right\rangle \\ &= \frac{1}{2L_xL_y} \left\langle \sum_{i<j}^{N} \beta u'(r_{ij}) \frac{r_{ij}^2 - 3z_{ij}^2}{r_{ij}} \right\rangle, \end{aligned} \tag{7.39}$$

where the final equality follows from the first by changing x to y and taking half the sum. The average that appears here may be written as an integral over space weighted by the inhomogeneous pair density. This result is known as the Kirkwood–Buff expression for the surface tension.

7.2.5 Equipartition Theorem

It will be recalled that the energy of an ideal gas, Eq. (7.23), is given by $\langle E^{\mathrm{id}} \rangle = \langle K \rangle = 3Nk_{\mathrm{B}}T/2$. The Gaussian integrals of the kinetic energy contribute $k_{\mathrm{B}}T/2$ for each component of momentum. More generally, for systems with intermolecular interactions there is a contribution of $k_{\mathrm{B}}T/2$ to the average energy for each quadratic degree of freedom in the Hamiltonian. This is called the equipartition theorem.

There is a generalised equipartition theorem that gives the temperature as a phase space average of a quantity related to the virial.[3] Recall from Ch. 5 that the partition function for an isolated system of energy E is given by Eq. (5.22),

$$Z(E) = \frac{\Delta_E}{N!h^{3N}} \int \mathrm{d}\boldsymbol{\Gamma}\, \delta(\mathcal{H}(\boldsymbol{\Gamma}) - E), \tag{7.40}$$

[3] A. Münster, *Statistical Thermodynamics*, Vol. 1, §1.9, Springer–Verlag, Berlin, 1969. K. Huang, *Statistical Mechanics*, §7.4, Wiley, New York, 1963.

where the number and volume dependence is suppressed for brevity. The partition function in essence gives the weight or volume of phase space within dE of the energy hypersurface. There is a related function that gives the weight or volume of phase space contained by the hypersurface,

$$\Omega(E) = \frac{1}{N!h^{3N}} \int \mathrm{d}\mathbf{\Gamma}\, \Theta(E - \mathcal{H}(\mathbf{\Gamma})), \tag{7.41}$$

where the integrand is the Heaviside step function. The two are related by

$$Z(E) = \Delta_E \frac{\partial \Omega(E)}{\partial E}. \tag{7.42}$$

Since the entropy, which is the logarithm of the partition function, is extensive, the partition function itself must grow exponentially with energy. This exceedingly rapid rate of increase means that almost all of the volume contained by the hypersurface is actually within close proximity to the hypersurface and one has

$$\begin{aligned} S(E) &= k_\mathrm{B} \ln Z(E) \\ &= k_\mathrm{B} \ln \Omega(E) + \mathcal{O} \ln E. \end{aligned} \tag{7.43}$$

In the thermodynamic limit the last term is negligible.

This result allows the evaluation of the desired average. Let x_i be a component of position or momentum of a particular molecule, and similarly for x_j, and consider the virial-like function $x_i \partial \mathcal{H}(\mathbf{\Gamma})/\partial x_j$. The average of this is given by

$$\begin{aligned} \left\langle x_i \frac{\partial \mathcal{H}(\mathbf{\Gamma})}{\partial x_j} \right\rangle &= \frac{\Delta_E}{N!h^{3N} Z(E)} \int \mathrm{d}\mathbf{\Gamma}\, \delta(\mathcal{H}(\mathbf{\Gamma}) - E) x_i \frac{\partial \mathcal{H}(\mathbf{\Gamma})}{\partial x_j} \\ &= \frac{\Delta_E}{N!h^{3N} Z(E)} \frac{\partial}{\partial E} \int \mathrm{d}\mathbf{\Gamma}\, \Theta(E - \mathcal{H}(\mathbf{\Gamma})) x_i \frac{\partial \mathcal{H}(\mathbf{\Gamma})}{\partial x_j} \\ &= \frac{\Delta_E}{N!h^{3N} Z(E)} \frac{\partial}{\partial E} \int \mathrm{d}\mathbf{\Gamma}\, \Theta(E - \mathcal{H}(\mathbf{\Gamma})) \\ &\quad \times \left[\frac{\partial}{\partial x_j} \left(x_i \left\{ \mathcal{H}(\mathbf{\Gamma}) - E \right\} \right) - \delta_{ij} \left\{ \mathcal{H}(\mathbf{\Gamma}) - E \right\} \right] \\ &= \frac{-\delta_{ij} \Delta_E}{N!h^{3N} Z(E)} \frac{\partial}{\partial E} \int \mathrm{d}\mathbf{\Gamma}\, \Theta(E - \mathcal{H}(\mathbf{\Gamma})) \left\{ \mathcal{H}(\mathbf{\Gamma}) - E \right\} \\ &= \frac{-\delta_{ij} \Delta_E}{N!h^{3N} Z(E)} \int \mathrm{d}\mathbf{\Gamma}\, \left[\delta(\mathcal{H}(\mathbf{\Gamma}) - E) \left\{ \mathcal{H}(\mathbf{\Gamma}) - E \right\} \right. \\ &\quad \left. - \Theta(E - \mathcal{H}(\mathbf{\Gamma})) \right] \\ &= \delta_{ij} \Delta_E \frac{\Omega(E)}{Z(E)} \\ &= \delta_{ij} \left[\frac{\partial \ln \Omega(E)}{\partial E} \right]^{-1} \\ &= \delta_{ij} k_\mathrm{B} T + \mathcal{O} E^{-1}. \end{aligned} \tag{7.44}$$

The third equality arises from manipulation of the partial derivative (E is independent of x_j, and $\partial x_i/\partial x_j = \delta_{ij}$). The fourth equality follows from the fact that $\mathcal{H} - E$ vanishes at the boundaries of the integration over x_j. The first integrand of the fifth equality vanishes on the energy hypersurface, and the integral over the second term is just the volume contained by the hypersurface. The final equality manifests the fact that the inverse temperature is the logarithmic energy derivative of the entropy, and that the contained volume and the hypersurface volume are equal in the thermodynamic limit. This result shows in particular that

$$\left\langle p_{i\alpha} \frac{\partial \mathcal{H}(\mathbf{\Gamma})}{\partial p_{i\alpha}} \right\rangle = \left\langle r_{i\alpha} \frac{\partial \mathcal{H}(\mathbf{\Gamma})}{\partial r_{i\alpha}} \right\rangle = k_{\mathrm{B}}T. \tag{7.45}$$

This is the generalised equipartition theorem. (The Hamiltonian may be replaced by the kinetic energy in the first equality, and by the potential energy in the second equality.) Notice that if $r_{i\alpha}$ is a quadratic term in the potential energy, then the averand is just twice the energy of this term. This confirms that the contribution of each harmonic term in the Hamiltonian to the average energy is $k_{\mathrm{B}}T/2$.

The configurational term above is obviously related to the virial of Clausius, Eq. (7.28). Summing over components and molecules gives

$$\left\langle \sum_{i\alpha} r_{i\alpha} \frac{\partial U^{\mathrm{total}}(\mathbf{r}^N)}{\partial r_{i\alpha}} \right\rangle = -\left\langle \mathcal{V}^{\mathrm{total}} \right\rangle = 3Nk_{\mathrm{B}}T. \tag{7.46}$$

This may be reconciled with the virial equation, by noting that the potential energy that appears here includes the external contribution from the walls, whereas only the internal part of the potential energy appears in Eq. (7.29). Combining the two results one has (with $\beta_s = 1/k_{\mathrm{B}}T$)

$$\begin{aligned} \langle p_s \rangle &= \frac{Nk_{\mathrm{B}}T}{V} + \frac{1}{3V}\left\langle \mathcal{V}^{\mathrm{int}} \right\rangle \\ &= \frac{1}{3V}\left\langle \mathcal{V}^{\mathrm{int}} - \mathcal{V}^{\mathrm{total}} \right\rangle \\ &= \frac{1}{3V}\left\langle \sum_{i\alpha} r_{i\alpha} \frac{\partial U^{\mathrm{ext}}(\mathbf{r}^N)}{\partial r_{i\alpha}} \right\rangle \\ &= \frac{-1}{3V}\left\langle \sum_{i\alpha} r_{i\alpha} f_{i\alpha}^{\mathrm{ext}} \right\rangle . \end{aligned} \tag{7.47}$$

This expresses the pressure in terms of the external forces acting on the system.[4]

7.2.6 Average Temperature

The preceding generalised equipartition theorem allows the temperature to be obtained as a phase space average. An alternative expression for the inverse

[4]M. P. Allen and D. J. Tildesley, *Computer Simulation of Liquids*, §2.4, Oxford Univ. Press, Oxford, 1987.

temperature may be obtained directly from the energy derivative of the entropy of the isolated system.[5] Recall that the temperature is defined as

$$T^{-1} = \frac{\partial S(E,N,V)}{\partial E}, \tag{7.48}$$

and that the entropy is

$$S(E,N,V) = k_{\rm B} \ln \frac{1}{N!h^{3N}} \int {\rm d}\mathbf{\Gamma} \delta(\mathcal{H}_N(\mathbf{\Gamma}) - E). \tag{7.49}$$

Now consider the transformation of phase space,

$$\mathbf{\Gamma}' = \mathbf{\Gamma} + \tilde{\mathbf{n}}(\mathbf{\Gamma})\Delta E, \tag{7.50}$$

where ΔE is an infinitesimal. The energy of the points transformed from the original energy hypersurface is

$$\mathcal{H}_N(\mathbf{\Gamma}') = \mathcal{H}_N(\mathbf{\Gamma}) + \Delta E \, \tilde{\mathbf{n}}(\mathbf{\Gamma}) \cdot \nabla \mathcal{H}_N(\mathbf{\Gamma}). \tag{7.51}$$

Accordingly the choice

$$\tilde{\mathbf{n}}(\mathbf{\Gamma}) = \frac{\mathbf{n}(\mathbf{\Gamma})}{\mathbf{n}(\mathbf{\Gamma}) \cdot \nabla \mathcal{H}_N(\mathbf{\Gamma})}, \tag{7.52}$$

represents, for arbitrary $\mathbf{n}(\mathbf{\Gamma})$, a new energy hypersurface parallel to the original,

$$\mathcal{H}_N(\mathbf{\Gamma}') = \mathcal{H}_N(\mathbf{\Gamma}) + \Delta E. \tag{7.53}$$

In view of this the entropy of the new energy hypersurface is

$$\begin{aligned} S(E+\Delta E, N, V) &= k_{\rm B} \ln \frac{1}{N!h^{3N}} \int {\rm d}\mathbf{\Gamma}' \delta(\mathcal{H}_N(\mathbf{\Gamma}') - E - \Delta E) \\ &= k_{\rm B} \ln \frac{1}{N!h^{3N}} \int {\rm d}\mathbf{\Gamma} \left| \frac{\partial \mathbf{\Gamma}'}{\partial \mathbf{\Gamma}} \right| \delta(\mathcal{H}_N(\mathbf{\Gamma}) - E). \end{aligned} \tag{7.54}$$

It is straightforward to show that[6]

$$\left| \frac{\partial \mathbf{\Gamma}'}{\partial \mathbf{\Gamma}} \right| = 1 + \Delta E \, \nabla \cdot \tilde{\mathbf{n}}(\mathbf{\Gamma}). \tag{7.55}$$

Accordingly, the inverse temperature may be written

$$\begin{aligned} \beta &= \frac{S(E+\Delta E,N,V) - S(E,N,V)}{k_{\rm B}\Delta E} \\ &= \frac{1}{\Delta E} \ln \langle 1 + \Delta E \, \nabla \cdot \tilde{\mathbf{n}}(\mathbf{\Gamma}) \rangle \\ &= \langle \nabla \cdot \tilde{\mathbf{n}}(\mathbf{\Gamma}) \rangle . \end{aligned} \tag{7.56}$$

---

[5] H. H. Rugh, Dynamical approach to temperature, *Phys. Rev. Lett.* **78** (1997), 772.

[6] In terms of the eigenvalues λ_i of the matrix $\underline{\underline{A}}$, $\left|\underline{\underline{I}} + \underline{\underline{A}}\right| = \prod_i (1+\lambda_i) = 1 + \sum_i \lambda_i + \sum_{i<j} \lambda_i \lambda_j + \ldots = 1 + {\rm Tr}\, \underline{\underline{A}} + [({\rm Tr}\, \underline{\underline{A}})^2 - {\rm Tr}(\underline{\underline{A}}^2)]/2 + \ldots$.

This result is Rugh's expression.

One may similarly show that the second derivative is

$$\frac{\partial^2 S(E,N,V)/k_{\rm B}}{\partial E^2} = -\left\langle \mathrm{Tr}\,[\nabla\tilde{\mathbf{n}}(\mathbf{\Gamma})]^2\right\rangle = -\left\langle \sum_{i,j=1}^{6N} \frac{\partial \tilde{n}_i}{\partial \Gamma_j}\frac{\partial \tilde{n}_j}{\partial \Gamma_i}\right\rangle. \tag{7.57}$$

This quantity gives the heat capacity. It also confirms that $S_{EE} < 0$ (because $\mathrm{Tr}(\underline{\underline{A}}^2) = \sum_i \lambda_i^2 > 0$).

These results hold for any $\mathbf{n}(\mathbf{\Gamma})$, provided only that $\mathbf{n}(\mathbf{\Gamma})\cdot\nabla\mathcal{H}_N(\mathbf{\Gamma}) \neq 0$. One possible choice is $\mathbf{n}(\mathbf{\Gamma}) = \nabla\mathcal{H}_N(\mathbf{\Gamma})$, in which case

$$\tilde{\mathbf{n}}(\mathbf{\Gamma}) = \frac{\nabla\mathcal{H}_N(\mathbf{\Gamma})}{\nabla\mathcal{H}_N(\mathbf{\Gamma})\cdot\nabla\mathcal{H}_N(\mathbf{\Gamma})}. \tag{7.58}$$

An alternative is $\mathbf{n}(\mathbf{\Gamma}) = \mathbf{r}_{i\alpha}$, in which case

$$\frac{1}{k_{\rm B}T} = \left\langle \frac{-\partial^2\mathcal{H}_N(\mathbf{\Gamma})/\partial r_{i\alpha}^2}{(\partial\mathcal{H}_N(\mathbf{\Gamma})/\partial r_{i\alpha})^2}\right\rangle. \tag{7.59}$$

7.2.7 Chemical Potential

The chemical potential is, in essence, the change in entropy upon adding a particle to a system. For an isothermal system it was shown in Section 4.4 that

$$\langle -\beta_s\mu_s\rangle = \ln Z(N,V,T) - \ln Z(N-1,V,T). \tag{7.60}$$

Exponentiating this and factorising the partition function into ideal and configurational parts one has

$$\begin{aligned} \exp\langle -\beta_s\mu_s\rangle &= \frac{\Lambda^{-3}V}{N}\frac{Q(N,V,T)}{VQ(N-1,V,T)} \\ &\equiv e^{-\beta\mu^{\rm id}}e^{-\beta\mu^{\rm ex}}. \end{aligned} \tag{7.61}$$

The first term is the ideal chemical potential, and the second term is the excess, which is the nontrivial part. In order to write the excess as an average, one needs to define the part of the potential energy that depends upon the Nth particle,

$$\Delta U(\mathbf{r}_N;\mathbf{r}^{N-1}) = U_N(\mathbf{r}^N) - U_{N-1}(\mathbf{r}^{N-1}). \tag{7.62}$$

With this one has

$$\begin{aligned} e^{-\beta\mu^{\rm ex}} &= \frac{\int_V \mathrm{d}\mathbf{r}_N \int_V \mathrm{d}\mathbf{r}^{N-1} e^{-\beta U_{N-1}(\mathbf{r}^{N-1})} e^{-\beta\Delta U(\mathbf{r}_N;\mathbf{r}^{N-1})}}{V\int_V \mathrm{d}\mathbf{r}^{N-1} e^{-\beta U_{N-1}(\mathbf{r}^{N-1})}} \\ &= \frac{1}{V}\int_V \mathrm{d}\mathbf{r}_N \left\langle e^{-\beta\Delta U(\mathbf{r}_N;\mathbf{r}^{N-1})}\right\rangle_{N-1}. \end{aligned} \tag{7.63}$$

In this expression the Nth particle is a ghost or virtual particle; it does not affect the other particles in the system as the configuration average is taken. In effect, it tests the value of the potential throughout the subsystem and measures how easy it would be to insert a particle into the subsystem. This is known as Widom's test particle expression.[7]

Alternatively, this may be turned upside down and written

$$\begin{aligned} e^{\beta\mu^{\mathrm{ex}}} &= \frac{\int_V \mathrm{d}\mathbf{r}^N e^{-\beta U_N(\mathbf{r}^N)} e^{\beta\Delta U(\mathbf{r}_N;\mathbf{r}^{N-1})}}{\int_V \mathrm{d}\mathbf{r}^N e^{-\beta U_N(\mathbf{r}^N)}} \\ &= \left\langle e^{\beta\Delta U(\mathbf{r}_N;\mathbf{r}^{N-1})} \right\rangle_N . \end{aligned} \tag{7.64}$$

In this expression the Nth particle is an actual particle of the subsystem and affects the configurations of it. The averand measures the potential at the position of an actual particle, and tells how easy it would be to delete a particle from the subsystem.

7.3 Particle Densities and Distributions

The positions of the atoms of a crystalline solid are obviously not randomly distributed; they are described by a spatial density. Although a fluid is disordered, there is, on average, local structure about each molecule, and this is characterised by the particle distribution functions. Each is the probability of finding a specified configuration of a few molecules in the fluid, irrespective of the positions of the rest, for example, the probability of finding a molecule a certain distance from a wall, or the probability of two molecules having a particular separation. Moreover, most averages can be written as integrals over the particle distribution functions, and there is a number of techniques for calculating the latter, as will be seen.

7.3.1 Particle Densities of a Closed Subsystem

In Section 6.5 the one-particle distribution was introduced. It gave the probability of finding a particular molecule, say the first, at a particular position, say $\mathbf{r}$, and it was given by $\wp(\mathbf{r}|N,V,T) = \langle \delta(\mathbf{r}-\mathbf{r}_1)\rangle$, or

$$\begin{aligned} \wp(\mathbf{r}|N,V,T) &= \int \mathrm{d}\mathbf{p}^N\, \mathrm{d}\mathbf{r}^N\, \wp(\mathbf{\Gamma}|N,V,T)\, \delta(\mathbf{r}-\mathbf{r}_1) \\ &= \int_V \mathrm{d}\mathbf{r}^N\, \wp(\mathbf{r}^N|N,V,T)\, \delta(\mathbf{r}-\mathbf{r}_1) \\ &= \frac{1}{Q(N,V,T)} \int_V \mathrm{d}\mathbf{r}^N e^{-\beta U_N(\mathbf{r}^N)} \delta(\mathbf{r}-\mathbf{r}_1) . \end{aligned} \tag{7.65}$$

This is normalised to unity, $\int_V \mathrm{d}\mathbf{r}\, \wp(\mathbf{r}|N,V,T) = 1$, since the atom *must* be somewhere in the volume.

[7] B. Widom, Some topics in the theory of fluids, *J. Chem. Phys.* **39** (1963), 2808.

The average number of atoms per unit volume at $\mathbf{r}$ is

$$\rho^{(1)}(\mathbf{r}) = \left\langle \sum_{i=1}^{N} \delta(\mathbf{r} - \mathbf{r}_i) \right\rangle . \tag{7.66}$$

For identical atoms each term in the sum is identical, and so it is clear that $\rho^{(1)}(\mathbf{r}) = N\wp(\mathbf{r}|N, V, T)$. This is called the singlet or one-particle density and one can write it as

$$\rho^{(1)}(\mathbf{r}) = \frac{N}{Q(N, V, T)} \int_V \mathrm{d}\mathbf{r}^N \, e^{-\beta U_N(\mathbf{r}^N)} \delta(\mathbf{r} - \mathbf{r}_N) , \tag{7.67}$$

or, since the Dirac delta makes $\mathbf{r}_N = \mathbf{r}$,

$$\rho^{(1)}(\mathbf{r}_N) = \frac{N}{Q(N, V, T)} \int \mathrm{d}\mathbf{r}^{N-1} \, e^{-\beta U_N(\mathbf{r}^N)} . \tag{7.68}$$

For a uniform fluid (i.e., in the absence of any external potential), the singlet density is constant, $\rho^{(1)}(\mathbf{r}) = N/V$. This follows from the above expression by choosing $\mathbf{r}_N$ as the origin for the various integrations.[8] This is called the density and is usually denoted ρ.

The two-particle density is essentially the average number of particles simultaneously at two positions, per unit volume squared. It is related to the probability of simultaneously finding a particle at $\mathbf{r}$ and a different one at $\mathbf{s}$, and it is defined as

$$\rho^{(2)}(\mathbf{r}, \mathbf{s}) = \left\langle \sum_{i=1}^{N} \sum_{j \neq i}^{N} \delta(\mathbf{r} - \mathbf{r}_i) \delta(\mathbf{s} - \mathbf{r}_j) \right\rangle , \tag{7.69}$$

and this is

$$\rho^{(2)}(\mathbf{r}_N, \mathbf{r}_{N-1}) = \frac{N(N-1)}{Q(N, V, T)} \int_V \mathrm{d}\mathbf{r}^{N-2} \, e^{-\beta U_N(\mathbf{r}^N)} . \tag{7.70}$$

The two particle density is proportional to the pair probability function, which is $\wp(\mathbf{r}, \mathbf{s}|N, V, T) = \langle \delta(\mathbf{r} - \mathbf{r}_1) \, \delta(\mathbf{s} - \mathbf{r}_2) \rangle$, and this gives the probability of simultaneously finding particle 1 at $\mathbf{r}$ and particle 2 at $\mathbf{s}$. The proportionality constant is the normalisation constant for the pair density, namely $N(N-1)$.

The *n-particle density* of an isothermal subsystem is

$$\rho_N^{(n)}(\mathbf{r}^n) = \frac{N!}{(N-n)! Q(N, V, T)} \int_V \mathrm{d}\mathbf{r}_{n+1} \ldots \mathrm{d}\mathbf{r}_N \, e^{-\beta U_N(\mathbf{r}^N)} . \tag{7.71}$$

This is proportional to the probability of simultaneously finding n particles within $\mathrm{d}\mathbf{r}_1, \ldots, \mathrm{d}\mathbf{r}_n$ of $\mathbf{r}_1, \ldots, \mathbf{r}_n$, irrespective of the positions of the remaining

[8] A system must display the same symmetries as its Hamiltonian. Since by definition a uniform fluid is translationally invariant, so must be the potential energy. Accordingly, the transformation $\mathbf{R}_i = \mathbf{r}_i - \mathbf{r}_N$ must leave the potential energy unchanged, $U_N(\mathbf{R}^N) = U_N(\mathbf{r}^N)$. Obviously the presence of an external one-body potential breaks this symmetry and creates an inhomogeneous fluid.

$N - n$ particles. That is, $\rho_N^{(n)}(\mathbf{r}^n) = (N!/(N-n)!)\wp(\mathbf{r}^n|N,V,T)$, since the density is normalised such that

$$\int_V \mathrm{d}\mathbf{r}^n \, \rho_N^{(n)}(\mathbf{r}^n) = \frac{N!}{(N-n)!}. \tag{7.72}$$

The subscript N has here been appended to make it clear that this is the result for a closed subsystem with N particles. It could similarly be appended to the singlet and pair densities given explicitly above.

7.3.2 Particle Densities of an Open Subsystem

The n-particle density of an open or grand canonical subsystem is simply the weighted sum over N. It is given by

$$\begin{aligned} \rho^{(n)}(\mathbf{s}^n) &= \left\langle \frac{N!}{(N-n)!} \delta(\mathbf{r}_1 - \mathbf{s}_1)\, \delta(\mathbf{r}_2 - \mathbf{s}_2) \ldots \delta(\mathbf{r}_n - \mathbf{s}_n) \right\rangle_\mu \\ &= \frac{1}{\Xi(\mu,V,T)} \sum_{N=n}^{\infty} \frac{\Lambda^{-3N} e^{\beta\mu N}}{(N-n)!} \\ &\quad \times \int \mathrm{d}\mathbf{r}^N \, e^{-\beta U_N(\mathbf{r}^N)} \delta(\mathbf{r}_1 - \mathbf{s}_1) \ldots \delta(\mathbf{r}_n - \mathbf{s}_n)\,, \end{aligned} \tag{7.73}$$

or

$$\begin{aligned} \rho^{(n)}(\mathbf{r}^n) &= \frac{1}{\Xi(\mu,V,T)} \sum_{N=n}^{\infty} \frac{\Lambda^{-3N} e^{\beta\mu N}}{(N-n)!} \int \mathrm{d}\mathbf{r}_{n+1} \ldots \mathrm{d}\mathbf{r}_N \, e^{-\beta U_N(\mathbf{r}^N)} \\ &= \frac{1}{\Xi(\mu,V,T)} \sum_{N=n}^{\infty} \frac{\Lambda^{-3N} e^{\beta\mu N}}{N!} Q(N,V,T) \rho_N^{(n)}(\mathbf{r}^n). \end{aligned} \tag{7.74}$$

The normalisation is evidently

$$\begin{aligned} \int \mathrm{d}\mathbf{r}^n \, \rho^{(n)}(\mathbf{r}^n) &= \frac{1}{\Xi(\mu,V,T)} \sum_{N=n}^{\infty} \frac{\Lambda^{-3N} e^{\beta\mu N}}{N!} Q(N,V,T) \frac{N!}{(N-n)!} \\ &= \left\langle \frac{N!}{(N-n)!} \right\rangle_\mu , \end{aligned} \tag{7.75}$$

as expected. In the case of the singlet density this is

$$\int \mathrm{d}\mathbf{r} \, \rho^{(1)}(\mathbf{r}) = \langle N \rangle_\mu \,, \tag{7.76}$$

and for the pair density it is

$$\int \mathrm{d}\mathbf{r}\, \mathrm{d}\mathbf{s} \, \rho^{(2)}(\mathbf{r},\mathbf{s}) = \left\langle N^2 \right\rangle_\mu - \langle N \rangle_\mu \,. \tag{7.77}$$

7.3.3 Ideal Gas and Asymptotic Limits

For a uniform ideal gas, which has 0 external potential, the configuration integral is $Q(N, V, T) = V^N$, and the multiparticle density for a closed system is

$$\rho_{N,\mathrm{id}}^{(n)}(\mathbf{r}^n) = \frac{N!}{(N-n)!V^n} = \frac{N}{V}\frac{N-1}{V}\cdots\frac{N-n+1}{V}. \tag{7.78}$$

If $n \ll N$, this is just ρ^n, where the singlet density is $\rho = N/V$. In the presence of an external potential this is multiplied by the product of the Boltzmann factors of the potential at the respective particle positions. Explicitly, the singlet density is

$$\rho_{N,\mathrm{id}}^{(1)}(\mathbf{r}) = \frac{N}{V}, \tag{7.79}$$

and the pair density is

$$\rho_{N,\mathrm{id}}^{(2)}(\mathbf{r}, \mathbf{s}) = \frac{N}{V}\frac{N-1}{V} = \rho^2(1 - N^{-1}). \tag{7.80}$$

Notice the factor of $N-1$ that occurs rather than N; this difference is negligible in the thermodynamic limit.

An ideal gas in an open system has

$$\begin{aligned} \rho_{\mu,\mathrm{id}}^{(n)}(\mathbf{r}^n) &= \frac{1}{\Xi(\mu, V, T)} \sum_{N=n}^{\infty} \frac{\Lambda^{-3N} e^{\beta\mu N}}{N!} V^N \frac{N!}{(N-n)!V^{N-n}} \\ &= \Lambda^{-3n} e^{\beta\mu n} \\ &= \rho^n, \end{aligned} \tag{7.81}$$

where the average density for the ideal gas is $\rho = \langle N \rangle_\mu / V = \Lambda^{-3} e^{\beta\mu}$. Notice that unlike the closed system, the open system multiparticle density is the strict product of singlet densities, even if the thermodynamic limit is not invoked.

Asymptotes

An ideal gas consists of noninteracting molecules. In this case the multiparticle densities are just the product of lower-order densities, strictly so for an open system. However, even real molecules, which interact at close separations, have negligible influence on each other when they are sufficiently far apart. In this asymptotic regime they are said to be uncorrelated, and the multiparticle density also factorises. For example, if the first m particles are far-separated from the remainder, so that $r_{ij} \to \infty$ for all $i \le m$ and $j > m$, then

$$\rho_\mu^{(m+n)}(\mathbf{r}_1, \ldots, \mathbf{r}_{m+n}) \sim \rho_\mu^{(m)}(\mathbf{r}_1, \ldots, \mathbf{r}_m)\rho_\mu^{(n)}(\mathbf{r}_{m+1}, \ldots, \mathbf{r}_{m+n}), \tag{7.82}$$

assuming $N \gg m + n$. This effectively says that for uncorrelated particles the joint probability is the product of independent probabilities. This is the leading asymptote; the primary correction to this is proportional to the potential energy of interaction between the two groups of particles, as will be discussed in a later chapter.

The opposite asymptotic limit is that of small separations. Real molecules have a finite size that prevents them overlapping (see the hardsphere potential or the r^{-12} repulsion of the Lennard–Jones potential discussed at the beginning of this chapter). Accordingly the particle densities must vanish for any configuration of overlapping particles,

$$\rho^{(n)}(\mathbf{r}^n) \to 0, \;\; r_{ij} \to 0, \;\text{ for any } i \in [1,n],\, j \in [1,n]. \tag{7.83}$$

Compressibility

As can be seen from the above results for the ideal gas the difference between the particle densities in an open and in a closed subsystem is negligible in the thermodynamic limit, where $N^{-1} \ll 1$. The important exception is the integral of the n-particle density itself, because then the correction is multiplied by the volume of the system; the canonical result will differ from the grand canonical result by terms of order V/N, which is not necessarily small compared to unity. The integral of the pair density $\rho^{(2)}_\mu(\mathbf{r},\mathbf{s})$ may be obtained by subtracting its asymptote $\rho^{(1)}_\mu(\mathbf{r})\rho^{(1)}_\mu(\mathbf{s})$, which makes the integral convergent. Also, in view of the normalisation Eq. (7.77), one may add the self-density, $\rho^{(1)}_\mu(\mathbf{r})\delta(\mathbf{r}-\mathbf{s})$, which gives

$$\begin{aligned}\int_V \mathrm{d}\mathbf{r}\,\mathrm{d}\mathbf{s}\, &\left[\rho^{(2)}_\mu(\mathbf{r},\mathbf{s}) - \rho^{(1)}_\mu(\mathbf{r})\rho^{(1)}_\mu(\mathbf{s}) + \rho^{(1)}_\mu(\mathbf{r})\delta(\mathbf{r}-\mathbf{s})\right] \\ &= \left\langle N^2\right\rangle_\mu - \left\langle N\right\rangle^2_\mu .\end{aligned} \tag{7.84}$$

This is just the fluctuation in particle number, which as seen in Section 4.3.3 is the second chemical potential derivative of the grand partition function, and which is related to the compressibility of the subsystem. The fluctuation in particle number is strictly 0 in a closed system, so one sees that the integral of the multiparticle densities do indeed differ between open and closed systems.

7.3.4 Particle Distribution Functions

The *n-particle distribution* functions are defined in terms of the corresponding densities by dividing by the singlet densities

$$g^{(n)}(\mathbf{r}^n) = \rho^{(n)}(\mathbf{r}^n) / \prod_{i=1}^{n} \rho^{(1)}(\mathbf{r}_i) \tag{7.85}$$

The product of singlet densities represents the *independent* distribution of molecules, as is the case for the ideal gas. Hence the n-particle distribution function measures the correlations between particles, and the departure from unity arises from their mutual interaction. As just discussed, $g^{(n)}(\mathbf{r}^n) = 0$ if any $r_{ij} \lesssim d$ (d is the molecular diameter and molecules don't overlap), and $g^{(n)}(\mathbf{r}^n) \to 1$, if all $r_{ij} \to \infty$ (because the interaction potential decays to 0 at large separations, and particles no longer feel each other and are hence uncorrelated).

Perhaps the most important particle distribution function is the pair one, since this is directly related to the structure factor, which is measurable by X-ray or neutron scattering. Also, when the configurational part of the Hamiltonian is pairwise additive, almost all thermodynamic properties can be written as integrals over the pair distribution function. For a uniform simple fluid (spherically symmetric molecules far from walls or external fields), the correlation between the particles depends only upon their separation In this case the two-particle distribution function is called the *radial distribution function* $g(r)$.

Excess Energy

For a uniform subsystem of spherical molecules with a pairwise additive potential, $U_N(\mathbf{r}^N) = \sum_{i<j}^{N} u(r_{ij})$, and the excess internal energy is

$$\begin{aligned}
E^{\text{ex}} &= \left\langle \sum_{i<j}^{N} u(r_{ij}) \right\rangle \\
&= \frac{N(N-1)}{2} \langle u(r_{N,N-1}) \rangle \\
&= \frac{N(N-1)}{2Q(N,V,T)} \int_V \mathrm{d}\mathbf{r}^N \, e^{-\beta U_N(\mathbf{r}^N)} u(r_{N,N-1}) \\
&= \frac{N(N-1)}{2Q(N,V,T)} \int_V \mathrm{d}\mathbf{r}_N \, \mathrm{d}\mathbf{r}_{N-1} \, u(r_{N,N-1}) \int_V \mathrm{d}\mathbf{r}^{N-2} \, e^{-\beta U_N(\mathbf{r}^N)} \\
&= \frac{1}{2} \int_V \mathrm{d}\mathbf{r}_N \, \mathrm{d}\mathbf{r}_{N-1} \, u(r_{N,N-1}) \rho_N^{(2)}(r_{N,N-1}) \\
&= \frac{\rho^2 V}{2} \int_V \mathrm{d}\mathbf{r} \, u(r) g(r). \qquad (7.86)
\end{aligned}$$

Note that the distinction between grand canonical and canonical subsystems is here unimportant because the decay of the pair potential makes the tail of the pair density, where the two differ, unimportant. The integrand is the energy of a pair of molecules at a separation r, weighted by the average number of molecules per unit volume simultaneously at that separation, $\rho^2 g(r)$. One particle is integrated over all of space with the other particle at the origin, and then the latter is also integrated over space, giving the volume factor. The factor of a half accounts for the double counting. The volume element is of course $\mathrm{d}\mathbf{r} = 4\pi r^2 \mathrm{d}r$.

Virial Pressure

The average of the virial of Clausius for a uniform system with pairwise additive potential is, using Eq. (7.31),

$$\langle \mathcal{V} \rangle = \frac{1}{Q(N,V,T)} \int_V \mathrm{d}\mathbf{r}^N \, e^{-\beta U_N^{\text{int}}(\mathbf{r}^N)} \frac{-1}{2} \sum_{i \neq j}^{N} r_{ij} u'(r_{ij})$$

$$
\begin{aligned}
&= \frac{-N(N-1)}{2Q(N,V,T)} \int_V \mathrm{d}\mathbf{r}^N \, e^{-\beta U_N^{\mathrm{int}}(\mathbf{r}^N)} r_{12} u'(r_{12}) \\
&= \frac{-1}{2} \int_V \mathrm{d}\mathbf{r}_1 \, \mathrm{d}\mathbf{r}_2 \, \rho^{(2)}(r_{12}) r_{12} u'(r_{12}) \\
&= \frac{-V}{2} \int_V \mathrm{d}\mathbf{r} \, \rho^{(2)}(r) r u'(r) \\
&= -2\pi\rho^2 V \int_0^\infty \mathrm{d}r \, g(r) r^3 u'(r). \qquad (7.87)
\end{aligned}
$$

Accordingly the virial equation, Eq. (7.29), gives the average pressure as

$$
\langle \beta_s p_s \rangle = \rho - \frac{2\pi\beta\rho^2}{3} \int_0^\infty \mathrm{d}r \, u'(r) g(r) r^3, \qquad (7.88)
$$

where the prime denotes differentiation with respect to argument. Again this result is the same, with relative error N^{-1}, for a closed and for an open subsystem.

If there is in addition a three-body potential, then using Eq. (7.32) it contributes to the average of the virial an amount,

$$
\begin{aligned}
\left\langle \mathcal{V}^{(3)} \right\rangle &= \frac{-2}{Q(N,V,T)} \int_V \mathrm{d}\mathbf{r}^N \, e^{-\beta U_N^{\mathrm{int}}(\mathbf{r}^N)} \sum_{i<j<k}^{N} r_{ij} \frac{\partial u(r_{ij}, r_{ik}, \theta_{jik})}{\partial r_{ij}} \\
&= \frac{-2N(N-1)(N-2)}{3! Q(N,V,T)} \int_V \mathrm{d}\mathbf{r}^N \, e^{-\beta U_N^{\mathrm{int}}(\mathbf{r}^N)} r_{12} \frac{\partial u(r_{12}, r_{13}, \theta_{213})}{\partial r_{12}} \\
&= \frac{-2}{3!} \int_V \mathrm{d}\mathbf{r}_1 \, \mathrm{d}\mathbf{r}_2 \, \mathrm{d}\mathbf{r}_3 \, \rho^{(3)}(r_{12}, r_{13}, \theta_{213}) r_{12} \frac{\partial u(r_{12}, r_{13}, \theta_{213})}{\partial r_{12}} \\
&= \frac{-8\pi^2\rho^3 V}{3} \int_0^\infty \mathrm{d}r \, r^2 \int_0^\infty \mathrm{d}s \, s^2 \int_0^\pi \mathrm{d}\theta \, \sin\theta \\
&\quad \times g^{(3)}(r,s,\theta) \, r \frac{\partial u(r,s,\theta)}{\partial r}. \qquad (7.89)
\end{aligned}
$$

This may be added to the ideal and the pair contributions.

Low-Density Limit

In the limit of vanishing density (e.g., when there are only two molecules in the system), the probability of them being at a particular separation is given by the Boltzmann distribution. Hence for a pairwise additive potential, the radial distribution function is $g(r) = \exp -\beta u(r)$, $\rho \to 0$. This limiting form for the radial distribution function can be used to obtain the low-density expansion of the internal energy and the virial pressure. As more particles are added to the system, there is an indirect contribution to the energy of a pair at a given separation, and this modifies the above limit. Notice that at large separations, the low-density limit gives the asymptotic expansion $g(r) \sim 1 - \beta u(r)$, $\quad r \to \infty$; this asymptote also applies at higher densities, except that the potential will then be multiplied by a density-dependent constant.

7.3.5 Contact Theorem

Planar Geometry

A system confined between two walls may be represented by an external one-body potential of the form $V(z;L)$, where L is the distance between the walls. In this case the configuration integrals have no explicit volume dependence; the z integrals are infinite in both directions as it is the potential that confines the particles. In terms of the volume of the system, $V = AL$, the derivative of the thermodynamic potential gives the pressure. Writing this as an L derivative gives the normal component,

$$\begin{aligned}
\langle \beta p_{zz} \rangle &= \frac{1}{Z(N,A,L,T)} \frac{\partial Z(N,A,L,T)}{A \partial L} \\
&= \frac{k_{\mathrm{B}}T}{N!AZ} \int \mathrm{d}\mathbf{r}^N \, \mathrm{d}\mathbf{p}^N e^{-\beta \mathcal{H}^*(\mathbf{r}^N,\mathbf{p}^N)} \frac{\partial}{\partial L} \prod_{i=1}^{N} e^{-\beta V(z_i;L)} \\
&= \frac{1}{N!AZ} \int_A \mathrm{d}\mathbf{R}_1 \int_{-\infty}^{\infty} \mathrm{d}z_1 \, N \frac{\partial[-\beta V(z_1;L)]}{\partial L} \\
&\quad \times \int \mathrm{d}\mathbf{r}^{N-1} \, \mathrm{d}\mathbf{p}^N e^{-\beta \mathcal{H}^*(\mathbf{r}^N,\mathbf{p}^N)} \prod_{i=1}^{N} e^{-\beta V(z_i;L)} \\
&= -\beta \int_{-\infty}^{\infty} \mathrm{d}z_1 \, \frac{\partial V(z;L)}{\partial L} \rho^{(1)}(z),
\end{aligned} \tag{7.90}$$

where $\mathcal{H}^*$ is that part of the energy excluding the external potential, and $\rho^{(1)}(z)$ is the singlet density. Due to the confining effects of the wall, $\rho^{(1)}(z) \to 0$, $|z| \to \infty$. (More precisely, $\rho^{(1)}(z) = 0$, $|z| \gtrsim L/2$.) This is a formally exact expression for the pressure of system in the presence of a one-body confining potential. It has been derived for a canonical system but it should be clear that the same expression holds for an open system.

A rather common case is when the external potential represents a hard wall, $V(z;L) = V^{\mathrm{hw}}(z;L)$, or a hard wall plus a soft potential, $V(z;L) = V^{\mathrm{hw}}(z;L) + V^{\mathrm{soft}}(z;L)$. The L-dependent part of the hard-wall potential is

$$V^{\mathrm{hw}}(z;L) = \begin{cases} \infty, & z > L \\ 0, & z < L. \end{cases} \tag{7.91}$$

There may or may not be a similar hard wall at $z = 0$. The singlet density vanishes beyond the hard wall and is discontinuous at the hard wall since it contains a factor of $e^{-\beta V(z;L)}$. Accordingly the (partial) cavity function $\rho_c^{(1)}(z) \equiv e^{\beta V^{\mathrm{hw}}(z;L)} \rho^{(1)}(z)$ is continuous across the wall. Since the exponential of the negative of the hard-wall potential is the Heaviside unit step function, and since the derivative of the step function is the Dirac δ function, the hard-wall contribution to the above may be written

$$\langle \beta p_{zz}^{\mathrm{hw}} \rangle = -\int_{-\infty}^{\infty} \mathrm{d}z \left[\frac{\partial \beta V^{\mathrm{hw}}(z;L)}{\partial L} e^{-\beta V^{\mathrm{hw}}(z;L)} \right] \left[e^{\beta V^{\mathrm{hw}}(z;L)} \rho^{(1)}(z) \right]$$

$$
\begin{aligned}
&= \int_{-\infty}^{\infty} \mathrm{d}z \, \frac{\partial e^{-\beta V^{\mathrm{hw}}(z;L)}}{\partial L} \rho_c^{(1)}(z) \\
&= \int_{-\infty}^{\infty} \mathrm{d}z \, \delta(z-L) \rho_c^{(1)}(z) \\
&= \rho_c^{(1)}(L) \\
&= \rho^{(1)}(L^-). \qquad (7.92)
\end{aligned}
$$

That is, the normal component of the pressure of the inhomogeneous system equals the contact density at the hard wall. The full expression is

$$
\langle \beta p_{zz} \rangle = \rho(L^-) - \int_{-\infty}^{L} \mathrm{d}z \, \frac{\partial V^{\mathrm{soft}}(z;L)}{\partial L} \rho(z). \qquad (7.93)
$$

This result holds as well for a semi-infinite system as for a slit–pore of finite width. In the former case, since the pressure must be constant throughout the system (since a system acts as a volume reservoir for itself), far from the wall p_{zz} may be identified with the bulk pressure. Otherwise, it is still constant throughout the pore but it is dependent on the separation of the walls.

Spherical Geometry

For the case of a spherical inhomogeneity of radius R, with external potential, $V(r;R)$, the derivative of the thermodynamic potential is similarly

$$
\begin{aligned}
\frac{\partial F(N,V,T;R)}{\partial R} &= \frac{-k_{\mathrm{B}}T}{Z(N,V,T;R)} \frac{\partial Z(N,V,T;R)}{\partial R} \\
&= 4\pi \int_{0}^{\infty} \mathrm{d}r \, r^2 \, \frac{\partial V(r;R)}{\partial R} \rho^{(1)}(r). \qquad (7.94)
\end{aligned}
$$

Strictly speaking, the left-hand side would be better expressed as the derivative of the total unconstrained entropy; in the thermodynamic limit this is $-\beta$ times the Helmholtz free energy. Again, if there is a hard cavity present, $V(r;R) = V^{\mathrm{hw}}(r;R) + V^{\mathrm{soft}}(r;R)$, this reduces to

$$
\begin{aligned}
&\frac{\partial F(N,V,T;R)}{\partial R} \\
&\quad = k_{\mathrm{B}}T 4\pi R^2 \rho^{(1)}(R^+) + 4\pi \int_{R}^{\infty} \mathrm{d}r \, r^2 \, \frac{\partial V^{\mathrm{soft}}(r;R)}{\partial R} \rho^{(1)}(r). \qquad (7.95)
\end{aligned}
$$

Integrating this with respect to radius gives the solvation free energy of the spherical solute. One expects that $F \sim 4\pi R^3 p/3 + 4\pi R^2 \gamma$, $R \to \infty$, where p is the pressure, and γ is the surface tension of the planar interface. Hence this expression can be used to obtain the curvature dependence of the surface tension.

Summary

- Intermolecular potentials include Coulomb, hard-sphere, Lennard–Jones, and a variety of other spherical pair potentials. In addition there exist orientation-dependent potentials, three- and many-body potentials, and external potentials.
- The partition function and its nontrivial part, the configuration integral, appear as normalising factors in the microstate and macrostate probabilities. Their logarithmic derivatives yield average values such as energy and pressure.
- The particle densities and the related distribution functions give the local structure of the molecules in a system. They may be used to obtain averages such as energy and pressure by spatial integration.
- The pressure of a system is given by $k_{\mathrm{B}}T$ times the contact density at a hard wall.

Chapter 8

Diagrammatic and Functional Expansions

8.1 Virial Expansion

8.1.1 Mayer f-function

It is easy to treat noninteracting particles because the $3N$-dimensional configuration integral factorises into independent integrals. At low densities, particles are on average so far apart that their interactions are negligible, and the ideal gas provides the correct limiting picture. For interacting particles, systematic treatments begin with an expansion in powers of density. This will yield, for example, the first correction to the ideal gas equation of state due to intermolecular interactions.

The simplest and most common system dealt with is one in which the potential energy consists solely of pair terms. That is, there is no density inhomogeneity due to a one-body external potential, and there are no three-body and higher interactions. Whilst it is relatively straightforward to include many body terms in the formalism, the complications that this introduces tends to obscure the concepts being described. In addition there are practical difficulties in implementing the full formalism, and this provides additional motivation for focussing on pair-wise additive interactions.

What is sought is an expansion for the thermodynamic potential and thence the pressure, which means that the configuration integral must first be dealt with. For a pair-wise additive potential, the integrand of the configuration integral may be written

$$\exp -\beta U_N(\mathbf{r}^N) = \exp -\beta \sum_{i<j}^{N} u(r_{ij}) = \prod_{i<j}^{N} \exp -\beta u(r_{ij}). \tag{8.1}$$

In general the pair potential $u(r)$ goes to infinity at small separations, which prevents molecular overlap, and goes to 0 at large separations, where molecules

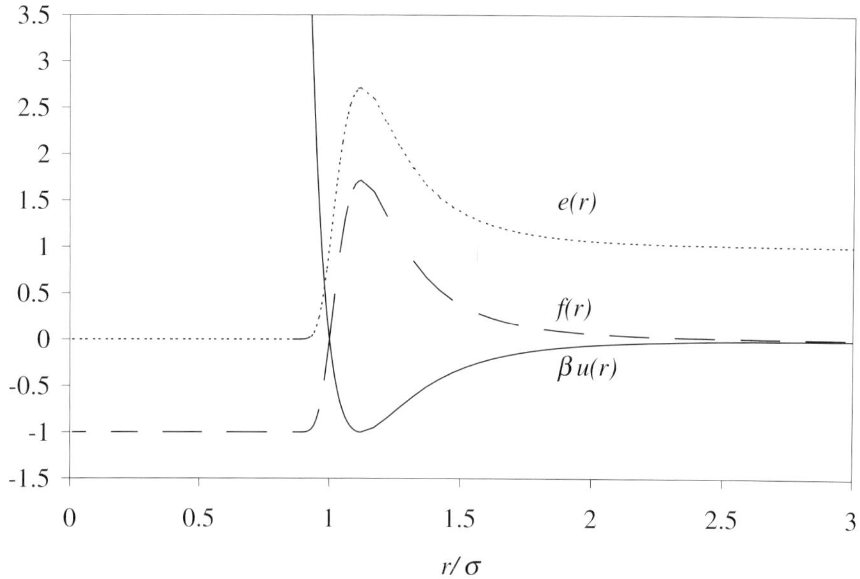

Figure 8.1: The Lennard–Jones potential divided by k_BT (full curve), its exponential (dotted curve), and the corresponding Mayer f-function (dashed curve), for $\epsilon = k_BT$.

no longer interact. Therefore its exponential goes to 0 at small separations and to unity at large separations. Since a dilute gas is characterised by large separations between the molecules, it is convenient to define a function that goes to 0 at large separations, thereby providing a basis for a well-behaved expansion. Subtracting the large separation asymptote, one defines the Mayer f-function,

$$f(r) \equiv e^{-\beta u(r)} - 1. \tag{8.2}$$

This has the convenient property that $f(r) \to 0$, $r \to \infty$, which facilitates the evaluation of the various integrals.

In Fig. 8.1 the behaviour of the various pair functions are shown. It may be seen that $f(r)$ is well behaved at small separations, $f(r) \approx -1$, $r \lesssim \sigma$, and that it decays to 0 at large separations at the same rate as the pair potential, $f(r) \sim -\beta u(r)$, $r \to \infty$. The figure corresponds to a temperature $T = \epsilon/k_B$. At temperatures higher than this the influence of the attractive tail of the Lennard–Jones potential is diminished, but the repulsive core is little changed (because the potential is so rapidly increasing in this region). Conversely, at low temperatures the attraction between the molecules, where $f(r) > 0$, becomes relatively more important.

With this definition, the exponential of the pair potential becomes $e(r) =$

$f(r)+1$, and the integrand of the configuration integral becomes

$$\exp -\beta U_N(\mathbf{r}^N) = \prod_{i<j}^{N} [1+f(r_{ij})]. \tag{8.3}$$

The product may be written out in full and the terms collected in groups of increasing powers of f. For example, for $N=3$ (writing $f_{ij} \equiv f(r_{ij})$),

$$\begin{aligned} &(1+f_{12})(1+f_{13})(1+f_{23}) \\ &= 1+f_{12}+f_{13}+f_{23}+f_{12}f_{13}+f_{12}f_{23}+f_{13}f_{23}+f_{12}f_{13}f_{23}. \end{aligned} \tag{8.4}$$

The number and variety of terms in the sum increases rapidly with N (for $N=4$ there are 64 terms), and techniques for dealing with them will be given below.

8.1.2 Low-Density Expansion

In view of the general expansion, the configuration integral may be written

$$\begin{aligned} Q(N,V,T) &= \int_V \mathrm{d}\mathbf{r}^N \exp -\beta U_N(\mathbf{r}^N) \\ &= \int_V \mathrm{d}\mathbf{r}^N \prod_{i<j}^{N} [1+f(r_{ij})] \\ &= \int_V \mathrm{d}\mathbf{r}^N \left[1 + \sum_{i<j} f(r_{ij}) + \ldots \right]. \end{aligned} \tag{8.5}$$

For the moment only these two explicit terms will be treated. The contribution of the zeroth term to the configuration integral is

$$Q_0 = \int_V \mathrm{d}\mathbf{r}_1 \ldots \mathrm{d}\mathbf{r}_N \, 1 = V^N. \tag{8.6}$$

This is the ideal gas contribution, and would have been obtained if $f(r)=0$, which is equivalent to $u(r)=0$, or to $T \rightarrow \infty$. It also arises as $N/V \rightarrow 0$.

There are no terms involving only a single atom (i.e., with only one index), but there are $N(N-1)/2$ terms involving only a pair of atoms (two indices). All of these terms give the same contribution because the indices are just dummy variables of integration. Hence

$$\begin{aligned} Q_2 &= \frac{N(N-1)}{2} \int_V \mathrm{d}\mathbf{r}_1 \ldots \mathrm{d}\mathbf{r}_N \, f(r_{12}) \\ &= \frac{N(N-1)}{2} V^{N-2} \int_V \mathrm{d}\mathbf{r} \, \mathrm{d}\mathbf{r}_2 \, f(r) \\ &= \frac{N(N-1)}{2V^2} V^N V \int_V \mathrm{d}\mathbf{r} \, f(r). \end{aligned} \tag{8.7}$$

The first factor is just $\rho^2/2$, the factor V^N is just Q_0, and the factor V comes from the integral over $\mathrm{d}\mathbf{r}_2$.

These two terms contribute to an expansion of the configuration integral, $Q(N,V,T) = Q_0(1+Q_2/Q_0+\ldots)$. From the relationship of the Helmholtz free energy with the logarithm of the partition function,

$$F(N,V,T) = -k_{\mathrm{B}}T\ln Z(N,V,T) = -k_{\mathrm{B}}T\ln\left[Z^{\mathrm{id}}V^{-N}Q\right], \tag{8.8}$$

and using the facts that $Z^{\mathrm{id}}(N,V,T) = V^N/N!\Lambda^{3N}$ and that $\ln[1+x] = x+\mathcal{O}x^2$, one has

$$\begin{aligned} F(N,V,T) &= -k_{\mathrm{B}}T\ln\left[V^N/N!\Lambda^{3N}\right] - k_{\mathrm{B}}T\ln\left[1+Q_2/Q_0+\ldots\right] \\ &= Nk_{\mathrm{B}}T(\ln[\rho\Lambda^3]-1) - \frac{\rho^2Vk_{\mathrm{B}}T}{2}\int_V \mathrm{d}\mathbf{r}\, f(r)+\ldots, \end{aligned} \tag{8.9}$$

where the first term is the ideal contribution, F^{id}. Note that all the terms scale with the size of the system. This is an expansion for the Helmholtz free energy in powers of density. The first neglected term is of order ρ^3V since it involves three molecules at a time. This result was obtained by retaining the leading two powers of f_{ij}, and hence it may also be loosely considered a high-temperature expansion, since $f \to 0$ as $T \to \infty$ (at least for the attractive tail).

The expansion for the pressure follows by differentiation with respect to volume (at constant N; the integral, if convergent, is independent of the volume),

$$p = \rho k_{\mathrm{B}}T - \frac{\rho^2k_{\mathrm{B}}T}{2}\int \mathrm{d}\mathbf{r}\, f(r)+\ldots. \tag{8.10}$$

Here one sees the first correction to the ideal gas equation of state due to intermolecular interactions. For interactions that are predominately attractive, $f(r) > 0$, one sees that the pressure is less than that of an ideal gas at the same density. The converse holds for repulsive interactions.

This result may be rederived from the virial equation, Eq. (7.88), by using the low-density limit for the radial distribution function, $g(r) = e^{-\beta u(r)}$.[1] One obtains

$$\begin{aligned} p &= \rho k_{\mathrm{B}}T - \frac{2\pi\rho^2}{3}\int_0^\infty \mathrm{d}r\, r^3\frac{\mathrm{d}u(r)}{\mathrm{d}r}e^{-\beta u(r)} \\ &= \rho k_{\mathrm{B}}T + \frac{2\pi\rho^2k_{\mathrm{B}}T}{3}\int_0^\infty \mathrm{d}r\, r^3\frac{\mathrm{d}f(r)}{\mathrm{d}r} \\ &= \rho k_{\mathrm{B}}T - 2\pi\rho^2k_{\mathrm{B}}T\int_0^\infty \mathrm{d}r\, r^2f(r), \end{aligned} \tag{8.11}$$

where $f'(r) = -\beta u'(r)e^{-\beta u(r)}$ has been used, as well as an integration by parts.

[1]This result follows since two molecules separated by r can be considered a macrostate whose energy $u(r)$ removes entropy $u(r)/T$ from the rest of the system. At low densities this is the main effect, and the Boltzmann factor is the exponential of this separation-dependent reservoir entropy. At higher densities the entropy of the remaining molecules becomes increasingly dependent upon the separation of these two, which directly determines the available configurations, and this effect becomes significant compared to the indirect effect on the entropy via the potential energy.

The expansion of the equation of state in powers of density is called the *virial expansion*

$$\beta p = \sum_{n=1}^{\infty} B_n \rho^n, \tag{8.12}$$

and the B_n are the virial coefficients. The first term, $B_1 = 1$, corresponds to the ideal gas expression. The first correction to the ideal gas law at low densities is the second virial coefficient, which from the above is

$$B_2 = \frac{-1}{2} \int \mathrm{d}\mathbf{r}\, f(r). \tag{8.13}$$

Its significance lies in the fact that it provides a direct relationship between the pressure and the intermolecular potential, at least at low densities. Hence *ab initio* quantum potentials for pairs of molecules can be inserted into the above, and the predicted coefficient can be compared to the measured one. Alternatively, parameters in the model potentials discussed in the preceding chapter can be optimised to provide the best fit to the measured coefficient over a range of temperatures. In both cases the subsequent pair potential should be applicable over the whole density regime, provided only that the assumption of pair-wise additivity holds.

8.2 Cluster Diagrams

The evaluation of the higher-order virial coefficients is greatly facilitated by the use of cluster diagrams.[2] These are a pictorial representation of multi-dimensional integrals, together with a few rules for their manipulation. A diagram consists of lines and circles, and a *simple* diagram has at most one line connecting any pair of circles. Although different meanings can be attached to these quantities, to begin with a circle represents a particle and a line represents the Mayer f function. A filled circle corresponds to an integration. Hence

$$= \int \mathrm{d}\mathbf{r}_2\, f(r_{12}) \tag{8.14}$$

represents the integral associated with the second virial coefficient. Filled circles are called field points and open circles are called root points. If both circles were filled, one would have

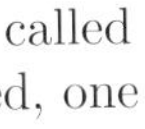

$$= \int \mathrm{d}\mathbf{r}_1\, \mathrm{d}\mathbf{r}_2\, f(r_{12}). \tag{8.15}$$

[2] G. Stell, Cluster expansions for classical systems in equilibrium, in *The Equilibrium Theory of Classical Fluids* (H. Frisch and J. L. Lebowitz, Eds.), p. II-171, Benjamin, New York, 1964. H. C. Anderson, Cluster methods in equilibrium statistical mechanics of fluids, in *Statistical Mechanics. Part A: Equilibrium Techniques* (B. J. Berne, Ed.), p. 1, Plenum, New York, 1977. J.-P. Hansen and I. R. McDonald, *Theory of Simple Liquids*, 2nd ed., Academic Press, London, 1986.

The two integrals associated with Q_3 are

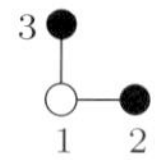

$$= \int d\mathbf{r}_2\, d\mathbf{r}_3\, f(r_{12}) f(r_{13}), \tag{8.16}$$

and

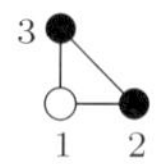

$$= \int d\mathbf{r}_2\, d\mathbf{r}_3\, f(r_{12}) f(r_{13}) f(r_{23}). \tag{8.17}$$

These diagrams are *labelled*, but it is more convenient to work with *unlabelled* diagrams. This is perhaps best illustrated using for example the configuration integral with $N = 3$ given above. In terms of labelled and unlabelled diagrams this may be written as

$$\int d\mathbf{r}_1\, d\mathbf{r}_2\, d\mathbf{r}_3 \prod_{i<j}^{3} [1 + f(r_{ij})]$$

$$= \int d\mathbf{r}_1\, d\mathbf{r}_2\, d\mathbf{r}_3\, [1 + f_{12} + f_{13} + f_{23} + f_{12}f_{13} + f_{23}f_{13} + f_{12}f_{23} + f_{12}f_{13}f_{23}]$$

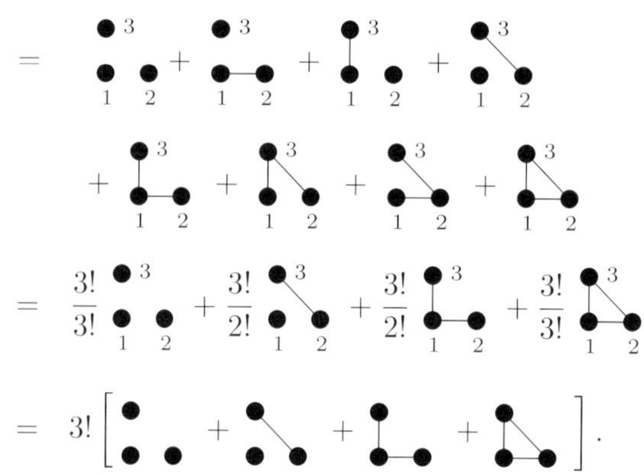

(8.18)

The passage from the second to the third equality follows because the variables of integration are dummy variables. There are only four distinct integrals, and the prefactor accounts for the number of integrals of each type. This is the total number of ways of rearranging the labels of the diagrams divided by the number of rearrangements that leave the diagram invariant. The numerator is $N!$, the number of permutations of the field points of the diagram.

The denominator is the symmetry number S, which is the number of permutations of the labels of an arbitrarily labelled diagram that leaves the topological structure (i.e., the actual labelled bonds) unchanged. If there is one set of m equivalent field points (two field points are *equivalent* if interchanging their labels does not change the labelled bonds of the diagram), the symmetry number is $S = m!$. If there are distinct sets, the ith one of which has m_i equivalent field points, then $S = \prod_i m_i!$. Equivalence sets that are themselves equivalent (i.e., simultaneously interchanging all their members leaves the structure unchanged) also contribute to S. If there are n_i equivalent sets of type i, then $S = \prod_i n_i!(m_i!)^{n_i}$. And so on, until all the permutations of the labels that preserve the topological structure of the diagram are counted. With reference to the above example, in the penultimate equality, the points 1, 2, and 3 are equivalent in the first and fourth diagrams, and hence $S = 3!$, and the points 2 and 3 are equivalent in the second and third diagrams, and hence $S = 2!$. The final equality defines an unlabelled diagram as $1/S$ times the same diagram with arbitrary labels attached to the field points. This result is true in general: the integral of the sum of all possible products of Mayer f-functions over N particles is $N!$ times the sum of all distinct unlabelled diagrams with N field points. The $N!/S$ that multiplies each corresponding arbitrarily labelled diagram accounts for the fact that each labelled arrangement of bonds occurs once and once only in the original expansion of the integrand.

As another example, the 64 terms for the configuration integral for $N = 4$ may be written

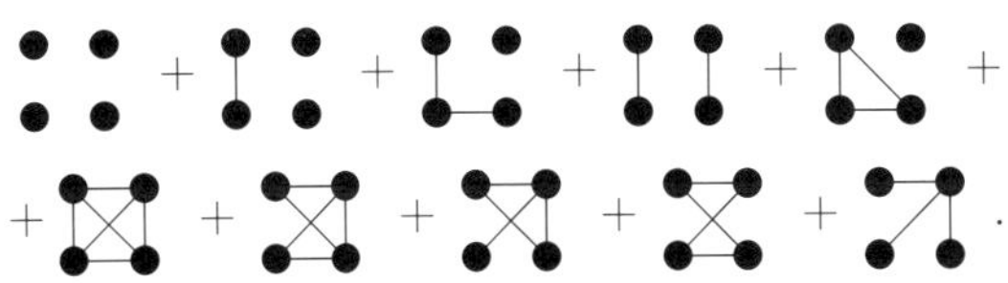

(8.19)

The symmetry numbers of the diagrams of the first row are $4!$, $2\times 2!$, $2!$, $2!\times(2!)^2$, $3!$, and $2!$, respectively. The same symmetry numbers belong to the respective diagrams of the second row, which are derived from those of the first row by interchanging bonds and nonbonds.

Here is a sum rule that provides the total number of diagrams with a certain number of bonds, and hence can be used to check the symmetry numbers. There are $N(N-1)/2$ different f_{ij} in diagrams with N field points. Hence the total number of diagrams with N field points and m bonds is ${}^{(N(N-1)/2)}C_m$. If each distinct unlabelled diagram with N field points and m bonds is indexed by i, and S_i is its symmetry number, then the sum rule is

$$\sum_i \frac{N!}{S_i} = {}^{(N(N-1)/2)}C_m = \frac{(N(N-1)/2)!}{m!(N(N-1)/2-m)!}. \tag{8.20}$$

For the case of $N = 4$, the right-hand side of this equals 1, 6, 15, and 20, for $m = 0$, 1, 2, and 3, respectively. These agree with the left-hand side using the symmetry numbers given above.

Some nomenclature and rules are now required. The multiplication of two diagrams, often called their *star-product*, consists of placing them side by side and considering them a single, possibly disconnected, diagram. Any root points with the same label are combined together, which has the effect of connecting the diagrams at the common root point. A *disconnected* diagram contains at least one pair of circles not joined by a path; a *path* is a sequence of adjacent circles; two circles are *adjacent* if they are directly connected by a bond. As an example, here are two products, the second of which results in an disconnected diagram:

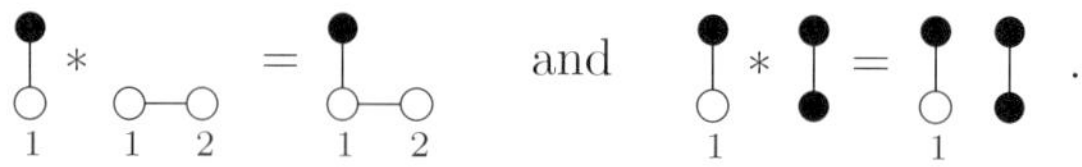

(8.21)

Star-irreducible diagrams cannot be decomposed into the product of two other diagrams. (The diagram consisting of a single root point is not regarded as star-irreducible.) All connected unlabelled diagrams consisting of field points only are star-irreducible. The star-product of such diagrams are disconnected. The exponential of a set of star-irreducible diagrams is one plus the sum of all possible star-products of the members of the set. For example,

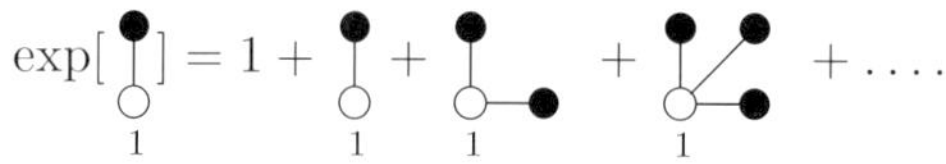

(8.22)

Compare this to the usual $e^x = 1 + x + x^2/2! + \ldots$, and note how the symmetry numbers take care of the factorials.

If removal of a circle (together with any bonds that touch it) causes a connected diagram to become disconnected, then that circle is said to be a *connecting* circle. If one of the pieces that became disconnected contains no root points, then the connecting circle is said to be an *articulation* circle. A diagram free of articulation circles is said to be *irreducible*. One can also speak of *connecting pairs* and *articulation pairs* of circles. If all paths connecting two root points pass through a circle, then that particular connecting circle is called a *nodal* point. These are readily illustrated:

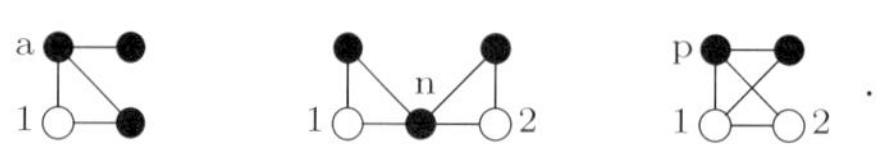

The first diagram contains an articulation point (a), the second a nodal point (n), and the third an articulation pair of points (1p).

In the above the field points have unit weight. However, one can attach a one-particle function to each field point, as is the case when an external potential causes a density inhomogeneity. This makes almost no change to the formalism. For example, in an open system with singlet potential $u^{(1)}(\mathbf{r})$ it is the generalised activity, $z(\mathbf{r}) = \Lambda^{-3} e^{\beta\mu} \exp -\beta u^{(1)}(\mathbf{r})$, that appears. A typical diagram is then

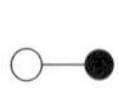

$$= \int \mathrm{d}\mathbf{r}_2\, z(\mathbf{r}_2) f(r_{12}), \tag{8.23}$$

and the diagram is described as having an f-bond, a 1-root point labelled 1, and a z-field point labelled 2.

8.3 Functional Differentiation

8.3.1 Basic Rules

The partition function depends upon the pair potential; it is said to be a *functional* of the pair potential. A functional is the continuum generalisation of a function of many variables. The total derivative of a function is the sum of its partial derivatives times the variation in each variable. The total derivative of a functional is the integral of its functional derivative times the variation at each point,

$$\delta\mathcal{F} = \int \mathrm{d}x \, \frac{\delta\mathcal{F}}{\delta u(x)} \delta u(x). \tag{8.24}$$

For example, if

$$\mathcal{F}[u] = \int \mathrm{d}x \, f(u(x)), \tag{8.25}$$

where $f(u)$ is an ordinary function, then

$$\begin{aligned} \delta\mathcal{F} &= \mathcal{F}[u+\delta u] - \mathcal{F}[u] \\ &= \int \mathrm{d}x \, [f(u(x)+\delta u(x)) - f(u(x))] \\ &= \int \mathrm{d}x \, f'(u(x))\delta u(x), \end{aligned} \tag{8.26}$$

and the functional derivative is defined as

$$\frac{\delta\mathcal{F}}{\delta u(x)} = f'(u(x)) \equiv \left. \frac{\mathrm{d}f(u)}{\mathrm{d}u} \right|_{u=u(x)}. \tag{8.27}$$

If

$$\mathcal{G}[u] = \int \mathrm{d}x \, \mathrm{d}y \, u(x)u(y)w(x,y) \tag{8.28}$$

where the kernel is assumed symmetric, $w(x,y) = w(y,x)$, and independent of $u(x)$, then

$$\frac{\delta\mathcal{G}}{\delta u(x)} = 2\int \mathrm{d}y \, u(y)w(x,y). \tag{8.29}$$

For the case of a functional of a function of several variables, for example,

$$\mathcal{H}[v] = \int \mathrm{d}\mathbf{x} \, \mathrm{d}\mathbf{y} \, f(v(\mathbf{x},\mathbf{y})), \tag{8.30}$$

the functional derivative is, as one would expect,

$$\frac{\delta\mathcal{H}}{\delta v(\mathbf{x},\mathbf{y})} = f'(v(\mathbf{x},\mathbf{y})) \equiv \left. \frac{\mathrm{d}f(v)}{\mathrm{d}v} \right|_{v=v(\mathbf{x},\mathbf{y})}. \tag{8.31}$$

In this example the functional derivative 'undoes' two integrals.

It is often necessary to take the functional derivative of a function rather than a functional. Since one can write

$$u(x) = \int \mathrm{d}y\, u(y)\delta(x-y), \tag{8.32}$$

one has the useful result that

$$\frac{\delta u(x)}{\delta u(y)} = \delta(x-y), \tag{8.33}$$

where it is the Dirac δ that appears on the right-hand side.

A function can also be a functional, $f(x,[g])$, which is to say that the value of f at x depends upon values of the function $g(x)$ over the whole interval, (e.g., $f(x,[g]) = \int \mathrm{d}y\, k(x,y)g(y)$). In these cases the functional derivative is a two-point function,

$$\frac{\delta f(x,[g])}{\delta g(y)} = h(x,y,[g]). \tag{8.34}$$

The functional chain rule or change of variable formula is

$$\frac{\delta \mathcal{F}}{\delta g(y)} = \int \mathrm{d}z\, \frac{\delta \mathcal{F}}{\delta f(z)} \frac{\delta f(z,[g])}{\delta g(y)}. \tag{8.35}$$

If one can invert the relationship and write the function g as a functional of f, $g(x,[f])$, then the two functional derivatives are inverses of each other. This may be seen by taking $\mathcal{F}[g] = g(x)$ in the above, in which case the left-hand side is just a Dirac δ and one has

$$\delta(x-y) = \int \mathrm{d}z\, \frac{\delta f(x,[g])}{\delta g(z)} \frac{\delta g(z,[f])}{\delta f(y)}. \tag{8.36}$$

8.3.2 Uniform Limit

In the event that the function of interest is a constant, $u(x) = u$, then the ordinary derivative is related to the functional derivative by the length of the relevant interval. This is because changing the value of the constant u is equivalent to changing the value of the function $u(x)$ over the whole interval. Specifically, the function used in the first example given above becomes $\mathcal{F}(u) = \int_L \mathrm{d}x\, f(u(x)) = Lf(u)$, and one has

$$\frac{\mathrm{d}\mathcal{F}(u)}{\mathrm{d}u} = Lf'(u) = L\frac{\delta \mathcal{F}(u)}{\delta u(x)} = \int \mathrm{d}x\, \left.\frac{\delta \mathcal{F}(u)}{\delta u(x)}\right|_{u(x)=u}. \tag{8.37}$$

In practice the most common system is a uniform one. However, in the development of the formalism it is often most convenient to obtain the results for an inhomogeneous system. The uniform limit of the latter may be taken by invoking a factor of volume as shown here.

For the case of the functional of the function of two variables given above, $\mathcal{H}[v]$, when the fluid is homogeneous the atomic pair functions become isotropic, $v(\mathbf{r}_1, \mathbf{r}_2) = v(r_{12})$. In this case one has

$$\mathcal{H}[v] = \int_V \mathrm{d}\mathbf{r}_1 \, \mathrm{d}\mathbf{r}_2 \, f(v(r_{12})) = V \int_V \mathrm{d}\mathbf{r} \, f(v(r)), \tag{8.38}$$

so that the functional derivative becomes

$$\frac{\delta \mathcal{H}}{\delta v(r)} = V f'(v(r)) \equiv V \left. \frac{\delta \mathcal{H}}{\delta v(\mathbf{r}_1, \mathbf{r}_2)} \right|_{r_{12}=r}. \tag{8.39}$$

Again one sees that a factor of volume appears in the isotropic case.

8.3.3 Derivatives of Diagrams

There are two rules for the functional derivative of diagrams. If $G[z, f]$ is a diagram of f-bonds and z-field points, then the functional derivative of G with respect to $z(\mathbf{r}_1)$ is the sum of all distinct diagrams obtained from G by replacing a z-field point by an unweighted root point labelled 1. The functional derivative with respect to $f(\mathbf{r}_1, \mathbf{r}_2)$ is half the sum of all distinct diagrams obtained from G by erasing an f-bond and turning the field points that it intersected into root points labelled 1 and 2.

For example, let

$G[z, f] =$ 

$$= \frac{1}{2!2!} \int \mathrm{d}\mathbf{r}_1 \, \mathrm{d}\mathbf{r}_2 \, \mathrm{d}\mathbf{r}_3 \, \mathrm{d}\mathbf{r}_4 \, z_1 z_2 z_3 z_4 f_{12} f_{13} f_{14} f_{23} f_{34}, \tag{8.40}$$

where $z_i \equiv z(\mathbf{r}_i)$ and $f_{ij} \equiv f(\mathbf{r}_i, \mathbf{r}_j)$. By considering the total variation δG with δz, one sees that one obtains four terms from the product rule. Writing these as labelled diagrams the functional derivative may be seen to be

$\dfrac{\delta G}{\delta z(\mathbf{r}_1)} =$ 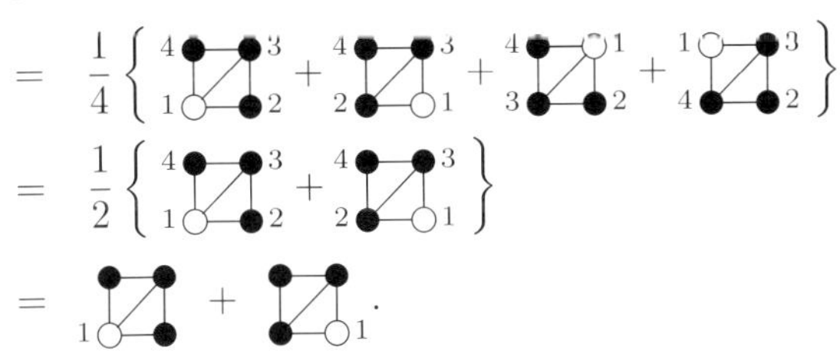 (8.41)

Hence one can see that the functional derivative with respect to the function attached to the field points is just the sum of the distinct unlabelled diagrams with a field point changed to a labelled 1-root point.

The total variation with respect to f is

$$\delta G = \frac{1}{4} \int \mathrm{d}\mathbf{r}_1 \, \mathrm{d}\mathbf{r}_2 \, \mathrm{d}\mathbf{r}_3 \, \mathrm{d}\mathbf{r}_4 \, z_1 z_2 z_3 z_4 \times [\delta f_{12} \, f_{13} f_{14} f_{23} f_{34} + \text{four other terms}\,]. \tag{8.42}$$

Hence one has

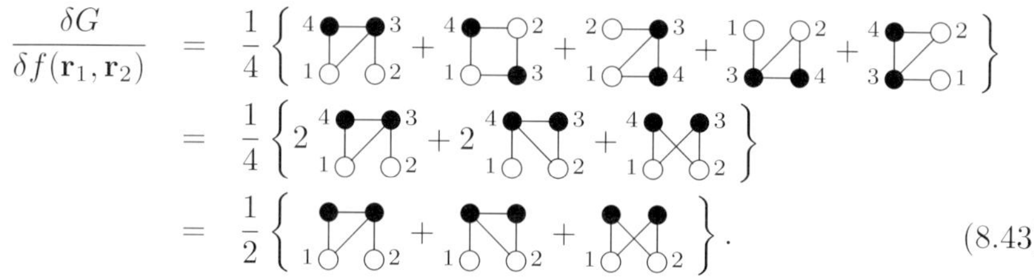

(8.43)

Note that for a homogeneous fluid, the first two diagrams in the bracket are numerically equal, but that they must still be counted as distinct due to the labelling on the root points. One sees indeed that the functional derivative with respect to the bond function is just half the sum of the distinct unlabelled diagrams obtained from the original by changing two of the adjacent field points to labelled root points and erasing their connecting bond.

8.4 Particle Densities

In Section 7.3, various particle densities were expressed as averages. Here the results are given in terms of the functional derivative of the partition function. For an open system with an external potential $u^{(1)}(\mathbf{r})$, the spatially varying activity is $z(\mathbf{r}) = \Lambda^{-3}e^{\beta\mu}\exp -\beta u^{(1)}(\mathbf{r})$. Defining the 'excess' part of the potential energy as $U_N^*(\mathbf{r}^N) = U_N(\mathbf{r}^N) - \sum_{i=1}^N u^{(1)}(\mathbf{r}_i)$, the grand partition function, Eq. (7.17), may be written

$$\Xi(\mu, V, T) = \sum_{N=0}^{\infty} \frac{1}{N!} \int \mathrm{d}\mathbf{r}^N \prod_{i=1}^{N} [z(\mathbf{r}_i)]\, e^{-\beta U_N^*(\mathbf{r}^N)}. \tag{8.44}$$

In terms of these, the n-particle density for an open system was given in Section 7.3 as

$$\begin{aligned} \rho^{(n)}(\mathbf{s}^n) &= \frac{1}{\Xi(\mu, V, T)} \sum_{N=n}^{\infty} \frac{1}{(N-n)!} \int \mathrm{d}\mathbf{r}^N \prod_{i=1}^{N} [z(\mathbf{r}_i)]\, e^{-\beta U_N^*(\mathbf{r}^N)} \\ & \quad \times \delta(\mathbf{r}_1 - \mathbf{s}_1) \ldots \delta(\mathbf{r}_n - \mathbf{s}_n). \end{aligned} \tag{8.45}$$

This expression for the n-particle density as an average in an open system will now be compared to the derivatives of the grand partition function in order to express the densities in terms of these derivatives.

The activity functional derivative of the grand partition function is

$$\frac{\delta \Xi}{\delta z(\mathbf{r})} = \sum_{N=1}^{\infty} \frac{N}{N!} \int \mathrm{d}\mathbf{r}_2 \ldots \mathrm{d}\mathbf{r}_N \prod_{i=2}^{N} [z(\mathbf{r}_i)]\, e^{-\beta U_N^*(\mathbf{r}^N)} \Big|_{\mathbf{r}_1 = \mathbf{r}}. \tag{8.46}$$

The factor of N in the numerator arises from the product rule and fact that the dummy variables of integration are all equivalent. Comparing this with the above average for $n = 1$, one concludes that the singlet particle density is given by

$$\rho^{(1)}(\mathbf{r}) = \frac{z(\mathbf{r})}{\Xi} \frac{\delta \Xi}{\delta z(\mathbf{r})}. \tag{8.47}$$

Similarly, the second functional derivative is

$$\frac{\delta^2 \Xi}{\delta z(\mathbf{r}) \delta z(\mathbf{s})} = \sum_{N=2}^{\infty} \int \frac{\mathrm{d}\mathbf{r}_3 \ldots \mathrm{d}\mathbf{r}_N}{(N-2)!} \prod_{i=3}^{N} [z(\mathbf{r}_i)] \, e^{-\beta U_N^*(\mathbf{r}^N)} \Big|_{\mathbf{r}_{1,2}=\mathbf{r},\mathbf{s}}, \tag{8.48}$$

and it is evident that the pair density is given by

$$\rho^{(2)}(\mathbf{r}, \mathbf{s}) = \frac{z(\mathbf{r}) z(\mathbf{s})}{\Xi} \frac{\delta^2 \Xi}{\delta z(\mathbf{r}) \delta z(\mathbf{s})}. \tag{8.49}$$

The general result is

$$\rho^{(n)}(\mathbf{s}^n) = \frac{z(\mathbf{s}_1) \ldots z(\mathbf{s}_n)}{\Xi} \frac{\delta^n \Xi}{\delta z(\mathbf{s}_1) \ldots \delta z(\mathbf{s}_n)}. \tag{8.50}$$

In the singlet case, one can write the right-hand side as a logarithmic derivative,

$$\rho^{(1)}(\mathbf{s}) = \frac{\delta \ln \Xi}{\delta \ln \Lambda^3 z(\mathbf{s})}. \tag{8.51}$$

(The constant thermal wavelength is included to make the argument of the logarithm dimensionless.) Since the numerator is the unconstrained total entropy, and since the denominator is essentially the one-particle potential, this shows that the singlet density and the singlet potential are conjugate variables. In the homogeneous limit, according to Eq. (8.37) this becomes

$$\rho = \frac{1}{V} \frac{\partial \ln \Xi}{\partial \ln \Lambda^3 z} = \frac{k_B T}{V} \frac{\partial \ln \Xi}{\partial \mu}, \tag{8.52}$$

which is of course one of the fundamental results of the open system, namely that the equilibrium number of particles is given by the negative of the chemical potential derivative of the grand potential, Eq. (3.10).

For the higher-order densities the logarithmic derivative gives additional terms. For the pair case, it is straightforward to show that

$$\begin{aligned}
&\frac{\delta^2 \ln \Xi}{\delta \ln \Lambda^3 z(\mathbf{r}) \delta \ln \Lambda^3 z(\mathbf{s})} \\
&= z(\mathbf{r}) \frac{\delta}{\delta z(\mathbf{r})} \left\{ \frac{1}{\Xi} \sum_{N=1}^{\infty} \frac{1}{(N-1)!} \int \mathrm{d}\mathbf{r}^N \prod_{i=1}^{N} [z(\mathbf{r}_i)] \, e^{-\beta U_N^*(\mathbf{r}^N)} \delta(\mathbf{r}_1 - \mathbf{s}) \right\} \\
&= \rho^{(2)}(\mathbf{r}, \mathbf{s}) - \rho^{(1)}(\mathbf{r}) \rho^{(1)}(\mathbf{s}) + \rho^{(1)}(\mathbf{r}) \delta(\mathbf{r}, \mathbf{s}).
\end{aligned} \tag{8.53}$$

The product of singlet densities that appears here would be the pair density if particles at $\mathbf{r}$ and $\mathbf{s}$ did not influence each other. Hence the difference between the first and second terms represents the degree to which these two positions are correlated. The final term is essentially a correction for the self-interaction that is counted in the product of singlet densities. It does not appear in the form

$$z(\mathbf{r})z(\mathbf{s})\frac{\delta^2 \ln \Xi}{\delta z(\mathbf{r})\delta z(\mathbf{s})} = \rho^{(2)}(\mathbf{r},\mathbf{s}) - \rho^{(1)}(\mathbf{r})\rho^{(1)}(\mathbf{s}). \tag{8.54}$$

The pair contribution to the potential energy may also be treated explicitly by writing $U_N^*(\mathbf{r}^N) = U_N(\mathbf{r}^N) - \sum_{i=1}^N u^{(1)}(\mathbf{r}_i) - \sum_{i<j} u^{(2)}(\mathbf{r}_i,\mathbf{r}_j)$. Defining the exponential function $e(\mathbf{r}_i,\mathbf{r}_j) = \exp -\beta u^{(2)}(\mathbf{r}_i,\mathbf{r}_j)$, the grand potential becomes

$$\Xi(\mu,V,T) = \sum_{N=0}^{\infty} \frac{1}{N!} \int \mathrm{d}\mathbf{r}^N \prod_{i=1}^{N} [z(\mathbf{r}_i)] \prod_{i<j}^{N} [e(\mathbf{r}_i,\mathbf{r}_j)]\, e^{-\beta U_N^*(\mathbf{r}^N)}. \tag{8.55}$$

The functional derivative of this is

$$\begin{aligned}
\frac{\delta \Xi}{\delta e(\mathbf{r}_1,\mathbf{r}_2)} &= \sum_{N=2}^{\infty} \int \frac{\mathrm{d}\mathbf{r}_3 \ldots \mathrm{d}\mathbf{r}_N}{2(N-2)!} \prod_{i=1}^{N} [z(\mathbf{r}_i)] \prod_{i<j}^{N}{}' \, [e(\mathbf{r}_i,\mathbf{r}_j)]\, e^{-\beta U_N^*(\mathbf{r}^N)} \\
&= \frac{\Xi}{2e(\mathbf{r}_1,\mathbf{r}_2)} \rho^{(2)}(\mathbf{r}_1,\mathbf{r}_2).
\end{aligned} \tag{8.56}$$

The differentiation gives $N(N-1)/2$ identical terms, and the prime on the second product indicates that the $ij = 12$ term is excluded. Writing this as the logarithmic derivative one has an alternative expression for the pair density,

$$\rho^{(2)}(\mathbf{r}_1,\mathbf{r}_2) = -2k_{\mathrm{B}}T \frac{\delta \ln \Xi}{\delta u^{(2)}(\mathbf{r}_1,\mathbf{r}_2)}. \tag{8.57}$$

Since $\ln \Xi$ is the unconstrained total entropy, this result may be interpreted as saying that the pair potential and the pair density are conjugate functions. In view of Eq. (8.39), in the homogeneous and isotropic limit the pair functions only depend upon particle separation and this result reduces to

$$\rho^{(2)}(r_{12}) = \frac{-2k_{\mathrm{B}}T}{V} \frac{\delta \ln \Xi}{\delta u^{(2)}(r_{12})}. \tag{8.58}$$

8.5 Expansions of the Density and Pressure

8.5.1 Grand Partition Function

The grand partition function for a fluid with one- and two-body potentials is

$$\Xi = \sum_{N=0}^{\infty} \frac{1}{N!} \int \prod_{i=1}^{N} [z(\mathbf{r}_i)] \prod_{i<j}^{N} [1 + f(\mathbf{r}_i,\mathbf{r}_j)]\, \mathrm{d}\mathbf{r}_1 \ldots \mathrm{d}\mathbf{r}_N. \tag{8.59}$$

Expanding the product of f-bonds, the first term has no bonds and is just

$$\sum_{N=0}^{\infty} \frac{1}{N!} \int \mathrm{d}\mathbf{r}^N \prod_{i=1}^{N} z_i = \sum_{N=0}^{\infty} \frac{1}{N!} (\bullet)^N$$

$$= \exp(\bullet)$$

$= 1 +$ 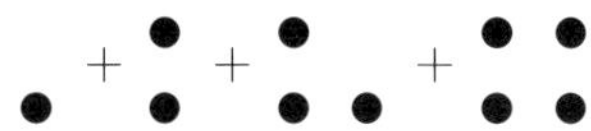 $+ \ldots.$ (8.60)

The second term has a single f-bond, and since there are $N(N-1)/2$ distinct pairs it is

$\sum_{N=2}^{\infty} \int \frac{\mathrm{d}\mathbf{r}^N}{2(N-2)!} f_{12} \prod_{i=1}^{N} z_i =$  $\sum_{N=2}^{\infty} \frac{(\bullet)^{(N-2)}}{(N-2)!}$

$=$  $+ \ldots.$ (8.61)

The third term has two f-bonds joined at a common point. There are $N(N-1)(N-2)/2$ distinct ways of doing this, and one obtains

$$\sum_{N=3}^{\infty} \int \frac{\mathrm{d}\mathbf{r}^N}{2(N-3)!} f_{12} f_{23} \prod_{i=1}^{N} z_i$$

$=$  $\sum_{N=3}^{\infty} \frac{(\bullet)^{(N-3)}}{(N-3)!}$

$=$ 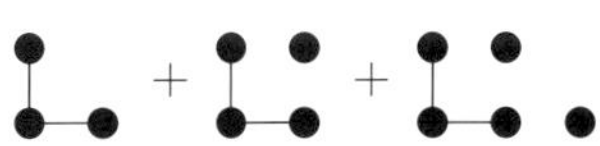 $+ \ldots.$ (8.62)

The fourth term has two unconnected f-bonds,

$$\sum_{N=4}^{\infty} \int \frac{\mathrm{d}\mathbf{r}^N}{8(N-4)!} f_{12} f_{34} \prod_{i=1}^{N} z_i$$

$=$  $\sum_{N=4}^{\infty} \frac{(\bullet)^{(N-4)}}{(N-4)!}$

$=$  $+ \ldots.$ (8.63)

This pattern continues and one obtains for the diagrammatic expansion of the grand partition function

$\Xi =$ {one plus the sum of all distinct simple diagrams with one or more z-field points and f-bonds}

$= 1 +$ 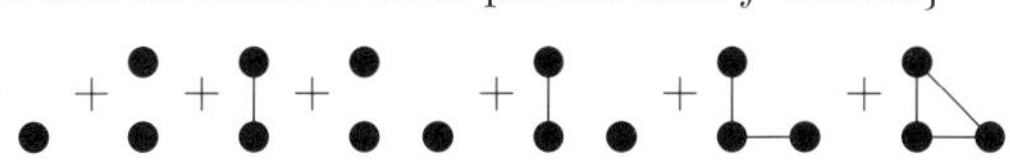 $+ \ldots.$ (8.64)

Recall that a simple diagram is one with at most one bond between any pair of points. Notice how the symmetry number in the denominator of the labelled diagrams was incorporated into the unlabelled diagrams.

Now the fact that disconnected diagrams appear in this sum indicates that it is the exponential of connected diagrams. Hence the logarithm of the grand partition function, which is the unconstrained total entropy and which is related to the grand potential, is

$$\ln \Xi = \{\text{the sum of all simple connected diagrams with one or more } z\text{-field points and } f\text{-bonds}\}$$

$= $ 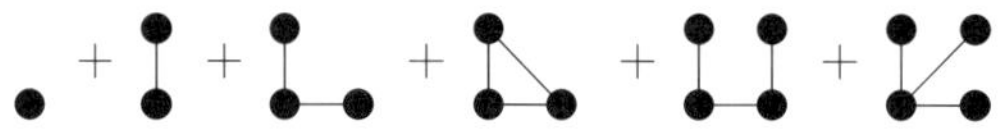 $+ \ldots .$ (8.65)

For a homogeneous system the pressure is $p = -\Omega/V = k_{\rm B}TV^{-1}\ln\Xi$, and this provides the expansion of the pressure in powers of the activity.

Three-Body Potential

The inclusion of many-body potentials is a formality. In the event that the potential energy consists of one-, two-, and three-body terms, in addition to the pair Mayer f-function, one defines a triplet function $f^{(3)}(\mathbf{r}_1, \mathbf{r}_2, \mathbf{r}_3) = -1 + \exp u^{(3)}(\mathbf{r}_1, \mathbf{r}_2, \mathbf{r}_3)$. It then follows in a straightforward manner that

$$\ln \Xi = \{\text{the sum of all simple connected diagrams with one or more } z\text{-field points, and } f^{(2)}\text{- and } f^{(3)}\text{-bonds.}\} \tag{8.66}$$

'Simple' means that there is at most one bond of a given type between a pair or triple of circles; it does not exclude simultaneous connection by bonds of different types. It is evident that this can be extended to all possible many-body interactions.

8.5.2 Density

The relation between activity and density follows from the above functional derivative expression for the singlet density, Eq. (8.47),[3]

$$\rho(\mathbf{r}) = z(\mathbf{r})\frac{\delta \ln \Xi}{\delta z(\mathbf{r})}$$

$$= \{\text{the sum of all simple connected diagrams with a } z\text{-root point labelled } \mathbf{r}\text{, and 0 or more } z\text{-field points, and } f\text{-bonds}\}$$

$= $ 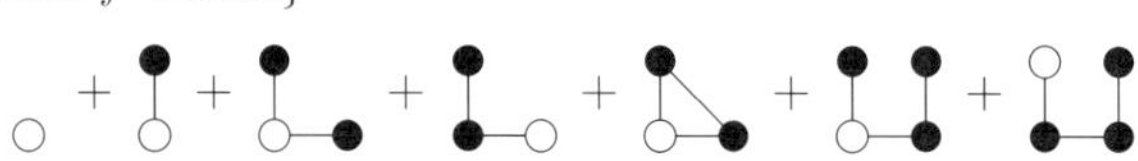 $\ldots .$ (8.67)

These diagrams may be divided into two sets: those in which the root point is an articulation circle and those in which it isn't. (Recall that an articulation

[3]In this section the label $\mathbf{r}$ attached to the root point in the diagrams has not been shown explicitly.

circle is the sole connection between a set of field points and the rest of the diagram.) Taking out the factor of $z(\mathbf{r})$ common to all the articulation root points, the exponential of the subset in which the 1-root point is not an articulation circle (called the star-irreducible diagrams) yields the diagrams of the entire set. Therefore the logarithm of the left-hand side divided by the common factor is just the star-irreducible diagrams,

$$\ln \frac{\rho(\mathbf{r})}{z(\mathbf{r})} = \{\text{the sum of all simple connected diagrams with a 1-root point labelled } \mathbf{r} \text{ that is not an articulation point, and one or more } z\text{-field points, and } f\text{-bonds}\}$$

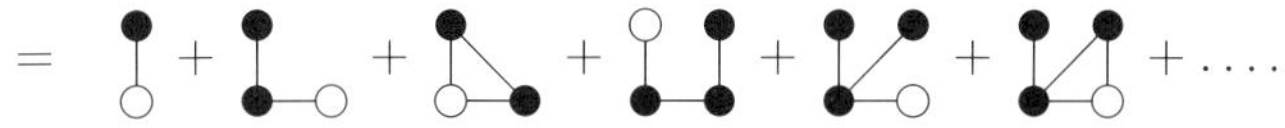

(8.68)

Now a subset of these diagrams has no articulation points at all; the remainder, in which some field points are articulation points, is just the former with the field points decorated with all the simple connected diagrams with one root point. (Decorate means to attach by the root point at that field point.) However, these decorations are just the diagrams of the singlet density, Eq. (8.67), and the procedure of removing the articulation field points corresponds to replacing the z-field points by ρ-field points. One has

$$\ln \frac{\rho(\mathbf{r})}{z(\mathbf{r})} = \{\text{the sum of all simple irreducible diagrams with a 1-root point labelled } \mathbf{r}, \text{ one or more } \rho\text{-field points, and } f\text{-bonds}\}$$

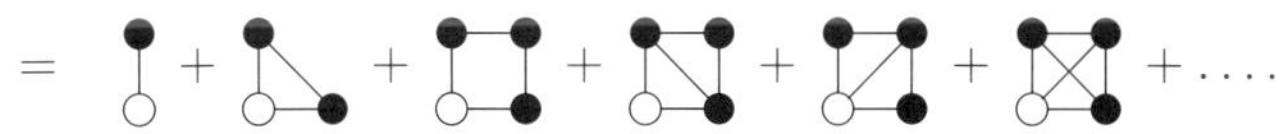

(8.69)

Here irreducible means free of articulation circles. This is an example of the very powerful procedure of *topological reduction*, whereby an infinite class of diagrams is reduced to a smaller number (albeit still infinite) of diagrams with simpler structure, by replacing the function attached to each circle or line.

This expression enables the activity of a homogeneous fluid to be given as an expansion in powers of density. From it the virial expansion of the equation of state, which is the expansion of the pressure in powers of density, can be derived. For a homogeneous fluid, the last expansion may be written

$$\ln \Lambda^3 z = \ln \Lambda^3 \rho - \sum_{n=1}^{\infty} \beta_n \rho^n, \tag{8.70}$$

where the β_n are the *irreducible Mayer cluster integrals*, as defined in Eq. (8.69), but with unit weight for all circles (β_n is the sum of all such diagrams with n field points). Now since $PV/k_{\mathrm{B}}T = \ln \Xi$, and since the homogeneous limit of Eq (8.47) is $\rho = (z/V)\partial \ln \Xi/\partial z$, it follows that $(\partial P/\partial z) = \rho k_{\mathrm{B}}T/z$, or $(\partial P/\partial \rho) = (\rho k_{\mathrm{B}}T/z)(\partial z/\partial \rho)$. Hence

$$\beta P = \int_0^{\rho} \mathrm{d}\rho' \, \rho' \frac{\mathrm{d} \ln \Lambda^3 z}{\mathrm{d}\rho'}$$

$$= \rho\left[1 - \sum_{n=1}^{\infty} \frac{n}{n+1}\beta_n \rho^n\right]$$

$=$ 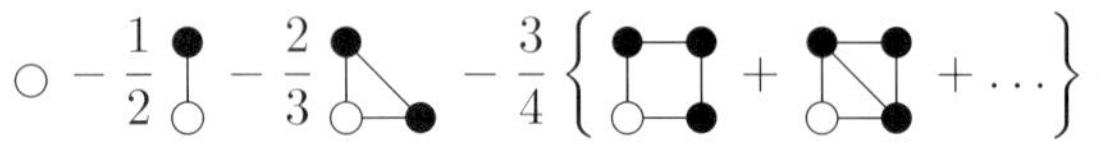 $- \ldots$. (8.71)

In these diagrams, the field points and the root points have weight ρ. This is the virial expansion, with the virial coefficients being $B_{n+1} = -\beta_n n/(n+1)$. It can be seen that the first correction to the ideal gas equation of state, B_2, agrees with that given above, Eq. (8.13).

Many-Body Potentials

As in the case of the activity expansion of the partition function, it is straightforward to include many-body potentials in the formalism. The idea of a connecting circle is essentially unchanged: when a circle is erased, it and all the bonds impinging upon it are removed from the diagram. Accordingly, the expression for the density in terms of irreducible diagrams, Eq. (8.69), is unchanged, with f-bonds interpreted to mean any of the pair and many-body Mayer f-functions. For example, for a homogeneous, isotropic fluid with a triplet potential, the first irreducible Mayer cluster integral is unchanged and the second is explicitly

$$\beta_2 = \frac{1}{2!}\int \mathrm{d}\mathbf{r}_2\,\mathrm{d}\mathbf{r}_3\, f(r_{12})f(r_{23})f(r_{13})\left[1 + f^{(3)}(r_{12}, r_{23}, r_{13})\right]. \tag{8.72}$$

8.5.3 Various Density Expansions

The virial expansion gives the grand partition function as

$$\ln \Xi = V \sum_{n=1}^{\infty} B_n \rho^n, \tag{8.73}$$

where $B_1 = 1$, and for $n > 1$, $B_{n+1} = -\beta_n(T)n/(n+1)$. (The irreducible Mayer cluster integrals are independent of volume if the integral of the Mayer f-function is convergent.) Since the grand partition function depends upon the chemical potential, or equivalently the activity, $\Xi(z, V, T)$, one concludes that the density that appears on the right-hand side is the equilibrium one, $\rho = \overline{\rho}(z, T)$. Explicit expressions for the density as a functional of the activity were given in Eqs. (8.67) and (8.68) above.

Equation (8.69) gives the activity as a function of density. It may be rewritten as

$$z(\mathbf{r}) = \rho(\mathbf{r}) \exp -c^{(1)}(\mathbf{r}; [\rho]), \tag{8.74}$$

where the *one-particle direct correlation function* is simply the right-hand side of Eq. (8.69),

$$c^{(1)}(\mathbf{r}; [\rho]) = \sum_{n=1}^{\infty} \beta_n(\mathbf{r})\rho^n. \tag{8.75}$$

For a homogeneous fluid, these represent the density expansion of the chemical potential,

$$\ln \Lambda^3 z = \beta\mu = \ln \Lambda^3 \rho - \sum_{n=1}^{\infty} \beta_n \rho^n. \tag{8.76}$$

Clearly the activity that appears here is the equilibrium one, $z = \overline{z}(\rho, T)$. In view of this density representation, one can regard the density as the independent variable in the virial expansion and write the grand partition function as $\Xi_\rho(\rho, V, T) = \Xi(\overline{z}(\rho, T), V, T)$.

The grand potential is the logarithm of the grand partition function, and for a homogeneous system this is $\Omega(\mu, V, T) = -k_B T \ln \Xi(\mu, V, T)$. The Helmholtz free energy is $F(N, V, T) = \Omega(\overline{\mu}, V, T) + \overline{\mu}N$. Accordingly, the above results can be used to obtain the density expansion of the Helmholtz free energy,

$$\beta F(N, V, T)/V = \rho \ln[\Lambda^3 \rho] - \rho - \sum_{n=1}^{\infty} \frac{\beta_n}{n+1} \rho^{n+1}. \tag{8.77}$$

As another example, the isothermal compressibility was given in Eq. (3.65) as the pressure derivative of the equilibrium volume. It may be inverted and rewritten in terms of the density derivative of the pressure, which lends itself to the virial expansion,

$$\chi_T^{-1} = -\rho^{-1} \left(\frac{\partial p}{\partial \rho} \right)_T = k_B T \sum_{n=1}^{\infty} n B_n \rho^{n-2}. \tag{8.78}$$

The excess energy can be expanded in powers of density by writing

$$\begin{aligned} E^{\text{ex}} &= -\left(\frac{\partial \ln \Xi}{\partial \beta} \right)_{z,V} \\ &= -\left(\frac{\partial \ln \Xi}{\partial \beta} \right)_{\rho,V} - \left(\frac{\partial \ln \Xi}{\partial \rho} \right)_{\beta,V} \left(\frac{\partial \rho}{\partial \beta} \right)_z . \end{aligned} \tag{8.79}$$

(This is the excess energy because the temperature-dependent contribution from the kinetic energy that appears as the thermal wavelength in the activity is held constant.) Using the virial expansion one obtains

$$\left(\frac{\partial \ln \Xi}{\partial \beta} \right)_{\rho,V} = -V \sum_{n=1}^{\infty} \frac{n}{n+1} \dot{\beta}_n \rho^{n+1}, \tag{8.80}$$

and

$$\left(\frac{\partial \ln \Xi}{\partial \rho} \right)_{\beta,V} = V - V \sum_{n=1}^{\infty} n \beta_n \rho^n, \tag{8.81}$$

where $\dot{\beta}_n \equiv \partial \beta_n / \partial \beta$. Differentiating the density expansion of $\ln z$, Eq. (8.70), one can show that

$$\left(\frac{\partial \rho}{\partial \beta} \right)_z = \frac{\sum_{n=1}^{\infty} \dot{\beta}_n \rho^{n+1}}{1 - \sum_{n=1}^{\infty} n \beta_n \rho^n}. \tag{8.82}$$

B_2	2.0944	B_5	2.1223
B_3	2.7415	B_6	1.5555
B_4	2.6362	B_7	1.1647

Table 8.1: Hard-Sphere Virial Coefficients (Units of d)

It therefore follows that

$$E^{\mathrm{ex}}/V = -\sum_{n=1}^{\infty} \frac{\dot{\beta}_n}{n+1} \rho^{n+1}. \tag{8.83}$$

This density expansion for the energy may be combined with the expansion for the Helmholtz free energy above to give the density expansion for the subsystem entropy density. Since $\overline{F} = \overline{E} - T\overline{S}$, one obtains

$$S/k_{\mathrm{B}}V = \rho - \ln[\Lambda^3 \rho] + \sum_{n=1}^{\infty} \frac{\beta_n - \beta\dot{\beta}_n}{n+1} \rho^{n+1}. \tag{8.84}$$

The temperature that appears on the right-hand side is the equilibrium one, $\overline{T}(E, N, V)$. Alternatively, one may regard the energy that appears as an argument of the entropy as the equilibrium one, $S(\overline{E}(N, V, T), N, V))$.

8.5.4 Hard-Sphere Fluid

The virial coefficients for a fluid interacting with a hard-sphere potential have been obtained analytically up to B_4, and numerically for higher orders.[4] For a hard-sphere diameter of d, so that $u^{(2)}(r) = \infty$, $r < d$, the Mayer f-function is $f(r) = -1$, $r < d$, and $f(r) = 0$, $r > d$. Because the potential is either infinite or 0, the Mayer f-function is independent of temperature, and all of the thermodynamic properties of the hard-sphere system have a trivial temperature dependence. The second virial coefficient is

$$B_2 = \frac{-1}{2}\beta_1 = \frac{-1}{2}\int \mathrm{d}\mathbf{r}\, f(r) = \frac{2}{3}\pi d^3. \tag{8.85}$$

The third virial coefficient is

$$B_3 = \frac{-2}{3}\beta_2 = \frac{-1}{3}\int \mathrm{d}\mathbf{r}\,\mathrm{d}\mathbf{r}'\, f(r) f(r') f(|\mathbf{r} - \mathbf{r}'|), \tag{8.86}$$

which can be evaluated to yield $B_3 = 5\pi^2 d^6/18$. Table 8.1 lists values for the first seven virial coefficients.

In the case of the hard-sphere fluid it is conventional to work with a dimensionless form of the density called the *packing fraction*, $\eta \equiv \pi \rho d^3/6$. Carnahan and Starling[5] observed that when the virial expansion was written in terms of

[4] J. -P. Hansen and I. R. McDonald, *Theory of Simple Liquids*, 2nd ed., Academic Press, London, 1986.

[5] N. F. Carnahan and K. E. Starling, Equation of state for noninteracting rigid spheres, *J. Chem. Phys.* **51** (1969), 635.

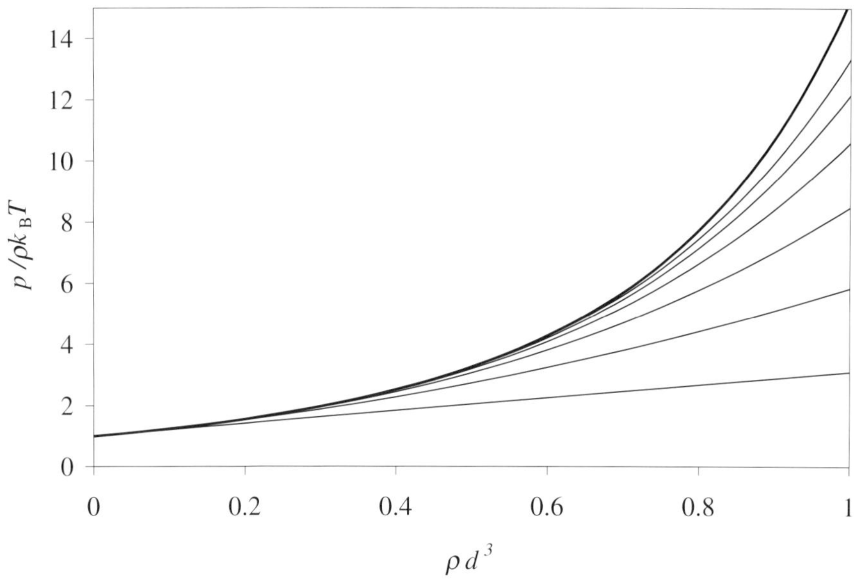

Figure 8.2: The pressure of a hard-sphere fluid relative to that of an ideal gas. The bold curve is the Carnahan–Starling equation of state, and the remaining curves use successively more terms in the virial expansion, truncating it at B_2 (bottom) and at B_7 (top).

the packing fraction, the first several coefficients were integers or close to integers. A quadratic function of the index was fitted to these three and shown to predict the remaining three known coefficients reasonably accurately. Carnahan and Starling summed the resultant series to obtain an analytic approximation for the hard-sphere equation of state,

$$\frac{P}{\rho k_B T} = \frac{1 + \eta + \eta^2 - \eta^3}{(1 - \eta)^3}. \tag{8.87}$$

It turns out that this approximation is almost exact over the whole fluid regime, and it has therefore proved extremely convenient in the study of hard-sphere fluids and of more general fluids that use the hard-sphere fluid as a reference.

Figure 8.2 compares the virial expansion truncated at successively higher orders with the Carnahan–Starling equation of state, which for present purposes can be regarded as exact. In general the pressure for the hard-sphere fluid increases at a rate faster than that for an ideal gas. (The ideal equation of state would be a horizontal line in the figure.) This is particularly marked approaching the fluid–solid transition ($\rho d^3 \approx 0.95$), as it becomes increasingly difficult to pack the spheres into the available volume. Because the hard-sphere potential is infinite when nonzero, the effective critical temperature of the system is 0, which is to say that the hard-sphere fluid is always supercritical and it displays no liquid–vapour transition. It can be seen that at low densities all the virial approximations agree. As the density is increased, one needs to retain more terms in the virial expansion to secure agreement with the exact results.

As an analytic expression, the Carnahan–Starling equation of state can be integrated or differentiated as required. For example, the pressure is the volume derivative of the Helmholtz free energy. In terms of excess quantities it is

$$p^{\mathrm{ex}} = -\frac{\partial F^{\mathrm{ex}}(N,V,T)}{\partial V} = \frac{\rho^2}{N}\frac{\partial F^{\mathrm{ex}}(N,V,T)}{\partial \rho}. \tag{8.88}$$

This can be integrated with respect to density to give

$$\begin{aligned} F^{\mathrm{ex}}(N,V,T) &= Nk_{\mathrm{B}}T\int_0^\rho \frac{\mathrm{d}\rho'}{\rho'}\frac{\beta p^{\mathrm{ex}}(\rho')}{\rho'} \\ &= Nk_{\mathrm{B}}T\int_0^\eta \frac{\mathrm{d}\eta'}{\eta'}\left[\frac{1+\eta+\eta^2-\eta^3}{(1-\eta)^3}-1\right] \\ &= Nk_{\mathrm{B}}T\frac{\eta(4-3\eta)}{(1-\eta)^2}. \end{aligned} \tag{8.89}$$

This is the Carnahan–Starling approximation for the excess Helmholtz free energy of the hard-sphere fluid. One can differentiate this with respect to number (at constant volume) to obtain the excess chemical potential. Alternatively, since at equilibrium $-\overline{p}V = \Omega = \overline{F} - \mu\overline{N}$, one has

$$\beta\mu^{\mathrm{ex}} = \frac{F^{\mathrm{ex}}(N,V,T)+p^{\mathrm{ex}}V}{Nk_{\mathrm{B}}T} = \frac{1+5\eta-6\eta^2+2\eta^3}{(1-\eta)^3}. \tag{8.90}$$

It is these closed form analytic expressions that make the Carnahan–Starling approximation particularly convenient. More accurate but less convenient fits to the hard-sphere equation of state also exist.[6]

8.5.5 Lennard–Jones Fluid

For the case of a Lennard–Jones potential, $u^{(2)}(r) = 4\epsilon[(\sigma/r)^{12} - (\sigma/r)^6]$, it is conventional to define dimensionless temperature, density, and pressure by $T^* = k_{\mathrm{B}}T/\epsilon$, $\rho^* = \rho\sigma^3$, and $p^* = p\sigma^3/\epsilon$, respectively. In this case the Mayer f-function is temperature-dependent, and the thermodynamic properties of the Lennard–Jones fluid are nontrivial functions of temperature. Unlike the hard-sphere fluid, because of the nonlinear nature of the potential one cannot evaluate even the second virial coefficient analytically. However, as a one-dimensional integral it is relatively straightforward to evaluate numerically, and the results have been fitted to an equation of the form[7]

$$B_2^* = x_1T^* + x_2 + x_3(T^*)^{1/2} + x_4/T^* + x_5/(T^*)^2, \tag{8.91}$$

where $B_2^* = B_2\sigma^{-6}$. The coefficients are listed in Table 8.2.

[6] J. J. Erpenbeck and W. W. Wood, Molecular dynamics calculations of the hard-sphere equation of state, *J. Stat. Phys.* **35** (1984), 321.

[7] J. J. Nicolas, K. E. Gubbins, W. B. Streett, and D. J. Tildesley, Equation of state for the Lennard–Jones fluid, *Mol. Phys.* **37** (1979), 1429.

x_1	-0.044481	x_4	3.8397
x_2	7.2738	x_5	-2.0058
x_3	-14.343		

Table 8.2: Coefficients for the Fit of the Lennard–Jones Second Virial Coefficient

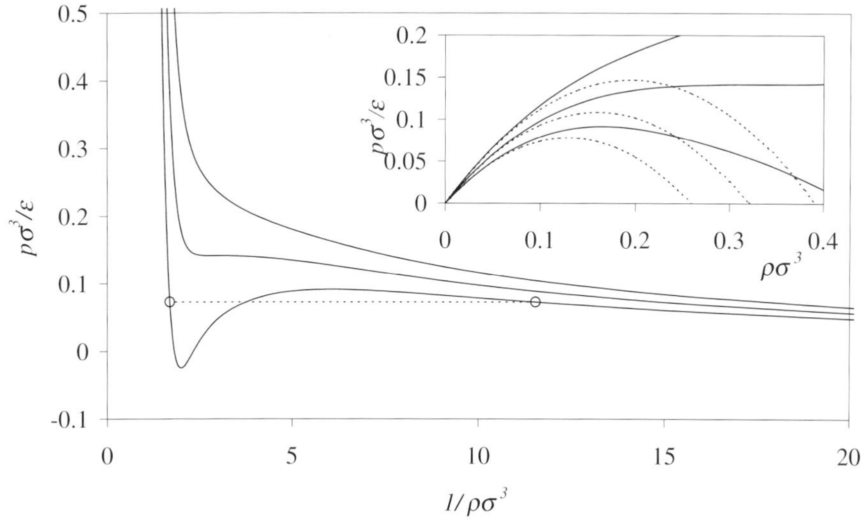

Figure 8.3: The equation of state of a Lennard–Jones fluid for, from bottom to top, $T^* = 1.2$, 1.35, and 1.5. The horizontal line ties the coexisting vapour and liquid densities. The inset tests the two-term virial expansion (dotted) at low densities. The critical temperature, density, and pressure are $T_c^* = 1.35$, $\rho_c^* = 0.35$, and $p_c^* = 0.142$, respectively.

The equation of state of the Lennard–Jones fluid has also been obtained and fitted to a polynomial in density and temperature by the aforementioned Nicolas *et al.* The result is shown in Fig. 8.3 for a subcritical, critical, and supercritical isotherm. For low temperatures the pressure is a nonmonotonic function of density. In fact, regions where $\partial p/\partial \rho < 0$ are unstable. This corresponds to the two-phase region of liquid–gas coexistence; the coexisting densities and the vapour pressure obtained by the Maxwell construction are shown (see below). The negative pressures on the liquid branch with positive compressibility are metastable states. Above the critical temperature the pressure is a monotonic increasing function of density. At a given reduced density the pressure of the Lennard–Jones fluid is much less than that of the hard-sphere fluid (cf. Fig. 8.2). This is in part because of the attractive r^{-6} tail of the Lennard–Jones potential, and in part because the effective core diameter of the Lennard–Jones potential (defined as when the potential becomes greater than $\approx k_{\mathrm{B}}T$) is less than σ, so that the reduced density is effectively smaller than $\rho\sigma^3$. In the inset to the figure it can be seen that the two-term virial expansion, $p^* = \rho^* + B_2^*\rho^{*2}$, works well at low densities but that it fails badly as the density gets larger.

Liquid and vapour phases coexisting in equilibrium must necessarily have equal pressure, $p_{\mathrm{l}} = p_{\mathrm{v}}$ and equal chemical potentials, $\mu_{\mathrm{l}} = \mu_{\mathrm{v}}$, since they can exchange number and volume with each other. The *Maxwell equal area construction* finds the coexisting densities by demanding that the (signed) area between the isotherm and the tie-line vanishes when plotted in terms of the volume per molecule, v. In terms of the density $\rho = 1/v$, denoting the coexistence pressure by p_0 and using the virial expansion for the pressure, this is

$$\begin{aligned} 0 &= \int_{\rho_{\mathrm{v}}}^{\rho_{\mathrm{l}}} \frac{\mathrm{d}\rho}{\rho^2}[p(\rho) - p_0] \\ &= \left[\ln\rho - \sum_{n=1}^{\infty}\beta_n \frac{\rho^n}{n+1} + \frac{p_0}{\rho}\right]_{\rho_{\mathrm{v}}}^{\rho_{\mathrm{l}}} \\ &= \left[1 + \ln\rho - \sum_{n=1}^{\infty}\beta_n\rho^n\right]_{\rho_{\mathrm{v}}}^{\rho_{\mathrm{l}}} \\ &= \ln z(\rho_{\mathrm{l}}) - \ln z(\rho_{\mathrm{v}}). \end{aligned} \tag{8.92}$$

The penultimate equality follows by writing $p(\rho)$ in place of p_0, which is valid at the respective limits, and then using the virial expansion. The final line, which follows from Eq. (8.70), shows that the Maxwell construction corresponds to equality of the chemical potentials of the two phases.

Summary

- The Mayer f-function goes to 0 at large separations, and it transforms the Boltzmann factor in the integrand of the partition function into a form suitable for expansion.

- Cluster diagrams are a convenient and graphic representation of the multidimensional integrals that appear in statistical mechanics, and they provide a picturesque method for deriving many results (see Table 8.3).

- Functional differentiation is another powerful procedure particularly suited for statistical mechanics. The various particle densities may be expressed as activity derivatives of the grand partition function.

- The activity expansion of the grand partition function may be expressed more compactly in terms of the density, and this provides explicit results for the coefficients that appear in the virial expansion of the pressure.

- The first three virial coefficients of the hard-sphere fluid are known exactly, and higher-order coefficients have been obtained numerically. The Carnahan–Starling equation of state is a convenient analytic approximation that is accurate over the whole fluid regime.

- The Lennard–Jones fluid displays a critical point and liquid–gas coexistence. Its virial coefficients are temperature-dependent and are not known analytically.

Adjacent	Directly connected by a bond
Articulation circle	A connecting circle whose removal creates at least one fragment composed only of field points
Bond	A line connecting two circles representing a pair function; higher-order bonds exist
Cluster diagram	A pictorial representation of a multidimensional integral
Connected diagram	A diagram with at least one path of bonds between any pair of circles
Connecting circle	A circle whose removal causes the diagram to become disconnected
Connecting pair	A pair of circles whose simultaneous removal disconnects the diagram
Field point	An integrated coordinate denoted by a filled circle
Irreducible diagram	A connected diagram with no articulation circles
Labelled diagram	An integral with all circles labelled with their coordinates
Nodal circle	A connecting circle whose removal severs all paths between two root points
Path	A linear arrangement of sequentially bonded circles
Removal	The erasure of a circle and its impinging bonds, or the erasure of a bond and rooting its terminal circles
Root point	A nonintegrated coordinate denoted by an open, labelled circle
Simple diagram	A diagram with at most one bond connecting any pair of circles
Star-irreducible	A diagram that is not the star-product of two other diagrams
Star-product	Two diagrams placed in parallel and sharing their common labelled root points, if any
Symmetry number	The number of permutations of field point labels that leave the diagram with the same labelled bonds
Unlabelled diagram	A representation of all the labelled diagrams with a particular structure that appear in the product expansion over the field points

Table 8.3: Diagrammatic Nomenclature

Chapter 9

Pair Functions

9.1 Density Expansions of the Pair Function

The pair probability function was defined in Section 7.3 as[1]

$$\wp(\mathbf{r}, \mathbf{s}) = \langle \delta(\mathbf{r}_1 - \mathbf{r})\delta(\mathbf{r}_2 - \mathbf{s}) \rangle . \tag{9.1}$$

This gives the probability of finding the particle labelled 1 at $\mathbf{r}$ and the particle labelled 2 at $\mathbf{s}$. The pair density in a closed system is simply proportional to this,

$$\rho(\mathbf{r}, \mathbf{s}) = N(N-1)\wp(\mathbf{r}, \mathbf{s}). \tag{9.2}$$

This is colloquially said to give the probability of finding any two particles simultaneously at the given positions. Strictly speaking this is an average and not a probability; integrating the pair density over $\mathbf{r}$ and $\mathbf{s}$ gives $N(N-1)$, not unity. One can say that the pair density is large when there is high probability of finding two particles simultaneously at the two positions, and that it is 0 when the two particles can't be there simultaneously.

When the positions are well separated (i.e., much beyond the range of the pair and many-body potentials), the pair probability function becomes the product of singlet functions. This is because whether particle 1 is at $\mathbf{r}$ can have no influence on particle 2 being at $\mathbf{s}$. The particles at these positions are said to be uncorrelated and one has $\wp(\mathbf{r}, \mathbf{s}) \sim \wp(\mathbf{r})\wp(\mathbf{s})$, $|\mathbf{r} - \mathbf{s}| \to \infty$. Since $\rho(\mathbf{r}) = N\wp(\mathbf{r})$, one sees that asymptotically the pair density equals the product of singlet densities,

$$\rho(\mathbf{r}, \mathbf{s}) \sim \rho(\mathbf{r})\rho(\mathbf{s}), \ |\mathbf{r} - \mathbf{s}| \to \infty. \tag{9.3}$$

[1] In this chapter, where ambiguity is removed by the number of arguments, the superscript denoting the order of the density and distribution functions is dropped.

9.1.1 Diagrammatic Definitions

The pair density can be expressed as

$$\begin{aligned}\rho(\mathbf{r},\mathbf{s}) &= \left\langle N(N-1)\delta(\mathbf{r}_1-\mathbf{r})\delta(\mathbf{r}_2-\mathbf{s})\right\rangle_\mu \\ &= \frac{z(\mathbf{r})z(\mathbf{s})}{\Xi}\frac{\delta^2\Xi}{\delta z(\mathbf{r})\delta z(\mathbf{s})}. \end{aligned} \tag{9.4}$$

For a homogeneous system, $z(\mathbf{r}) = z$ and $\rho(\mathbf{r}) = \rho$, and for an isotropic system $f(\mathbf{r}_1, \mathbf{r}_2) = f(r_{12})$ and $\rho(\mathbf{r}_1, \mathbf{r}_2) = \rho^{(2)}(r_{12})$. The pair density can be normalised by the product of singlet densities to give the pair distribution function, Eq. (7.85),

$$g(\mathbf{r}_1,\mathbf{r}_2) = \frac{\rho(\mathbf{r}_1,\mathbf{r}_2)}{\rho(\mathbf{r}_1)\rho(\mathbf{r}_2)}. \tag{9.5}$$

As mentioned above, at large separations the density at $\mathbf{r}_1$ is not influenced by the density at $\mathbf{r}_2$, and the two positions are said to be uncorrelated. The joint density is just the product of the pair density, and hence the normalisation ensures that the pair distribution function goes to unity, $g(\mathbf{r}_1, \mathbf{r}_2) \to 1$, $r_{12} \to \infty$. For the case of an isotropic fluid, this is called the radial distribution function and it is denoted simply $g(r_{12})$. In view of this asymptotic behaviour it is convenient to define the *total correlation function*,

$$h(\mathbf{r}_1,\mathbf{r}_2) = g(\mathbf{r}_1,\mathbf{r}_2) - 1. \tag{9.6}$$

This measures the degree of correlation of the two positions. Where the particles exert no influence on each other, they are said to be uncorrelated, and $h = 0$. This obviously occurs at large separations in all systems, and at all separations in the ideal gas. Real particles become increasingly correlated with each other as their separation decreases. As they approach each other, $r_{12} \to 0$, then $g(r_{12}) \to 0$ and $h(r_{12}) \to -1$, because there is 0 probability of molecular overlap. Positive values of the total correlation function correspond to an enhanced density at that separation (due, for example, to a well in the intermolecular potential or to favourable geometric packing of the molecules). Conversely, when the joint density is reduced, the total correlation function is negative.

Equation (8.57) gave the pair density as a logarithmic derivative of the grand partition function, and it may be written as

$$\rho(\mathbf{r}_1,\mathbf{r}_2) = 2e(\mathbf{r}_1,\mathbf{r}_2)\frac{\delta \ln \Xi}{\delta f(\mathbf{r}_1,\mathbf{r}_2)}, \tag{9.7}$$

where the fact that $\delta f = \delta e$ has been used. Using the diagrammatic definition, Eq. (8.65), and recalling that differentiation by a pair function is half the sum of distinct diagrams obtained by erasing a bond, this gives

$\rho(\mathbf{r}_1, \mathbf{r}_2)$ = {the sum of all simple connected diagrams with z-field points and f-bonds, and two z-root points labelled 1 and 2 connected by an e-bond}

= 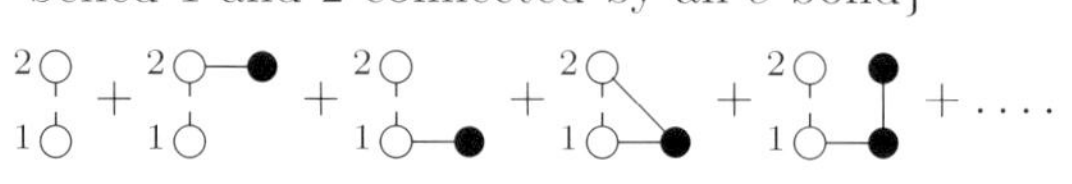 (9.8)

The broken line here signifies an e-bond. The root points and the field points are articulation points in many of these diagrams, and, as above, topological reduction can be used to remove them. This replaces the activity by the singlet density,

$$\rho(\mathbf{r}_1, \mathbf{r}_2) = \{\text{the sum of all simple irreducible diagrams with } \rho^{(1)}\text{-field points and } f\text{-bonds, and two } \rho^{(1)}\text{-root points labelled 1 and 2 connected by an } e\text{-bond}\}$$

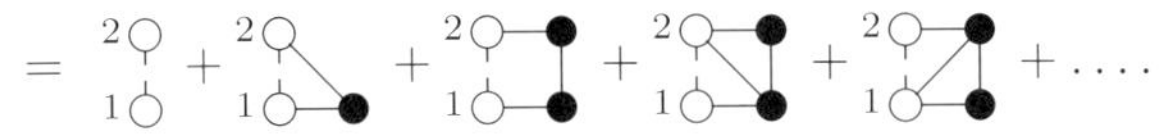

$= \ldots + \ldots$ (9.9)

The pair distribution function g has the same definition as this, except that it has 1-root points. If one replaces the e-bond by $f+1$, one doubles the number of diagrams: each of the above diagrams gives rise to a pair of diagrams, one with an f-bond between the root points, and the other with no direct connection between the root points. All of the diagrams except for one are connected diagrams. The disconnected diagram consists only of the two root points. In the representation of g this has value unity, so that the total correlation function, $h = g - 1$, is just the set of connected diagrams,

$$h(\mathbf{r}_1, \mathbf{r}_2) = \{\text{the sum of all simple irreducible diagrams with } \rho^{(1)}\text{-field points and } f\text{-bonds, and two 1-root points labelled 1 and 2}\}$$

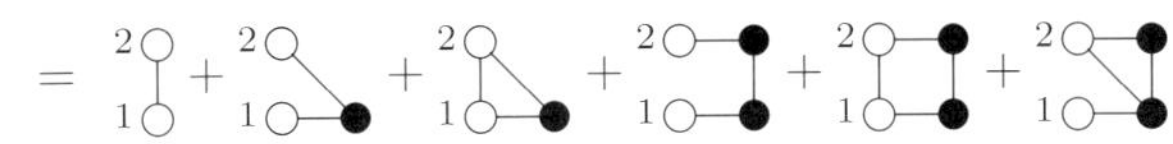

$= \ldots + \ldots$ (9.10)

Note that irreducible necessarily implies connected.

9.1.2 Density Expansion

For the homogeneous, isotropic system, one can expand the total correlation function in powers of density as $h(r) = \sum_{n=0}^{\infty} h_n(r)\rho^n$. The zeroth-order term is

$$h_0(r) = f(r) = -1 + \exp -\beta u^{(2)}(r). \tag{9.11}$$

This result is the same as the low-density expression used for the radial distribution function in the preceding chapter, $g(r) \sim \exp -\beta u^{(2)}(r)$, $\rho \to 0$. From the behaviour of the pair potential for physically realistic systems $u^{(2)}(r) \to 0$, $r \to \infty$, and $u^{(2)}(r) \to \infty$, $r \to 0$, it is clear that $h_0(r) \sim -\beta u^{(2)}(r)$, $r \to \infty$, and that $h_0(r) \to -1$, $r \to 0$.

The term linear in density is the sum of the second and third diagrams above, and this gives

$$h_1(r_{12}) = e(r_{12}) \int_V \mathrm{d}\mathbf{r}_3\, f(r_{13}) f(r_{32}). \tag{9.12}$$

Due to the exponential prefactor, which vanishes at small separations, one has $h_1(r) \to 0$, $r \to 0$. In fact, for every diagram of $h(r)$ with no bond between the root points, there is an identical diagram with an $f(r)$ bond. These can always be added together to give an $e(r)$ bond, and so one concludes that all the density-dependent terms go to 0 at small separations. The convolution integral complicates the analysis of the large-separation asymptotic behaviour of $h_2(r)$. Briefly however, for a power law pair potential whose integral converges, $u^{(2)}(r) \sim Ar^{-n}$, $r \to \infty$, where n is greater than the dimension of the relevant space, one has

$$h_1(r) \sim -2\beta u^{(2)}(r) \int_V \mathrm{d}\mathbf{r}'\, f(r'),\ r \to \infty. \tag{9.13}$$

This is a general result: $h(r) \sim \mathcal{O}[u^{(2)}(r)]$, $r \to \infty$, where the notation (read order of) means $\lim_{r\to\infty} h(r)/u^{(2)}(r) = \text{const}$. For potentials that decay quicker than power law (e.g., the hard-sphere potential) the total correlation function decays exponentially to 0. For potentials that decay so slowly that their integral is not convergent (e.g., the Coulomb potential), the individual terms in the density expansion diverge. In these cases a resummation must be carried out.

9.1.3 Mixtures

To date only a one-component system has been dealt with. The extension to a mixture is straightforward. The components may be indexed by a Greek subscript, so that $u_\alpha(\mathbf{r})$ is the one-body potential acting on species α, and $u_{\alpha\gamma}(\mathbf{r}, \mathbf{s})$ is the pair potential between particles of type α and γ. For a m-component mixture interacting with one- and two-body potentials the grand partition function is

$$\begin{aligned}\Xi(\boldsymbol{\mu}, V, T) &= \sum_{N_1=0}^{\infty} \cdots \sum_{N_m=0}^{\infty} \prod_{\alpha=1}^{m} \left[\frac{e^{\beta\mu_\alpha}}{\Lambda_\alpha^{3N_\alpha} N_\alpha!}\right] \int_V \mathrm{d}\mathbf{r}^{N_1} \ldots \mathrm{d}\mathbf{r}^{N_m} \\ &\quad \times \prod_{i\alpha} e^{-\beta u_\alpha(\mathbf{r}_{i\alpha})} \prod_{i\alpha, j\gamma}{}' e^{-\beta u_{\alpha\gamma}(\mathbf{r}_{i\alpha}, \mathbf{r}_{j\gamma})},\end{aligned} \tag{9.14}$$

where $\mathbf{r}_{i\alpha}$ is the position of the ith particle of type α, and where the prime on the product indicates that each pair interaction occurs once only so that there are no self-interaction terms.

The singlet density is

$$\begin{aligned}\rho_\alpha(\mathbf{r}) &= \langle N_\alpha \delta(\mathbf{r}_{1\alpha} - \mathbf{r})\rangle_\mu \\ &= \frac{z_\alpha(\mathbf{r})}{\Xi} \frac{\delta \Xi}{\delta z_\alpha(\mathbf{r})},\end{aligned} \tag{9.15}$$

where the activity is $z_\alpha(\mathbf{r}) = \Lambda_\alpha^{-3} \exp -\beta u_\alpha(\mathbf{r})$. The pair density can be expressed as

$$\begin{aligned}\rho_{\alpha\gamma}(\mathbf{r}, \mathbf{s}) &= \langle N_\alpha (N_\gamma - \delta_{\alpha\gamma}) \delta(\mathbf{r}_{1\alpha} - \mathbf{r}) \delta(\mathbf{r}_{1\gamma} - \mathbf{s})\rangle_\mu \\ &= \frac{z_\alpha(\mathbf{r}) z_\gamma(\mathbf{s})}{\Xi} \frac{\delta^2 \Xi}{\delta z_\alpha(\mathbf{r}) \delta z_\gamma(\mathbf{s})}.\end{aligned} \tag{9.16}$$

In the event that $\alpha = \gamma$, the coordinate $\mathbf{r}_{1\gamma}$ should be replaced by $\mathbf{r}_{2\alpha}$, because this is a distinct particle density. The pair function is clearly symmetric, $\rho_{\alpha\gamma}(\mathbf{r}, \mathbf{s}) = \rho_{\gamma\alpha}(\mathbf{s}, \mathbf{r})$.

The diagrammatic definitions are unchanged, except of course for the facts that the type of the root points must be specified, that z-points means z_α-points, and that f-bonds means $f_{\alpha\gamma}$-bonds. Also, a field point corresponds to integration over space *and* to summation over the species index.

9.2 Ornstein–Zernike Equation

9.2.1 Direct Correlation Function

In the preceding chapter it was shown that the derivative of the logarithm of the grand partition function with respect to the logarithm of the activity gave

$$\frac{\delta^2 \ln \Xi}{\delta \ln \Lambda^3 z(\mathbf{r}) \delta \ln \Lambda^3 z(\mathbf{s})} = \rho(\mathbf{r}, \mathbf{s}) - \rho(\mathbf{r})\rho(\mathbf{s}) + \rho(\mathbf{r})\delta(\mathbf{r}, \mathbf{s}). \tag{9.17}$$

One can define $\rho_\delta(\mathbf{r}, \mathbf{s})$ as the right-hand side of this, and in view of the fact that the singlet density is a similar first derivative, $\rho_\delta^{(1)} = \rho^{(1)}$, one has

$$\rho_\delta(\mathbf{r}, \mathbf{s}) = \frac{\delta \rho(\mathbf{r})}{\delta \ln \Lambda^3 z(\mathbf{s})}. \tag{9.18}$$

Successive differentiation yields a hierarchy of such densities. Thus $\ln \Xi$ is a generating functional for the many-particle densities. At the pair level, $\rho_\delta^{(2)}$ is obviously related to the total correlation function,

$$\rho_\delta(\mathbf{r}, \mathbf{s}) = \rho(\mathbf{r})\delta(\mathbf{r}, \mathbf{s}) + \rho(\mathbf{r})\rho(\mathbf{s})h(\mathbf{r}, \mathbf{s}). \tag{9.19}$$

In the preceding chapter the one-particle direct correlation function, $c(\mathbf{r})$, was defined in terms of the Mayer irreducible cluster integrals, Eq. (8.75). One can define a generating functional for $c^{(1)}$,

$\beta\mathcal{F}^{\rm ex}[\rho]$ = {the sum of all simple irreducible diagrams with ρ-field points and one or more f-bonds}

= 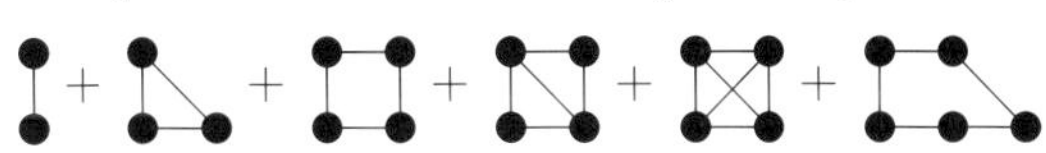 + ... (9.20)

It is clear that

$$c(\mathbf{r}) = \frac{\delta \beta \mathcal{F}^{\rm ex}}{\delta \rho(\mathbf{r})}. \tag{9.21}$$

The functional $\mathcal{F}^{\rm ex}$ acts as a generator for the hierarchy of direct correlation functions. The second member of the hierarchy is the *pair direct correlation function*,

$$c(\mathbf{r}, \mathbf{s}) = \frac{\delta c(\mathbf{r})}{\delta \rho(\mathbf{s})}. \tag{9.22}$$

In terms of diagrams this is just the irreducible pair diagrams without a nodal point (because a nodal point would have been an articulation point before differentiation). One has

$c(\mathbf{r}_1, \mathbf{r}_2)$ = {the sum of all simple irreducible diagrams with two 1-root points labelled $\mathbf{r}_1$ and $\mathbf{r}_2$, ρ-field points, f-bonds, and no nodal points}

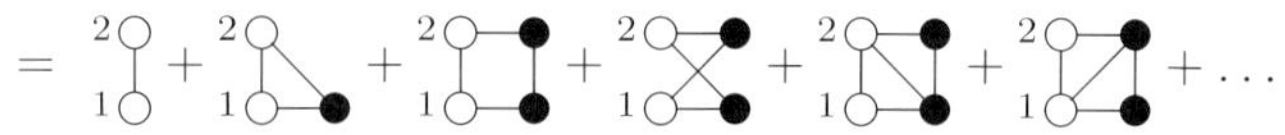

(9.23)

It is evident that the diagrams of $c^{(2)}$ form a subset of the diagrams of $h^{(2)}$.

A related generating functional is $\mathcal{F}[\rho] = \mathcal{F}^{\rm id}[\rho] - \mathcal{F}^{\rm ex}[\rho]$, where

$$\beta\mathcal{F}^{\rm id}[\rho] = \int \mathrm{d}\mathbf{s}\, \rho(\mathbf{s}) \left[-1 + \ln \Lambda^3 \rho(\mathbf{s})\right]. \tag{9.24}$$

Its first derivative is

$$\begin{aligned} \frac{\delta \beta \mathcal{F}[\rho]}{\delta \rho(\mathbf{r})} &= \ln \Lambda^3 \rho(\mathbf{r}) - c(\mathbf{r}) \\ &= \ln \Lambda^3 z(\mathbf{r}, [\rho]). \end{aligned} \tag{9.25}$$

Accordingly the second derivative is

$$\frac{\delta \ln \Lambda^3 z(\mathbf{r}, [\rho])}{\delta \rho(\mathbf{s})} = \frac{\delta(\mathbf{r}, \mathbf{s})}{\rho(\mathbf{s})} - c(\mathbf{r}, \mathbf{s}) \equiv c_\delta(\mathbf{r}, \mathbf{s}). \tag{9.26}$$

9.2.2 Functional Inverse

This last equation shows that $c_\delta^{(2)}$ is the functional inverse of $\rho_\delta^{(2)}$. Hence the two satisfy

$$\int \mathrm{d}\mathbf{t}\, \rho_\delta(\mathbf{r}, \mathbf{t}) c_\delta(\mathbf{t}, \mathbf{s}) = \delta(\mathbf{r}, \mathbf{s}). \tag{9.27}$$

Writing these in full and rearranging gives the *Ornstein–Zernike* equation,

$$h(\mathbf{r}, \mathbf{s}) = c(\mathbf{r}, \mathbf{s}) + \int \mathrm{d}\mathbf{t}\, h(\mathbf{r}, \mathbf{t})\, \rho(\mathbf{t})\, c(\mathbf{t}, \mathbf{s}). \tag{9.28}$$

This equation is formally exact, and it plays a central role in approaches to the pair distribution function.

As mentioned above, the pair direct correlation function is the subset of diagrams of the pair total correlation function that have no nodal points. The remaining diagrams of $h^{(2)}$ can be split into classes that contain one, two, ... nodal points connected in series. A little thought shows that the function that connects each pair of nodal points is the direct correlation function, since it is composed of all diagrams free of nodal points. One has

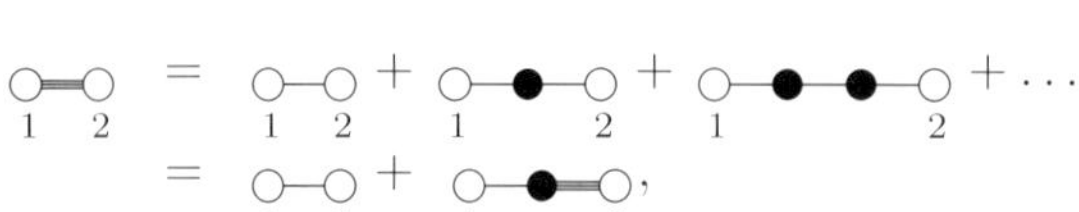

(9.29)

where the heavy line represents an h-bond and the single line a c-bond, the empty circles have unit weight, and the filled circles are weighted with the singlet density. This is just the Ornstein–Zernike equation, and one sees that it represents an expansion of the total correlation function in terms of the number of nodal points.

Since the convolution integral in the Ornstein–Zernike equation has the effect of summing series of increasing powers of c-bonds, the order of the arguments does not matter. One has

$$\int \mathrm{d}\mathbf{t}\, h(\mathbf{r},\mathbf{t})\,\rho(\mathbf{t})\,c(\mathbf{t},\mathbf{s}) = \int \mathrm{d}\mathbf{t}\, c(\mathbf{r},\mathbf{t})\,\rho(\mathbf{t})\,h(\mathbf{t},\mathbf{s}). \tag{9.30}$$

The Ornstein–Zernike equation was postulated early last century by two eponymous scientists concerned with the scattering of light by fluids at their critical point.[2] The two pair functions continue to bear the names 'total' and 'direct', which were historically ascribed to them in part because of the physical basis originally used to justify the equation. The total correlation between two molecules is the sum of the direct correlation between them plus the indirect correlation mediated by a third. The latter is the direct correlation between one and three times the total correlation between three and two, integrated over all of space. Alternatively, the total correlation between two molecules is the sum of direct correlations between by zero, one, two, ... intermediate molecules. Modern-day usage of the Ornstein–Zernike equation far exceeds anything originally envisaged.

For a homogeneous fluid the Ornstein–Zernike equation is

$$h(r_{12}) = c(r_{12}) + \rho \int \mathrm{d}\mathbf{r}_3\, c(r_{13})h(r_{31}). \tag{9.31}$$

The convolution integral that appears here is most naturally treated by means of Fourier transforms. Using the fact that $\int \mathrm{d}\mathbf{r}\, e^{i\mathbf{k}\cdot\mathbf{r}} = (2\pi)^3\delta(\mathbf{k})$, and defining the Fourier transform to be $\hat{f}(k) = \int \mathrm{d}\mathbf{r}\, e^{-i\mathbf{k}\cdot\mathbf{r}} f(r)$, it follows that

$$\begin{aligned} \hat{h}(k) &= \hat{c}(k) + \rho\hat{c}(k)\hat{h}(k) \\ &= \frac{\hat{c}(k)}{1-\rho\hat{c}(k)} \\ &= \hat{c}(k) + \rho\hat{c}(k)^2 + \rho^2\hat{c}(k)^3 + \ldots . \end{aligned} \tag{9.32}$$

The last expansion shows once more that the total correlation function is a series of terms of successively higher powers of the direct correlation function. The convolution integrals of real space have become products in Fourier space. This reduction or factorisation of the Ornstein–Zernike equation occurs using the Fourier transform in arbitrary dimension for an homogeneous system. In the case of a planar inhomogeneity, such that $\rho(\mathbf{r}) = \rho(z)$ and $h(\mathbf{r}_1,\mathbf{r}_2) = h(z_1,z_2,R_{12})$, a Hankel transform of order 0 factorises the equation, and in the case of a spherical inhomogeneity, $\rho(\mathbf{r}) = \rho(r)$ and $h(\mathbf{r}_1,\mathbf{r}_2) = h(r_1,r_2,\theta_{12})$, a Legendre transform accomplishes the same thing.

[2] L. S. Ornstein and F. Zernike, Accidental deviations of density and opalescence at the critical point of a single substance, *Proc. Akad. Sci. Amsterdam* **17** (1914), 793.

Mixtures

For a multicomponent system the Ornstein–Zernike equation becomes

$$h_{\alpha\gamma}(\mathbf{r},\mathbf{s}) = c_{\alpha\gamma}(\mathbf{r},\mathbf{s}) + \sum_{\lambda} \int \mathrm{d}\mathbf{t}\, h_{\alpha\lambda}(\mathbf{r},\mathbf{t})\, \rho_\lambda(\mathbf{t})\, c_{\lambda\gamma}(\mathbf{t},\mathbf{s}). \tag{9.33}$$

Because the field points now correspond to integration over space and summation over species, one sees that the nodal point embodied in the Ornstein–Zernike integral also has a summation over the components of the mixture.

The form of the equation is suggestive of matrix multiplication. One can define the total correlation function matrix $\underline{\underline{h}}(\mathbf{r},\mathbf{s})$, with components $h_{\alpha\gamma}(\mathbf{r},\mathbf{s})$, and similarly for the direct correlation function matrix, $\underline{\underline{c}}(\mathbf{r},\mathbf{s})$. With the diagonal density matrix, $\underline{\underline{\rho}}(\mathbf{r})$, which has components $\rho_\alpha(\mathbf{r})\delta_{\alpha\gamma}$, the Ornstein–Zernike equation for a mixture may be written

$$\underline{\underline{h}}(\mathbf{r},\mathbf{s}) = \underline{\underline{c}}(\mathbf{r},\mathbf{s}) + \int \mathrm{d}\mathbf{t}\, \underline{\underline{h}}(\mathbf{r},\mathbf{t})\, \underline{\underline{\rho}}(\mathbf{t})\, \underline{\underline{c}}(\mathbf{t},\mathbf{s}). \tag{9.34}$$

For a homogeneous system this is simply written in Fourier space as

$$\underline{\underline{\hat{h}}}(k) = \underline{\underline{\hat{c}}}(k) + \underline{\underline{\hat{h}}}(k)\underline{\underline{\rho\hat{c}}}(k), \tag{9.35}$$

which has solution

$$\underline{\underline{\hat{h}}}(k) = \underline{\underline{\hat{c}}}(k) \left(\underline{\underline{I}} - \underline{\underline{\rho\hat{c}}}(k)\right)^{-1}. \tag{9.36}$$

An alternative and slightly neater formulation emerges by defining the matrix $\underline{\underline{H}}(\mathbf{r},\mathbf{s})$ with components $\rho_\alpha(\mathbf{r})^{1/2} h_{\alpha\gamma}(\mathbf{r},\mathbf{s})\rho_\gamma(\mathbf{s})^{1/2}$ and similarly for $\underline{\underline{C}}(\mathbf{r},\mathbf{s})$. In this case for a homogeneous system the Ornstein–Zernike equation in Fourier space has solution

$$\underline{\underline{\hat{H}}}(k) = \underline{\underline{\hat{C}}}(k) \left(\underline{\underline{I}} - \underline{\underline{\hat{C}}}(k)\right)^{-1}. \tag{9.37}$$

It should be noted that these are symmetric matrices, since $h_{\alpha\gamma}(r) = h_{\gamma\alpha}(r)$, and $c_{\alpha\gamma}(r) = c_{\gamma\alpha}(r)$. Also, the two matrices $\underline{\underline{\hat{H}}}(k)$ and $\underline{\underline{\hat{C}}}(k)$ commute since the former is a sum of products of the latter.

9.3 Closure Relations

9.3.1 Potential of Mean Force

The Ornstein–Zernike equation represents a relation between two (unknown) pair correlation functions, and a second equation is required to close the set.[3] One can proceed in a formally exact fashion by noting that many of the diagrams

[3]G. Stell, Cluster expansions for classical systems in equilibrium, in *The Equilibrium Theory of Classical Fluids* (H. Frisch and J. L. Lebowitz, Eds.), p. II-171, Benjamin, New York, 1964.

that compose $h(\mathbf{r}_1, \mathbf{r}_2)$ are the star-product of simpler diagrams connected in parallel at the root points. That is, when both root points are removed, these diagrams become disconnected; the root points form an *articulation pair*. Hence the total correlation function can be decomposed into successively higher-order star-products of diagrams free of such an articulation pair of root points. One defines

$$v(\mathbf{r}_1, \mathbf{r}_2) = \{\text{the sum of all irreducible diagrams composed of two nonadjacent 1-root points labelled } \mathbf{r}_1 \text{ and } \mathbf{r}_2\text{, and one or more } \rho^{(1)}\text{-field points and } f\text{-bonds, such that the root points do not form an articulation pair}\}$$

$=$ 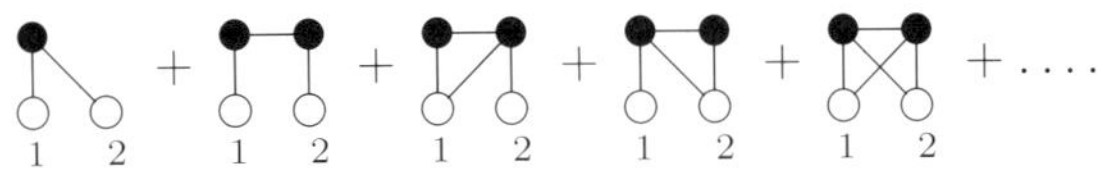 (9.38)

Exponentiating these connects them in parallel, and generates almost all of the diagrams of $h^{(2)}$. What is missing are those diagrams with an f-bond between the root points, and so one needs to multiply the exponential by $1 + f(\mathbf{r}_1, \mathbf{r}_2)$. Doing this gives rise to the diagram with two disconnected field points, so that one recognises that one has generated $g^{(2)}$ rather than $h^{(2)}$. That is,

$$g(\mathbf{r}_1, \mathbf{r}_2) = [1 + f(\mathbf{r}_1, \mathbf{r}_2)] \exp v(\mathbf{r}_1, \mathbf{r}_2). \tag{9.39}$$

In view of the fact that the prefactor is just the exponential of the pair potential, one can define the *potential of mean force*,

$$w(\mathbf{r}_1, \mathbf{r}_2) = u(\mathbf{r}_1, \mathbf{r}_2) - k_{\mathrm{B}} T v(\mathbf{r}_1, \mathbf{r}_2), \tag{9.40}$$

in terms of which the pair distribution function is simply

$$g(\mathbf{r}_1, \mathbf{r}_2) = \exp -\beta w(\mathbf{r}_1, \mathbf{r}_2). \tag{9.41}$$

The potential of mean force for a multicomponent mixture is essentially the same as that for a single component system. This is because the star-product connects diagrams at the root points that have the same label (i.e., the same position and the same species). Accordingly the radial distribution function for a mixture is just

$$g_{\alpha\gamma}(\mathbf{r}_1, \mathbf{r}_2) = \exp -\beta w_{\alpha\gamma}(\mathbf{r}_1, \mathbf{r}_2). \tag{9.42}$$

As in the derivation of the Ornstein–Zernike equation, one splits the diagrams composing $v^{(2)}$ into two classes; those with nodal points sum to the *series function*, $s(\mathbf{r}_1, \mathbf{r}_2)$, and the remainder form the *bridge function*, $d(\mathbf{r}_1, \mathbf{r}_2)$. One has

$$g(\mathbf{r}_1, \mathbf{r}_2) = e^{-\beta u(\mathbf{r}_1, \mathbf{r}_2)} \exp[s(\mathbf{r}_1, \mathbf{r}_2) + d(\mathbf{r}_1, \mathbf{r}_2)]. \tag{9.43}$$

Now since the pair diagrams without nodes form the pair direct correlation function, subtracting it from the total correlation function must yield the diagrams with nodes,

$$s(\mathbf{r}_1, \mathbf{r}_2) = h(\mathbf{r}_1, \mathbf{r}_2) - c(\mathbf{r}_1, \mathbf{r}_2)$$

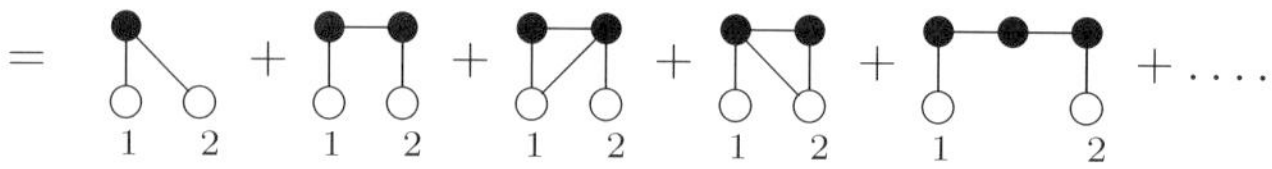

(9.44)

These are called the series diagrams because, in view of the Ornstein–Zernike equation, they represent the sum of convolution products of the direct correlation function, which is to say that they are connected in series.

The bridge function is defined as

$d(\mathbf{r}_1, \mathbf{r}_2)$ = {the sum of all irreducible diagrams composed of two nonadjacent 1-root points labelled $\mathbf{r}_1$ and $\mathbf{r}_2$, and two or more $\rho^{(1)}$-field points and f-bonds, such that the root points do not form an articulation pair and that there are no nodal points}

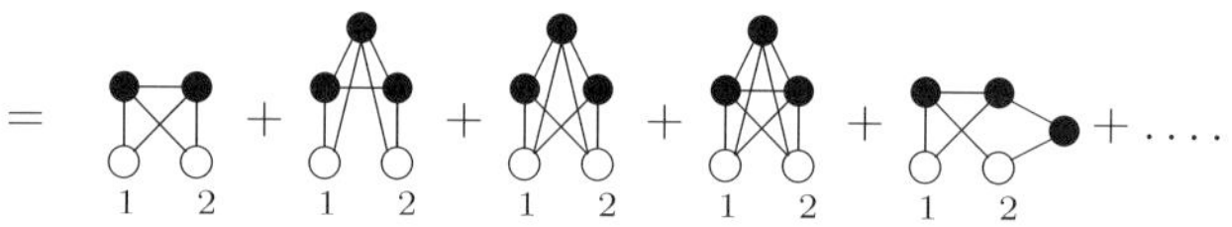

(9.45)

These diagrams are evidently more compact and more highly connected than the series diagrams.

In view of the above, one has two specified functions (ρ and $u^{(2)}$), two equations (the Ornstein–Zernike Eq. (9.28) and the potential of mean force Eq. (9.43)), and three unknown functions (the total, h, and direct, c, correlation functions, and the bridge function, d). In order to solve numerically this system of equations one needs a closure equation.

9.3.2 Approximate Closures

There is no simple expression for the bridge function. It is in fact quite difficult to evaluate, and the *hypernetted chain* (HNC) approximation neglects it entirely,

$$d^{\mathrm{HNC}}(\mathbf{r}_1, \mathbf{r}_2) = 0. \tag{9.46}$$

It is emphasised that this neglect of the bridge function is the only approximation that has been introduced into the system of formally exact equations. For an homogeneous fluid, the hypernetted chain closure may be written as

$$h(r) = -1 + \exp[-\beta u(r) + h(r) - c(r)]. \tag{9.47}$$

This and the Ornstein–Zernike equation gives two equations for the two unknown pair correlation functions. One usually sets up an iterative procedure that solves them alternately in real and Fourier space, utilising the fast Fourier transform. The HNC closure usually provides relatively accurate numerical results. It also can be used to give various analytic results, as will be seen below.

The spirit of the Ornstein–Zernike approach differs from a simple density expansion. The hypernetted chain closure includes diagrams of all orders in density. It also neglects an infinite class of diagrams (the bridge diagrams) but it does not introduce any additional approximations. The first neglected diagram is of order ρ^2, and so this closure gives the exact results in the low-density limit. Conversely, because it includes diagrams of all orders it is also applicable to dense fluids and to solids. In general, one finds that the scheme becomes less accurate in highly coupled (low-temperature, high-density) systems.

The reason that the bridge function is difficult to evaluate is that the bonds between the field points make them highly connected and prevent the reduction of the multidimensional integrals. One of the oldest closures, the *Percus–Yevick approximation*, can be derived by replacing some of these f-bonds by -1. The rationale for this is that the high connectivity of the bridge diagrams means that the integrals are dominated by regions where the field points are close together, and in the core region, $r \lesssim d$, $f(r) = -1$. The first bridge diagram becomes

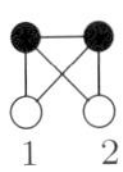 $\Rightarrow -$ 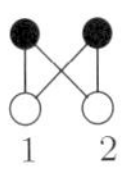

$$= \frac{-1}{2}\left[\rho \int \mathrm{d}\mathbf{r}_3\, f(r_{13}) f(r_{32})\right]^2 . \tag{9.48}$$

The right-hand side is the square of the first series diagram. By judicious choice of the replacement bonds (i.e., those bonds that make the root points an articulation pair *and* create a nodal point in the fragments), all of the bridge diagrams can be written as star-products of series diagrams with an appropriate prefactor. One ends up with

$$\begin{aligned} d^{\mathrm{PY}}(r) &= \frac{-1}{2} s(r)^2 + \frac{1}{3} s(r)^3 - \ldots \\ &= \ln[1 + s(r)] - s(r). \end{aligned} \tag{9.49}$$

With this approximation, and the Ornstein–Zernike relation, $s(r) = h(r) - c(r)$, the Percus–Yevick closure is

$$g(r) = [g(r) - c(r)]e^{-\beta u(r)}. \tag{9.50}$$

For the case of the hard-sphere fluid, this corresponds to setting the radial distribution function to 0 within the hard core, which is exact, and to setting the direct correlation function to 0 beyond it. The Percus–Yevick closure can be solved analytically for the hard-sphere fluid, and turns out to be slightly more accurate than the HNC at higher densities. For fluids with long-ranged potentials, the HNC is preferable.

A closely related closure is the *mean spherical approximation*. This closure, which reduces to the Percus–Yevick for hard spheres, sets $g(r) = 0, r < d$, and $c(r) = -\beta u(r), r > d$, which is exact at large r as will be seen. Analytic solutions exist for a variety of hard-sphere plus electrostatic multipolar fluids,[4] and it provides a relatively accurate description of the primitive model electrolyte.[5]

[4] M.S. Wertheim, Exact solution of the mean spherical model for fluids of hard spheres with permanent electric dipole moments, *J. Chem. Phys.* **55** (1971), 4291.

[5] L. Blum, Mean spherical model for a mixture of charged spheres and hard dipoles, *Chem. Phys. Lett.* **26** (1974), 200. J. S. Høye and G. Stell, New self-consistent approximations for ionic and polar fluids, *J. Chem. Phys.* **67** (1977), 524.

9.3.3 First Bridge Diagram

The hypernetted chain approximation may be systematically improved by including successive bridge diagrams. The first bridge diagram is

$$d^{(2)}(r_{12}) = \frac{1}{2}\rho^2 \int \mathrm{d}\mathbf{r}_3\, \mathrm{d}\mathbf{r}_4\, f(r_{13})f(r_{14})f(r_{34})f(r_{23})f(r_{24}). \tag{9.51}$$

This first bridge diagram alone significantly improves the hypernetted chain closure for more highly coupled systems, and including it extends the regime of applicability of the approximation. The numerical evaluation of this bridge diagram is facilitated by expansion in Legendre polynomials

$$\begin{aligned} d^{(2)}(r_{12}) &= 2\pi^2\rho^2 \sum_{n=0}^{\infty} \left(\frac{2}{2n+1}\right)^2 \int_0^{\infty} \mathrm{d}r_3\, r_3^2 \int_0^{\infty} \mathrm{d}r_4\, r_4^2 \\ &\quad \times f(r_3)f(r_4)\hat{f}^{(n)}(r_1, r_3)\hat{f}^{(n)}(r_1, r_4)\hat{f}^{(n)}(r_3, r_4), \end{aligned} \tag{9.52}$$

where the molecule 2 has been placed at the origin, $\mathbf{r}_2 = \mathbf{0}$, and the Legendre coefficients are

$$\hat{f}^{(n)}(r, s) = \frac{2n+1}{2} \int_{-1}^{1} \mathrm{d}x\, P_n(x) f\left(\sqrt{r^2 + s^2 - 2rsx}\right), \tag{9.53}$$

where $P_n(x)$ is the Legendre polynomial of order n. The evaluation of multidimensional integrals by expansion in Legendre polynomials is a powerful technique.[6] It is convenient to evaluate the Legendre coefficients by a discrete, orthogonal technique,[7] which corresponds to a Gaussian quadrature; typically only 10–20 terms are needed in the expansion. Expressions for the Legendre expansion of the bridge diagrams with three field points have been given.[8]

This first f-bond bridge diagram is evaluated at the beginning, and it is then fixed during the iteration of the hypernetted chain closure. Alternatively one may replace the f-bonds by h-bonds, which corresponds to resumming an infinite class of bridge diagrams. In this case the resummed bridge diagram must be reevaluated during the iteration procedure, typically every 50–100 iterations of the total correlation function. The number of radial grid points used in the Legendre procedure is generally less than that used in the Fourier transform (because it is numerically intensive, and because the bridge diagram is more short-ranged and smoother than the total correlation function itself). However, in the case of a hard core, the f-bonds have a discontinuity at contact, and one is well advised to subract this discontinuity in real space and to add back its analytic transform in Legendre space, thereby reducing the need for a fine grid and a high number of Legendre coefficients.

[6] A. D. J. Haymet, S. A. Rice, and W. G. Madden, An accurate integral equation for the pair and triplet distribution functions of a simple liquid, *J. Chem. Phys.* **74** (1981), 3033.

[7] P. Attard, Spherically inhomogeneous fluids. I. Percus–Yevick hard spheres: Osmotic coefficients and triplet correlations, *J. Chem. Phys.* **91** (1989), 3072.

[8] P. Attard and G. N. Patey, Hypernetted chain closure with bridge diagrams. Asymmetric hard-sphere mixtures, *J. Chem. Phys.* **92** (1990), 4970.

9.3.4 Functional Taylor Expansion

These closure approximations can also be derived by truncating various functional Taylor series expansions at the first term. The key idea is to consider an inhomogeneous fluid in which the external one-body potential equals the pair potential due to a fixed particle. In this case the singlet density is just proportional to the radial distribution function of the homogeneous fluid. One considers the reference state of no fixed particle, and assumes that changes in the singlet density and in the external potential due to introducing the particle are small.

One starts with a homogeneous fluid of density ρ with isotropic pair functions due to a pair potential $u^{(2)}(r_{12})$, and one considers the effect of a perturbing external potential $\Delta u^{(1)}(\mathbf{r})$. In order to approximate the pair functions, a relationship between the perturbed singlet density and the pair function of the reference system is required. As just mentioned, if the perturbing potential is taken to represent a particle of the system, fixed at $\mathbf{r}_1$ say, then

$$\Delta u^{(1)}(\mathbf{r}_2) = u^{(2)}(r_{12}). \tag{9.54}$$

In this case the final singlet density, $\rho + \Delta\rho^{(1)}(\mathbf{r}_2)$, represents, in essence, the probability of finding a particle at $\mathbf{r}_2$ in the presence of the particle fixed at $\mathbf{r}_1$, and as such it must be proportional to the pair probability function of the reference fluid, $\rho + \Delta\rho^{(1)}(\mathbf{r}_2) = \rho g^{(2)}(r_{12})$, or

$$\Delta\rho^{(1)}(\mathbf{r}_2) = \rho h^{(2)}(r_{12}). \tag{9.55}$$

Mean Spherical Approximation

The formal expression for the change in density due to the change in external potential is

$$\begin{aligned}
\Delta\rho^{(1)}(\mathbf{r}_2) &= \int \mathrm{d}\mathbf{r}_3 \, \frac{\delta\rho^{(1)}(\mathbf{r}_2)}{\delta u^{(1)}(\mathbf{r}_3)} \Delta u^{(1)}(\mathbf{r}_3) \\
&= -\beta \int \mathrm{d}\mathbf{r}_3 \, \frac{\delta\rho^{(1)}(\mathbf{r}_2)}{\delta \ln \Lambda^3 z(\mathbf{r}_3)} \Delta u^{(1)}(\mathbf{r}_3) \\
&= -\beta \int \mathrm{d}\mathbf{r}_3 \, \rho_\delta^{(2)}(\mathbf{r}_2, \mathbf{r}_3) \Delta u^{(1)}(\mathbf{r}_3) \\
&= -\beta\rho^{(1)}(\mathbf{r}_2)\Delta u^{(1)}(\mathbf{r}_2) \\
&\quad - \beta \int \mathrm{d}\mathbf{r}_3 \, \rho^{(1)}(\mathbf{r}_2)\rho^{(1)}(\mathbf{r}_3) h^{(2)}(\mathbf{r}_2, \mathbf{r}_3) \Delta u^{(1)}(\mathbf{r}_3).
\end{aligned} \tag{9.56}$$

This is called the Yvon equation, and it shows that the pair function $\rho_\delta^{(2)}$ gives the linear response of the density to an external field. The density and total correlation function in the integrand are evaluated in the reference system, as befits the functional Taylor expansion; they are ρ and $h^{(2)}(r_{23})$, respectively. When the perturbing singlet potential is the reference pair potential due to a particle at $\mathbf{r}_1$, the left-hand side is just density times the reference total correlation function. Inserting all of these and dividing by density one has

$$h(r_{12}) = -\beta u(r_{12}) - \beta\rho \int \mathrm{d}\mathbf{r}_3\, u(r_{13}) h(r_{32}). \tag{9.57}$$

Comparing this with the Ornstein–Zernike equation, one sees that this is equivalent to the closure for the pair direct correlation function,

$$c^{(2)}(r) = -\beta u^{(2)}(r). \tag{9.58}$$

This is the *mean spherical approximation* given above.

What is approximate in this derivation is the truncation of the Taylor expansion for the change in density at the linear term. If the perturbing potential were vanishingly small, the final result would be exact. However, in general the pair potential of interest is not particularly small, and hence the whole procedure is approximate. In particular, in the core region where $r \to 0$, the pair potential goes to infinity. Accordingly, in practice one invokes the mean spherical closure for the direct correlation function beyond the core, and one utilises the fact that the total correlation is exactly -1 inside the core. That is,

$$\begin{array}{ll} h^{(2)}(r) = -1, & r < d, \\ c^{(2)}(r) = -\beta u^{(2)}(r), & r > d, \end{array} \tag{9.59}$$

where d is the core diameter. It is this version of the closure that is generally known as the mean spherical approximation.

Percus–Yevick Approximation

The density is not the only function that one can expand, and it is not only the external potential that one need expand with respect to. Consider instead the perturbation in the density divided by the activity, due to a change in the density itself,

$$\begin{aligned} \Delta[\rho^{(1)}(\mathbf{r}_2)/z(\mathbf{r}_2)] &= \int \mathrm{d}\mathbf{r}_3\, \frac{\delta[\rho^{(1)}(\mathbf{r}_2)/z(\mathbf{r}_2)]}{\delta\rho^{(1)}(\mathbf{r}_3)} \Delta\rho^{(1)}(\mathbf{r}_3) \\ &= \int \mathrm{d}\mathbf{r}_3\, \frac{\rho^{(1)}(\mathbf{r}_2)}{z(\mathbf{r}_2)} \frac{\delta \ln[\rho^{(1)}(\mathbf{r}_2)/z(\mathbf{r}_2)]}{\delta\rho^{(1)}(\mathbf{r}_3)} \Delta\rho^{(1)}(\mathbf{r}_3) \\ &= \int \mathrm{d}\mathbf{r}_3\, \frac{\rho^{(1)}(\mathbf{r}_2)}{z(\mathbf{r}_2)} \frac{\delta c^{(1)}(\mathbf{r}_2)}{\delta\rho^{(1)}(\mathbf{r}_3)} \Delta\rho^{(1)}(\mathbf{r}_3) \\ &= \int \mathrm{d}\mathbf{r}_3\, \frac{\rho^{(1)}(\mathbf{r}_2)}{z(\mathbf{r}_2)} c^{(2)}(\mathbf{r}_2, \mathbf{r}_3) \Delta\rho^{(1)}(\mathbf{r}_3). \end{aligned} \tag{9.60}$$

Again the density, activity, and pair direct correlation function in the integrand are those of the reference fluid, ρ, z, and $c^{(2)}(r_{23})$, respectively. The change in density in the integrand is $\rho h^{(2)}(r_{13})$, and the left-hand side is

$$\frac{\rho g^{(2)}(r_{12})}{\Lambda^{-3} e^{\beta\mu} e^{-\beta u^{(2)}(r_{12})}} - \frac{\rho}{\Lambda^{-3} e^{\beta\mu}}.$$

Hence one has

$$\begin{aligned} g^{(2)}(r_{12}) &= e^{-\beta u^{(2)}(r_{12})} + e^{-\beta u^{(2)}(r_{12})}\rho \int \mathrm{d}\mathbf{r}_3\, c^{(2)}(r_{23})h^{(2)}(r_{13}) \\ &= e^{-\beta u^{(2)}(r_{12})}\left[g^{(2)}(r_{12}) - c^{(2)}(r_{12})\right]. \end{aligned} \tag{9.61}$$

This is the *Percus–Yevick approximation* derived above by diagrammatic means. In combination with the Orstein–Zernike equation, it provides a second equation for the two unknown pair functions.

In the case of the hard-sphere potential, the Percus–Yevick closure is the same as the mean spherical approximation. Inside the core the right-hand side vanishes due to the exponential prefactor and one has the exact result that $g^{(2)}(r) = 0$, $r < d$. Beyond the core the potential vanishes and the closure gives $g^{(2)}(r) = g^{(2)}(r) - c^{(2)}(r)$, with solution $c^{(2)}(r) = -\beta u^{(2)}(r) = 0$, $r > d$.

Hypernetted Chain

If instead of expanding ρ/z one expands $\ln \rho/z$ with respect to the density, one obtains the hypernetted chain closure approximation given above. Following closely the derivation of the Percus–Yevick approximation one has

$$\begin{aligned} \Delta \ln[\rho^{(1)}(\mathbf{r}_2)/z(\mathbf{r}_2)] &= \int \mathrm{d}\mathbf{r}_3\, \frac{\delta \ln[\rho^{(1)}(\mathbf{r}_2)/z(\mathbf{r}_2)]}{\delta\rho^{(1)}(\mathbf{r}_3)} \Delta\rho^{(1)}(\mathbf{r}_3) \\ &= \int \mathrm{d}\mathbf{r}_3\, c^{(2)}(\mathbf{r}_2, \mathbf{r}_3)\Delta\rho^{(1)}(\mathbf{r}_3). \end{aligned} \tag{9.62}$$

The left-hand side is

$$\ln \frac{\rho g^{(2)}(r_{12})}{\Lambda^{-3}e^{\beta\mu}e^{-\beta u^{(2)}(r_{12})}} - \ln \frac{\rho}{\Lambda^{-3}e^{\beta\mu}} = \beta u^{(2)}(r_{12}) + \ln g^{(2)}(r_{12}), \tag{9.63}$$

so that one has

$$\ln g^{(2)}(r_{12}) = -\beta u^{(2)}(r_{12}) + \rho \int \mathrm{d}\mathbf{r}_3\, c^{(2)}(r_{23})h^{(2)}(r_{31}). \tag{9.64}$$

The Ornstein–Zernike convolution integral is just the series function, so that this may be written

$$g^{(2)}(r_{12}) = \exp\left[-\beta u^{(2)}(r_{12}) + h^{(2)}(r_{12}) - c^{(2)}(r_{12})\right]. \tag{9.65}$$

Clearly this corresponds to neglecting the bridge function in the potential of mean force, which is just the *hypernetted chain approximation* derived by diagrammatic means above.

If one were to linearise the exponential in the hypernetted chain approximation with respect to the series function, one would obtain the Percus–Yevick approximation. Hence an alternative interpretation of the latter is that it neglects the bridge diagrams and all the diagrams with an articulation pair of points.

9.4 Born–Green–Yvon Hierarchy

An alternative to the Ornstein–Zernike equation for calculating the pair distribution function is based upon the so-called Born–Green–Yvon hierarchy. This is a general approach that relates the gradient of the many-particle density to integrals of the gradient of the potential and higher-order densities. An approximate closure is required to relate these higher-order densities to the one of interest, which is usually the pair density.

Suppose that the potential energy consists of singlet, pair, triplet, and higher-order terms,

$$U_N(\mathbf{r}^N) = \sum_{i=1}^{N} u^{(1)}(\mathbf{r}_i) + \sum_{i<j}^{N} u^{(2)}(\mathbf{r}_i, \mathbf{r}_j) + \sum_{i<j<k}^{N} u^{(3)}(\mathbf{r}_i, \mathbf{r}_j, \mathbf{r}_k) + \ldots. \tag{9.66}$$

Its derivative with respect to $\mathbf{r}_1$ is

$$\begin{aligned} \nabla_1 U_N(\mathbf{r}^N) &= \nabla_1 u^{(1)}(\mathbf{r}_1) + \sum_{j=2}^{N} \nabla_1 u^{(2)}(\mathbf{r}_1, \mathbf{r}_j) \\ &\quad + \sum_{1<j<k}^{N} \nabla_1 u^{(3)}(\mathbf{r}_1, \mathbf{r}_j, \mathbf{r}_k) + \ldots. \end{aligned} \tag{9.67}$$

The first member of the hierarchy is obtained by multiplying both sides of this by $Z(N,V,T)^{-1} N e^{-\beta U_N}$ and integrating over $\mathrm{d}\mathbf{r}_2 \ldots \mathrm{d}\mathbf{r}_N$. Using the definition of the many-particle density, Eq. (7.71), the left-hand side is

$$\begin{aligned} & \frac{N}{Z(N,V,T)} \int_V \mathrm{d}\mathbf{r}_2 \ldots \mathrm{d}\mathbf{r}_N \, e^{-\beta U_N(\mathbf{r}^N)} \nabla_1 U_N(\mathbf{r}^N) \\ &= \frac{-N k_\mathrm{B} T}{Z(N,V,T)} \nabla_1 \int_V \mathrm{d}\mathbf{r}_2 \ldots \mathrm{d}\mathbf{r}_N \, e^{-\beta U_N(\mathbf{r}^N)} \\ &= -k_\mathrm{B} T \nabla_1 \rho^{(1)}(\mathbf{r}_1). \end{aligned} \tag{9.68}$$

The right-hand side is

$$\begin{aligned} & \frac{N}{Z(N,V,T)} \int_V \mathrm{d}\mathbf{r}_2 \ldots \mathrm{d}\mathbf{r}_N \, e^{-\beta U_N(\mathbf{r}^N)} \\ & \times \left[\nabla_1 u^{(1)}(\mathbf{r}_1) + \sum_{j=2}^{N} \nabla_1 u^{(2)}(\mathbf{r}_1, \mathbf{r}_j) + \sum_{1<j<k}^{N} \nabla_1 u^{(3)}(\mathbf{r}_1, \mathbf{r}_j, \mathbf{r}_k) + \ldots \right] \\ &= \rho^{(1)}(\mathbf{r}_1) \nabla_1 u^{(1)}(\mathbf{r}_1) + \int_V \mathrm{d}\mathbf{r}_2 \, \rho^{(2)}(\mathbf{r}_1, \mathbf{r}_2) \nabla_1 u^{(2)}(\mathbf{r}_1, \mathbf{r}_2) \\ &\quad + \frac{1}{2} \int_V \mathrm{d}\mathbf{r}_2 \, \mathrm{d}\mathbf{r}_3 \, \rho^{(3)}(\mathbf{r}_1, \mathbf{r}_2, \mathbf{r}_3) \nabla_1 u^{(3)}(\mathbf{r}_1, \mathbf{r}_2, \mathbf{r}_2) + \ldots. \end{aligned} \tag{9.69}$$

The middle term follows because the $N-1$ terms involving $\mathbf{r}_j$ are equivalent, and the third term follows because there are $(N-1)(N-2)/2$ equivalent terms.

Equating these the first member of the hierarchy is

$$\begin{aligned}
&\nabla_1 \rho^{(1)}(\mathbf{r}_1) \\
&= -\rho^{(1)}(\mathbf{r}_1)\nabla_1\beta u^{(1)}(\mathbf{r}_1) - \int_V \mathrm{d}\mathbf{r}_2\, \rho^{(2)}(\mathbf{r}_1,\mathbf{r}_2)\nabla_1\beta u^{(2)}(\mathbf{r}_1,\mathbf{r}_2) \\
&\quad - \frac{1}{2}\int_V \mathrm{d}\mathbf{r}_2\,\mathrm{d}\mathbf{r}_3\, \rho^{(3)}(\mathbf{r}_1,\mathbf{r}_2,\mathbf{r}_3)\nabla_1\beta u^{(3)}(\mathbf{r}_1,\mathbf{r}_2,\mathbf{r}_3) - \ldots. \qquad (9.70)
\end{aligned}$$

This says that the gradient of the singlet density at a particular position is equal to the average force acting on a particle at that position.

The second member of the hierarchy follows by multiplying the gradient of the potential with an appropriate factor and integrating over $\mathrm{d}\mathbf{r}_3 \ldots \mathrm{d}\mathbf{r}_N$. The left-hand side is

$$\begin{aligned}
&\frac{N(N-1)}{Z(N,V,T)}\int_V \mathrm{d}\mathbf{r}_3 \ldots \mathrm{d}\mathbf{r}_N\, e^{-\beta U_N(\mathbf{r}^N)}\nabla_1 U_N(\mathbf{r}^N) \\
&= \frac{-N(N-1)k_{\mathrm{B}}T}{Z(N,V,T)}\nabla_1 \int_V \mathrm{d}\mathbf{r}_3 \ldots \mathrm{d}\mathbf{r}_N\, e^{-\beta U_N(\mathbf{r}^N)} \\
&= -k_{\mathrm{B}}T\nabla_1\rho^{(2)}(\mathbf{r}_1,\mathbf{r}_2). \qquad (9.71)
\end{aligned}$$

The right-hand side is

$$\begin{aligned}
&\frac{N(N-1)}{Z(N,V,T)}\int_V \mathrm{d}\mathbf{r}_3 \ldots \mathrm{d}\mathbf{r}_N\, e^{-\beta U_N(\mathbf{r}^N)} \\
&\times \Big[\nabla_1 u^{(1)}(\mathbf{r}_1) + \nabla_1 u^{(2)}(\mathbf{r}_1,\mathbf{r}_2) + (N-2)\nabla_1 u^{(2)}(\mathbf{r}_1,\mathbf{r}_3) \\
&+(N-2)\nabla_1 u^{(3)}(\mathbf{r}_1,\mathbf{r}_2,\mathbf{r}_3) + \frac{1}{2}(N-2)(N-3)\nabla_1 u^{(3)}(\mathbf{r}_1,\mathbf{r}_3,\mathbf{r}_4) + \ldots\Big] \\
&= \rho^{(2)}(\mathbf{r}_1,\mathbf{r}_2)\left[\nabla_1 u^{(1)}(\mathbf{r}_1) + \nabla_1 u^{(2)}(\mathbf{r}_1,\mathbf{r}_2)\right] \\
&\quad + \int_V \mathrm{d}\mathbf{r}_3\, \rho^{(3)}(\mathbf{r}_1,\mathbf{r}_2,\mathbf{r}_3)\left[\nabla_1 u^{(2)}(\mathbf{r}_1,\mathbf{r}_3) + \nabla_1 u^{(3)}(\mathbf{r}_1,\mathbf{r}_2,\mathbf{r}_3)\right] \\
&\quad + \frac{1}{2}\int_V \mathrm{d}\mathbf{r}_3\,\mathrm{d}\mathbf{r}_4\, \rho^{(4)}(\mathbf{r}_1,\mathbf{r}_2,\mathbf{r}_3,\mathbf{r}_4)\nabla_1 u^{(3)}(\mathbf{r}_1,\mathbf{r}_3,\mathbf{r}_4) + \ldots. \qquad (9.72)
\end{aligned}$$

Equating these, ones sees that the gradient at $\mathbf{r}_1$ of the pair probability for particles at $\mathbf{r}_1$ and $\mathbf{r}_2$, is the average force at $\mathbf{r}_1$ due to a particle at $\mathbf{r}_2$, taking into account the other particles in the fluid. Higher members in the hierarchy may be generated in this fashion.

This second member of the hierarchy forms the basis of an approximation for the radial distribution function. For a homogeneous fluid, $\rho^{(1)}(\mathbf{r}) = \rho$, with isotropic pair potential, $u^{(2)}(\mathbf{r}_1,\mathbf{r}_2) = u^{(2)}(r_{12})$, the pair density is $\rho^{(2)}(\mathbf{r}_1,\mathbf{r}_2) = \rho^2 g^{(2)}(r_{12})$. With the only potential being the pair one, $u^{(n)} = 0$, $n \neq 2$, the second member of the hierarchy is

$$\begin{aligned}
-k_{\mathrm{B}}T\nabla_1 g^{(2)}(r_{12}) &= g^{(2)}(r_{12})\nabla_1 u^{(2)}(r_{12}) \\
&\quad + \rho\int_V \mathrm{d}\mathbf{r}_3\, g^{(3)}(r_{12},r_{23},r_{31})\nabla_1 u^{(2)}(r_{13}). \qquad (9.73)
\end{aligned}$$

Dropping the superscripts and using a prime to signify the derivative, this may be written as

$$\begin{aligned} g'(r_{12}) &= -g(r_{12})\beta u'(r_{12}) \\ &\quad - 2\pi\rho \int_0^\pi \mathrm{d}\theta \int_0^\infty \mathrm{d}r_3\, r_3^2 g(r_{12}, r_{23}, r_{31})\beta u'(r_{13}) \cos\theta, \end{aligned} \tag{9.74}$$

where $r_{23}^2 = r_{12}^2 + r_{13}^2 - r_{12}r_{13}\cos\theta$.

This equation relates the pair and triplet distribution functions, both of which are unknown. A second relationship is required. One seeks to express the triplet function as a superposition of pair functions, and in view of the symmetry the most general expression is

$$g(r_{12}, r_{13}, r_{23}) = \Gamma(r_{12})\Gamma(r_{13})\Gamma(r_{23}), \tag{9.75}$$

with Γ to be determined. Now the probability of finding three particles at certain positions, in the limit that the third particle is removed from the rest, must equal that of finding the remaining two particles times that of finding the third,

$$\wp(r_{12}, r_{13}, r_{23}) \sim \wp(r_{12})\wp(\mathbf{r}_3), \ r_3 \rightarrow \infty. \tag{9.76}$$

That is, the third particle is uninfluenced by the other two, and vice versa, and they are therefore uncorrelated; the joint probability is just the product of the independent probabilities. Since the density is proportional to the probability and the distribution is proportional to the density, this is the same as

$$g(r_{12}, r_{13}, r_{23}) \sim g(r_{12}), \ r_3 \rightarrow \infty. \tag{9.77}$$

If the two remaining particles are further separated, then since $g(r_{12}) \sim 1$, $r_{12} \rightarrow \infty$, one has

$$g(r_{12}, r_{13}, r_{23}) \sim 1, \ r_{12} \rightarrow \infty, \ r_{23} \rightarrow \infty, \ r_{31} \rightarrow \infty. \tag{9.78}$$

With $\gamma = \lim_{r\rightarrow\infty} \Gamma(r)$, the asymptote of the superposition approximation yields

$$\begin{aligned} g(r_{12}, r_{13}, r_{23}) &\sim \Gamma(r_{12})\gamma^2, \ r_3 \rightarrow \infty \\ &\sim \gamma^3, \ r_{12} \rightarrow \infty. \end{aligned} \tag{9.79}$$

Hence one concludes that $\gamma = 1$ and that $\Gamma(r) = g(r)$. The only pair function whose product gives the triplet distribution function *and* which satisfies the correct asymptotic limit is the radial distribution function,

$$g^{(3)}(r_{12}, r_{13}, r_{23}) = g(r_{12})g(r_{13})g(r_{23}). \tag{9.80}$$

This is known as the Kirkwood superposition approximation.

Using the Kirkwood superposition approximation, the Born–Green–Yvon equation becomes

$$g'(r) = -g(r)\beta u'(r) - 2\pi\rho g(r) \int_{-1}^{1} \mathrm{d}x \int_0^\infty \mathrm{d}s\, s^2\, g(s)g(t)\beta u'(s), \tag{9.81}$$

where $t^2 = r^2 + s^2 - rsx$. This is an integro-differential equation for the radial distribution function.

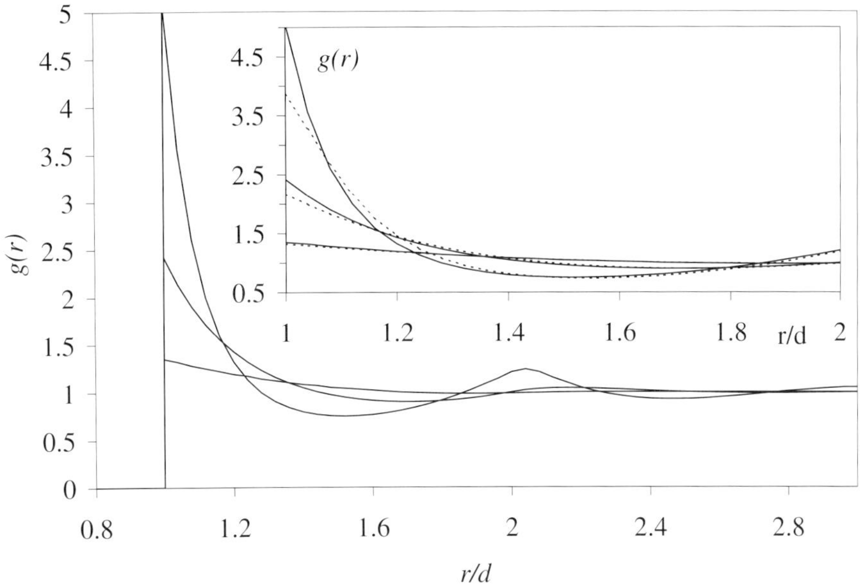

Figure 9.1: The radial distribution function for a hard-sphere fluid, as given by the hypernetted chain approximation. From bottom to top at contact the densities are $\rho d^3 = 0.2$, 0.5, and 0.8, respectively. The inset shows the effect of including the bridge diagrams of second and third order in density (dotted curves).

9.5 Thermodynamic Properties

9.5.1 Radial Distribution Function

Hard Spheres

Figure 9.1 shows the radial distribution function of the hard-sphere fluid at several densities. These were obtained by solving the Ornstein–Zernike equation with the hypernetted chain closure. The fact that $g(r) = 0$, $r < d$, manifests the impossibility of overlap of the molecular cores. At large separations the radial distribution function goes to unity, due to the decreasing correlation between the molecules. The low-density fluid becomes uncorrelated more quickly than the high-density one. The peak in $g(r)$ at contact indicates that there is a high probability of finding touching molecules. The occurrence of such configurations grows markedly with density. The oscillations evident at the highest densities have a period slightly greater than the molecular diameter and indicate regularities in the molecular packing. The second peak corresponds to a high probability of finding a second neighbour from the central molecule, and the trough at about 1.5 diameters is due to the exclusion of molecules from this region (a molecule can only fit here if there is *not* one already in contact with the central molecule, which is unlikely).

The inset of Fig. 9.1 compares the bare hypernetted chain approximation with the closure that includes the bridge function. The latter was approximated as the sum of all bridge diagrams of second and third order in density.[9] It can be seen that the bridge function is most important at higher densities and close to contact, where its effect is to decrease the hypernetted chain contact value.

There is evidently a discontinuity in the radial distribution function at contact, which reflects the discontinuity in the hard-sphere potential. From the derivation of the pair density and the total correlation function, it is evident that there is an e-bond between the root points. The e-bond behaves as a unit step function,

$$e^{-\beta u^{\rm hs}(r)} = \begin{cases} 1, & r > d \\ 0, & r < d, \end{cases} \tag{9.82}$$

and it is this that causes the core discontinuity in $g(r)$ (because the discontinuities in the remaining f-bonds between field points are all integrated out). In view of this one defines the *cavity function*,

$$y(r) \equiv e^{\beta u(r)} g(r), \tag{9.83}$$

which cancels the discontinuous e-bond in $g(r)$ and hence is a continuous function.

The virial expression for the pressure, Eq. (7.88), involves the derivative of the pair potential, the derivative of which is problematic for the hard-sphere potential. By changing to the cavity function one can convert this to the derivative of the exponential of the potential,

$$\begin{aligned} \frac{{\rm d}u^{\rm hs}(r)}{{\rm d}r} g(r) &= \frac{{\rm d}u^{\rm hs}(r)}{{\rm d}r} e^{-\beta u^{\rm hs}(r)} e^{\beta u^{\rm hs}(r)} g(r) \\ &= -k_{\rm B}T \frac{{\rm d}e^{-\beta u^{\rm hs}(r)}}{{\rm d}r} y(r) \\ &= -k_{\rm B}T \delta(r-d) y(r). \end{aligned} \tag{9.84}$$

The last line follows because the derivative of the unit step function is the Dirac δ function. With this result the pressure of the hard-sphere system is

$$\begin{aligned} \beta p &= \rho + \frac{2\pi\rho^2}{3} \int_0^\infty {\rm d}r\, r^3 y(r) \delta(r-d) \\ &= \rho + \frac{2\pi\rho^2 d^3 y(d)}{3} \\ &= \rho + \frac{2\pi\rho^2 d^3 g(d^+)}{3}. \end{aligned} \tag{9.85}$$

The penultimate equality hold because the cavity function is continuous across the core, and the final equality follows because for the hard-sphere system the

[9] P. Attard and G. N. Patey, Hypernetted chain closure with bridge diagrams. Asymmetric hard-sphere mixtures, *J. Chem. Phys.* **92** (1990), 4970.

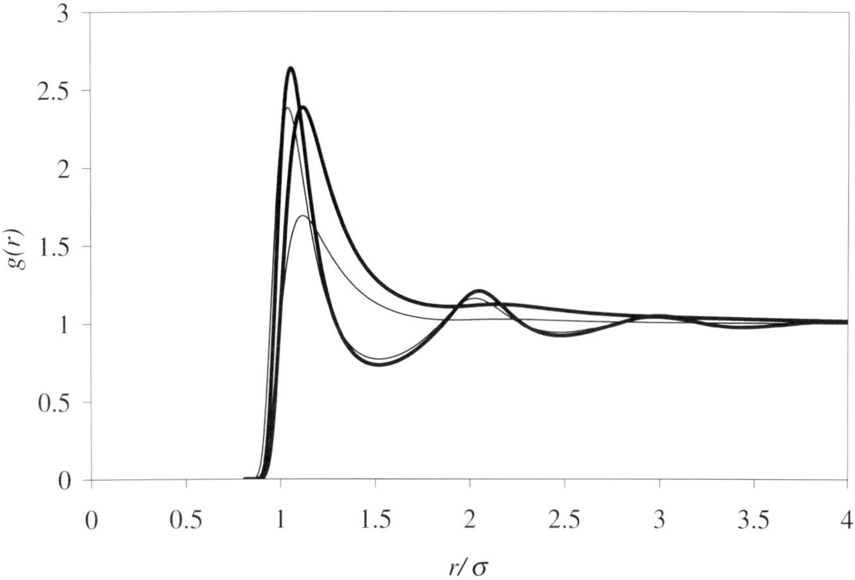

Figure 9.2: The radial distribution function for a Lennard–Jones fluid, as given by the hypernetted chain approximation. The bold curves are for a temperature of $T^* = 1.2$, and the ordinary curves are for $T^* = 1.9$. The oscillatory curves are for a density of $\rho^* = 0.8$, and the smooth curves are for $\rho^* = 0.1$.

cavity function coincides with the radial distribution function beyond the core. This result says that the pressure of the hard-sphere system is determined by the contact density.

To lowest order in density, $g(d^+) = 1$, and this gives

$$\frac{\beta p}{\rho} \sim 1 + \frac{2\pi \rho d^3}{3}, \quad \rho \to 0. \tag{9.86}$$

The leading correction to the ideal gas law exhibited here is just the hard-sphere second virial coefficient, Eq. (8.85), namely $B_2 = 2\pi d^3/3$.

Lennard–Jones

Figure 9.2 shows the radial distribution function of a Lennard–Jones fluid in hypernetted chain approximation. There is a peak at contact, more pronounced at high densities, but less sharp than in the hard-sphere system. This indicates that molecular packing is important here, as confirmed by the oscillations in $g(r)$, and the relative insensitivity to temperature. In contrast at low densities the peak is markedly higher at the subcritical temperature shown than at the supercritical temperature, although there is no qualitative difference between the two temperatures. That the peak at this low-density is higher in

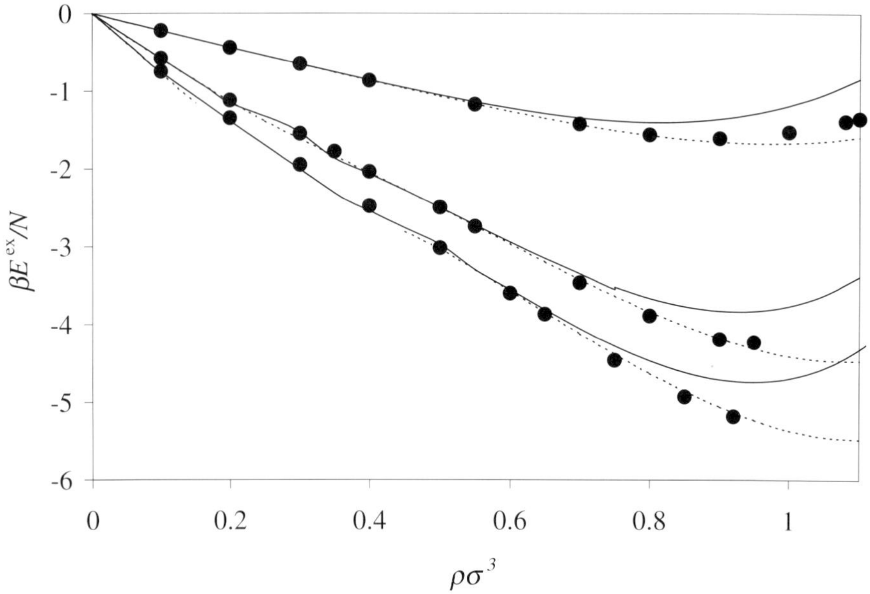

Figure 9.3: The average excess energy per Lennard–Jones molecule as a function of density for, from bottom to top, temperatures of $T^* = 1.15$, 1.35, and 2.74. The symbols are simulation results, the solid curve is the hypernetted chain approximation, and the dotted curve is the Percus–Yevick approximation.

the Lennard–Jones system than in the hard-sphere system indicates the effect of the attractive nature of the Lennard–Jones potential, the adhesion of which enhances the probability of molecular contact.

9.5.2 Average Energy

The average excess energy for a homogenous system, Eq. (7.86), may be written (but with $u^{(1)}(\mathbf{r}) = 0$)

$$\begin{aligned} E^{\mathrm{ex}} &= \frac{\rho^2 V}{2} \int_V \mathrm{d}\mathbf{r}\, u(r) g(r) \\ &= \frac{1}{2} \int \mathrm{d}\mathbf{r}_1\, \mathrm{d}\mathbf{r}_2\, \rho^{(2)}(\mathbf{r}_1, \mathbf{r}_2)\, u^{(2)}(\mathbf{r}_1, \mathbf{r}_2) \end{aligned}$$

$$= \frac{1}{2} \Big[\quad \Big], \qquad (9.87)$$

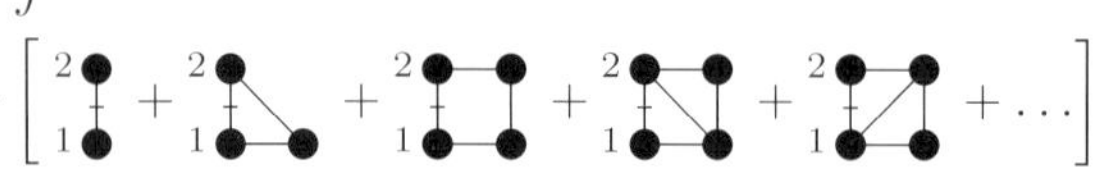

where the crossed-line represents $u(r_{12}) \exp -\beta u(r_{12})$. It may be shown that explicit evaluation of the first few terms of the virial expansion of the energy, Eq. (8.83), agree with this expansion.

Figure 9.3 shows the average energy for a Lennard–Jones fluid. The simulation data may be regarded as exact.[10] At low densities the energy per molecule becomes more negative with increasing density due to the increasing numbers of molecules that experience the Lennard–Jones tail. The curves are not monotonic however; at higher densities molecules are forced into the repulsive soft-core region and this causes the energy to increase. At a given density the magnitude of the energy decreases with increasing temperature, $T^* = k_{\mathrm{B}}T/\epsilon$, as the potential between the molecules becomes relatively less important. Indeed, to a reasonable approximation the energy per molecule is inversely proportional to T^*. Two of the closure approximations are tested against the simulation results in the figure. At low densities they agree with each other and with the simulation data. At high densities the Percus–Yevick performs better than the hypernetted chain approximation; the latter increasingly underestimates the magnitude of the energy in this region. On the subcritical isothermal both the simulations and the hypernetted chain yielded homogeneous solutions in what should be the two-phase region. It was not possible to obtain a uniform solution of the Percus–Yevick equation here, as signified by the break in the curve. The compressibility in the hypernetted chain approximation diverged, just at one would expect at the spinodal line, and was negative beyond this, which indicates an unstable fluid; in these two aspects it is physically realistic. The Percus–Yevick compressibility also diverged just prior to the region in which solutions were unobtainable.

9.5.3 Compressibility

The isothermal compressibility is related to the density fluctuations of a system, and these in turn are given by the integral of the pair density. The homogeneous limit of the definition of $\rho_\delta^{(2)}$, Eq. (9.17), gives

$$\begin{aligned}
\frac{\partial^2 \ln \Xi}{(\partial \beta\mu)^2} &= \int_V \mathrm{d}\mathbf{r}_1\, \mathrm{d}\mathbf{r}_2\, \left. \frac{\delta^2 \ln \Xi}{\delta \ln \Lambda^3 z(\mathbf{r}_1) \delta \ln \Lambda^3 z(\mathbf{r}_2)} \right|_{u^{(1)}=0} \\
&= \int_V \mathrm{d}\mathbf{r}_1\, \mathrm{d}\mathbf{r}_2\, [\rho^2 h(r_{12}) + \rho\delta(\mathbf{r}_{12})] \\
&= \rho V + \rho^2 V \int_V \mathrm{d}\mathbf{r}\, h(r) \\
&= \left\langle N^2 \right\rangle - \langle N \rangle^2 . \qquad (9.88)
\end{aligned}$$

This was given above as Eq. (7.84); the final step follows from the normalisation of the pair density, Eq. (7.77). Equating the left-hand side to the final right-hand side also follows from direct differentiation of the homogeneous partition function; the left-hand side equals $\partial \langle N \rangle / \partial \beta\mu$. Note that the density that appears above is the equilibrium one, $\overline{\rho}(\mu, T)$.

---

[10] J. J. Nicolas, K. E. Gubbins, W. B. Streett, and D. J. Tildesley, Equation of state for the Lennard–Jones fluid, *Mol. Phys.* **37** (1979), 1429.

The isothermal compressibility, Eq. (3.65), is

$$\begin{aligned}
\chi_T &= \frac{-1}{\overline{V}} \left(\frac{\partial \overline{V}}{\partial p} \right)_{T,N} \\
&= \rho^{-1} \left(\frac{\partial \rho}{\partial p} \right)_T \\
&= \rho^{-1} \left(\frac{\partial \rho}{\partial \mu} \right)_T \left(\frac{\partial \mu}{\partial p} \right)_T \\
&= \frac{\beta}{V \rho^2} \left(\frac{\partial \langle N \rangle}{\partial \beta \mu} \right)_T .
\end{aligned} \tag{9.89}$$

The second equality follows because at equilibrium an intensive variable is a function of only two other intensive variables, so that $(\partial \rho / \partial V)_{p,T} = 0$, and it doesn't matter whether N or V is held fixed. The final line utilises the Gibbs–Duhem relation, Eq. (3.77), $(\partial p / \partial \mu)_T = \rho$, which also follows, of course, from direct differentiation of the grand potential. Combining these two results one obtains

$$\begin{aligned}
1 + \rho \int_V \mathrm{d}\mathbf{r}\, h(r) &= \frac{\langle N^2 \rangle - \langle N \rangle^2}{\langle N \rangle} \\
&= \frac{1}{\langle N \rangle} \left(\frac{\partial \langle N \rangle}{\partial \beta \mu} \right)_T \\
&= \rho k_{\mathrm{B}} T \chi_T .
\end{aligned} \tag{9.90}$$

Hence the integral of the total correlation function gives the isothermal compressibility of the system.

The integral on the left is just the Fourier transform of the total correlation function evaluated at $k = 0$, $\hat{h}(0)$. Using the Ornstein–Zernike equation in Fourier space the isothermal compressibility may therefore be written

$$\rho k_{\mathrm{B}} T \chi_T = 1 + \rho \hat{h}(0) = \frac{1}{1 - \rho \hat{c}(0)} . \tag{9.91}$$

For vanishing density, the right-hand sides go to unity, which gives the ideal gas result, $\chi_T^{\mathrm{id}} = 1/\rho k_{\mathrm{B}} T$. Conversely at the critical point and the spinodal line the compressibility becomes infinite, which corresponds to a divergence of the integral of the total correlation function. The mechanism by which this occurs is that the range of $h(r)$ diverges, which is to say that it decays increasingly slowly as the critical point or spinodal line is approached. In contrast in these regions $\hat{c}(0) \rightarrow \rho^{-1}$, which is to say that its integral is finite. So the direct correlation function is well-behaved near-criticality, which is to say that it is of shorter range than the total correlation function, $h(r)/c(r) \rightarrow \infty$, $r \rightarrow \infty$. This is not true in general; at arbitrary points on the phase diagram both pair correlation functions have precisely the same range, $h(r)/c(r) \rightarrow$ const., $r \rightarrow \infty$.

The divergence of χ_T and $\hat{h}(0)$ near criticality corresponds to increasing fluctuations in the local density in the system, and these give rise to the phenomenon of *critical opalescence*. As the size of the density inhomogeneities become comparable to the wavelength of light, a near-critical system scatters light strongly and it appears turbid. It was this phenomenon that motivated Ornstein and Zernike to introduce the direct correlation function in the form of the equation that bears their name.

Mixtures

For a mixture one may write the compressibility as

$$\begin{aligned}
\chi_T^{-1} &= \frac{-1}{V}\left(\frac{\partial \overline{p}}{\partial V}\right)_{T,\underline{N}} \\
&= \frac{1}{V}\left(\frac{\partial \overline{p}}{\partial V^{-1}}\right)_{T,\underline{N}} \\
&= \frac{1}{V}\sum_\alpha \left(\frac{\partial \rho_\alpha}{\partial V^{-1}}\right)_{T,\underline{N}} \left(\frac{\partial \overline{p}}{\partial \rho_\alpha}\right)_{T,\rho_{\gamma\neq\alpha}} \\
&= \sum_\alpha \rho_\alpha \sum_\gamma \left(\frac{\partial \beta\mu_\gamma}{\partial \rho_\alpha}\right)_{T,\rho_{\lambda\neq\alpha}} \left(\frac{\partial k_{\mathrm{B}} T \ln \Xi}{V \partial \beta\mu_\gamma}\right)_{T,\mu_{\lambda\neq\gamma}} \\
&= \sum_{\alpha,\gamma} \rho_\alpha \hat{c}^\delta_{\alpha\gamma}(0) \rho_\gamma k_{\mathrm{B}} T \\
&= k_{\mathrm{B}} T \sum_{\alpha,\gamma} \left[\rho_\alpha \delta_{\alpha\gamma} - \rho_\alpha \rho_\gamma \hat{c}_{\alpha\gamma}(0)\right].
\end{aligned} \tag{9.92}$$

The homogeneous limit of

$$\frac{\delta \ln z_\gamma(\mathbf{s})\Lambda_\gamma^3}{\delta \rho_\alpha(\mathbf{r})} = c^\delta_{\alpha\gamma}(\mathbf{r},\mathbf{s}) = \rho_\alpha(\mathbf{r})^{-1}\delta_{\alpha\gamma}(\mathbf{r}-\mathbf{s}) - c_{\alpha\gamma}(\mathbf{r},\mathbf{s}) \tag{9.93}$$

has been used to obtain the penultimate line.

Defining the vector $\underline{r}$ with components $\{\underline{r}\}_\alpha = \rho_\alpha^{1/2}$, one can write this in a relatively compact form,

$$\begin{aligned}
\chi_T^{-1} &= k_{\mathrm{B}} T \underline{r}^{\mathrm{T}} \left[\underline{\underline{I}} - \underline{\underline{\hat{C}}}(0)\right] \underline{r} \\
&= k_{\mathrm{B}} T \underline{r}^{\mathrm{T}} \left[\underline{\underline{I}} + \underline{\underline{\hat{H}}}(0)\right]^{-1} \underline{r}.
\end{aligned} \tag{9.94}$$

The final form could be written as a sum over the components of the cofactor matrix of the total correlation function. The fact that this quadratic form must yield a positive compressibility places certain restrictions upon the matrices of the correlation functions that appear (see below). The relationship between the fluctuations in composition, the derivatives of the chemical potential, and the integral of the total correlation function is often called Kirkwood–Buff theory.[11]

[11] J. G. Kirkwood and F. P. Buff, The statistical mechanical theory of solutions. I, *J. Chem. Phys.* **19** (1951), 774.

9.5.4 Chemical Potential

According to Eq. (7.61), the average excess chemical potential is the number difference of the logarithm of the configuration integral,

$$\langle -\beta_s \mu_s^{\mathrm{ex}} \rangle = \ln Q(N,V,T) - \ln V Q(N-1,V,T). \tag{9.95}$$

One can interpolate between the two systems by introducing a coupling parameter λ for the Nth particle. One is now dealing with a two-component mixture, with $N-1$ particles of the first type (the solvent) and one particle of the second type (the solute). For a homogeneous, isotropic system with pair potential $u(r)$ between particles of the first type, the interaction potential for the partially coupled solute particle may be written $u(r;\lambda)$, with $u(r;0)=0$ and $u(r;1)=u(r)$. The simplest coupling is the linear one, $u(r;\lambda)=\lambda u(r)$, but it is not necessary to restrict the analysis that follows to this case.

In view of these definitions the excess chemical potential may formally be rewritten as

$$\begin{aligned}
\langle -\beta_s \mu_s^{\mathrm{ex}} \rangle &= \int_0^1 \mathrm{d}\lambda \frac{\partial}{\partial \lambda} \ln Q_\lambda(N,V,T) \\
&= \int_0^1 \mathrm{d}\lambda \frac{\partial}{\partial \lambda} \ln \int_V \frac{\mathrm{d}\mathbf{r}^N}{(N-1)!} \prod_{i<j}^{N-1} e^{-\beta u(r_{ij})} \prod_{i=1}^{N-1} e^{-\beta u(r_{iN};\lambda)} \\
&= \int_0^1 \mathrm{d}\lambda \frac{1}{Q_\lambda(N,V,T)} \int_V \frac{\mathrm{d}\mathbf{r}^N}{(N-1)!} \prod_{i<j}^{N-1} e^{-\beta u(r_{ij})} \\
&\quad \times \prod_{i=1}^{N-1} e^{-\beta u(r_{iN};\lambda)} \frac{\partial[-\beta u(r_{iN};\lambda)]}{\partial \lambda} \\
&= -\beta \int_0^1 \mathrm{d}\lambda \int_V \mathrm{d}\mathbf{r}\, \rho g(r;\lambda) \frac{\partial u(r;\lambda)}{\partial \lambda}.
\end{aligned} \tag{9.96}$$

In the final result the quantity $\rho g(r;\lambda)$ represents the density of solvent at $\mathbf{r}$ when the partially coupled solute particle is at the origin. It is important for what follows to understand that in the diagrammatic representation of $g(r;\lambda)$, only the bonds to the solute root particle depend upon the coupling constant; all the remaining particles and bonds between them are fully coupled and independent of λ.

The radial distribution function may be written in terms of the potential of mean force, $g(r;\lambda) = e^{-\beta u(r;\lambda)} e^{v(r;\lambda)}$, and rearrangement gives

$$\begin{aligned}
\langle -\beta_s \mu_s^{\mathrm{ex}} \rangle &= \rho \int_V \mathrm{d}\mathbf{r} \int_0^1 \mathrm{d}\lambda\, e^{v(r;\lambda)} \frac{\partial e^{-\beta u(r;\lambda)}}{\partial \lambda} \\
&= \rho \int_V \mathrm{d}\mathbf{r} \left[g(r;\lambda)\Big|_0^1 - \int_0^1 \mathrm{d}\lambda\, g(r;\lambda) \frac{\partial v(r;\lambda)}{\partial \lambda} \right] \\
&= \rho \int_V \mathrm{d}\mathbf{r} \left[h(r) - v(r) - \int_0^1 \mathrm{d}\lambda\, h(r;\lambda) \frac{\partial v(r;\lambda)}{\partial \lambda} \right].
\end{aligned} \tag{9.97}$$

The second equality follows from an integration by parts, and the third equality follows since $g(r;1) = g(r)$ and $g(r;0) = 1$, and similarly $v(r;1) = v(r)$ and $v(r;0) = 0$.

Now the quantity $v(r)$, which in essence is the excess potential of mean force, is the sum of the series diagrams and the bridge diagrams. The series diagrams can be expressed as the convolution product of an h-bond and a c-bond; the h-bond can be taken to be connected to the solute and hence dependent upon λ, whilst the direct correlation function depends solely upon the solvent particles and is independent of the coupling constant.

One can use topological reduction to eliminate articulation pairs from the diagrammatic definition of the bridge function. This corresponds to replacing the f-bonds by h-bonds, since h is the sum of all the connected pair diagrams. The more compact set of diagrams that results is

$$d(\mathbf{r}_1, \mathbf{r}_2) = \{\text{the sum of all irreducible diagrams composed of two nonadjacent 1-root points labelled } \mathbf{r}_1 \text{ and } \mathbf{r}_2\text{, and two or more } \rho^{(1)}\text{-field points and } h\text{-bonds, such that there are no nodal points nor any articulation pairs of points}\}. \tag{9.98}$$

With this resummation it is possible to classify the bridge diagrams according to how many h-bonds impinge upon one of the root points. One can write

$$v(r;\lambda) = \sum_{n=1}^{\infty} v^{(n)}(r;\lambda), \tag{9.99}$$

where the series diagram is just $v^{(1)}(r;\lambda) = s(r;\lambda)$, and the bridge diagrams start at $n = 2$. The function $v^{(n)}(r;\lambda)$ contains the convolution of n partially coupled h-bonds with an $(n+1)$-body function independent of λ. This latter function is the generalisation of the singlet and pair direct correlation function.[12] It is important to note that this function is symmetric in its arguments (because it belongs to a functional derivative hierarchy). Accordingly one has

$$\begin{aligned}
&\rho \int_V \mathrm{d}\mathbf{r}_2\, h(r_{12};\lambda) \frac{\partial v^{(n)}(r_{12};\lambda)}{\partial \lambda} \\
&= \rho \int_V \mathrm{d}\mathbf{r}_2\, h(r_{12};\lambda) \frac{\partial}{\partial \lambda} \rho^n \int_V \frac{\mathrm{d}\mathbf{r}_3 \ldots \mathrm{d}\mathbf{r}_{n+2}}{n!} c(\mathbf{r}_2, \mathbf{r}_3 \ldots \mathrm{d}\mathbf{r}_{n+2}) \\
&\quad \times h(r_{13};\lambda) \ldots h(r_{1,n+2};\lambda) \\
&= \rho^{n+1} \frac{n}{n+1} \frac{\partial}{\partial \lambda} \int_V \frac{\mathrm{d}\mathbf{r}_2 \ldots \mathrm{d}\mathbf{r}_{n+2}}{n!} c(\mathbf{r}_2, \mathbf{r}_3 \ldots \mathrm{d}\mathbf{r}_{n+2}) \\
&\quad \times h(r_{12};\lambda) \ldots h(r_{1,n+2};\lambda) \\
&= \rho \frac{n}{n+1} \frac{\partial}{\partial \lambda} \int_V \mathrm{d}\mathbf{r}_2 h(r_{12};\lambda) v^{(n)}(r_{12};\lambda).
\end{aligned} \tag{9.100}$$

[12] This function is composed of the irreducible diagrams with $n+1$ root points, ρ-field points and no articulation points, no nodal points, and, for $n > 1$, h-bonds, and no articulation pairs of points. It cannot contain a node because such would form half of an articulation pair with one of the root points of the bridge diagram, and these have just been eliminated. A little thought shows that the singlet and pair direct correlation functions satisfy these criteria.

The factor of $n/(n+1)$ arises because the original form gave rise to n identical terms involving the derivative of the partially coupled total correlation function, whereas the final form gives rise to $n+1$ such terms.

With this the coupling constant integral can be performed term by term, with the final result[13]

$$\begin{aligned} & \langle -\beta_s \mu_s^{\rm ex} \rangle \\ &= \rho \int_V d\mathbf{r} \left[h(r) - v(r) - h(r) \sum_{n=1}^{\infty} \frac{n}{n+1} v^{(n)}(r) \right] \\ &= \rho \int_V d\mathbf{r} \left[h(r) - v(r) - h(r) v(r) + h(r) \sum_{n=1}^{\infty} \frac{v^{(n)}(r)}{n+1} \right], \end{aligned} \tag{9.101}$$

where $v(r) \equiv v(r;1)$ is in essence the excess potential of mean force.

Approximations

This is a formally exact expression for the chemical potential. The sum on the right-hand side can be evaluated for the various closure approximations to give approximate expressions for the chemical potential. In the case of the hypernetted chain approximation, the bridge function is neglected, $v^{(n)}(r) = 0$, $n \geq 2$. In this case $v(r) = s(r) = h(r) - c(r)$ and one has[14]

$$\langle -\beta_s \mu_s^{\rm ex} \rangle_{\rm HNC} = \frac{\rho}{2} \int_V d\mathbf{r}\ [2c(r) - h(r)\{h(r) - c(r)\}]. \tag{9.102}$$

In the case of the Percus–Yevick approximation, the potential of mean force function is $v^{\rm PY}(r) = \ln[1 + s(r)]$, which corresponds to

$$v_{\rm PY}^{(n)}(r) = \frac{(-1)^{n+1}}{n} s(r)^n. \tag{9.103}$$

Hence

$$\begin{aligned} & \langle -\beta_s \mu_s^{\rm ex} \rangle_{\rm PY} \\ &= \rho \int_V d\mathbf{r} \left[h(r) - \ln[1+s(r)] - h(r) \sum_{n=1}^{\infty} \frac{(-1)^{n+1}}{n+1} s(r)^n \right] \\ &= \rho \int_V d\mathbf{r} \left[h(r) - \ln[1+s(r)] + \frac{h(r)}{s(r)} \{\ln[1+s(r)] - s(r)\} \right] \\ &= \rho \int_V d\mathbf{r}\, \frac{c(r)}{h(r)-c(r)} \ln[1 + h(r) - c(r)]. \end{aligned} \tag{9.104}$$

---

[13] L. L. Lee, Correlation functions of classical fluids. III. The method of partition function variation applied to the chemical potential: Cases of PY and HNC2, *J. Chem. Phys.* **60** (1974), 1197. P. Attard, The chemical potential in terms of n-particle direct correlation functions, *J. Chem. Phys.* **94** (1991), 2379.

[14] T. Morita, Theory of classical fluids: Hypernetted chain approximation. III, *Progr. Theor. Phys.* **23** (1960), 829.

The mean spherical approximation in its most elementary form corresponds to $c(r) = -\beta u(r)$, and $h(r)$ is the sum of Ornstein–Zernike chains of c-bonds in series. (By elementary here is meant the model with 0 core diameter.) Only the first c-bond depends upon the coupling constant, so that one may write symbolically,

$$h_\lambda = c_\lambda * [1 + c + c*c + c*c*c + \ldots]. \tag{9.105}$$

Accordingly in this approximation one has

$$\begin{aligned}
\langle -\beta_s \mu_s^{\text{ex}} \rangle_{\text{MSA}} &= -\beta \int_V \mathrm{d}\mathbf{r} \int_0^1 \mathrm{d}\lambda \, \frac{\partial u(r;\lambda)}{\partial \lambda} g(r;\lambda) \\
&= -\beta \int_V \mathrm{d}\mathbf{r} \left[u(r) + \int_0^1 \mathrm{d}\lambda \, \frac{\partial u(r;\lambda)}{\partial \lambda} h(r;\lambda) \right] \\
&= -\beta \int_V \mathrm{d}\mathbf{r} \left[u(r) + \int_0^1 \mathrm{d}\lambda \, \frac{1}{2} \frac{\partial}{\partial \lambda} [u(r;\lambda) h(r;\lambda)] \right] \\
&= -\beta \int_V \mathrm{d}\mathbf{r} \left[u(r) + \frac{1}{2} u(r) h(r) \right].
\end{aligned} \tag{9.106}$$

The factor of one-half arises because the two c_λ bonds connected to the solute (one is the $u(r;\lambda)$ that appears explicitly, and the other is the $c_\lambda *$ that occurs in all the $h(r;\lambda)$ diagrams) are identical due to the integral over $\mathbf{r}$.

9.6 Asymptotic Analysis

9.6.1 Exact Results

The asymptotic behaviour of the pair correlation function is important because it can be obtained explicitly and analytically, at least formally, and it therefore provides a useful check on the approximations and numerical procedures used to obtain the full functions. In this section it is assumed that the system is homogeneous and isotropic, and that the pair potential has asymptotic behaviour

$$u(r) \sim \mathcal{O}(r^{-\eta}), \; r \to \infty, \; \eta > d, \tag{9.107}$$

where d is the dimensionality of the space. This ensures that the Mayer f-function, which goes like $f(r) \sim -\beta u(r)$, $r \to \infty$, is integrable. In three dimensions it precludes the Coulomb and dipole potentials from further consideration.

The Mayer cluster diagrams, and the Ornstein–Zernike equation, contain convolution integrals, and so it is important to understand the asymptotic behaviour of these. The integral $\int \mathrm{d}\mathbf{r}_3 \, f(r_{13}) f(r_{23})$ is dominated by the two regions around the root points in the limit $r_{12} \to \infty$. In these regions there is one short bond and one stretched bond, so that the integrand has value proportional to $r_{12}^{-\eta}$. In contrast, when the field point is halfway between the two root points, both bonds are stretched, and the value of the integrand is proportional to

$(r_{12}/2)^{-2\eta}$. In the limit $r_{12} \to \infty$ the latter is much smaller than the former. Accordingly one has

$$\begin{aligned} \int_V \mathrm{d}\mathbf{r}_3\, f(r_{13}) f(r_{23}) &\approx \int_{\approx \mathbf{r}_1} \mathrm{d}\mathbf{r}_3\, f(r_{13}) f(r_{23}) + \int_{\approx \mathbf{r}_2} \mathrm{d}\mathbf{r}_3\, f(r_{13}) f(r_{23}) \\ &\approx \int_{\approx \mathbf{r}_1} \mathrm{d}\mathbf{r}_3\, f(r_{13}) f(r_{12}) + \int_{\approx \mathbf{r}_2} \mathrm{d}\mathbf{r}_3\, f(r_{12}) f(r_{23}) \\ &\approx f(r_{12}) \int_V \mathrm{d}\mathbf{r}_3\, f(r_{13}) + f(r_{12}) \int_V \mathrm{d}\mathbf{r}_3\, f(r_{23}) \\ &= 2 f(r_{12}) \hat{f}(0),\ r_{12} \to \infty. \end{aligned} \tag{9.108}$$

The second line follows from a Taylor expansion of the appropriate function; the first neglected term is proportional to $f'(r_{12})$. One may extend the integrations to the entire volume in the penultimate line if the function is integrable.

This result may be derived directly in Fourier space. The Fourier transform of a power law potential of the above form has a $|k|^{\eta-d}$ singularity at the origin. Hence the small-k expansion goes like

$$\hat{f}(k) \sim f_\eta |k|^{\eta-d} + \hat{f}(0) + f_2 k^2 + \ldots,\ k \to 0. \tag{9.109}$$

Since the convolution becomes a product in Fourier space one has

$$\hat{f}(k)^2 \sim 2\hat{f}(0) f_\eta |k|^{\eta-d} + f_\eta^2 |k|^{2\eta-2d} + f_2 f_\eta |k|^{\eta-d+2} + \ldots,\ k \to 0. \tag{9.110}$$

Because $\eta - d > 0$, the first exhibited term is the strongest singularity (i.e., the one with the lowest power of k), and it dominates the large-separation asymptotic behaviour. The inverse Fourier transform yields the result given above.

The convolution integral just discussed is a nodal one. For diagrams without nodes, at large separations the integral is still dominated by regions where the field points are clustered around the root points, and the asymptote is proportional to the number of stretched bonds that result. For example, the first bridge diagram goes like

$$\begin{aligned} &\frac{1}{2} \int_V \mathrm{d}\mathbf{r}_3\, \mathrm{d}\mathbf{r}_4\, f(r_{13}) f(r_{23}) f(r_{14}) f(r_{24}) f(r_{34}) \\ &\quad \sim\ f(r_{12})^2 \int_V \mathrm{d}\mathbf{r}_3\, \mathrm{d}\mathbf{r}_4\, f(r_{13}) f(r_{14}) f(r_{34}),\ r_{12} \to \infty. \end{aligned} \tag{9.111}$$

Obviously $\lim_{r\to\infty} f(r)^2/f(r) = 0$, and hence this and the higher-order bridge diagrams and other multiply connected diagrams can be neglected in the asymptotic limit compared to the singly connected diagrams.

In view of these general considerations, the asymptotic behaviour of the various pair correlation functions may now be extracted. The direct correlation function lacks any nodes, and consists of a single diagram that is $f(r_{12})$, and multiply connected diagrams. Since the latter decay at least as fast as $f(r_{12})^2$, one has

$$\begin{aligned} c(r) &\sim f(r) + \mathcal{O}(f(r)^2) \\ &\sim -\beta u(r) + \mathcal{O}(u(r)^2),\ r \to \infty. \end{aligned} \tag{9.112}$$

What is remarkable about this exact result is that the asymptote is directly given by the pair potential, and apart from the temperature it is independent of the thermodynamic properties of the system.

The easiest way to obtain the asymptotic behaviour of the total correlation function is via the Ornstein–Zernike equation. Provided that attention is restricted to regions where the total correlation function decays as an integrable power law, one has

$$\begin{aligned} h(r) &\sim c(r) + \rho\hat{h}(0)c(r) + \rho\hat{c}(0)h(r) \\ &\sim \frac{1+\rho\hat{h}(0)}{1-\rho\hat{c}(0)}c(r) \\ &\sim -(k_{\mathrm{B}}T\rho\chi_T)^2\beta u(r) + \mathcal{O}(u(r)^2),\ r \to \infty. \end{aligned} \tag{9.113}$$

Here the expression for the compressibility, Eq. (9.91), has been used. This shows that the total and the direct correlation function decay at the same rate, $\lim_{r\to\infty} h(r)/c(r) = \text{const}$. However, unlike the direct correlation function, the asymptote of the total correlation function depends upon the thermodynamic properties of the system, notably its compressibility.

The physical interpretation of this asymptote is straightforward. At large separations each of the particles has an adsorption excess of particles, $1+\rho\hat{h}(0) = k_{\mathrm{B}}T\rho\chi_T$. Hence the total interaction energy between the two particles is effectively the sum of the potentials between the two adsorbed clouds. The asymptote is equivalent to the radial distribution function of an almost-ideal gas of such clouds. That the isothermal compressibility should determine the magnitude of the adsorption excess is reasonable because it is a measure of the ease with which the density of the system may be changed; incompressible systems have a small excess. If the potential at long-range is attractive, then $h(r) > 0$, which is to say that there is a positive correlation between the particles. In other words, the compressibility may magnify the asymptotic interaction of the particles, but it does not change its nature.

Second-Order Effects

The correction to the primary asymptote for the direct correlation function is[15]

$$c(r) + \beta u(r) \sim \frac{1}{2}\left(\beta\rho^2\frac{\partial^2\mu}{\partial\rho^2}\right)^2 h(r)^2,\ r \to \infty. \tag{9.114}$$

The derivative that appears here may be related to the volume integral of the triplet direct correlation function.

[15] G. Stell, Fluids with long-range forces: Toward a simple analytic theory, in *Statistical Mechanics. Part A: Equilibrium Techniques* (B. J. Berne, Ed.), p. 47, Plenum, New York, 1977. P. Attard, Integral equations and closure relations for the bridge function and for the triplet correlation function, *J. Chem. Phys.* **93** (1990), 7301.

9.6.2 Near Criticality

In some cases the preceding analysis does not apply as such, although the general features still hold. The exceptions are when the solvent pair potentials decay faster than any power law (e.g., hard sphere) or they decay so slowly as to be nonintegrable (e.g., Coulomb), or else in the vicinity of the critical point or line of spinodal decomposition. The latter is characterised by the divergence of the compressibility of the system, which, in view of Eq. (9.90), occurs when the total correlation function becomes so long-ranged as to be nonintegrable. One can see that the preceding asymptotic analysis breaks down in this case because the amplitudes of the asymptotes depend upon the $\hat{h}(0)$, which is infinite. In these cases one can no longer assert that the asymptotic expansion is dominated by the nodal diagrams; their very divergence indicates that all orders of connectivity contribute, and a resummation must be carried out.

In contrast, when the compressibility diverges, the direct correlation function remains integrable, $\hat{c}(0) \to \rho^{-1}$. Hence it is believed that even at the critical point the state-independent asymptote found above, $c(r) \sim -\beta u(r)$, still holds. The direct correlation function is in this sense better behaved than the total correlation function, and it is part of the reason why Ornstein and Zernike introduced it in their studies of criticality.

The simplest way to analyse the asymptotic behaviour of the pair correlation functions near the critical point and in the vicinity of the spinodal line is to assume the classical Ornstein–Zernike form for the correlation function,

$$h(r) \sim A\frac{e^{-\xi r}}{r}, \quad r \to \infty. \tag{9.115}$$

This form is not exact (the factor of r in the denominator should be replaced by $r^{d-2+\eta}$, with $d = 3$ and $\eta \approx 0.06$),[16] but it is satisfactory for the present purposes. The decay length diverges as one approaches the critical point, $\xi^{-1} \to \infty$, $\rho \to \rho_{\rm c}$, $T \to T_{\rm c}^{+}$. For the compressibility to be positive, one must have $A > 0$. The Fourier transform of the Yukawa form is

$$\hat{h}(k) \sim \frac{4\pi A}{k^2 + \xi^2}, \quad k \to 0. \tag{9.116}$$

Since the Ornstein–Zernike equation gives $\hat{c}(k) = \hat{h}(k)/[1 + \rho\hat{h}(k)]$, one has

$$\hat{c}(k) \sim \frac{4\pi A}{k^2 + \chi^2}, \quad k \to 0, \ \chi^2 \equiv \xi^2 + 4\pi\rho A. \tag{9.117}$$

Since $\chi > \xi$, this shows that near the critical point the direct correlation function is more short-ranged than the total correlation function.

For future reference it is worth noting that the asymptotic decay of h is determined by the 0 of $1 - \rho\hat{c}(k)$. In this case this is $0 = (k^2 + \chi^2 - \rho 4\pi A)/(k^2 + \chi^2) = (k^2 + \xi^2)/(k^2 + \chi^2)$.

[16] G. Stell, Extension of the Ornstein–Zernike theory of the critical region, *Phys. Rev. Lett.* **20** (1968), 533; Part II, *Phys. Rev. B* **1** (1970), 2265.

9.6.3 Mixtures

The direct correlation function of a mixture has asymptote that is the negative of the pair potential, $c_{\alpha\gamma}(r) \sim -\beta u_{\alpha\gamma}(r)$. In view of the Ornstein–Zernike equation for mixtures, the asymptotic behaviour of the total correlation function is

$$\begin{aligned} \underline{\underline{H}}(r) &\sim \underline{\underline{C}}(r) + \underline{\underline{\hat{H}}}(0)\underline{\underline{C}}(r) + \underline{\underline{H}}(r)\underline{\underline{\hat{C}}}(0) \\ &= \left[\underline{\underline{I}} + \underline{\underline{\hat{H}}}(0)\right] \underline{\underline{C}}(r) \left[\underline{\underline{I}} + \underline{\underline{\hat{H}}}(0)\right]. \end{aligned} \tag{9.118}$$

Note that the elements of the total correlation function matrix are given by $\{\underline{\underline{H}}\}_{\alpha\gamma} = \rho_\alpha^{1/2}\rho_\gamma^{1/2}h_{\alpha\gamma}$, and similarly for the direct correlation function matrix. Here a representation of the Ornstein–Zernike equation has been used, namely, $\left[\underline{\underline{I}} - \underline{\underline{\hat{C}}}(0)\right]^{-1} = \left[\underline{\underline{I}} + \underline{\underline{\hat{H}}}(0)\right]$. In component form this is

$$\begin{aligned} &\rho_\alpha\rho_\gamma h_{\alpha\gamma}(r) \\ &\quad \sim -\beta \sum_{\lambda\chi} \left[\rho_\alpha\delta_{\alpha\lambda} + \rho_\alpha\hat{h}_{\alpha\lambda}(0)\rho_\lambda\right] u_{\lambda\chi}(r) \left[\rho_\gamma\delta_{\chi\gamma} + \rho_\chi\hat{h}_{\chi\gamma}(0)\rho_\gamma\right]. \end{aligned} \tag{9.119}$$

Because all of the $u_{\lambda\chi}(r)$ contribute to each $h_{\alpha\gamma}(r)$, in the asymptotic regime the most long-ranged of the pair potentials dominates, and all of the total correlation functions decay at the same rate.

In view of the expressions for the singlet and pair densities given earlier for a mixture, one has

$$\begin{aligned} &\frac{\delta^2 \ln \Xi}{\delta \ln \Lambda_\alpha^3 z_\alpha(\mathbf{r}) \delta \ln \Lambda_\gamma^3 z_\gamma(\mathbf{s})} \\ &\quad = \frac{\delta \rho_\gamma(\mathbf{s})}{\delta \ln \Lambda_\alpha^3 z_\alpha(\mathbf{r})} \\ &\quad = \rho_\alpha(\mathbf{r})\delta_{\alpha\gamma}(\mathbf{r},\mathbf{s}) + \rho_\alpha(\mathbf{r})\rho_\gamma(\mathbf{s})h_{\alpha\gamma}(\mathbf{r},\mathbf{s}). \end{aligned} \tag{9.120}$$

In the homogeneous limit, with $\ln \Lambda_\gamma^3 z_\gamma = \beta\mu_\gamma$, one has

$$\begin{aligned} \frac{\partial^2 \ln \Xi}{\partial\beta\mu_\alpha\, \partial\beta\mu_\gamma} &= \int_V \mathrm{d}\mathbf{r}\, \mathrm{d}\mathbf{s}\, \left. \frac{\delta^2 \ln \Xi}{\delta \ln \Lambda_\alpha^3 z_\alpha(\mathbf{r}) \delta \ln \Lambda_\gamma^3 z_\gamma(\mathbf{s})} \right|_{z_\alpha, z_\gamma} \\ &= \int_V \mathrm{d}\mathbf{r}\, \mathrm{d}\mathbf{s}\, \left[\rho_\alpha\delta_{\alpha\gamma}(\mathbf{r},\mathbf{s}) + \rho_\alpha\rho_\gamma h_{\alpha\gamma}(|\mathbf{r}-\mathbf{s}|)\right] \\ &= V\rho_\alpha\delta_{\alpha\gamma} + V\rho_\alpha\rho_\gamma\hat{h}_{\alpha\gamma}(0) \\ &= \langle N_\alpha N_\gamma\rangle - \langle N_\alpha\rangle \langle N_\gamma\rangle . \end{aligned} \tag{9.121}$$

The final line may be derived by direct differentiation of the partition function. Similarly, $\partial \ln \Xi / \partial\beta\mu_\alpha = V\rho_\alpha$. It therefore follows that the coefficients in the asymptote may be written

$$\left[\rho_\alpha\delta_{\alpha\lambda} + \rho_\alpha\hat{h}_{\alpha\lambda}(0)\rho_\lambda\right] = \frac{\partial\rho_\alpha}{\partial\beta\mu_\lambda} = \frac{\partial\rho_\lambda}{\partial\beta\mu_\alpha}. \tag{9.122}$$

Hence the asymptote may be written compactly as

$$h_{\alpha\gamma}(r) \sim \frac{-\beta}{\rho_\alpha \rho_\gamma} \sum_{\lambda\chi} \frac{\partial \rho_\alpha}{\partial \beta\mu_\lambda} u_{\lambda\chi}(r) \frac{\partial \rho_\chi}{\partial \beta\mu_\gamma}. \tag{9.123}$$

In the event that the pair potential is dyadic, $u_{\alpha\gamma}(r) = a_\alpha a_\gamma u(r)$, then the asymptote assumes the particularly simple form,

$$h_{\alpha\gamma}(r) \sim \tilde{a}_\alpha \tilde{a}_\gamma u(r),\ r \to \infty, \tag{9.124}$$

where the effective 'charges' are

$$\tilde{a}_\alpha = \sum_\lambda \frac{\partial \rho_\alpha}{\partial \beta\mu_\lambda} a_\lambda. \tag{9.125}$$

There is reason to believe that in the real world *all* pair potentials are dyadic in the asymptotic regime. The triplet potential is believed to be asymptotically triadic, and so on for the higher-order many-body potentials.

Near Criticality

Near the critical point and along the line of spinodal decomposition, the compressibility diverges. As in the single-component case, this occurs because the $h_{\alpha\gamma}(r)$ become so long-ranged as to be nonintegrable. The zeroth Fourier components, $\hat{h}_{\alpha\gamma}(0)$, were given above as Eq. (9.121). In terms of the matrix of second derivatives of the grand potential, this may be written

$$\frac{-1}{V} \underline{\underline{\Omega}} = \underline{\underline{\rho}} \left[\underline{\underline{I}} + \underline{\underline{\hat{H}}}(0) \right] \underline{\underline{\rho}}, \tag{9.126}$$

where the density matrix is diagonal with components $\{\underline{\underline{\rho}}\}_{\alpha\gamma} = \rho_\alpha^{1/2} \delta_{\alpha\gamma}$. Now the grand potential matrix is concave, as shown by Eq. (4.26), and hence the matrix product on the right is convex with all the eigenvalues being positive. This means that the diagonal elements are positive,

$$\rho_\alpha + \rho_\alpha^2 \hat{h}_{\alpha\alpha}(0) = \frac{1}{V} \left[\left\langle N_\alpha^2 \right\rangle - \left\langle N_\alpha \right\rangle^2 \right] > 0. \tag{9.127}$$

Denoting the fluctuation in number by $\Delta_{\alpha\gamma} \equiv \langle N_\alpha N_\gamma \rangle - \langle N_\alpha \rangle \langle N_\gamma \rangle$, one also has

$$\Delta_{\alpha\alpha} \Delta_{\gamma\gamma} - \Delta_{\alpha\gamma}^2 > 0. \tag{9.128}$$

In addition the determinant is positive,

$$\left| \underline{\underline{I}} + \underline{\underline{\hat{H}}}(0) \right| = \left| \underline{\underline{I}} - \underline{\underline{\hat{C}}}(0) \right|^{-1} > 0, \tag{9.129}$$

since $|\underline{\underline{\rho}}| > 0$. From this convexity, the isothermal compressibility of the multicomponent mixture is positive, as follows from the quadratic form, Eq. (9.94).

The critical point or line of spinodal decomposition corresponds to the divergence of the the compressibility, and to the nonintegrability of the total correlation functions. Accordingly, approaching criticality the determinant of the central matrix must also diverge, and hence that of its inverse must vanish,

$$\left| \underline{\underline{I}} - \underline{\underline{\hat{C}}}(0) \right| \to 0^+. \tag{9.130}$$

In the vicinity of criticality, the $\hat{h}_{\alpha\gamma}(0)$ are finite but large. At some $k = i\xi$ off the real axis they diverge, with $\xi \to 0$ approaching criticality. (Note that since the pair correlation functions are even functions of r, they are real functions of k throughout the complex plane.) Let $\underline{x}_j(k)$ be one of the orthonormal eigenvectors of the symmetric matrix $\underline{\underline{\hat{C}}}(k)$ with corresponding eigenvalue $\lambda_j(k)$. In view of the convexity established above, one must have $\lambda_j(k) > 1$. If the first eigenvalue is the smallest, then it must approach unity from above at $k = i\xi$. If it has the functional form

$$\lambda_1(k) \sim \frac{A}{A + k^2 + \xi^2}, \ k \to i\xi, \tag{9.131}$$

with $A > 0$ to ensure that $\lambda_j(0) > 1$, then the determinant goes like

$$\left| \underline{\underline{I}} - \underline{\underline{\hat{C}}}(k) \right| = \prod_{j=1}^{m} [1 - \lambda_j(k)] \to \frac{k^2 + \xi^2}{A + k^2 + \xi^2} \times \text{ const.} \to 0^+. \tag{9.132}$$

The matrix $\underline{\underline{\hat{C}}}(k)$ is diagonal in the eigenvector basis, as is $\underline{\underline{\hat{H}}}(k)$, since it is the sum of products of the former. Denoting the elements of the latter in this basis by $\eta_{\alpha\alpha}(k)$, one sees that the first element has a simple pole at $k = i\xi$,

$$\eta_{11}(k) = \frac{\lambda_1(k)}{1 - \lambda_1(k)} \sim \frac{A}{k^2 + \xi^2}, \ k \to i\xi. \tag{9.133}$$

All the other elements are regular. Accordingly one has

$$\underline{\underline{\hat{H}}}(k) \sim \frac{A}{k^2 + \xi^2} \underline{x}\,\underline{x}^{\mathrm{T}} + \ldots, \ k \to i\xi, \tag{9.134}$$

where $\underline{x} = \underline{x}_1(i\xi)$, and the regular terms have not been shown here. Hence asymptotically the total correlation function goes like

$$h_{\alpha\gamma}(r) \sim x_\alpha x_\gamma \frac{A e^{-\xi r}}{4\pi r}, \ r \to \infty. \tag{9.135}$$

The dyadic nature of the asymptote near criticality is similar to that found in the rest of the phase diagram.

9.6.4 Approximations

Since the total correlation function is small at large separations, so must be the exponent that appears in the hypernetted chain closure approximation, Eq. (9.65). Expanding to second order one obtains

$$h(r) \sim h(r) - c(r) - \beta u(r) + \frac{1}{2} \left[h(r) - c(r) - \beta u(r) \right]^2, \ r \to \infty. \tag{9.136}$$

The linear order result yields $c(r) \sim -\beta u(r)$, which is exact. The second-order result gives

$$c(r) + \beta u(r) \sim \frac{1}{2} h(r)^2, \ r \to \infty. \tag{9.137}$$

That this is incorrect results from neglecting the bridge function, which is asymptotically of order $u(r)^2$.

Similarly expanding the Percus–Yevick closure, Eq. (9.61), yields

$$\begin{aligned} c(r) &\sim -\beta u(r) - \frac{1}{2}\beta^2 u(r)^2 - \beta u(r) h(r) \\ &\sim -\beta u(r) - \frac{1}{2}\beta^2 u(r)^2 + (k_B T \rho \chi_T)^2 \beta^2 u(r)^2, \ r \to \infty. \end{aligned} \tag{9.138}$$

Again the approximation is correct to leading order but fails at the second-order term. The mean spherical approximation is of course $c(r) \sim -\beta u(r)$, there being no higher-order terms.

Summary

- The pair distribution function gives the joint probability of finding two particles at particular positions in the system, and as such gives structural and thermodynamic information about the system. The structure factor, which is measured in scattering experiments, is essentially the Fourier transform of the pair distribution function.

- For a system interacting with only a pair potential, the pair distribution function gives the average energy, pressure, and, in certain approximations, the chemical potential. More generally, its integral gives the number fluctuations of the system and its compressibility.

- The Ornstein–Zernike equation is an exact relation between the direct and total correlation functions that corresponds to connecting diagrams in series.

- The pair distribution function is the exponential of the potential of mean force, which corresponds to connecting diagrams in parallel.

- The closure that completes the set of equations for the pair functions generally correspond to neglecting (hypernetted chain) or to approximating (Percus–Yevick) the bridge function. Approaches based on the iterative numerical solution of these equations are more powerful than straight density expansions, and they are relatively accurate even for dense systems.

- The pair total and direct correlation functions decay to 0 at the same rate as the pair potential (provided it is an integrable power law potential). The direct correlation function shows a universal behaviour asymptotically, whereas the magnitude of the decay of the total correlation function scales with the square of the compressibility of the system.

Chapter 10

Functional and Perturbation Theory

In the previous chapter several techniques for dealing with systems of interacting particles were presented: a density expansion, the Born–Green–Yvon equation, and the Ornstein–Zernike equation. The advantage of the latter is that it is applicable to dense systems and solutions can be obtained by straightforward iterative techniques. The main disadvantage is that in practice factorisation of the Ornstein–Zernike equation is essential, and this restricts its application to uniform systems or to the simplest inhomogeneous system. In addition, for certain systems (e.g., highly coupled systems, or systems in the vicinity of a phase transition), the iterative procedure may not converge and the results can become unacceptably sensitive to the approximations made. Alternative methods for treating both homogeneous and inhomogeneous systems have been developed. These are based on constrained entropy maximisation (equivalently, constrained thermodynamic potential minimisation). Such density functional theories form the content of this chapter.

10.1 Uniform System

10.1.1 Constrained Thermodynamic Potential

In the third chapter it was shown that for an open subsystem that could exchange energy and particles with a reservoir, the constrained total entropy when the subsystem had $\tilde{N}$ particles and energy $\tilde{E}$ was

$$S(\tilde{E}, \tilde{N}|\mu, V, T) = S(\tilde{E}, \tilde{N}, V) - \frac{\tilde{E}}{T} + \frac{\mu \tilde{N}}{T}. \tag{10.1}$$

This neglects a constant independent of the presence of the subsystem. The left-hand side is the constrained total entropy, the first term on the right is the entropy of the subsystem in the particular macrostate, and the remainder is the subsystem-dependent part of the reservoir entropy in the macrostate. The quantities with a tilde refer to the subsystem; the tilde is used to emphasise that these are constrained or nonequilibrium quantities. The constrained thermodynamic potential is the negative of the temperature times the constrained total entropy,

$$\begin{aligned}\Omega(\tilde{E},\tilde{N}|\mu,V,T) &= -TS(\tilde{E},\tilde{N}|\mu,V,T) \\ &= \tilde{E}-\mu\tilde{N}+S(\tilde{E},\tilde{N},V). \end{aligned} \tag{10.2}$$

The constrained thermodynamic potential is a nonequilibrium quantity and the probability of a macrostate occurring due to random fluctuations about equilibrium is the exponential of the total entropy,

$$\wp(\tilde{E},\tilde{N}|\mu,V,T) = \frac{e^{-\Omega(\tilde{E},\tilde{N}|\mu,V,T)/k_{\mathrm{B}}T}}{\Xi(\mu,V,T)}. \tag{10.3}$$

The logarithm of the partition function is the total unconstrained entropy, $S_\mu(\mu,V,T) = k_{\mathrm{B}}\ln\Xi(\mu,V,T)$. The grand potential, which is the equilibrium thermodynamic potential for the open system, is the minimum value of the constrained thermodynamic potential, $\Omega(\mu,V,T) = \Omega(\overline{E},\overline{N}|\mu,V,T)$. Here the equilibrium values satisfy

$$\left.\frac{\partial S(\tilde{E},\tilde{N},V)}{\partial\tilde{E}}\right|_{\overline{E},\overline{N}} = \frac{1}{T}, \quad \left.\frac{\partial S(\tilde{E},\tilde{N},V)}{\partial\tilde{N}}\right|_{\overline{E},\overline{N}} = \frac{-\mu}{T}. \tag{10.4}$$

This says that at equilibrium the temperatures are equal and the chemical potentials are equal between the subsystem and the reservoir, $T(\overline{E},\overline{N},V) = T$, and $\mu(\overline{E},\overline{N},V) = \mu$. An overline is used to denote the equilibrium state.

The logarithm of the grand partition function gives the unconstrained total entropy. The partition function is the weighted sum over the macrostates, and if fluctuations are negligible, as they are in the thermodynamic limit, this sum is dominated by its largest term, which is the equilibrium macrostate. Hence one has

$$\begin{aligned}k_{\mathrm{B}}\ln\Xi(\mu,V,T) &= k_{\mathrm{B}}\ln\sum_{\tilde{E},\tilde{N}} e^{S(\tilde{E},\tilde{N}|\mu,V,T)/k_{\mathrm{B}}} \\ &\approx k_{\mathrm{B}}\ln e^{S(\overline{E},\overline{N}|\mu,V,T)/k_{\mathrm{B}}} \\ &= -\Omega(\mu,V,T)/T. \end{aligned} \tag{10.5}$$

This is the conventional result, that the thermodynamic potential is in essence the logarithm of the partition function. It is emphasised that this conventional equality only holds in the thermodynamic limit, whereas equating the logarithm of the partition function to the total unconstrained entropy holds in general.

10.1.2 Meta-reservoir

One can introduce an imaginary reservoir that has temperature $\tilde{T}$ and chemical potential $\tilde{\mu}$ that would be in equilibrium with the constrained subsystem, $\tilde{E} =$

$\overline{E}(\tilde{\mu}, V, \tilde{T})$, and $\tilde{N} = \overline{N}(\tilde{\mu}, V, \tilde{T})$. In terms of this meta-reservoir, one can rewrite the expression for the constrained total entropy, Eq. (10.1), as

$$\begin{aligned} S(\tilde{E}, \tilde{N}|\mu, V, T) &= \left[S(\tilde{E}, \tilde{N}, V) - \frac{\tilde{E}}{\tilde{T}} + \frac{\tilde{\mu}\tilde{N}}{\tilde{T}} \right] \\ &\quad - \tilde{E}\left\{ \frac{1}{T} - \frac{1}{\tilde{T}} \right\} + \tilde{N}\left\{ \frac{\mu}{T} - \frac{\tilde{\mu}}{\tilde{T}} \right\}. \end{aligned} \tag{10.6}$$

The term in brackets represents the constrained total entropy of the equilibrium macrostate of the subsystem and meta-reservoir; it is $S(\tilde{E}, \tilde{N}|\tilde{\mu}, V, \tilde{T})$. The two terms in braces represent the difference between the entropy of the actual reservoir and that of the meta-reservoir, when the subsystem is in the macrostate $\tilde{E}$, $\tilde{N}$. Hence the constrained total entropy may be written

$$S(\tilde{E}, \tilde{N}|\mu, V, T) = S(\tilde{E}, \tilde{N}|\tilde{\mu}, V, \tilde{T}) + \Delta S_{\text{res}}(\tilde{E}, \tilde{N}). \tag{10.7}$$

The invocation of the meta-reservoir represents a departure from the attitude that has underpinned the book to date. So far a physical approach has been taken, as exemplified by the reservoir formalism wherein both the constrained and the unconstrained thermodynamic potentials were shown to represent the actual exchange of measurables with a reservoir. The meta-reservoir is an artificial construction that has no physical basis. It is a mathematical convenience that allows certain formal thermodynamic relationships for constrained quantities to be derived, but it has no physical realisation beyond this. The reason for introducing the meta-reservoir is that because the constrained state is in equilibrium with it, the various equilibrium results that have been established are applicable, whereas they cannot be directly used for the nonequilibrium system itself.

Notwithstanding the metaphysical nature of the meta-reservoir, this particular result may be stated more generally: the constrained total entropy of a subsystem and actual reservoir may be written as the constrained total entropy of the subsystem in equilibrium with the meta-reservoir, plus the difference between the constrained entropy of the actual reservoir and that of the meta-reservoir.

10.2 Singlet Functionals

10.2.1 Density Functional

In the case that the reservoir imposes an external potential $u(\mathbf{r})$ that causes a density inhomogeneity $\rho(\mathbf{r})$, one divides the energy of the subsystem into an internal and an external part. If the subsystem is in the macrostate $\tilde{\rho}(\mathbf{r})$ (i.e., it is constrained to have the density profile $\tilde{\rho}(\mathbf{r})$), then the particle number and the external energy are given by

$$\tilde{N} = \int \mathrm{d}\mathbf{r}\, \tilde{\rho}(\mathbf{r}), \quad \tilde{E}^{\text{ext}} = \int \mathrm{d}\mathbf{r}\, \tilde{\rho}(\mathbf{r}) u(\mathbf{r}). \tag{10.8}$$

The internal energy of the subsystem is still denoted by $\tilde{E}$, and the energy of the reservoir is $E_0 - \tilde{E} - \tilde{E}^{\rm ext}$. Writing the local chemical potential of the reservoir as $\mu(\mathbf{r}) = \mu - u(\mathbf{r})$, the constrained total entropy of the subsystem and reservoir is

$$\begin{aligned} & S(\tilde{E}, \tilde{\rho}|\mu, V, T) \\ &= S(\tilde{E}, \tilde{\rho}, V) - \frac{\tilde{E}}{T} + \frac{1}{T}\int \mathrm{d}\mathbf{r}\, \tilde{\rho}(\mathbf{r})\mu(\mathbf{r}) \\ &= S(\tilde{E}, \tilde{\rho}, V) - \frac{\tilde{E}}{\tilde{T}} + \frac{1}{\tilde{T}}\int \mathrm{d}\mathbf{r}\, \tilde{\rho}(\mathbf{r})\tilde{\mu}(\mathbf{r}) \\ &\quad - \tilde{E}\left\{\frac{1}{T} - \frac{1}{\tilde{T}}\right\} + \int \mathrm{d}\mathbf{r}\, \tilde{\rho}(\mathbf{r})\left\{\frac{\mu(\mathbf{r})}{T} - \frac{\tilde{\mu}(\mathbf{r})}{\tilde{T}}\right\} \\ &= S(\tilde{E}, \tilde{\rho}|\tilde{\mu}, V, \tilde{T}) - \tilde{E}\left\{\frac{1}{T} - \frac{1}{\tilde{T}}\right\} + k_{\rm B}\int \mathrm{d}\mathbf{r}\, \tilde{\rho}(\mathbf{r}) \ln \frac{z(\mathbf{r})}{\tilde{z}(\mathbf{r})}. \end{aligned} \tag{10.9}$$

The definition of the activity, $z(\mathbf{r}) = \Lambda^{-3} \exp \beta\mu(\mathbf{r})$, gives the final equality. It should be clear that the above are functionals, not functions, of the density and of the local chemical potential; for typographical clarity the latter arguments are not here enclosed in brackets. In the case that the subsystem is thermally equilibrated, $\tilde{T} = T$, one has

$$S(\tilde{\rho}|z, V, T) = S(\tilde{\rho}|\tilde{z}, V, T) + k_{\rm B}\int \mathrm{d}\mathbf{r}\, \tilde{\rho}(\mathbf{r}) \ln \frac{z(\mathbf{r})}{\tilde{z}(\mathbf{r})}, \tag{10.10}$$

where the activity rather than the chemical potential is now used. Since in general the constrained thermodynamic potential is the negative of the temperature times the constrained entropy, multiplying by $-T$ one may write this as

$$\Omega[\tilde{\rho}|z] = \Omega[\tilde{z}] + k_{\rm B}T\int \mathrm{d}\mathbf{r}\, \tilde{\rho}(\mathbf{r}) \ln \frac{\tilde{z}(\mathbf{r})}{z(\mathbf{r})}. \tag{10.11}$$

The grand potential of the meta-system appears as the first term on the right-hand side because the given macrostate is the equilibrium state for the subsystem in contact with the meta-reservoir; its constrained thermodynamic potential is at its minimum value, $-TS(\tilde{\rho}|\tilde{z}, V, T) = \Omega([\tilde{z}], V, T)$.

This functional of the density is minimised by the equilibrium density profile, $\tilde{\rho}(\mathbf{r}) = \overline{\rho}(\mathbf{r}, [z])$. At equilibrium $\tilde{z}(\mathbf{r}) = z(\mathbf{r})$, and the integrand vanishes, which shows that the minimum value of the functional equals $\Omega[z]$, the grand potential of the subsystem plus reservoir.

This functional is an implicit functional of the density. The activity of the meta-reservoir that appears is the one that gives $\tilde{\rho}(\mathbf{r})$ as the equilibrium density profile, and hence it is a functional of the constrained density, $\tilde{z}(\mathbf{r}) = \overline{z}(\mathbf{r}, [\tilde{\rho}])$. An explicit formula for the activity as a functional of the density was given as Eq. (8.69), $\tilde{z}(\mathbf{r}) = \tilde{\rho}(\mathbf{r}) \exp -c(\mathbf{r}; [\tilde{\rho}])$. The grand potential Ω is an explicit functional of the activity, and it is therefore an implicit function of the density. Hence one can define $\Omega_\rho[\rho] \equiv \Omega[\overline{z}[\rho]]$, which has derivative

$$\frac{-\delta\beta\Omega_\rho[\rho]}{\delta\rho(\mathbf{r})} = \int \mathrm{d}\mathbf{s}\, \frac{-\delta\beta\Omega[z]}{\delta z(\mathbf{s})}\frac{\delta z(\mathbf{s})}{\delta\rho(\mathbf{r})}$$

$$
\begin{aligned}
&= \int \mathrm{d}\mathbf{s}\, \frac{\rho(\mathbf{s})}{z(\mathbf{s})} \frac{\delta z(\mathbf{s})}{\delta \rho(\mathbf{r})} \\
&= \int \mathrm{d}\mathbf{s}\, \left[\delta(\mathbf{r}, \mathbf{s}) - \rho(\mathbf{s}) c(\mathbf{r}, \mathbf{s}) \right], \qquad (10.12)
\end{aligned}
$$

where Eq. (9.26) has been used. The function $\Omega_\rho[\rho]$ is well defined, at least implicitly. (One could write it simply as $\Omega[\rho]$, provided that one kept in mind that this is a different function of its argument to $\Omega[z]$ due to the convention that a function is distinguished by the symbol used for its argument.)

In view of the fact that this derivative gives an integral of the pair direct correlation function, and the fact that the function $\mathcal{F}^{\mathrm{ex}}[\rho]$ introduced in the previous chapter is a generating function for the direct correlation functions, one is motivated to consider the explicit functional of density

$$
\Omega'_\rho[\rho] = k_{\mathrm{B}} T \int \mathrm{d}\mathbf{s}\, \rho(\mathbf{s}) \left[-1 + \ln \frac{\rho(\mathbf{s})}{z(\mathbf{s}; [\rho])} \right] - \mathcal{F}^{\mathrm{ex}}[\rho]. \qquad (10.13)
$$

It will be recalled from Eq. (9.20) that

$\beta \mathcal{F}^{\mathrm{ex}}[\rho]$ = {the sum of all simple irreducible diagrams with ρ-field points and one or more f-bonds}

= 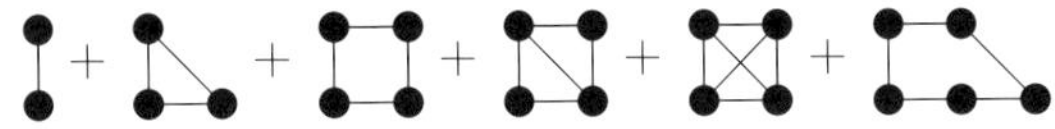 + ... (10.14)

and from Eq. (9.21) that

$$
\frac{\delta \beta \mathcal{F}^{\mathrm{ex}}[\rho]}{\delta \rho(\mathbf{r})} = c(\mathbf{r}) = \ln \frac{\rho(\mathbf{r})}{z(\mathbf{r})}. \qquad (10.15)
$$

To show that $\Omega'_\rho = \Omega_\rho$, one takes its derivative,

$$
\begin{aligned}
\frac{\delta \beta \Omega'_\rho[\rho]}{\delta \rho(\mathbf{r})} &= \left[-1 + \ln \frac{\rho(\mathbf{r})}{z(\mathbf{r})} + 1 - \int \mathrm{d}\mathbf{s}\, \frac{\rho(\mathbf{s})}{z(\mathbf{s})} \frac{\delta z(\mathbf{s})}{\delta \rho(\mathbf{r})} \right] - \ln \frac{\rho(\mathbf{r})}{z(\mathbf{r})} \\
&= \frac{\delta \beta \Omega_\rho[\rho]}{\delta \rho(\mathbf{r})}. \qquad (10.16)
\end{aligned}
$$

Hence the two expressions are equal up to an arbitrary constant independent of density. In the limit of vanishing density, where $z(\mathbf{r}) = \rho(\mathbf{r})$, it is clear by explicit evaluation that $-\beta\Omega_\rho[\rho] = -\beta\Omega'_\rho[\rho] = \bullet$. This then is the explicit expression for the grand potential in terms of the density.[1]

In view of this result, the density constrained thermodynamic potential is explicitly

$$
\begin{aligned}
\Omega[\tilde{\rho}|z] &= \Omega_\rho[\tilde{\rho}] + k_{\mathrm{B}} T \int \mathrm{d}\mathbf{r}\, \tilde{\rho}(\mathbf{r}) \ln \frac{\tilde{z}(\mathbf{r})}{z(\mathbf{r})} \\
&= k_{\mathrm{B}} T \int \mathrm{d}\mathbf{r}\, \tilde{\rho}(\mathbf{r}) \left[-1 + \ln \frac{\tilde{\rho}(\mathbf{r})}{z(\mathbf{r})} \right] - \mathcal{F}^{\mathrm{ex}}[\tilde{\rho}], \qquad (10.17)
\end{aligned}
$$

[1] G. Stell, Cluster expansions for classical systems in equilibrium, in *The Equilibrium Theory of Classical Fluids* (H. Frisch and J. L. Lebowitz, Eds.), p. II-171, Benjamin, New York, 1964. P. Attard, The explicit density functional and its connection with entropy maximisation, *J. Stat. Phys.* **100** (2000), 445.

where $\tilde{z}(\mathbf{r}) = \overline{z}(\mathbf{r}, [\tilde{\rho}])$. The minimum value of this functional occurs at $\tilde{\rho}(\mathbf{r}) = \overline{\rho}(\mathbf{r}, [z])$, and is equal to the grand potential, $\Omega[\overline{\rho}|z] = \Omega_\rho[\overline{\rho}] = \Omega[z]$.

Probability of Density Fluctuations

In view of the present interpretation of statistical mechanics, a physical meaning can be given to density functional theory. The density functional $\Omega[\tilde{\rho}|z]$ is nothing but the constrained thermodynamic potential, which is (the negative of the temperature times) the entropy of the subsystem when it has a particular density profile, plus the entropy of the reservoir in the corresponding macrostate. Minimisation of the density functional corresponds to maximisation of the constrained total entropy, which is the procedure for establishing the equilibrium macrostate. Furthermore, the density functional given explicitly here is unique: whereas there are many functionals that are mathematically minimised at equilibrium, it is the present one that determines the density fluctuations of the ensemble. Because of the relationship with the constrained total entropy, the probability of such fluctuations is given by

$$\wp[\tilde{\rho}|z] = \frac{1}{\Xi[z]} e^{-\beta\Omega[\tilde{\rho}|z]}. \tag{10.18}$$

The left-hand side may be visualised by subdividing space into cells $\mathbf{r}_i$ of size Δ_i, so that for a given density $\tilde{\rho}(\mathbf{r})$ the cell occupancy is $\tilde{n}_i = \Delta_i \tilde{\rho}(\mathbf{r}_i)$, and the entire density function is represented by the set of occupancies $\underline{\tilde{n}}$. The left-hand side then represents the joint probability for the cells to be so occupied.

Cross–Entropy

There is an alternative way of viewing the density functional. One invokes the so-called cross-entropy of the configuration probabilities of the reservoir and the meta-reservoir. The configuration probability of the actual system is given by

$$\wp(\mathbf{r}^N, N|\mu, V, T) = \frac{e^{-\beta U_N^*(\mathbf{r}^N)}}{\Lambda^{3N}\Xi(\mu, V, T)} \prod_{i=1}^{N} z(\mathbf{r}_i), \tag{10.19}$$

where U_N^* is the potential energy without the singlet contribution. The normalisation factor is the grand partition function,

$$\Xi(\mu, V, T) = \sum_{N=0}^{\infty} \int \frac{\mathrm{d}\mathbf{r}^N}{\Lambda^{3N} N!} e^{-\beta U_N^*(\mathbf{r}^N)} \prod_{i=1}^{N} z(\mathbf{r}_i), \tag{10.20}$$

and the grand potential is, in the thermodynamic limit, $\Omega(\mu, V, T) = -k_{\mathrm{B}} T \ln \Xi$. Analogous quantities, denoted by a tilde, are defined for the meta-system.

The cross-entropy is defined to be

$$S[\tilde{\wp}, \wp]/k_{\mathrm{B}} = \sum_{N=0}^{\infty} \frac{1}{N!} \int \mathrm{d}\mathbf{r}^N \, \tilde{\wp}(\mathbf{r}^N) \ln \frac{\tilde{\wp}(\mathbf{r}^N)}{\wp(\mathbf{r}^N)}. \tag{10.21}$$

In view of the fact that the configuration probabilities are normalised to unity, this may be written

$$S[\tilde{\wp}, \wp]/k_{\rm B} = \sum_{N=0}^{\infty} \int \frac{{\rm d}\mathbf{r}^N}{N!} \wp(\mathbf{r}^N) \left\{ \frac{\tilde{\wp}(\mathbf{r}^N)}{\wp(\mathbf{r}^N)} \ln \frac{\tilde{\wp}(\mathbf{r}^N)}{\wp(\mathbf{r}^N)} - \frac{\tilde{\wp}(\mathbf{r}^N)}{\wp(\mathbf{r}^N)} + 1 \right\}. \quad (10.22)$$

The integrand is nonnegative (since $x \ln x \geq x - 1$), and one concludes that $S[\tilde{\wp}, \wp] \geq 0$. Obviously the cross-entropy vanishes if and only if $\tilde{\wp} = \wp$.

In view of the nonnegativity of the cross-entropy, one can add it to the grand potential of the subsystem plus actual reservoir to give a functional that is minimised at equilibrium, and that has minimum value equal to the actual grand potential. This functional is

$$\begin{aligned} \beta\Omega''[\tilde{\rho}|z] &= \beta\Omega[z] + S[\tilde{\wp}, \wp]/k_{\rm B} \\ &= \beta\Omega[z] + \sum_{N=0}^{\infty} \frac{1}{N!} \int {\rm d}\mathbf{r}^N \, \tilde{\wp}(\mathbf{r}^N) \ln \frac{\tilde{\wp}(\mathbf{r}^N)}{\wp(\mathbf{r}^N)} \\ &= \beta\Omega[z] + \sum_{N=0}^{\infty} \int \frac{{\rm d}\mathbf{r}^N}{N!} \tilde{\wp}(\mathbf{r}^N) \ln \frac{\Xi}{\tilde{\Xi}} \prod_{i=1}^{N} \frac{\tilde{z}(\mathbf{r}_i)}{z(\mathbf{r}_i)} \\ &= \beta\Omega[\tilde{z}] + \int {\rm d}\mathbf{r} \, \tilde{\rho}(\mathbf{r}) \ln \frac{\tilde{z}(\mathbf{r})}{z(\mathbf{r})}. \end{aligned} \quad (10.23)$$

This is just the density functional given above. Hence the cross-entropy of the reservoir and meta-reservoir plays a rôle rather similar to that of the constrained entropy of the general formalism.

10.2.2 HNC Approximation

The utility of density functional theory lies in the availability of robust and tractable approximations for the nontrivial part of the functional, $\mathcal{F}^{\rm ex}$. Here it will be shown how it can be used to obtain the hypernetted chain approximation.

One does a coupling integral from an initial density, $\rho_0(\mathbf{r})$, to the final density, $\rho_1(\mathbf{r})$, along a linear path, $\rho_\lambda(\mathbf{r}) = \rho_0(\mathbf{r}) + \lambda\Delta\rho(\mathbf{r})$, where $\Delta\rho(\mathbf{r}) = \rho_1(\mathbf{r}) - \rho_0(\mathbf{r})$. One assumes that the initial density is the equilibrium one, $\rho_0(\mathbf{r}) = \overline{\rho}(\mathbf{r}; z_0)$, but that the final density does not correspond to the final activity, $\rho_1(\mathbf{r}) \neq \overline{\rho}(\mathbf{r}; z_1)$. One further assumes that the chemical potential is fixed and that only the external potential depends upon the coupling constant, $\mu_\lambda(\mathbf{r}) = \mu - u_\lambda(\mathbf{r})$.

The change in $\mathcal{F}^{\rm ex}$ is

$$\begin{aligned} \mathcal{F}^{\rm ex}[\rho_1] - \mathcal{F}^{\rm ex}[\rho_0] &= \int_0^1 {\rm d}\lambda \, \frac{\partial \mathcal{F}^{\rm ex}[\rho_\lambda]}{\partial\lambda} \\ &= \int_0^1 {\rm d}\lambda \int_V {\rm d}\mathbf{r} \, \frac{\delta \mathcal{F}^{\rm ex}[\rho_\lambda]}{\delta\rho_\lambda(\mathbf{r})} \frac{\partial\rho_\lambda(\mathbf{r})}{\partial\lambda} \\ &= k_{\rm B}T \int_0^1 {\rm d}\lambda \int_V {\rm d}\mathbf{r} \, c(\mathbf{r}; \rho_\lambda)\Delta\rho(\mathbf{r}). \end{aligned} \quad (10.24)$$

Similarly,

$$\begin{aligned} c(\mathbf{r};\rho_\lambda) - c(\mathbf{r};\rho_0) &= \int_0^\lambda \mathrm{d}\lambda' \, \frac{\partial c(\mathbf{r};\rho_{\lambda'})}{\partial \lambda'} \\ &= \int_0^\lambda \mathrm{d}\lambda' \int_V \mathrm{d}\mathbf{s}\, c(\mathbf{r},\mathbf{s};\rho_{\lambda'})\Delta\rho(\mathbf{s}). \end{aligned} \tag{10.25}$$

Using the fact that

$$\int_0^1 \mathrm{d}\lambda \int_0^\lambda \mathrm{d}\lambda'\, f(\lambda') = \int_0^1 \mathrm{d}\lambda' \,(1-\lambda') f(\lambda'), \tag{10.26}$$

the change in the thermodynamic functional is

$$\begin{aligned} &\beta\mathcal{F}^{\mathrm{ex}}[\rho_1] - \beta\mathcal{F}^{\mathrm{ex}}[\rho_0] \\ &= \int_0^1 \mathrm{d}\lambda \int_V \mathrm{d}\mathbf{r}\, \Delta\rho(\mathbf{r}) \left[c(\mathbf{r};\rho_0) + \int_0^\lambda \mathrm{d}\lambda' \int_V \mathrm{d}\mathbf{s}\, \Delta\rho(\mathbf{s}) c(\mathbf{r},\mathbf{s};\rho_{\lambda'}) \right] \\ &= \int_V \mathrm{d}\mathbf{r}\, \Delta\rho(\mathbf{r}) c(\mathbf{r};\rho_0) \\ &\quad + \int_0^1 \mathrm{d}\lambda\, (1-\lambda) \int_V \mathrm{d}\mathbf{r}\, \mathrm{d}\mathbf{s}\, \Delta\rho(\mathbf{r})\Delta\rho(\mathbf{s}) c(\mathbf{r},\mathbf{s};\rho_\lambda). \end{aligned} \tag{10.27}$$

From Eq. (10.17), and using the fact that $c(\mathbf{r};\rho_0) = \ln[\Lambda^3\rho_0(\mathbf{r})] - \beta\mu + \beta u_0(\mathbf{r})$, the change in the density functional is

$$\begin{aligned} &\beta\Omega[\rho_1|z_1] - \beta\Omega[\rho_0|z_0] \\ &= \int_V \mathrm{d}\mathbf{r}\, \rho_1(\mathbf{r}) \left[-1 + \ln\frac{\rho_1(\mathbf{r})}{z_1(\mathbf{r})} \right] - \int_V \mathrm{d}\mathbf{r}\, \rho_0(\mathbf{r}) \left[-1 + \ln\frac{\rho_0(\mathbf{r})}{z_0(\mathbf{r})} \right] \\ &\quad - \int_V \mathrm{d}\mathbf{r}\, \Delta\rho(\mathbf{r}) c(\mathbf{r};\rho_0) \\ &\quad - \int_0^1 \mathrm{d}\lambda\, (1-\lambda) \int_V \mathrm{d}\mathbf{r}\, \mathrm{d}\mathbf{s}\, \Delta\rho(\mathbf{r})\Delta\rho(\mathbf{s}) c(\mathbf{r},\mathbf{s};\rho_\lambda) \\ &= \int_V \mathrm{d}\mathbf{r} \left[\rho_1(\mathbf{r}) \ln\frac{\rho_1(\mathbf{r})}{\rho_0(\mathbf{r})} - \{\rho_1(\mathbf{r}) - \rho_0(\mathbf{r})\} + \rho_1(\mathbf{r}) \{\beta u_1(\mathbf{r}) - \beta u_0(\mathbf{r})\} \right] \\ &\quad - \int_0^1 \mathrm{d}\lambda\, (1-\lambda) \int_V \mathrm{d}\mathbf{r}\, \mathrm{d}\mathbf{s}\, \Delta\rho(\mathbf{r})\Delta\rho(\mathbf{s}) c(\mathbf{r},\mathbf{s};\rho_\lambda). \end{aligned} \tag{10.28}$$

This is an exact expression for density functional that can be minimised with respect to ρ_1 to yield the equilibrium density profile that corresponds to u_1. The terms corresponding to the initial state are constant with respect to ρ_1.

Now take the pair direct correlation function to be constant along the integration path and equal to its initial value,

$$c(\mathbf{r},\mathbf{s};\rho_\lambda) \approx c(\mathbf{r},\mathbf{s};\rho_0). \tag{10.29}$$

This is of course an approximation that allows the coupling integral to be performed. For the initial state a uniform system in equilibrium is chosen, $u_0(\mathbf{r}) = 0$, and $\rho_0(\mathbf{r}) = \overline{\rho}(\mathbf{r};\mu) = \rho_0$. With this the density functional becomes

$$\begin{aligned} & \beta\Omega[\rho_1|z_1] \\ = \ & \beta\Omega(\mu, V, T) + \int_V \mathrm{d}\mathbf{r} \left[\rho_1(\mathbf{r}) \ln \frac{\rho_1(\mathbf{r})}{\rho_0} - \{\rho_1(\mathbf{r}) - \rho_0\} + \rho_1(\mathbf{r})\beta u_1(\mathbf{r}) \right] \\ & - \frac{1}{2} \int_V \mathrm{d}\mathbf{r}\, \mathrm{d}\mathbf{s}\, \{\rho_1(\mathbf{r}) - \rho_0\} \{\rho_1(\mathbf{s}) - \rho_0\} c(\mathbf{r}, \mathbf{s}; \rho_0). \end{aligned} \tag{10.30}$$

The functional derivative is

$$\frac{\delta \beta\Omega[\rho_1|z_1]}{\delta \rho_1(\mathbf{r})} = \ln \frac{\rho_1(\mathbf{r})}{\rho_0} + \beta u_1(\mathbf{r}) - \int_V \mathrm{d}\mathbf{s}\, \{\rho_1(\mathbf{s}) - \rho_0\} c(\mathbf{r}, \mathbf{s}; \rho_0), \tag{10.31}$$

which vanishes at the equilibrium density $\overline{\rho}_1(\mathbf{r}; u_1, \mu)$. Rearranging this gives

$$\overline{\rho}_1(\mathbf{r}) = \rho_0 e^{-\beta u_1(\mathbf{r})} \exp\left[\int_V \mathrm{d}\mathbf{s}\, \{\overline{\rho}_1(\mathbf{s}) - \rho_0\} c(\mathbf{r}, \mathbf{s}; \rho_0) \right]. \tag{10.32}$$

For the case that the reference fluid is isotropic and has pair-wise additive potential $u^{(2)}(r_{12})$, the pair direct correlation function depends only upon the separation between the particles, $c(\mathbf{r}_3, \mathbf{r}_2; \rho_0) = c_0(r_{32})$. Further, if one takes the external potential to represent a particle fixed at $\mathbf{r}_1$, so that $u_1(\mathbf{r}_2) = u^{(2)}(r_{12})$, then the density profile about the fixed particle is simply related to the radial distribution function of the uniform reference system,

$$\overline{\rho}_1(\mathbf{r}_2) = \rho_0 g_0(r_{12}) = \rho_0 h_0(r_{12}) + \rho_0. \tag{10.33}$$

In this case the equation for the equilibrium density profile may be written

$$\rho_0 g_0(r_{12}) = \rho_0 e^{-\beta u^{(2)}(r_{12})} \exp\left[\rho_0 \int_V \mathrm{d}\mathbf{r}_3\, h_0(r_{13}) c_0(r_{32}) \right]. \tag{10.34}$$

One recognises the second exponent as the Ornstein–Zernike series function, and hence this is just the hypernetted chain closure approximation, Eq. (9.65).

At equilibrium, the change in grand potential in this approximation is

$$\begin{aligned} & \beta\Omega[\overline{\rho}_1|z_1] - \beta\Omega(\mu, V, T) \\ = \ & \rho \int_V \mathrm{d}\mathbf{r}_2 \left[\{1 + h(r_{12})\} \left(\beta u^{(2)}(r_{12}) + \ln\{1 + h(r_{12})\} \right) - h(r_{12}) \right] \\ & - \frac{\rho^2}{2} \int_V \mathrm{d}\mathbf{r}_2\, \mathrm{d}\mathbf{r}_3\, h(r_{12}) c(r_{23}) h(r_{31}). \end{aligned} \tag{10.35}$$

For clarity the subscript 0 has been dropped since all densities and functions on the right-hand side refer to the uniform reference system. The difference between the final and the initial system is the particle fixed at $\mathbf{r}_1$. Hence the difference in the grand potentials represents the excess chemical potential of the reference system. (The ideal contributions are not included because the particle is fixed.) The excess chemical potential in hypernetted chain approximation was given above as Eq. (9.102). Rearranging it gives

$$\begin{aligned}
\langle -\beta_s \mu_s^{\text{ex}} \rangle_{\text{HNC}} & \\
= & \frac{\rho}{2} \int_V \mathrm{d}\mathbf{r}_2 \, [2c(r_{12}) - h(r_{12})\{h(r_{12}) - c(r_{12})\}] \\
= & -\rho \int_V \mathrm{d}\mathbf{r}_2 \, [\{h(r_{12}) - c(r_{12})\} \{1 + h(r_{12})\} - h(r_{12})] \\
& + \frac{\rho}{2} \int_V \mathrm{d}\mathbf{r}_2 \, [\{h(r_{12}) - c(r_{12})\} h(r_{12})] \\
= & -\rho \int_V \mathrm{d}\mathbf{r}_2 \, \left[\left(\beta u^{(2)}(r_{12}) + \ln\{1 + h(r_{12})\} \right) \{1 + h(r_{12})\} - h(r_{12}) \right] \\
& + \frac{\rho^2}{2} \int_V \mathrm{d}\mathbf{r}_2 \, \mathrm{d}\mathbf{r}_3 \, h(r_{12}) c(r_{23}) h(r_{31}).
\end{aligned} \tag{10.36}$$

The hypernetted chain closure has been used to obtain the first term, and the series function has been replaced by the Ornstein–Zernike convolution integral to obtain the second. This is the same as that given by the difference in the grand potentials, which confirms indeed that the hypernetted chain approximation is equivalent to taking the pair direct correlation function to be given by its bulk value and independent of the partially coupled particle in the density functional.

10.2.3 Approximation Schemes

The hypernetted chain approximation derived above is one example of a density functional approximation. It consisted of writing the excess part of the Helmholtz free energy functional as an integral involving the inhomogeneous pair direct correlation function evaluated at an intermediate density, and then approximating the latter as that of a reference system, usually taken as a uniform bulk system.

In order to go beyond the hypernetted chain, one may go to the next order in the coupling constant expansion, which involves the triplet direct correlation function. One approximates the latter by that of the uniform reference system. This level of approximation is equivalent to retaining a certain infinite class of bridge diagrams, namely those that can be written as

$$d_2(\mathbf{r}, \mathbf{s}) = \frac{1}{2} \int \mathrm{d}\mathbf{u} \, \mathrm{d}\mathbf{t} \, c(\mathbf{r}, \mathbf{u}, \mathbf{t}) \rho(\mathbf{u}) \rho(\mathbf{t}) h(\mathbf{u}, \mathbf{s}) h(\mathbf{t}, \mathbf{s}). \tag{10.37}$$

The actual bridge function is the series of d_n, where d_n is the convolution of $c^{(n+1)}$ with n pair total correlation functions. Hence going to higher orders in the coupling constant expansion is equivalent to retaining successive classes of bridge diagrams in the hypernetted chain closure. This then is a systematic way of improving the density functional approximation.

Other approximate approaches have been used, based upon some explicit formula for $\mathcal{F}^{\rm ex}[\rho]$.[2] There is a degree of arbitrariness in the choice of the latter and the accuracy of the final density profile depends to a large extent on the appropriateness of the chosen functional for the problem at hand.

The simplest approach is to invoke the Helmholtz free energy density of a uniform bulk system with density equal to the local density,

$$\mathcal{F}^{\rm ex}[\rho] = -\int {\rm d}\mathbf{r}\, f_{\rm ex}(\rho(\mathbf{r})). \tag{10.38}$$

There are at least three limitations of this approximation. First, in the case that one is describing two-phase coexistence, the local density may belong to the unstable part of the phase diagram, and hence the free energy density may be undefined or unphysical. Second, the free energy density may not be known for the particular system of interest. And third, this local density approximation corresponds to taking the pair direct correlation function to be a Dirac δ, which is a poor approximation to real systems. This may be seen by taking the functional derivative,

$$c^{(1)}(\mathbf{r}) = \frac{\delta \mathcal{F}^{\rm ex}[\rho]}{\delta \rho(\mathbf{r})} = -f'_{\rm ex}(\rho(\mathbf{r})), \tag{10.39}$$

and

$$c^{(2)}(\mathbf{r},\mathbf{s}) = \frac{\delta c^{(1)}(\mathbf{r})}{\delta \rho(\mathbf{s})} = -f''_{\rm ex}(\rho(\mathbf{r}))\delta(\mathbf{r},\mathbf{s}), \tag{10.40}$$

where the primes denote differentiation with respect to density.

In order to go beyond this local density approximation, one can add to it terms involving the density gradient. Due to the isotropy of the bulk uniform system, one has

$$f[\rho] = f_0(\rho(\mathbf{r})) + f_2(\rho(\mathbf{r}))\,(\nabla\rho(\mathbf{r}))^2 + \ldots. \tag{10.41}$$

The second coefficient here is essentially the second moment of the bulk pair direct correlation function. The utility of such gradient expansions are restricted to systems that are only weakly nonuniform.

A higher-order approach, essentially a resummation of the gradient expansion, is to add a nonlocal term quadratic in the density,

$$\mathcal{F}^{\rm ex}[\rho] = -\int {\rm d}\mathbf{r}\, f_{\rm ex}(\rho(\mathbf{r})) - \frac{k_{\rm B}T}{4}\int {\rm d}\mathbf{r}\,{\rm d}\mathbf{s}\, c(\mathbf{r},\mathbf{s};\rho^*)\,[\rho(\mathbf{r})-\rho(\mathbf{s})]^2. \tag{10.42}$$

To leading order in density the second density derivative of this yields a pair direct correlation function. The latter is often taken to be that of a bulk uniform system at density ρ^*, which can be chosen as some suitable average of the densities at the two positions (e.g., $\rho^* = [\rho(\mathbf{r})+\rho(\mathbf{s})]/2$).

[2] J. -P. Hansen and I. R. McDonald, *Theory of Simple Liquids*, 2nd ed., Academic Press, London, 1986. R. Evans, The nature of the liquid–vapour interface and other topics in the statistical mechanics of nonuniform classical fluids, *Adv. Phys.* **28** (1979), 143.

Perturbative approximations have also been pursued, using the hard-sphere system as a reference. One treats the attractive long-range tail of the pair potential, $v(r)$, in a mean field fashion, and one supposes that the structure is mainly determined by the repulsive core, for which one can use a hard-sphere density functional,

$$\mathcal{F}^{\mathrm{ex}}[\rho] = \mathcal{F}_d^{\mathrm{ex}}[\rho] - \frac{1}{2}\int \mathrm{d}\mathbf{r}\,\mathrm{d}\mathbf{s}\, v(|\mathbf{r}-\mathbf{s}|)\rho(\mathbf{r})\rho(\mathbf{s}). \tag{10.43}$$

For the hard-sphere functional of the inhomogeneous density one often invokes the excess free energy density of a uniform hard-sphere system at some local weighted density $\rho^*(\mathbf{r})$. That is,

$$\mathcal{F}_d^{\mathrm{ex}}[\rho] = -\int \mathrm{d}\mathbf{r}\,\rho(\mathbf{r})\frac{f_d(\rho^*(\mathbf{r}))}{\rho^*(\mathbf{r})}, \tag{10.44}$$

with

$$\rho^*(\mathbf{r}) = \int \mathrm{d}\mathbf{s}\,\rho(\mathbf{s})w(\mathbf{r}-\mathbf{s}), \tag{10.45}$$

with normalised weight function, $\int \mathrm{d}\mathbf{s}\, w(\mathbf{r}-\mathbf{s}) = 1$. In the local density version, $w(\mathbf{r}-\mathbf{s}) = \delta(\mathbf{r}-\mathbf{s})$, the second density functional derivative gives for the pair direct correlation function

$$c(\mathbf{r},\mathbf{s}) = -\beta f_d''(\rho^*(\mathbf{r}))\delta(\mathbf{r}-\mathbf{s}) - \beta v(|\mathbf{r}-\mathbf{s}|), \tag{10.46}$$

which has the correct asymptote, but is again a poor approximation at short range. Averaging the density instead over the neighbourhood of a hard-sphere diameter yields better behaviour at short range. One has a certain flexibility in the choice of the hard-sphere diameter, and in the weighting function. Although specific recipes may work well for particular systems, the approach is somewhat ad hoc and it is not clear how systematic improvements can be made.

10.2.4 Activity Functional

Uniform System

In the early chapters it was shown that an open subsystem in thermal equilibrium had constrained thermodynamic potential

$$\Omega(\tilde{N}|\mu, V, T) = F(\tilde{N}, V, T) - \mu\tilde{N}, \tag{10.47}$$

where the Helmholtz free energy appears. The minimum value of this was the grand potential,

$$\Omega(\mu, V, T) = \Omega(\overline{N}|\mu, V, T) = F(\overline{N}, V, T) - \mu\overline{N}, \tag{10.48}$$

where the equilibrium number is $\tilde{N} = \overline{N}(\mu, V, T)$.

Now a variational principle for the chemical potential will be developed. This is more a mathematical convenience than a physical representation of a real system. Consider the function

$$\mathcal{F}(\tilde{\mu}|N,V,T) = \Omega(\tilde{\mu},V,T) + \tilde{\mu}N. \tag{10.49}$$

Its derivative is

$$\frac{\partial \mathcal{F}(\tilde{\mu}|N,V,T)}{\partial \tilde{\mu}} = -\overline{N}(\tilde{\mu},V,T) + N, \tag{10.50}$$

which obviously vanishes when $\tilde{\mu} = \overline{\mu}(N,V,T)$. Taking the second derivative yields

$$\frac{\partial^2 \mathcal{F}(\tilde{\mu}|N,V,T)}{\partial \tilde{\mu}^2} = \frac{-\partial \overline{N}(\tilde{\mu},V,T)}{\partial \tilde{\mu}} < 0, \tag{10.51}$$

as follows from Eq. (4.68). This shows that the stationary point is a maximum, which is in contrast to the previous constrained thermodynamic potentials where a minimum characterises equilibrium. Nevertheless $\mathcal{F}(\tilde{\mu}|N,V,T)$ represents a variational principle for the chemical potential, and its maximum value is the Helmholtz free energy

$$F(N,V,T) = \mathcal{F}(\overline{\mu}|N,V,T) = \Omega(\overline{\mu},V,T) + \overline{\mu}N. \tag{10.52}$$

Inhomogeneous System

For an inhomogeneous system, one can generalise this to a functional of the activity,

$$\mathcal{F}(\tilde{z}|\rho,V,T) = \Omega(\tilde{z},V,T) + k_{\mathrm{B}}T \int \mathrm{d}\mathbf{s}\; \rho(\mathbf{s}) \ln \Lambda^3 \tilde{z}(\mathbf{s}). \tag{10.53}$$

The functional derivative of this is

$$\frac{\delta \beta \mathcal{F}(\tilde{z}|\rho,V,T)}{\delta \ln \Lambda^3 \tilde{z}(\mathbf{r})} = -\tilde{\rho}(\mathbf{r}) + \rho(\mathbf{r}), \tag{10.54}$$

which, since $\tilde{\rho}(\mathbf{r}) = \overline{\rho}(\mathbf{r},[\tilde{z}])$, vanishes when $\tilde{z}(\mathbf{r}) = \overline{z}(\mathbf{r},[\rho])$. Hence $\mathcal{F}(\tilde{z}|\rho,V,T)$ provides a variational principle that yields the activity that corresponds to a given density profile. Its maximum value is

$$\begin{aligned} \mathcal{F}(\rho,V,T) &= \mathcal{F}(\overline{z}|\rho,V,T) \\ &= \Omega_\rho(\rho,V,T) + k_{\mathrm{B}}T \int \mathrm{d}\mathbf{s}\; \rho(\mathbf{s}) \ln \Lambda^3 \overline{z}(\mathbf{s};[\rho]). \end{aligned} \tag{10.55}$$

Since the logarithm of the activity gives the local chemical potential, $\mu(\mathbf{s}) = \mu - u(\mathbf{s})$, one sees that this is similar to the Helmholtz free energy, $F = \Omega + \mu N$. In the Helmholtz free energy the chemical potential is effectively removed from Ω. The fact that both it and the external potential are effectively removed from the grand potential in this expression is the reason that $\mathcal{F}$ is often called the intrinsic Helmholtz free energy.

This density functional is the same as the functional denoted by the same symbol given previously, Eqs. (9.20) and (9.24). This follows by direct substitution into the explicit representation of Ω_ρ, Eq.(10.13),

$$\begin{aligned}\mathcal{F}(\rho, V, T) &= k_{\mathrm{B}}T \int \mathrm{d}\mathbf{s}\, \rho(\mathbf{s}) \left[-1 + \ln \frac{\rho(\mathbf{s})}{z(\mathbf{s}; [\rho])} \right] - \mathcal{F}^{\mathrm{ex}}[\rho] \\ &\quad + k_{\mathrm{B}}T \int \mathrm{d}\mathbf{s}\; \rho(\mathbf{s}) \ln \Lambda^3 z(\mathbf{s}; [\rho]) \\ &= \int \mathrm{d}\mathbf{s}\, \rho(\mathbf{s})[-1 + \ln \Lambda^3 \rho(\mathbf{s})] - \mathcal{F}^{\mathrm{ex}}[\rho] \\ &= \mathcal{F}^{\mathrm{id}}[\rho] - \mathcal{F}^{\mathrm{ex}}[\rho]. \end{aligned} \tag{10.56}$$

It will be recalled that this thermodynamic potential was the generating functional for the direct correlation functions, c_δ.

Cross–Entropy

In terms of the cross-entropy, consider the functional maximised at equilibrium with maximum value equal to the actual intrinsic Helmholtz free energy,

$$\begin{aligned} &\beta\mathcal{F}'[\tilde{z}|\rho] \\ &= \beta\mathcal{F}(\rho, V, T) - S[\wp, \tilde{\wp}] \\ &= \beta\mathcal{F}(\rho, V, T) - \sum_{N=0}^{\infty} \frac{1}{N!} \int \mathrm{d}\mathbf{r}^N\; \wp(\mathbf{r}^N) \ln \frac{\wp(\mathbf{r}^N)}{\tilde{\wp}(\mathbf{r}^N)} \\ &= \beta\mathcal{F}(\rho, V, T) - \sum_{N=0}^{\infty} \int \frac{\mathrm{d}\mathbf{r}^N}{N!} \wp(\mathbf{r}^N) \ln \frac{\tilde{\Xi}}{\Xi} \prod_{i=1}^{N} \frac{z(\mathbf{r}_i)}{\tilde{z}(\mathbf{r}_i)} \\ &= \beta\mathcal{F}(\rho, V, T) + \beta\Omega(\tilde{z}, V, T) - \beta\Omega(z, V, T) - \int \mathrm{d}\mathbf{r}\, \rho(\mathbf{r}) \ln \frac{z(\mathbf{r})}{\tilde{z}(\mathbf{r})}. \end{aligned} \tag{10.57}$$

Using the fact that

$$\Omega(z, V, T) = \mathcal{F}(\overline{\rho}, V, T) - k_{\mathrm{B}}T \int \mathrm{d}\mathbf{r}\, \overline{\rho}(\mathbf{r}) \ln \Lambda^3 z(\mathbf{r}), \tag{10.58}$$

and that $\rho(\mathbf{r}) = \overline{\rho}(\mathbf{r}, [z])$, this is

$$\beta\mathcal{F}'[\tilde{z}|\rho] = \beta\Omega(\tilde{z}, V, T) + \int \mathrm{d}\mathbf{r}\, \rho(\mathbf{r}) \ln \Lambda^3 \tilde{z}(\mathbf{r}), \tag{10.59}$$

which is just the activity functional given above.

10.3 Pair Functionals

10.3.1 Total Correlation Function

In this section a variational principle for the pair correlation function is developed. One again considers an open subsystem in contact with a reservoir that creates a local chemical potential $\mu(\mathbf{r}) = \mu - u(\mathbf{r})$. As for the density functional, with the singlet density constrained to be $\tilde{\rho}(\mathbf{r})$, the corresponding number of particles is $\tilde{N}$ and external energy is $\tilde{E}^{\rm ext}$. One supposes that the particles of the actual subsystem interact with pair potential $u(\mathbf{r}, \mathbf{s})$, and that the subsystem is in the macrostate with pair density $\tilde{\rho}(\mathbf{r}, \mathbf{s})$. Then the pair part of the subsystem energy is

$$\tilde{E}^{(2)} = \frac{1}{2} \int \mathrm{d}\mathbf{r}\, \mathrm{d}\mathbf{s}\, \tilde{\rho}(\mathbf{r}, \mathbf{s}) u(\mathbf{r}, \mathbf{s}). \tag{10.60}$$

The internal nonpair energy of the subsystem is denoted by $\tilde{E}$, and the energy of the reservoir is therefore $E_0 - \tilde{E} - \tilde{E}^{\rm ext} - \tilde{E}^{(2)}$. The constrained total entropy of the subsystem and reservoir is

$$\begin{aligned}
& S(\tilde{E}, \tilde{\rho}^{(1)}, \tilde{\rho}^{(2)} | \mu, u^{(1)}, u^{(2)}, V, T) \\
&= S(\tilde{E}, \tilde{\rho}^{(1)}, \tilde{\rho}^{(2)}, V) - \frac{\tilde{E}}{T} + \frac{1}{T} \int \mathrm{d}\mathbf{r}\, \tilde{\rho}(\mathbf{r}) \mu(\mathbf{r}) \\
&\quad - \frac{1}{2T} \int \mathrm{d}\mathbf{r}\, \mathrm{d}\mathbf{s}\, \tilde{\rho}(\mathbf{r}, \mathbf{s}) u(\mathbf{r}, \mathbf{s}) \\
&= S(\tilde{E}, \tilde{\rho}^{(1)}, \tilde{\rho}^{(2)}, V) - \frac{\tilde{E}}{\tilde{T}} + \frac{1}{\tilde{T}} \int \mathrm{d}\mathbf{r}\, \tilde{\rho}(\mathbf{r}) \tilde{\mu}(\mathbf{r}) \\
&\quad - \frac{1}{2\tilde{T}} \int \mathrm{d}\mathbf{r}\, \mathrm{d}\mathbf{s}\, \tilde{\rho}(\mathbf{r}, \mathbf{s}) \tilde{u}(\mathbf{r}, \mathbf{s}) \\
&\quad - \tilde{E} \left\{ \frac{1}{T} - \frac{1}{\tilde{T}} \right\} + \int \mathrm{d}\mathbf{r}\, \tilde{\rho}(\mathbf{r}) \left\{ \frac{\mu(\mathbf{r})}{T} - \frac{\tilde{\mu}(\mathbf{r})}{\tilde{T}} \right\} \\
&\quad - \frac{1}{2} \int \mathrm{d}\mathbf{r}\, \mathrm{d}\mathbf{s}\, \tilde{\rho}(\mathbf{r}, \mathbf{s}) \left\{ \frac{1}{T} u(\mathbf{r}, \mathbf{s}) - \frac{1}{\tilde{T}} \tilde{u}(\mathbf{r}, \mathbf{s}) \right\} \\
&= S(\tilde{E}, \tilde{\rho}^{(1)}, \tilde{\rho}^{(2)} | \tilde{\mu}, \tilde{u}^{(1)}, \tilde{u}^{(2)}, V, T) - \tilde{E} \left\{ \frac{1}{T} - \frac{1}{\tilde{T}} \right\} \\
&\quad + k_{\rm B} \int \mathrm{d}\mathbf{r}\, \tilde{\rho}(\mathbf{r}) \ln \frac{z(\mathbf{r})}{\tilde{z}(\mathbf{r})} + \frac{k_{\rm B}}{2} \int \mathrm{d}\mathbf{r}\, \mathrm{d}\mathbf{s}\, \tilde{\rho}(\mathbf{r}, \mathbf{s}) \ln \frac{e(\mathbf{r}, \mathbf{s})}{\tilde{e}(\mathbf{r}, \mathbf{s})}.
\end{aligned} \tag{10.61}$$

The definitions of the activity $z(\mathbf{r}) = \Lambda^{-3} \exp \beta\mu(\mathbf{r})$ and the exponential pair function $e(\mathbf{r}, \mathbf{s}) = \exp -\beta u(\mathbf{r}, \mathbf{s})$ give the final equality.

In the case that the subsystem is thermally equilibrated, $\tilde{T} = T$, one has

$$\begin{aligned}
S(\tilde{\rho}^{(1)}, \tilde{\rho}^{(2)} | z, e, V, T) &= S(\tilde{\rho}^{(1)}, \tilde{\rho}^{(2)} | \tilde{z}, \tilde{e}, V, T) + k_{\rm B} \int \mathrm{d}\mathbf{r}\, \tilde{\rho}(\mathbf{r}) \ln \frac{z(\mathbf{r})}{\tilde{z}(\mathbf{r})} \\
&\quad + \frac{k_{\rm B}}{2} \int \mathrm{d}\mathbf{r}\, \mathrm{d}\mathbf{s}\, \tilde{\rho}(\mathbf{r}, \mathbf{s}) \ln \frac{e(\mathbf{r}, \mathbf{s})}{\tilde{e}(\mathbf{r}, \mathbf{s})},
\end{aligned} \tag{10.62}$$

where the activity rather than the chemical potential is now used. Since in general the constrained thermodynamic potential is the negative of the temperature times the constrained entropy, multiplying by $-T$ one may write this as

$$\Omega[\tilde{\rho}^{(1)}, \tilde{\rho}^{(2)} | z, e] = \Omega[\tilde{z}, \tilde{e}] + k_{\rm B}T \int \mathrm{d}\mathbf{r}\, \tilde{\rho}(\mathbf{r}) \ln \frac{\tilde{z}(\mathbf{r})}{z(\mathbf{r})}$$

$$+\frac{k_{\rm B}T}{2}\int {\rm d}\mathbf{r}\,{\rm d}\mathbf{s}\,\tilde\rho(\mathbf{r},\mathbf{s})\ln\frac{\tilde e(\mathbf{r},\mathbf{s})}{e(\mathbf{r},\mathbf{s})}. \tag{10.63}$$

The grand potential of the meta-system appears as the first term on the right-hand side because the given macrostate is the equilibrium state for the subsystem in contact with the meta-reservoir; its constrained thermodynamic potential is at its minimum value, $-TS(\tilde\rho^{(1)},\tilde\rho^{(2)}|\tilde z,\tilde e,V,T)=\Omega(\tilde z,\tilde e,V,T)$.

This functional of the density is minimised by the equilibrium densities, $\tilde\rho(\mathbf{r})=\overline{\rho}(\mathbf{r};[z],[e])$, and $\tilde\rho(\mathbf{r},\mathbf{s})=\overline{\rho}(\mathbf{r},\mathbf{s};[z],[e])$. At equilibrium $\tilde z(\mathbf{r})=z(\mathbf{r})$, and $\tilde e(\mathbf{r},\mathbf{s})=e(\mathbf{r},\mathbf{s})$ and the integrands vanish, which shows that the minimum value of the functional equals $\Omega[z,e]$, the actual grand potential of the open subsystem.

The pair density is related to the pair total correlation function by $\rho(\mathbf{r},\mathbf{s})=\rho(\mathbf{r})\rho(\mathbf{s})[1+h(\mathbf{r},\mathbf{s})]$. Henceforth the analysis is particularised to the case of fixed density, $\tilde\rho(\mathbf{r})=\rho(\mathbf{r})$, and the above functional is written $\Omega[\tilde h|e,z]$. This means that the activity of the meta-system is now a dependent functional, $\tilde z(\mathbf{r})=z(\mathbf{r};[\tilde e],[\rho])$.

The grand potential is an implicit functional of the total correlation function. The function $\tilde e(\mathbf{r},\mathbf{s})$, which is the exponential function of the meta-system, is the one that gives $\tilde h(\mathbf{r},\mathbf{s})$ as the equilibrium total correlation function of the meta-system, and hence it is a functional of the constrained total correlation function, $\tilde e(\mathbf{r},\mathbf{s})=\overline{e}(\mathbf{r},\mathbf{s};[\tilde h])$.

One now requires the derivative of the total correlation functional $\Omega[\tilde h|e,z]$. For this one uses the fact that

$$\begin{aligned}
&\frac{\delta\beta\Omega[\tilde z,\tilde e]}{\delta\tilde h(\mathbf{r},\mathbf{s})}\\
&= \int {\rm d}\mathbf{u}\,{\rm d}\mathbf{t}\,\frac{\delta\beta\tilde\Omega}{\delta\ln\tilde e(\mathbf{u},\mathbf{t})}\frac{\delta\ln\tilde e(\mathbf{u},\mathbf{t})}{\delta\tilde h(\mathbf{r},\mathbf{s})}+\int{\rm d}\mathbf{t}\,\frac{\delta\beta\tilde\Omega}{\delta\ln\tilde z(\mathbf{t})}\frac{\delta\ln\tilde z(\mathbf{t})}{\delta\tilde h(\mathbf{r},\mathbf{s})}\\
&= \frac{-1}{2}\int{\rm d}\mathbf{u}\,{\rm d}\mathbf{t}\,\tilde\rho(\mathbf{u},\mathbf{t})\frac{\delta\ln\tilde e(\mathbf{u},\mathbf{t})}{\delta\tilde h(\mathbf{r},\mathbf{s})}-\int{\rm d}\mathbf{t}\,\tilde\rho(\mathbf{t})\frac{\delta\ln\tilde z(\mathbf{t})}{\delta\tilde h(\mathbf{r},\mathbf{s})}.
\end{aligned} \tag{10.64}$$

Here Eq. (8.57), which gives the pair density as the pair potential derivative of the grand potential, has been used. Accordingly, for fixed density, $\tilde\rho(\mathbf{r})=\rho(\mathbf{r})$, these two terms terms cancel with the derivatives of some of the remaining terms in $\Omega[\tilde h|e,z]$, and one is left with

$$\frac{\delta\Omega[\tilde h|e,z]}{\delta\tilde h(\mathbf{r},\mathbf{s})}=\frac{1}{2}\rho(\mathbf{r})\rho(\mathbf{s})\ln\frac{\tilde e(\mathbf{r},\mathbf{s})}{e(\mathbf{r},\mathbf{s})}. \tag{10.65}$$

Explicit Functional

The total correlation functional $\Omega[\tilde h|e,z]$ may be written as an explicit functional of the the total correlation function. One requires the closure relationship and two generating functions for the classes of diagrams that appear therein. The closure was given in the preceding chapter as Eq. (9.43), and it may be written

$$h(\mathbf{r},\mathbf{s})=-1+e(\mathbf{r},\mathbf{s})e^{h(\mathbf{r},\mathbf{s})-c(\mathbf{r},\mathbf{s})+d(\mathbf{r},\mathbf{s})}, \tag{10.66}$$

where $d^{(2)}$ is the bridge function.

Now one needs to define two sets of diagrams that will yield the series function $h - c$, and the bridge function d, when differentiated with respect to the total correlation function. A ring is a linear chain closed on itself, and one defines the ring diagrams as

$$\mathcal{R}[h] = \{\text{the sum of all simple connected diagrams with } (-\rho^{(1)})\text{-field points and } h^{(2)}\text{-bonds, such that each field point is intersected by exactly two bonds}\}$$

$=$ 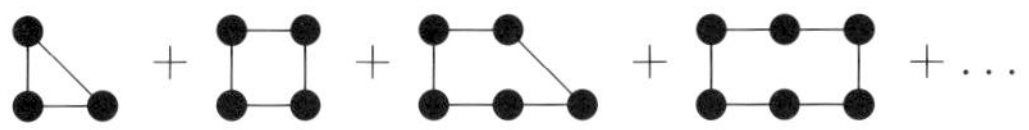 (10.67)

Note that the sign of successive ring diagrams alternates, since the circles represent $(-\rho^{(1)})$-field points. Differentiation yields the series function,

$$\begin{aligned} \frac{\delta \mathcal{R}[h]}{\delta h(\mathbf{r},\mathbf{s})} &= \frac{1}{2}\rho(\mathbf{r})\rho(\mathbf{s})\left[-h*h+h*h*h-h*h*h*h+\ldots\right](\mathbf{r},\mathbf{s}) \\ &= \frac{-1}{2}\rho(\mathbf{r})\rho(\mathbf{s})\left[h(\mathbf{r},\mathbf{s})-c(\mathbf{r},\mathbf{s})\right]. \end{aligned} \tag{10.68}$$

The asterisks here denote a density-weighted convolution integral. This result follows by using successive substitution in the Ornstein–Zernike equation written in the form

$$\begin{aligned} c &= h - h*c \\ &= h - h*h + h*h*c \\ &= h - h*h + h*h*h - h*h*h*h + \ldots. \end{aligned} \tag{10.69}$$

The second function is defined as

$$\mathcal{B}[h] = \{\text{the sum of all simple connected diagrams with four or more } \rho^{(1)}\text{-field points and no articulation points, and with } h^{(2)}\text{-bonds and no articulation pairs of points}\}$$

$=$ 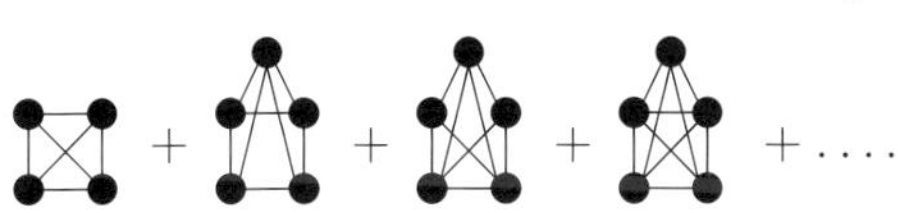 (10.70)

Differentiation yields the bridge function,

$$\frac{\delta \mathcal{B}[\rho,h]}{\delta h(\mathbf{r},\mathbf{s})} = \frac{1}{2}\rho(\mathbf{r})\rho(\mathbf{s})d(\mathbf{r},\mathbf{s}). \tag{10.71}$$

Using these definitions, the explicit functional of the total correlation function is

$$\begin{aligned} &\beta\Omega[\tilde{h}|e,z] \\ &= \int \mathrm{d}\mathbf{r}\,\rho(\mathbf{r})[\ln\frac{\rho(\mathbf{r})}{z(\mathbf{r})} - 1] + \mathcal{R}[\tilde{h}] - \mathcal{B}[\tilde{h}] \end{aligned}$$

$$+\frac{1}{2}\int \mathrm{d}\mathbf{r}\,\mathrm{d}\mathbf{s}\,\rho(\mathbf{r})\rho(\mathbf{s})\left\{[1+\tilde{h}(\mathbf{r},\mathbf{s})]\ln[1+\tilde{h}(\mathbf{r},\mathbf{s})]-\tilde{h}(\mathbf{r},\mathbf{s})\right\}$$
$$-\frac{1}{2}\int \mathrm{d}\mathbf{r}\,\mathrm{d}\mathbf{s}\,\rho(\mathbf{r})\rho(\mathbf{s})[1+\tilde{h}(\mathbf{r},\mathbf{s})]\ln e(\mathbf{r},\mathbf{s}). \tag{10.72}$$

Using the closure equation, differentiation yields

$$\frac{\delta\beta\Omega[\tilde{h}|e,z]}{\delta\tilde{h}(\mathbf{r},\mathbf{s})}=\frac{1}{2}\rho(\mathbf{r})\rho(\mathbf{s})\ln\frac{\tilde{e}(\mathbf{r},\mathbf{s})}{e(\mathbf{r},\mathbf{s})}, \tag{10.73}$$

which shows that up to a constant independent of $\tilde{h}$ this explicit functional is the same as the implicit one, Eq. (10.63).[3]

The value of the constant may be determined by taking the high-temperature limit, $\tilde{f}\to 0$. It is straightforward in both cases to show that

$$\beta\Omega[\tilde{h}|e,z]$$
$$\to \int \mathrm{d}\mathbf{r}\,\rho(\mathbf{r})[-1+\ln\frac{\rho(\mathbf{r})}{z(\mathbf{r})}]-\frac{1}{2}\int \mathrm{d}\mathbf{r}\,\mathrm{d}\mathbf{s}\,\rho(\mathbf{r})\rho(\mathbf{s})\ln e(\mathbf{r},\mathbf{s}), \tag{10.74}$$

and one concludes that the two functionals are in fact identical. One also concludes that the minimum value of both functionals is $\Omega[h|e,z]=\Omega[e,z]$.

It is obvious that the pair potential determines the pair correlation function. The fact that an explicit functional was given above indicates that the obverse holds: the pair correlation function determines the pair potential. In other words, there is a unique pair potential for each pair density, which means that well may one write $e(\mathbf{r},\mathbf{s};[h])$.

Fluctuations in the Pair Correlation Function

Because of the relationship of the pair correlation functional with the constrained total entropy, the probability of fluctuations in the pair correlation function is given by

$$\wp[\tilde{h}|z,e]=\frac{1}{\Xi[z,e]}e^{-\beta\Omega[\tilde{h}|z,e]}. \tag{10.75}$$

Conceptually, the probability of the pair density is a straightforward extension of the probability of the singlet density.

Cross–Entropy

As in the singlet case, one can use the cross-entropy to derive the variational functional. The normalised configuration probability is

$$\wp(\mathbf{r}^N,N|\mu,V,T)=\frac{e^{-\beta U_N^*(\mathbf{r}^N)}}{\Lambda^{3N}\Xi(\mu,V,T)}\prod_{i=1}^{N}z(\mathbf{r}_i)\prod_{i<j}^{N}e(\mathbf{r}_i,\mathbf{r}_j), \tag{10.76}$$

[3]This result was obtained by G. Stell, Cluster expansions for classical systems in equilibrium, in *The Equilibrium Theory of Classical Fluids* (H. Frisch and J. L. Lebowitz, Eds.), p. II-171, Benjamin, New York, 1964. The present derivation follows that of P. Attard, The explicit density functional and its connection with entropy maximisation, *J. Stat. Phys.* **100** (2000), 445.

where U_N^* is the potential energy without the singlet and pair contributions. Analogous quantities, denoted by a tilde, are defined for the meta-system.

In view of the nonnegativity of the cross-entropy, $S[\tilde{\wp}, \wp]$, defined exactly as above, one can add it to the grand potential of the subsystem plus actual reservoir to give a functional that is minimised at equilibrium and that has minimum value equal to the actual grand potential. This functional is

$$\begin{aligned}
&\beta\Omega[\tilde{\rho}^{(1)}, \tilde{\rho}^{(2)}|z, e] \\
&= \beta\Omega[z, e] + S[\tilde{\wp}, \wp]/k_{\rm B} \\
&= \beta\Omega[z, e] + \sum_{N=0}^{\infty} \frac{1}{N!} \int {\rm d}\mathbf{r}^N\, \tilde{\wp}(\mathbf{r}^N) \ln \frac{\tilde{\wp}(\mathbf{r}^N)}{\wp(\mathbf{r}^N)} \\
&= \beta\Omega[z, e] + \sum_{N=0}^{\infty} \int \frac{{\rm d}\mathbf{r}^N}{N!} \tilde{\wp}(\mathbf{r}^N) \ln \frac{\Xi}{\tilde{\Xi}} \prod_{i=1}^{N} \frac{\tilde{z}(\mathbf{r}_i)}{z(\mathbf{r}_i)} \prod_{i<j}^{N} \frac{\tilde{e}(\mathbf{r}_i, \mathbf{r}_j)}{e(\mathbf{r}_i, \mathbf{r}_j)} \\
&= \beta\Omega[\tilde{z}, \tilde{e}] + \int {\rm d}\mathbf{r}\, \tilde{\rho}(\mathbf{r}) \ln \frac{\tilde{z}(\mathbf{r})}{z(\mathbf{r})} + \frac{1}{2} \int {\rm d}\mathbf{r}\, {\rm d}\mathbf{s}\, \tilde{\rho}(\mathbf{r}, \mathbf{s}) \ln \frac{\tilde{e}(\mathbf{r}, \mathbf{s})}{e(\mathbf{r}, \mathbf{s})}, \qquad (10.77)
\end{aligned}$$

which is just the implicit pair density functional given above.

10.3.2 HNC Approximation

The ring diagrams are readily evaluated, at least in the uniform density case, $\rho(\mathbf{r}) = \rho$, $h(\mathbf{r}, \mathbf{s}) = h(|\mathbf{r} - \mathbf{s}|)$. One has $\mathcal{R}[h] = \sum_{n=3}^{\infty} \mathcal{R}_n$, with

$$\begin{aligned}
\mathcal{R}_n &= \frac{(-\rho)^n}{2n} \int {\rm d}\mathbf{r}_1 \ldots {\rm d}\mathbf{r}_n\, h(r_{12})h(r_{23}) \ldots h(r_{n-1,n})h(r_{n,1}) \\
&= \frac{(-\rho)^n V}{2n} \int {\rm d}\mathbf{r}_{12}\, h(r_{12})h_{n-1}(r_{21}) \\
&= \frac{(-\rho)^n V}{2n(2\pi)^3} \int {\rm d}\mathbf{k}\, \hat{h}(k)\hat{h}_{n-1}(k) \\
&= \frac{(-\rho)^n V}{2n(2\pi)^3} \int {\rm d}\mathbf{k}\, \hat{h}(k)^n, \qquad (10.78)
\end{aligned}$$

where the circumflex denotes the Fourier transform and Parsevaal's theorem and the convolution theorem have been used. It follows that

$$\begin{aligned}
\mathcal{R}[h] &= \frac{V}{2(2\pi)^3} \int {\rm d}\mathbf{k}\, \left\{ \ln[1 + \rho\hat{h}(k)] - \rho\hat{h}(k) + \rho^2 \hat{h}(k)^2/2 \right\} \\
&= \frac{V}{2(2\pi)^3} \int {\rm d}\mathbf{k}\, \left\{ \ln[1 + \rho\hat{h}(k)] - \rho\hat{c}(k) \right\} \\
&\quad - \frac{V\rho^2}{2} \int {\rm d}\mathbf{r}\, \left\{ c(r)h(r) - \frac{h(r)^2}{2} \right\}, \qquad (10.79)
\end{aligned}$$

where the Ornstein–Zernike equation has been used, $\hat{c}(k) = \hat{h}(k)/[1 + \rho\hat{h}(k)]$.

The hypernetted chain approximation consists of the neglect of the bridge diagrams $\mathcal{B}[h] = 0$. This gives

$$\begin{aligned}
&\beta\Omega^{\rm HNC}[\tilde{h}|e,z]/V \\
&= \rho\ln[\Lambda^3\rho] - \rho - \rho\beta\mu + \mathcal{R}[\tilde{h}]/V - \frac{\rho^2}{2}\int \mathrm{d}\mathbf{r}\,[1+\tilde{h}(r)]\ln e(r) \\
&\quad + \frac{\rho^2}{2}\int \mathrm{d}\mathbf{r}\,\left\{[1+\tilde{h}(r)]\ln[1+\tilde{h}(r)] - \tilde{h}(r)\right\}. \qquad (10.80)
\end{aligned}$$

This HNC functional to be minimised is the same as that given by Olivares and McQuarrie.[4] Using the HNC closure, $h(r) = -1 + e(r)\exp[h(r) - c(r)]$, this may be rewritten as

$$\begin{aligned}
&\beta\Omega^{\rm HNC}[\tilde{h}|e,z]/V \\
&= \rho\ln[\Lambda^3\rho] - \rho - \rho\beta\mu + \mathcal{R}[\tilde{h}]/V + \frac{\rho^2}{2}\int \mathrm{d}\mathbf{r}\,[1+\tilde{h}(r)]\ln\frac{\tilde{e}(r)}{e(r)} \\
&\quad + \frac{\rho^2}{2}\int \mathrm{d}\mathbf{r}\,\left\{\tilde{h}(r)^2 - \tilde{h}(r)\tilde{c}(r) - \tilde{c}(r)\right\}, \qquad (10.81)
\end{aligned}$$

from which it is apparent that its minimum value, which is the grand potential, is

$$\begin{aligned}
\frac{\beta\Omega^{\rm HNC}}{V} &= \rho\ln[\Lambda^3\rho] - \rho - \rho\beta\mu + \frac{\rho^2}{2}\int \mathrm{d}\mathbf{r}\,\left\{\frac{h(r)^2}{2} - c(r)\right\} \\
&\quad - \frac{1}{16\pi^3}\int \mathrm{d}\mathbf{k}\left\{\ln[1+\rho\hat{h}(k)] - \rho\hat{c}(k)\right\}. \qquad (10.82)
\end{aligned}$$

Since the grand potential is related to the Helmholtz free energy by $\Omega = F - \mu N$, one may recognise in this the HNC result for the excess Helmholtz free energy[5]

$$\begin{aligned}
\beta F_{\rm ex}^{\rm HNC}/V &= \frac{\rho^2}{2}\int \mathrm{d}\mathbf{r}\,\left\{h(r)^2/2 - c(r)\right\} \\
&\quad - \frac{1}{16\pi^3}\int \mathrm{d}\mathbf{k}\left\{\ln[1+\rho\hat{h}(k)] - \rho\hat{c}(k)\right\}. \qquad (10.83)
\end{aligned}$$

The HNC expression for the excess chemical potential was given above, and also in the preceding chapter, Eq. (9.102).

Inhomogeneous System

It is straightforward to obtain the HNC result for a nonuniform system. One recognises that one has an infinite dimensional vector space with the density providing the weighting for the inner product,

$$(f,g) = \int \mathrm{d}\mathbf{r}\,\rho(\mathbf{r})f(\mathbf{r})g(\mathbf{r}). \qquad (10.84)$$

[4] W. Olivares and D. A. McQuarrie, A variational approach to the theory of ionic solutions, *J. Chem. Phys.* **65** (1976), 3604.

[5] T. Morita, Theory of classical fluids: Hypernetted chain approximation. III, *Progr. Theor. Phys.* **23** (1960), 829.

The total correlation function is an operator,

$$Hf = \int \mathrm{d}\mathbf{s}\, \rho(\mathbf{s}) h(\mathbf{r},\mathbf{s}) f(\mathbf{s}). \tag{10.85}$$

Since this is symmetric, $(Hf, g) = (f, Hg)$, a basis of orthonormal eigenfunctions exist, and one has $Hf_{\mathbf{k}} = \lambda_{\mathbf{k}} f_{\mathbf{k}}$. The convolution products that appear in the ring diagrams are simply the trace of powers of H,

$$\mathrm{Tr}\, H^n = \sum_{\mathbf{k}} \lambda_{\mathbf{k}}^n, \tag{10.86}$$

and it follows that the ring diagrams sum to

$$\mathcal{R}[h] = \frac{1}{2} \sum_{\mathbf{k}} \left\{ \ln[1+\lambda_{\mathbf{k}}] - \lambda_{\mathbf{k}} + \lambda_{\mathbf{k}}^2/2 \right\}. \tag{10.87}$$

The remainder of the HNC expression for the thermodynamic potential may be evaluated directly.

10.3.3 Pair Potential Functional

The variational functional of the singlet and pair potential is

$$\begin{aligned} \mathcal{F}^{(2)}(\tilde{z},\tilde{e}|\rho^{(1)},\rho^{(2)},V,T) &= \Omega(\tilde{z},\tilde{e},V,T) + k_{\mathrm{B}}T \int \mathrm{d}\mathbf{r}\, \rho(\mathbf{r}) \ln \Lambda^3 \tilde{z}(\mathbf{r}) \\ &\quad + \frac{k_{\mathrm{B}}T}{2} \int \mathrm{d}\mathbf{r}\, \mathrm{d}\mathbf{s}\, \rho(\mathbf{r},\mathbf{s}) \ln \tilde{e}(\mathbf{r},\mathbf{s}). \end{aligned} \tag{10.88}$$

The functional derivatives of this are

$$\frac{\delta \beta \mathcal{F}^{(2)}(\tilde{z},\tilde{e}|\rho^{(1)},\rho^{(2)},V,T)}{\delta \ln \Lambda^3 \tilde{z}(\mathbf{r})} = -\tilde{\rho}(\mathbf{r}) + \rho(\mathbf{r}) \tag{10.89}$$

and

$$\frac{\delta \beta \mathcal{F}^{(2)}(\tilde{z},\tilde{e}|\rho^{(1)},\rho^{(2)},V,T)}{\delta \ln \tilde{e}(\mathbf{r},\mathbf{s})} = -\frac{1}{2}\tilde{\rho}(\mathbf{r},\mathbf{s}) + \frac{1}{2}\rho(\mathbf{r},\mathbf{s}), \tag{10.90}$$

which vanish when $\tilde{z}(\mathbf{r}) = \overline{z}(\mathbf{r};\rho^{(1)},\rho^{(2)})$ and $\tilde{e}(\mathbf{r},\mathbf{s}) = \overline{e}(\mathbf{r},\mathbf{s};\rho^{(1)},\rho^{(2)})$. Hence $\mathcal{F}^{(2)}(\tilde{z},\tilde{e}|\rho^{(1)},\rho^{(2)},V,T)$ provides a variational principle that yields the activity and pair potential that corresponds to a given density profile. Its maximum value is

$$\begin{aligned} &\mathcal{F}^{(2)}(\rho^{(1)},\rho^{(2)},V,T) \\ &= \mathcal{F}^{(2)}(\overline{z},\overline{e}|\rho^{(1)},\rho^{(2)},V,T) \\ &= \Omega_\rho(\rho^{(1)},\rho^{(2)},V,T) + k_{\mathrm{B}}T \int \mathrm{d}\mathbf{r}\, \rho(\mathbf{r}) \ln \Lambda^3 \overline{z}(\mathbf{r};\rho^{(1)},\rho^{(2)}) \\ &\quad + \frac{k_{\mathrm{B}}T}{2} \int \mathrm{d}\mathbf{r}\, \mathrm{d}\mathbf{s}\, \rho(\mathbf{r},\mathbf{s}) \ln \overline{e}(\mathbf{r},\mathbf{s};\rho^{(1)},\rho^{(2)}) \\ &= \mathcal{F}^{(1)}(\rho^{(1)},h,V,T) \\ &\quad + \frac{k_{\mathrm{B}}T}{2} \int \mathrm{d}\mathbf{r}\, \mathrm{d}\mathbf{s}\, \rho(\mathbf{r},\mathbf{s}) \ln \overline{e}(\mathbf{r},\mathbf{s};\rho^{(1)},\rho^{(2)}). \end{aligned} \tag{10.91}$$

Here $\mathcal{F}^{(1)}$ is the intrinsic Helmholtz free energy, which has the singlet potential removed; it was analysed several times above, $\mathcal{F}^{(1)} = \mathcal{F}^{\mathrm{id}} - \mathcal{F}^{\mathrm{ex}}$. The quantity $\mathcal{F}^{(2)}$ removes in addition the pair potential; it may be called the intrinsic pair Helmholtz free energy.

Cross–Entropy

In terms of the cross entropy, consider the functional maximised at equilibrium, and that has maximum value equal to the intrinsic pair Helmholtz free energy,

$$\begin{aligned}
&\beta\mathcal{F}_2'[\tilde{z},\tilde{e}|\rho^{(1)},\rho^{(2)}] \\
&= \beta\mathcal{F}^{(2)}(\rho^{(1)},\rho^{(2)},V,T) - S[\wp,\tilde{\wp}] \\
&= \beta F(\rho^{(1)},\rho^{(2)},V,T) - \sum_{N=0}^{\infty}\frac{1}{N!}\int \mathrm{d}\mathbf{r}^N\,\wp(\mathbf{r}^N)\ln\frac{\wp(\mathbf{r}^N)}{\tilde{\wp}(\mathbf{r}^N)} \\
&= \beta\mathcal{F}^{(2)}(\rho^{(1)},\rho^{(2)},V,T) - \sum_{N=0}^{\infty}\int\frac{\mathrm{d}\mathbf{r}^N}{N!}\wp(\mathbf{r}^N)\ln\frac{\tilde{\Xi}}{\Xi}\prod_{i=1}^{N}\frac{z(\mathbf{r}_i)}{\tilde{z}(\mathbf{r}_i)}\prod_{i<j}^{N}\frac{e(\mathbf{r}_i,\mathbf{r}_j)}{\tilde{e}(\mathbf{r}_i,\mathbf{r}_j)} \\
&= \beta\mathcal{F}^{(2)}(\rho^{(1)},\rho^{(2)},V,T) + \beta\Omega(\tilde{z},\tilde{e},V,T) - \beta\Omega(z,e,V,T) \\
&\quad - \int \mathrm{d}\mathbf{r}\,\rho(\mathbf{r})\ln\frac{z(\mathbf{r})}{\tilde{z}(\mathbf{r})} - \frac{1}{2}\int \mathrm{d}\mathbf{r}\,\mathrm{d}\mathbf{s}\,\rho(\mathbf{r},\mathbf{s})\ln\frac{e(\mathbf{r},\mathbf{s})}{\tilde{e}(\mathbf{r},\mathbf{s})}. \qquad (10.92)
\end{aligned}$$

Using the definition of $\mathcal{F}^{(2)}$ given in the penultimate equation,

$$\begin{aligned}
\mathcal{F}^{(2)}(\rho^{(1)},\rho^{(2)},V,T) &= \Omega(z,e,V,T) + k_{\mathrm{B}}T\int \mathrm{d}\mathbf{r}\,\rho(\mathbf{r})\ln\Lambda^3 z(\mathbf{r}) \\
&\quad + \frac{k_{\mathrm{B}}T}{2}\int \mathrm{d}\mathbf{r}\,\mathrm{d}\mathbf{s}\,\rho(\mathbf{r},\mathbf{s})\ln e(\mathbf{r},\mathbf{s}), \qquad (10.93)
\end{aligned}$$

where $\rho(\mathbf{r}) = \overline{\rho}(\mathbf{r}; z, e)$ and $\rho(\mathbf{r},\mathbf{s}) = \overline{\rho}(\mathbf{r},\mathbf{s}; z, e)$, this is

$$\begin{aligned}
\beta\mathcal{F}_2'[\tilde{z},\tilde{e}|\rho^{(1)},\rho^{(2)}] &= \beta\Omega(\tilde{z},\tilde{e},V,T) + \int \mathrm{d}\mathbf{r}\,\rho(\mathbf{r})\ln\Lambda^3\tilde{z}(\mathbf{r}) \\
&\quad + \frac{1}{2}\int \mathrm{d}\mathbf{r}\,\mathrm{d}\mathbf{s}\,\rho(\mathbf{r},\mathbf{s})\ln\tilde{e}(\mathbf{r},\mathbf{s}). \qquad (10.94)
\end{aligned}$$

In view of the singlet and pair potential functional given above, one concludes that $\mathcal{F}_2' \equiv \mathcal{F}^{(2)}$.

10.4 Perturbation Theory

Frequently one is interested in the perturbative effects of a weak but long-ranged potential. This can be an external potential applied to the system, or it may arise from the division of the pair potential into a short-range repulsive part (the reference system) and a long-range attractive tail. The mean field theory of Section 6.5 was applicable to just such a system. An alternative is perturbation theory, which has also been developed to handle these types of problems.

10.4.1 External Perturbation

Let the potential energy of the reference system consist of one- and two-body potentials,

$$U(\mathbf{r}^N) = \sum_{i=1}^{N} \psi_0^{\rm ext}(\mathbf{r}_1) + \sum_{i<j}^{N} u(\mathbf{r}_i, \mathbf{r}_j), \tag{10.95}$$

and suppose that a small change is made to the external potential,

$$\psi^{\rm ext}(\mathbf{r}) = \psi_0^{\rm ext}(\mathbf{r}) + \Delta\psi^{\rm ext}(\mathbf{r}). \tag{10.96}$$

Making a functional Taylor expansion, the change in the singlet density is

$$\begin{aligned} \rho(\mathbf{r}) - \rho_0(\mathbf{r}) &= \int \mathrm{d}\mathbf{s}\, \left.\frac{\delta\rho(\mathbf{r})}{\delta\beta\psi^{\rm ext}(\mathbf{s})}\right|_0 \left[\beta\psi^{\rm ext}(\mathbf{s}) - \beta\psi_0^{\rm ext}(\mathbf{s})\right] \\ &= -\int \mathrm{d}\mathbf{s}\, \rho_0^{(\delta)}(\mathbf{r}, \mathbf{s}) \left[\beta\psi^{\rm ext}(\mathbf{s}) - \beta\psi_0^{\rm ext}(\mathbf{s})\right], \end{aligned} \tag{10.97}$$

where Eq. (9.17) for the pair density, $\rho^{(\delta)}(\mathbf{r}, \mathbf{s}) = \rho(\mathbf{r})\delta(\mathbf{r}, \mathbf{s}) + \rho(\mathbf{r})\rho(\mathbf{s})h(\mathbf{r}, \mathbf{s})$, has been used. This says that $\rho^{(\delta)}$ is the linear response function linking the change in singlet density to the change in singlet potential. Note that this change in density occurs at constant chemical potential, and hence this expression is appropriate for an open system.

The mean potential at $\mathbf{r}$ is the sum of the external potential and the average pair potential at that position,

$$\psi^{\rm mean}(\mathbf{r}) = \psi^{\rm ext}(\mathbf{r}) + \int \mathrm{d}\mathbf{s}\, \rho(\mathbf{s}) u(\mathbf{r}, \mathbf{s}). \tag{10.98}$$

A similar definition holds for the mean potential of the reference system, with the subscript 0 appended everywhere. Just as $\rho^{(\delta)}$ gives the density response to the external potential, so may one define a function τ that gives the response to the mean potential,

$$\rho(\mathbf{r}) - \rho_0(\mathbf{r}) = \int \mathrm{d}\mathbf{s}\, \tau_0(\mathbf{r}, \mathbf{s}) \left[\beta\psi^{\rm mean}(\mathbf{s}) - \beta\psi_0^{\rm mean}(\mathbf{s})\right]. \tag{10.99}$$

The response function τ will turn out to be short-ranged and well behaved, in essence because the long-range tail of the pair potential causes changes in the far density by the external potential to affect the local density, whereas such far field effects are already incorporated into the mean potential and so it is only local changes in the latter that are important.

In the event that the reference system is homogeneous and isotropic, so that the pair functions only depend upon separation, these equations may be written in Fourier space

$$\Delta\hat{\rho}(\mathbf{k}) = -\beta\hat{\rho}_0^{(\delta)}(k)\Delta\hat{\psi}^{\rm ext}(\mathbf{k}), \tag{10.100}$$

$$\Delta\hat{\psi}^{\rm mean}(\mathbf{k}) = \Delta\hat{\psi}^{\rm ext}(\mathbf{k}) + \hat{u}(k)\Delta\hat{\rho}(\mathbf{k}), \tag{10.101}$$

and

$$\Delta\hat{\rho}(\mathbf{k}) = -\beta\hat{\tau}_0(k)\Delta\hat{\psi}^{\mathrm{mean}}(\mathbf{k}). \tag{10.102}$$

These may be solved to give the relationship between the two response functions,

$$\hat{\rho}_0^{(\delta)}(k) = \frac{\hat{\tau}_0(k)}{1+\beta\hat{u}(k)\hat{\tau}_0(k)}. \tag{10.103}$$

Comparing this to the Ornstein–Zernike equation,

$$\hat{\rho}_0^{(\delta)}(k) = \left[\hat{c}_0^{(\delta)}(k)\right]^{-1}, \tag{10.104}$$

where $c^{(\delta)}(r) = \delta(\mathbf{r})/\rho - c(r)$, one concludes that the mean potential response function is related to the direct correlation function by

$$\hat{\tau}_0(k)^{-1} = \hat{c}_0^{(\delta)}(k) - \beta\hat{u}(k) = \rho^{-1} - \left[\hat{c}_0(k) + \beta\hat{u}(k)\right]. \tag{10.105}$$

In so far as $c(r) \sim -\beta u(r)$, $r \to \infty$, one concludes from this that $\tau(r)$ is a short-ranged function, as asserted above.

Its worth mentioning that even if the reference system is not homogeneous and isotropic, one can show that

$$\rho_0^{(\delta)}(\mathbf{r},\mathbf{s}) = \tau_0(\mathbf{r},\mathbf{s}) - \beta\int \mathrm{d}\mathbf{t}\,\mathrm{d}\mathbf{v}\,\tau_0(\mathbf{r},\mathbf{t})u(\mathbf{t},\mathbf{v})\rho_0^{(\delta)}(\mathbf{v},\mathbf{s}). \tag{10.106}$$

10.4.2 Truncated Potential

Suppose now that the pair potential is split into a truncated short-range part and a long-range tail,

$$u(r) = u^{\mathrm{tr}}(r) + u^{\mathrm{tail}}(r). \tag{10.107}$$

The intention now is to use as the reference system the system interacting only with the truncated potential. The contribution from the tail will be treated as a mean potential and will be added to the external potential. A molecule at $\mathbf{r}$ in the truncated system with this augmented external potential feels the same mean potential as a molecule in the full system with the bare external potential. Hence the density profiles in the two systems are the same, $\rho_0^{\mathrm{tr}}(\mathbf{r}) = \rho_0(\mathbf{r})$, and similarly $\rho^{\mathrm{tr}}(\mathbf{r}) = \rho(\mathbf{r})$. Accordingly, the requisite augmented external potential is

$$\psi^{\mathrm{tail}}(\mathbf{r}) = \psi^{\mathrm{ext}}(\mathbf{r}) + \int \mathrm{d}\mathbf{s}\,u^{\mathrm{tail}}(\mathbf{r},\mathbf{s})\rho(\mathbf{s}), \tag{10.108}$$

and similarly for $\psi_0^{\mathrm{tail}}(\mathbf{r})$. Hence the change in this potential is

$$\Delta\psi^{\mathrm{tail}}(\mathbf{r}) = \Delta\psi^{\mathrm{ext}}(\mathbf{r}) + \int \mathrm{d}\mathbf{s}\,u^{\mathrm{tail}}(\mathbf{r},\mathbf{s})\Delta\rho(\mathbf{s}). \tag{10.109}$$

The density response function of the truncated system is $\rho_{\rm tr}^{(\delta)}$. A Taylor expansion for the change in density profile of the truncated system due to changing the augmented external potential therefore yields

$$\Delta\rho(\mathbf{r}) = -\beta \int \mathrm{d}\mathbf{s}\, \rho_{\rm tr}^{(\delta)}(\mathbf{r},\mathbf{s})\Delta\psi^{\rm tail}(\mathbf{s}). \tag{10.110}$$

For a homogeneous, isotropic system, these may be solved in Fourier space to yield

$$\begin{aligned} \Delta\hat{\rho}(\mathbf{k}) &= -\beta\hat{\rho}_{\rm tr}^{(\delta)}(k)\left[\Delta\psi^{\rm ext}(\mathbf{k}) + \beta\hat{u}^{\rm tail}(k)\Delta\hat{\rho}(\mathbf{k})\right] \\ &= \frac{-\beta\hat{\rho}_{\rm tr}^{(\delta)}(k)}{1+\beta\hat{u}^{\rm tail}(k)\hat{\rho}_{\rm tr}^{(\delta)}(k)}\Delta\psi^{\rm ext}(\mathbf{k}). \end{aligned} \tag{10.111}$$

Comparing this with the result obtained in the previous section, one can relate the pair density in the full and in the truncated systems,

$$\rho_0^{(\delta)}(k) = \frac{\hat{\rho}_{\rm tr}^{(\delta)}(k)}{1+\beta\hat{u}^{\rm tail}(k)\hat{\rho}_{\rm tr}^{(\delta)}(k)}. \tag{10.112}$$

For the case that the external field represents an atom fixed at the origin, the left-hand side is essentially the total correlation function of the system interacting with the full potential, and the numerator of the right-hand side is that of the system with truncated potential. This result for dealing with the perturbative effects on the pair correlation functions of long-range tails was given by Lado, who derived it from the hypernetted chain closure approximation.[6]

For the general system the result is

$$\rho_0^{(\delta)}(\mathbf{r},\mathbf{s}) = \rho_{\rm tr}^{(\delta)}(\mathbf{r},\mathbf{s}) - \beta\int \mathrm{d}\mathbf{t}\,\mathrm{d}\mathbf{v}\,\rho_{\rm tr}^{(\delta)}(\mathbf{r},\mathbf{t})u^{\rm tail}(\mathbf{t},\mathbf{v})\rho_0^{(\delta)}(\mathbf{v},\mathbf{s}). \tag{10.113}$$

This may be confirmed by successive substitution.

The mean potential is the same in the full system and in the truncated system with augmented external potential. For the full system it was given in the preceding section; for the truncated system it is

$$\Delta\psi^{\rm mean}(\mathbf{r}) = \Delta\psi^{\rm tail}(\mathbf{r}) + \int \mathrm{d}\mathbf{s}\, u^{\rm tr}(\mathbf{r},\mathbf{s})\Delta\rho(\mathbf{s}). \tag{10.114}$$

In the preceding section the density response function for the mean potential was defined as τ_0. Combining that definition with this representation of the mean potential, in Fourier space one obtains

$$\begin{aligned} \Delta\hat{\rho}(\mathbf{k}) &= -\beta\hat{\tau}_0(k)\Delta\psi^{\rm mean}(\mathbf{k}) \\ &= -\beta\hat{\tau}_0(k)\left[\Delta\psi^{\rm tail}(\mathbf{k}) + \beta\hat{u}^{\rm tr}(k)\Delta\hat{\rho}(\mathbf{k})\right] \\ &= \frac{-\beta\hat{\tau}_0(k)}{1+\beta\hat{u}^{\rm tr}(k)\hat{\tau}_0(k)}\Delta\psi^{\rm tail}(\mathbf{k}). \end{aligned} \tag{10.115}$$

[6] F. Lado, Perturbation correction to the radial distribution function, *Phys. Rev. A* **135** (1964), 1013.

Hence one concludes that

$$\hat{\rho}_{\rm tr}^{(\delta)}(k) = \frac{\hat{\tau}_0(k)}{1+\beta\hat{u}^{\rm tr}(k)\hat{\tau}_0(k)}, \tag{10.116}$$

or

$$\hat{\tau}_0(k) = \frac{\hat{\rho}_{\rm tr}^{(\delta)}(k)}{1-\beta\hat{u}^{\rm tr}(k)\hat{\rho}_{\rm tr}^{(\delta)}(k)}. \tag{10.117}$$

This is to say that the quantity τ depends only upon the properties of the truncated system; it is entirely insensitive to the tail of the pair potential. In view of the relationship between τ and the nontrivial part of the direct correlation function given above, it follows that the latter is also independent of the long-range tail of the pair potential and that it can be obtained from a truncated system.

10.4.3 Perturbation Grand Potential

The truncated potential is often taken to be the short-range, repulsive part of the potential, for example, a hard-sphere potential. The grand potential of such a reference system may be known, and one is interested in the change in it due to turning on the tail of the potential. If the tail is characterised by a coupling constant, $u^{\rm tail}(r;\lambda)$, with $\lambda = 0$ corresponding to the reference system, and $\lambda = 1$ corresponding to the full system, then in the thermodynamic limit when the thermodynamic potential is equal to the logarithm of the partition function, the grand potential for the partially coupled system is given by

$$\Omega(\lambda) = -k_{\rm B}T \ln \sum_{N=0}^{\infty} \frac{e^{\beta\mu N}}{\Lambda^{3N} N!} \int {\rm d}\mathbf{r}^N \, e^{-\beta U^{\rm tr}(\mathbf{r}^N)} e^{-\beta U^{\rm tail}(\mathbf{r}^N;\lambda)}. \tag{10.118}$$

The reason that the grand potential is sought rather than the Helmholtz free energy is that the expressions for the change in density found above were at constant chemical potential (i.e., an open system). Differentiating this yields the average of the tail potential,

$$\frac{\partial \Omega(\lambda)}{\partial \lambda} = \left\langle \dot{U}^{\rm tail}(\lambda) \right\rangle_\lambda, \tag{10.119}$$

where the dot signifies differentiation with respect to λ. Hence the change in grand potential due to the partially coupled tail is

$$\begin{aligned}
\Omega(\lambda) - \Omega^{\rm tr} &= \int_0^\lambda {\rm d}\lambda' \left\langle \dot{U}^{\rm tail}(\lambda') \right\rangle_{\lambda'} \\
&= \frac{1}{2}\int_0^\lambda {\rm d}\lambda' \int {\rm d}\mathbf{r}\,{\rm d}\mathbf{s}\, \rho^{(2)}(\mathbf{r},\mathbf{s};\lambda')\dot{u}^{\rm tail}(\mathbf{r},\mathbf{s};\lambda') \\
&= \frac{1}{2}\int_0^\lambda {\rm d}\lambda' \int {\rm d}\mathbf{r}\,{\rm d}\mathbf{s}\, \rho(\mathbf{r};\lambda')\rho(\mathbf{s};\lambda')\dot{u}^{\rm tail}(\mathbf{r},\mathbf{s};\lambda') \\
&\quad + \frac{1}{2}\int_0^\lambda {\rm d}\lambda' \int {\rm d}\mathbf{r}\,{\rm d}\mathbf{s}\, \rho^{(\delta)}(\mathbf{r},\mathbf{s};\lambda')\dot{u}^{\rm tail}(\mathbf{r},\mathbf{s};\lambda').
\end{aligned} \tag{10.120}$$

(The self-interaction, which is the Dirac δ introduced in the final line, should strictly be subtracted. However, it does not contribute here because the tail of the potential is 0 when the molecules occupy the same position.) In view of this one sees that the grand potential is the sum of a mean term and a correlation term, $\Omega(\lambda) - \Omega^{\rm tr} = \Omega^{\rm tail}_{\rm mean}(\lambda) + \Omega^{\rm tail}_{\rm corr}(\lambda)$.

Mean Contribution

In order to evaluate the mean grand potential, an expression for the density of the partially coupled system, $\rho(\mathbf{r};\lambda)$, is required. In the previous section the density was the same in the full and in the truncated system because in the latter the external potential included the mean tail of the pair potential. Now hold the external potential fixed and find the change in the singlet density as the coupling of the tail is changed. Let $\rho_{\rm tr}(\mathbf{r})$ be the density in the truncated system for the external potential $\psi^{\rm ext}(\mathbf{r})$. The mean potential due to the tail and the external potential in the partially coupled system is

$$\psi^{\rm tail}(\mathbf{r};\lambda) = \psi^{\rm ext}(\mathbf{r}) + \int {\rm d}\mathbf{s}\, u^{\rm tail}(\mathbf{r},\mathbf{s};\lambda)\rho(\mathbf{s};\lambda). \tag{10.121}$$

Now if in the truncated system the external potential $\psi^{\rm ext}(\mathbf{r})$ were replaced by $\psi^{\rm tail}(\mathbf{r};\lambda)$, then the density would change from $\rho_{\rm tr}(\mathbf{r})$ to $\rho(\mathbf{r};\lambda)$. Hence using the density response function of the truncated reference system one has

$$\begin{aligned} \rho(\mathbf{r};\lambda) - \rho_{\rm tr}(\mathbf{r}) &= -\beta \int {\rm d}\mathbf{s}\, \rho^{(\delta)}_{\rm tr}(\mathbf{r},\mathbf{s}) \left[\psi^{\rm tail}(\mathbf{r};\lambda) - \psi^{\rm ext}(\mathbf{r})\right] \\ &= -\beta \int {\rm d}\mathbf{s}\,{\rm d}\mathbf{t}\, \rho^{(\delta)}_{\rm tr}(\mathbf{r},\mathbf{s}) u^{\rm tail}(\mathbf{s},\mathbf{t};\lambda)\rho(\mathbf{t};\lambda). \end{aligned} \tag{10.122}$$

For an isotropic system, in Fourier space this is

$$\begin{aligned} \hat\rho(\mathbf{k};\lambda) &= \hat\rho_{\rm tr}(\mathbf{k}) - \beta\hat\rho^{(\delta)}_{\rm tr}(k)\hat u^{\rm tail}(k;\lambda)\hat\rho(\mathbf{k};\lambda) \\ &= \frac{\hat\rho_{\rm tr}(\mathbf{k})}{1+\beta\hat\rho^{(\delta)}_{\rm tr}(k)\hat u^{\rm tail}(k;\lambda)}. \end{aligned} \tag{10.123}$$

For a given density of the truncated system this enables the change in density to be obtained as the tail coupling is turned on.

The mean part of the grand potential due to the partial tail is

$$\begin{aligned} &\Omega^{\rm tail}_{\rm mean}(\lambda) \\ &= \frac{1}{2}\int_0^\lambda {\rm d}\lambda' \int {\rm d}\mathbf{r}\,{\rm d}\mathbf{s}\, \rho(\mathbf{r};\lambda')\dot u^{\rm tail}(\mathbf{r},\mathbf{s};\lambda')\rho(\mathbf{s};\lambda') \\ &= \frac{1}{2(2\pi)^3}\int {\rm d}\mathbf{k} \int_0^\lambda {\rm d}\lambda'\, \hat\rho(\mathbf{k};\lambda')\hat\rho(-\mathbf{k};\lambda')\frac{\partial}{\partial\lambda'}\hat u^{\rm tail}(k;\lambda') \\ &= \frac{1}{2(2\pi)^3}\int {\rm d}\mathbf{k} \int_0^\lambda {\rm d}\lambda'\, \frac{\hat\rho_{\rm tr}(\mathbf{k})\hat\rho_{\rm tr}(-\mathbf{k})}{\left[1+\beta\hat\rho^{(\delta)}_{\rm tr}(k)\hat u^{\rm tail}(k;\lambda')\right]^2}\frac{\partial}{\partial\lambda'}\hat u^{\rm tail}(k;\lambda') \end{aligned}$$

$$
\begin{aligned}
&= \frac{1}{2(2\pi)^3}\int \mathrm{d}\mathbf{k}\frac{\hat{\rho}_{\mathrm{tr}}(\mathbf{k})\hat{\rho}_{\mathrm{tr}}(-\mathbf{k})}{\beta\hat{\rho}^{(\delta)}_{\mathrm{tr}}(k)}\left.\frac{-1}{1+\beta\hat{\rho}^{(\delta)}_{\mathrm{tr}}(k)\hat{u}^{\mathrm{tail}}(k;\lambda')}\right|_0^{\lambda} \\
&= \frac{1}{2(2\pi)^3}\int \mathrm{d}\mathbf{k}\frac{\hat{\rho}_{\mathrm{tr}}(\mathbf{k})\hat{u}^{\mathrm{tail}}(k;\lambda)\hat{\rho}_{\mathrm{tr}}(-\mathbf{k})}{1+\beta\hat{\rho}^{(\delta)}_{\mathrm{tr}}(k)\hat{u}^{\mathrm{tail}}(k;\lambda)}. \qquad (10.124)
\end{aligned}
$$

Correlation Contribution

Using the result established in the preceding section that related the full pair density to the pair density of the truncated system, Eq. (10.112), the correlation contribution to the grand potential is

$$
\begin{aligned}
&\Omega^{\mathrm{tail}}_{\mathrm{corr}}(\lambda) \\
&= \frac{1}{2(2\pi)^3}\int \mathrm{d}\mathbf{k}\int_0^{\lambda}\mathrm{d}\lambda'\,\hat{\rho}^{(\delta)}(k;\lambda')\frac{\partial}{\partial\lambda'}\hat{u}^{\mathrm{tail}}(k;\lambda') \\
&= \frac{1}{2(2\pi)^3}\int \mathrm{d}\mathbf{k}\int_0^{\lambda}\mathrm{d}\lambda'\,\frac{\hat{\rho}^{(\delta)}_{\mathrm{tr}}(k)}{1+\beta\hat{\rho}^{(\delta)}_{\mathrm{tr}}(k)\hat{u}^{\mathrm{tail}}(k;\lambda')}\frac{\partial}{\partial\lambda'}\hat{u}^{\mathrm{tail}}(k;\lambda') \\
&= \frac{k_{\mathrm{B}}T}{2(2\pi)^3}\int \mathrm{d}\mathbf{k}\ln\left[1+\beta\hat{\rho}^{(\delta)}_{\mathrm{tr}}(k)\hat{u}^{\mathrm{tail}}(k;\lambda)\right]. \qquad (10.125)
\end{aligned}
$$

This result is in essential details similar to that obtained by Lado.[7] Note the close relationship with the ring diagrams used in obtaining the hypernetted chain grand potential above (cf. Eq. (10.67)). This indicates that the present approach is equivalent to including all orders of the many-body correlation function of the reference system, approximating them by superpositions of pair correlation functions. In contrast, conventional perturbation theories represent truncation of the expansion at the pair term.[8]

Helmholtz Free Energy

At equilibrium, the Helmholtz free energy is related to the the grand potential by $F(\overline{N}) = \Omega(\mu) + \mu\overline{N}$. Hence the change in Helmholtz free energy due to turning on the tail of the pair potential is

$$
F(\lambda) - F^{\mathrm{tr}} = \Omega(\lambda) - \Omega^{\mathrm{tr}} + \mu_{\mathrm{tr}}\int \mathrm{d}\mathbf{r}\,\left[\rho(\mathbf{r};\lambda) - \rho_{\mathrm{tr}}(\mathbf{r})\right], \qquad (10.126)
$$

where the integrand was given above in Eq. (10.122). Since the chemical potential is constant it may be evaluated in the truncated system.

[7] F. Lado, Perturbation correction for the free energy and structure of simple fluids, *Phys. Rev. A* **8** (1973), 2548.

[8] J. A. Barker and D. Henderson, Perturbation theory and equation of state for fluids. II. A successful theory of liquids, *J. Chem. Phys.* **47** (1967), 4714. J. D. Weeks, D. Chandler, and H. C. Andersen, Role of repulsive forces in determining the equilibrium structure of simple liquids, *J. Chem. Phys.* **54** (1971), 5237. J.-P. Hansen and I. R. McDonald, *Theory of Simple Liquids*, 2nd ed., Academic Press, London, 1986.

10.4.4 Interaction Grand Potential

A particularly interesting and useful application of perturbation theory is to the interaction between two systems, for example, the interaction between two separated bodies, or between two fluids separated by a membrane. The systems do not have to be spatially separated; the theory applies as well to two weakly interacting species of a given system. In this case the tail potential can be considered to be the pair potential between the two systems, and the truncated part corresponds to the two bodies isolated from each other. One introduces a coupling parameter λ, which could correspond to the separation between the bodies. The potential energy may be written $U(\lambda) = U_1 + U_2 + U^{\rm int}(\lambda)$, where the first two terms on the right-hand side refer to the two isolated bodies. One is interested in that part of the thermodynamic potential due to the coupling between the bodies. The interaction grand potential is the difference in grand potential between the coupled and the decoupled system. In the case that one is dealing with separated systems, the interaction grand potential goes to 0 at large separations.

The analysis is essentially identical to that given above for the tail of the potential, with the result that the interaction grand potential is

$$\begin{aligned}
\Omega^{\rm int}(\lambda) &= \int_0^\lambda {\rm d}\lambda' \left\langle \dot{U}^{\rm int}(\lambda') \right\rangle_{\lambda'} \\
&= \int_0^\lambda {\rm d}\lambda' \int_1 {\rm d}\mathbf{r} \int_2 {\rm d}\mathbf{s}\, \rho_{12}^{(2)}(\mathbf{r},\mathbf{s};\lambda') \dot{u}^{\rm int}(\mathbf{r},\mathbf{s};\lambda') \\
&= \int_0^\lambda {\rm d}\lambda' \int_1 {\rm d}\mathbf{r} \int_2 {\rm d}\mathbf{s}\, \rho_1(\mathbf{r};\lambda')\rho_2(\mathbf{s};\lambda') \dot{u}^{\rm int}(\mathbf{r},\mathbf{s};\lambda') \\
&\quad + \int_0^\lambda {\rm d}\lambda' \int_1 {\rm d}\mathbf{r} \int_2 {\rm d}\mathbf{s}\, \rho_{12}^{(\delta)}(\mathbf{r},\mathbf{s};\lambda') \dot{u}^{\rm int}(\mathbf{r},\mathbf{s};\lambda'),
\end{aligned} \tag{10.127}$$

where the subscript 12 indicates that this is the cross pair density. (The Dirac δ does not contribute here because the species linked in the cross term are distinct.) In view of this one sees that the interaction grand potential is the sum of a mean term and a correlation term, $\Omega^{\rm int}(\lambda) = \Omega^{\rm int}_{\rm mean}(\lambda) + \Omega^{\rm int}_{\rm corr}(\lambda)$.

Mean Contribution

The mean term may vanish, as, for example, occurs with interacting charge systems due to electro-neutrality. More generally, however, the singlet density is nonzero, and the interaction potential contributes to the external potential. As in the preceding section the singlet density may be written

$$\begin{aligned}
\rho_1(\mathbf{r};\lambda) &= \rho_1(\mathbf{r};0) - \beta \int_1 {\rm d}\mathbf{s}\, \rho_0^{(\delta)}(\mathbf{r},\mathbf{s}) \left[\psi_1^{\rm tail}(\mathbf{s};\lambda) - \psi_1^{\rm tail}(\mathbf{s};0)\right] \\
&= \rho_1(\mathbf{r};0) - \beta \int_1 {\rm d}\mathbf{s} \int_2 {\rm d}\mathbf{t}\, \rho_0^{(\delta)}(\mathbf{r},\mathbf{s}) u^{\rm int}(\mathbf{s},\mathbf{t};\lambda)\rho_2(\mathbf{t};\lambda).
\end{aligned} \tag{10.128}$$

Here $\rho_0^{(\delta)}(\mathbf{r},\mathbf{s}) \equiv \rho_{11}^{(\delta)}(\mathbf{r},\mathbf{s};0)$ is the same-body pair density for the isolated or uncoupled body. An analogous equation holds for $\rho_2(\mathbf{r};\lambda)$. However, for simplicity only the symmetric case, $\rho_2 = \rho_1$ and $\rho_{22}^{(\delta)} = \rho_{11}^{(\delta)}$ will henceforth be considered. Using a dot to denote an integrated coordinate the density equation may be written symbolically as

$$\rho_1(\mathbf{r};\lambda) = \rho_1(\mathbf{r};0) - \beta\rho_0^{(\delta)}(\mathbf{r},\cdot)u^{\text{int}}(\cdot,\cdot;\lambda)\rho_1(\cdot;\lambda), \tag{10.129}$$

with solution

$$\rho_1(\mathbf{r};\lambda) = \delta(\mathbf{r},\cdot)\frac{\rho_1(\cdot;0)}{1+\beta\rho_0^{(\delta)}(\cdot,\cdot)u^{\text{int}}(\cdot,\cdot;\lambda)}. \tag{10.130}$$

The last equation acquires meaning as a series expansion where the products correspond to iterated convolution integrals.

Using this result, the mean contribution to the interaction grand potential is

$$\begin{aligned}
&\Omega^{\text{int}}_{\text{mean}}(\lambda)\\
&= \int_0^\lambda \mathrm{d}\lambda'\, \rho_1(\cdot;\lambda')\dot{u}^{\text{int}}(\cdot,\cdot;\lambda')\rho_1(\cdot;\lambda')\\
&= \int_0^\lambda \mathrm{d}\lambda'\, \frac{\rho_1(\cdot;0)}{1+\beta\rho_0^{(\delta)}(\cdot,\cdot)u^{\text{int}}(\cdot,\cdot;\lambda')}\dot{u}^{\text{int}}(\cdot,\cdot;\lambda')\frac{\rho_1(\cdot;0)}{1+\beta\rho_0^{(\delta)}(\cdot,\cdot)u^{\text{int}}(\cdot,\cdot;\lambda')}\\
&= \int_0^\lambda \mathrm{d}\lambda'\, \frac{-\rho_1(\cdot;0)^2}{\beta\rho_0^{(\delta)}(\cdot,\cdot)}\frac{\partial}{\partial\lambda'}\frac{1}{1+\beta\rho_0^{(\delta)}(\cdot,\cdot)u^{\text{int}}(\cdot,\cdot;\lambda')}\\
&= \left.\frac{-1}{\beta\rho_0^{(\delta)}(\cdot,\cdot)}\frac{\rho_1(\cdot;0)^2}{1+\beta\rho_0^{(\delta)}(\cdot,\cdot)u^{\text{int}}(\cdot,\cdot;\lambda')}\right|_0^\lambda\\
&= \frac{\rho_1(\cdot;0)u^{\text{int}}(\cdot,\cdot;\lambda)\rho_1(\cdot;0)}{1+\beta\rho_0^{(\delta)}(\cdot,\cdot)u^{\text{int}}(\cdot,\cdot;\lambda)}.
\end{aligned} \tag{10.131}$$

This is the contribution to the interaction grand potential ignoring the correlations between the two surfaces (although these are included in the calculation of $\rho_1(\cdot;\lambda)$.

In the event that one can apply the Fourier transform, this result can be written in closed form. For the case of two interacting planar bodies, assuming homogeneity, $\rho_1(\mathbf{r}) = \rho_1$, and isotropy, $\rho_0^{(\delta)}(\mathbf{r},\mathbf{s}) = \rho_0^{(\delta)}(|\mathbf{r}-\mathbf{s}|)$, the density equation may be written

$$\begin{aligned}
\rho_1(\lambda) &= \rho_1(0) - \beta\rho_1(\lambda)\int \mathrm{d}\mathbf{s}\int \mathrm{d}\mathbf{t}\, \rho_0^{(\delta)}(s)u^{\text{int}}(|\mathbf{s}-\mathbf{t}|;\lambda)\\
&= \rho_1(0) - \beta\rho_1(\lambda)\hat{\rho}_0^{(\delta)}(0)\hat{u}^{\text{int}}(0;\lambda)\\
&= \frac{\rho_1(0)}{1+\beta\hat{\rho}_0^{(\delta)}(0)\hat{u}^{\text{int}}(0;\lambda)}.
\end{aligned} \tag{10.132}$$

Hence the mean interaction grand potential per unit area is

$$\begin{aligned}
\Omega^{\rm int}_{\rm mean}(\lambda)/A
&= \int_0^\lambda {\rm d}\lambda' \int {\rm d}\mathbf{r}\, \rho_1(\lambda')^2 \dot{u}^{\rm int}(r;\lambda') \\
&= \int_0^\lambda {\rm d}\lambda' \frac{\rho_1(0)^2}{\left[1+\beta\hat{\rho}_0^{(\delta)}(0)\hat{u}^{\rm int}(0;\lambda')\right]^2} \frac{\partial}{\partial\lambda'}\hat{u}^{\rm int}(0;\lambda') \\
&= \frac{-\rho_1(0)^2}{\beta\hat{\rho}_0^{(\delta)}(0)} \left.\frac{1}{1+\beta\hat{\rho}_0^{(\delta)}(0)\hat{u}^{\rm int}(0;\lambda')}\right|_0^\lambda \\
&= \frac{\rho_1(0)\hat{u}^{\rm int}(0;\lambda)\rho_1(0)}{1+\beta\hat{\rho}_0^{(\delta)}(0)\hat{u}^{\rm int}(0;\lambda)}. \qquad (10.133)
\end{aligned}$$

This could also have been written down directly from the general result.

The net pressure between the walls is the negative derivative of the interaction grand potential per unit area. Assuming that λ represents the separation, the mean contribution is

$$\begin{aligned}
p^{\rm net}_{\rm mean}(\lambda) &\equiv \frac{-1}{A}\frac{\partial\Omega^{\rm int}_{\rm mean}(\lambda)}{\partial\lambda} \\
&= \frac{-\rho_1(0)^2}{\left[1+\beta\hat{\rho}_0^{(\delta)}(0)\hat{u}^{\rm int}(0;\lambda)\right]^2}\frac{\partial\hat{u}^{\rm int}(0;\lambda)}{\partial\lambda} \\
&= -\rho_1(\lambda)^2\frac{\partial\hat{u}^{\rm int}(0;\lambda)}{\partial\lambda} \\
&= -\rho_1(\lambda)^2\int {\rm d}\mathbf{r}\,\frac{\partial u^{\rm int}(r;\lambda)}{\partial\lambda}. \qquad (10.134)
\end{aligned}$$

This is just the mean force per unit area between the walls, as one would expect.

Correlation Contribution

The cross pair density $\rho_{12}^{(\delta)}$ was defined above in connection with the correlation contribution to the average of the derivative of the interaction potential, as were the same-system pair densities, $\rho_{11}^{(\delta)}$ and $\rho_{22}^{(\delta)}$. Again attention is restricted to the symmetric case, so that the latter two are equal. In the uncoupled system only the same pair density is nonzero, and it may be denoted by $\rho_0^{(\delta)}$. Using Eq. (10.113) above, one has two equations for the pair densities,

$$\begin{aligned}
&\rho_{11}^{(\delta)}(\mathbf{r},\mathbf{s};\lambda) \\
&\quad = \rho_0^{(\delta)}(\mathbf{r},\mathbf{s}) - \beta\int {\rm d}\mathbf{t}\,{\rm d}\mathbf{v}\,\rho_0^{(\delta)}(\mathbf{r},\mathbf{t})u^{\rm int}(\mathbf{t},\mathbf{v};\lambda)\rho_{12}^{(\delta)}(\mathbf{v},\mathbf{s};\lambda) \qquad (10.135)
\end{aligned}$$

and

$$\rho_{12}^{(\delta)}(\mathbf{r},\mathbf{s};\lambda) = -\beta\int {\rm d}\mathbf{t}\,{\rm d}\mathbf{v}\,\rho_0^{(\delta)}(\mathbf{r},\mathbf{t})u^{\rm int}(\mathbf{t},\mathbf{v};\lambda)\rho_{11}^{(\delta)}(\mathbf{v},\mathbf{s};\lambda). \qquad (10.136)$$

These represent two linked integral equations for the coupled pair correlation functions that only require as input the pair correlation function of the decoupled system. Again denoting an integration by a dot, these may be written symbolically as

$$\rho_{11}^{(\delta)}(\mathbf{r},\mathbf{s};\lambda) = \rho_0^{(\delta)}(\mathbf{r},\mathbf{s}) - \beta\rho_0^{(\delta)}(\mathbf{r},\cdot)u^{\rm int}(\cdot,\cdot;\lambda)\rho_{12}^{(\delta)}(\cdot,\mathbf{s};\lambda) \tag{10.137}$$

and

$$\rho_{12}^{(\delta)}(\mathbf{r},\mathbf{s};\lambda) = -\beta\rho_0^{(\delta)}(\mathbf{r},\cdot)u^{\rm int}(\cdot,\cdot;\lambda)\rho_{11}^{(\delta)}(\cdot,\mathbf{s};\lambda), \tag{10.138}$$

with solution

$$\rho_{11}^{(\delta)}(\mathbf{r},\mathbf{s};\lambda) = \delta(\mathbf{r},\cdot)\frac{\rho_0^{(\delta)}(\cdot,\cdot)}{1-\left[\beta\rho_0^{(\delta)}(\cdot,\cdot)u^{\rm int}(\cdot,\cdot;\lambda)\right]^2}\delta(\cdot,\mathbf{s}) \tag{10.139}$$

and

$$\rho_{12}^{(\delta)}(\mathbf{r},\mathbf{s};\lambda) = \delta(\mathbf{r},\cdot)\frac{-\beta\rho_0^{(\delta)}(\cdot,\cdot)u^{\rm int}(\cdot,\cdot;\lambda)\rho_0^{(\delta)}(\cdot,\cdot)}{1-\left[\beta\rho_0^{(\delta)}(\cdot,\cdot)u^{\rm int}(\cdot,\cdot;\lambda)\right]^2}\delta(\cdot,\mathbf{s}). \tag{10.140}$$

Accordingly, the coupling constant integration may be performed to give the correlation part of the interaction grand potential as

$$\begin{aligned}\Omega_{\rm corr}^{\rm int}(\lambda) &= \int_0^\lambda {\rm d}\lambda'\,\rho_{12}^{(\delta)}(\cdot,\cdot;\lambda')\dot{u}^{\rm int}(\cdot,\cdot;\lambda') \\ &= \int_0^\lambda {\rm d}\lambda'\,\frac{-\beta\rho_0^{(\delta)}(\cdot,\cdot)u^{\rm int}(\cdot,\cdot;\lambda')\rho_0^{(\delta)}(\cdot,\cdot)}{1-\left[\beta\rho_0^{(\delta)}(\cdot,\cdot)u^{\rm int}(\cdot,\cdot;\lambda')\right]^2}\frac{\partial}{\partial\lambda'}u^{\rm int}(\cdot,\cdot;\lambda') \\ &= \frac{k_{\rm B}T}{2}\ln\left(1-\left[\beta\rho_0^{(\delta)}(\cdot,\cdot)u^{\rm int}(\cdot,\cdot;\lambda')\right]^2\right).\end{aligned} \tag{10.141}$$

The logarithm is defined by its series expansion. Again one can recognise these integrals as essentially the ring diagrams defined in Eq. (10.67), and one can see that one has effectively performed a superposition approximation for the many-body, same-surface correlation functions.

A closed form for this result may be obtained when one can invoke the Fourier transform. For two interacting planar surfaces, the two-dimensional Fourier transform of the equations for the same and cross pair densities yield

$$\hat{\rho}_{11}^{(\delta)}(k;\lambda) = \hat{\rho}_0^{(\delta)}(k) - \beta\hat{\rho}_0^{(\delta)}(k)\hat{u}^{\rm int}(k;\lambda)\hat{\rho}_{12}^{(\delta)}(k;\lambda) \tag{10.142}$$

and

$$\hat{\rho}_{12}^{(\delta)}(k;\lambda) = -\beta\hat{\rho}_0^{(\delta)}(k)\hat{u}^{\rm int}(k;\lambda)\hat{\rho}_{11}^{(\delta)}(k;\lambda), \tag{10.143}$$

assuming homogeneity and isotropy. Hence the cross pair density is

$$\hat{\rho}_{12}^{(\delta)}(k;\lambda) = \frac{-\beta\hat{\rho}_0^{(\delta)}(k)^2\hat{u}^{\rm int}(k;\lambda)}{1-\beta^2\hat{\rho}_0^{(\delta)}(k)^2\hat{u}^{\rm int}(k;\lambda)^2}. \tag{10.144}$$

Using Parsevaal's theorem, the interaction grand potential per unit area becomes

$$\begin{aligned}
\Omega^{\rm int}_{\rm corr}(\lambda)/A &= \frac{1}{(2\pi)^2}\int_0^\lambda {\rm d}\lambda' \int {\rm d}\mathbf{k}\, \hat\rho^{(\delta)}_{12}(k;\lambda')\frac{\partial}{\partial\lambda'}\hat u^{\rm int}(k;\lambda') \\
&= \frac{1}{(2\pi)^2}\int {\rm d}\mathbf{k} \int_0^\lambda {\rm d}\lambda' \left[1-\beta^2\hat\rho^{(\delta)}_0(k)^2\hat u^{\rm int}(k;\lambda')^2\right]^{-1} \\
&\quad \times \frac{\partial}{2\partial\lambda'}\left[-\beta\hat\rho^{(\delta)}_0(k)^2\hat u^{\rm int}(k;\lambda')^2\right] \\
&= \frac{k_{\rm B}T}{2(2\pi)^2}\int {\rm d}\mathbf{k}\,\ln\left[1-\beta^2\hat\rho^{(\delta)}_0(k)^2\hat u^{\rm int}(k;\lambda)^2\right]. \qquad (10.145)
\end{aligned}$$

As mentioned above, the net pressure between these two interacting walls is an important quantity. If λ is the separation, the correlation contribution is the negative derivative of the correlation part of the interaction grand potential per unit area,

$$\begin{aligned}
p^{\rm net}_{\rm corr}(\lambda) &\equiv \frac{-1}{A}\frac{\partial\Omega^{\rm int}_{\rm corr}(\lambda)}{\partial\lambda} \\
&= \frac{-1}{(2\pi)^2}\int {\rm d}\mathbf{k}\,\hat\rho^{(\delta)}_{12}(k;\lambda)\frac{\partial\hat u^{\rm int}(k;\lambda)}{\partial\lambda} \\
&= -\int {\rm d}\mathbf{r}\,\rho^{(\delta)}_{12}(r;\lambda)\frac{\partial u^{\rm int}(r;\lambda)}{\partial\lambda}. \qquad (10.146)
\end{aligned}$$

The integrand is the density-weighted normal component of the force between two molecules, as one would expect on physical grounds. This final result holds in general and could have been written down by directly differentiating Eq. (10.127); this would give the total net force between the bodies rather than the pressure.

Summary

- A nonequilibrium subsystem may be considered to be in equilibrium with a meta-reservoir, and the standard equilibrium results may be applied to the meta-system. The total entropy of the constrained state is the equilibrium entropy of the subsystem and meta-reservoir, plus the difference between the entropies of the actual reservoir and the meta-reservoir.

- The meta-reservoir gives rise to a density functional minimised by the equilibrium density profile. Its minimum value is the grand potential, which may be expressed explicitly in terms of the intrinsic Helmholtz free energy density functional. The density functional may also be expressed implicitly in terms of the cross-entropy.

- Various approximation schemes for the intrinsic Helmholtz free energy density functional, including using the pair direct correlation function of the bulk, which is equivalent to the hypernetted chain approximation, and of local and various weighted densities, as well as perturbative approaches, have been developed.

- A functional maximised by the equilibrium activity corresponding to the specified density may also be obtained, explicitly from the meta-reservoir or implicitly from the cross-entropy. Its maximum value is the intrinsic Helmholtz free energy.

- Variational principles for the pair total correlation function and for the pair potential may also be derived. Retaining only the ring diagrams in the former is equivalent to the hypernetted chain approximation, and hence an expression for the thermodynamic potential follows in that approximation.

- Perturbation theory is based upon the linear density response to a weak field. The latter can be the external field, the mean field of the tail of the pair potential, or the interaction with another body. The coupling integral can be performed to give the change in the density and in the grand potential due to the tail or interaction. The response function for the mean field (equivalently the short-range part of the direct correlation function) is independent of the long-range tail.

Chapter 11

Inhomogeneous Systems

Nonuniform systems arise due to the presence of external fields, such as those due to gravity, or to an applied electric potential. Such density inhomogeneities also occur in the vicinity of the walls of the container of the system, at the surfaces of cavities or inclusions, or around and between solute particles. The density functional theory described in the preceding chapter represents a variational approach to inhomogeneous systems based upon constrained entropy maximisation. This chapter explores alternative approaches based upon the Ornstein–Zernike equation and the approximate closures derived in Ch. 9. There are basically two ways of proceeding: one can regard the inhomogeneity as due to a solute particle and solve the Ornstein–Zernike equation for a uniform mixture, or one can invoke the inhomogeneous Ornstein–Zernike equation. The first part of this chapter is concerned with the former approach, using the so-called singlet Ornstein–Zernike equation; the latter technique is discussed in the concluding sections.

11.1 Spherical Inhomogeneity

11.1.1 Ornstein–Zernike Equation

The Ornstein–Zernike equation for a multicomponent mixture was given in Eq. (9.33). In the case of a uniform, isotropic system this is

$$h_{\alpha\gamma}(r_{12}) = c_{\alpha\gamma}(r_{12}) + \sum_{\lambda} \rho_\lambda \int \mathrm{d}\mathbf{r}_3\, h_{\alpha\lambda}(r_{13})\, c_{\lambda\gamma}(r_{32}), \tag{11.1}$$

where the Greek indices label the species.

Consider one of the species, say $\alpha = 0$, to be a solute at infinite dilution, $\rho_0 = 0$. The remaining species, $\alpha > 0$, comprise the solvent. In this case one has

$$h_{\alpha\gamma}(r_{12}) = c_{\alpha\gamma}(r_{12}) + \sum_{\lambda>0} \rho_\lambda \int \mathrm{d}\mathbf{r}_3\, h_{\alpha\lambda}(r_{13})\, c_{\lambda\gamma}(r_{32}). \tag{11.2}$$

For $\alpha > 0$, $\gamma > 0$, only solvent pair functions appear, and this equation is just that of the pure solvent in the absence of the solute. For $\alpha = 0$, $\gamma > 0$ (or vice versa), solute–solvent and solvent–solvent correlation functions are present. The quantity $\rho_\gamma[1+h_{0\gamma}(r)]$ is the density of species γ a distance r from the spherical solute. Obtaining the density inhomogeneity from the solute Ornstein–Zernike equation is called a singlet method because the solute–solvent total correlation function is equivalent to the (singlet) density profile.

For $\alpha = \gamma = 0$ this is an equation for solute–solute correlation functions in terms of the solute–solvent ones. The logarithm of $h_{00}(r)$ is essentially the potential of mean force or constrained thermodynamic potential for two solutes separated by r.

For the common case of a single-component solvent, one has the solvent–solvent Ornstein–Zernike equation,

$$h_{11}(r_{12}) = c_{11}(r_{12}) + \rho_1 \int \mathrm{d}\mathbf{r}_3\, h_{11}(r_{13})\, c_{11}(r_{32}), \tag{11.3}$$

the solute–solvent equation,

$$h_{01}(r_{12}) = c_{01}(r_{12}) + \rho_1 \int \mathrm{d}\mathbf{r}_3\, h_{01}(r_{13})\, c_{11}(r_{32}), \tag{11.4}$$

and the solute–solute equation,

$$h_{00}(r_{12}) = c_{00}(r_{12}) + \rho_1 \int \mathrm{d}\mathbf{r}_3\, h_{01}(r_{13})\, c_{01}(r_{32}). \tag{11.5}$$

By successive substitution in the solute–solvent Ornstein–Zernike equation, one sees that the solute–solvent total correlation function is the sum of chains of solvent–solvent direct correlation functions, terminated by a solute–solvent direct correlation function,

$$h_{01} = c_{01} + c_{01} * c_{11} + c_{01} * c_{11} * c_{11} + c_{01} * c_{11} * c_{11} * c_{11} + \ldots. \tag{11.6}$$

Hence it makes no difference whether the integrand of the solute–solvent equation is written $h_{01} * c_{11}$ or $c_{01} * h_{11}$. Similarly, the solute–solute equation may be written

$$h_{00} = c_{00} + c_{01} * c_{10} + c_{01} * c_{11} * c_{10} + c_{01} * c_{11} * c_{11} * c_{10} + \ldots. \tag{11.7}$$

This is an important point: the Ornstein–Zernike series function shows that the indirect solute–solvent and solute–solute correlations propagate through solvent–solvent correlations unaffected by the presence of the solute.

11.1.2 Closure

The solute–solvent potential may be denoted $V_{0\alpha}(r)$. Any spherical inhomogeneity can be represented in this fashion. A quite common system is that of a spherical cavity of radius R,

$$V_{0\alpha}(r) = \begin{cases} \infty, & r < R + d_\alpha/2, \\ 0, & r > R + d_\alpha/2, \end{cases} \tag{11.8}$$

where $d_\alpha/2$ represents the distance of closest approach of the centre of the solvent of type α to the surface of the cavity. As in the hard-sphere systems discussed in Section 9.5.1, the solute–solvent density vanishes inside the cavity, $g_{0\alpha}(r) = 0$, or $h_{0\alpha}(r) = -1$, $r < R + d_\alpha/2$. One can can add to this core part of the solute potential other spherically symmetric potentials, such as the Coulomb potential, or one can replace it with, for example, a Lennard–Jones potential.

In this example of a spherical cavity, the interaction potential between two such solutes would be represented by

$$V_{00}(r) = \begin{cases} \infty, & r < 2R, \\ 0, & r > 2R. \end{cases} \tag{11.9}$$

Again one may add to this other solute–solute potentials.

The formally exact closure equation has been given previously, Eq. (9.43). For a mixture it reads

$$h_{\alpha\gamma}(r) = -1 + e^{-\beta u_{\alpha\gamma}(r)} e^{s_{\alpha\gamma}(r) + d_{\alpha\gamma}(r)}, \tag{11.10}$$

where $s_{\alpha\gamma} = h_{\alpha\gamma} - c_{\alpha\gamma}$ is the series function and $d_{\alpha\gamma}$ is the bridge function. Hence the solute–solvent closure is just this with $\alpha = 0$, and $u_{0\gamma} = V_{0\gamma}$, and the solute–solute closure has $\alpha = \gamma = 0$, and $u_{00} = V_{00}$.

The closure approximations that one can make are the same as for a simple system. The hypernetted chain approximation neglects the bridge function, $d_{\alpha\gamma}^{(2)} = 0$, the Percus–Yevick approximates it as a sum of powers of the series function, and the mean spherical approximation invokes the asymptotic behaviour of the direct correlation function.

One generally uses the same closure for all the pair correlation functions, although there is in fact no logical necessity to do this. For example, in some cases there may be some advantage to using the Percus–Yevick bulk total correlation functions in the solute–solvent Ornstein–Zernike equation, together with the hypernetted chain closure for the solute–solvent correlation functions.

11.1.3 Numerical Solution

As in the case of the homogeneous, isotropic single-component system, one solves the present solute equations by using a Fourier transform to factorise the Ornstein–Zernike equation. For a mixture a matrix formulation is convenient, and according to Eq. (9.37) one has

$$\underline{\underline{\hat{H}}}(k) = \underline{\underline{\hat{C}}}(k) \left(\underline{\underline{I}} - \underline{\underline{\hat{C}}}(k) \right)^{-1}. \tag{11.11}$$

In the present context this is the solvent–solvent equation, with the components of the matrices being $\{\underline{\underline{H}}\}_{\alpha\gamma} = \rho_\alpha^{1/2} h_{\alpha\gamma} \rho_\gamma^{1/2}$ and $\{\underline{\underline{C}}\}_{\alpha\gamma} = \rho_\alpha^{1/2} c_{\alpha\gamma} \rho_\gamma^{1/2}$, for $\alpha > 0$ and $\gamma > 0$.

The usual iterative algorithm for solving the solvent equations, and, for example, the hypernetted chain closure, is to Fourier transform the nth iterate of the direct correlation function, $C^{(n)} \Rightarrow \hat{C}^{(n)}$, and to use matrix inversion and multiplication to obtain $\hat{H}^{(n)}$ from the above equation. One can inverse transform this, $\hat{H}^{(n)} \Rightarrow H^{(n)}$, and form the series function $S^{(n)} = H^{(n)} - C^{(n)}$ (or form this in Fourier space before transforming), which upon subtraction of the pair potential and exponentiation of the components gives the next iterate of the components of the total correlation function, $H^{(n+1)}$. This may be combined with the series function to give the new direct correlation function, $C^{(n+1)} = H^{(n+1)} - S^{(n)}$. Quite often one mixes the new and old iterates to improve the reliability of the convergence of this simple Picard scheme. Nonlinear methods based upon the Jacobean of the iteration procedure have been used successfully to accelerate the procedure and to make it more robust.[1]

In the case of the solute–solvent correlations, one can define vectors with components $\{\underline{H}\}_\alpha = \rho_\alpha^{1/2} h_{0\alpha}$ and $\{\underline{C}\}_\alpha = \rho_\alpha^{1/2} c_{0\alpha}$. In this notation the Fourier transform of the solute–solvent Ornstein–Zernike equation is

$$\begin{aligned} \underline{\hat{H}}(k) &= \underline{\hat{C}}(k) + \underline{\underline{\hat{C}}}(k)\underline{\hat{H}}(k) \\ &= \left(\underline{\underline{I}} - \underline{\underline{\hat{C}}}(k)\right)^{-1} \underline{\hat{C}}(k). \end{aligned} \tag{11.12}$$

One can either solve this equation in parallel with the solvent–solvent equations or simply invoke a converged solution for the solvent direct correlation function matrix that appears. One still must iterate even in the latter case because both the direct and the total solute–solvent pair correlation functions appear here. The iteration procedure is identical to the solvent case with successive transformations between Fourier and real space.

For the solute–solute equation one has

$$\begin{aligned} \hat{h}_{00}(k) &= \hat{c}_{00}(k) + \underline{\hat{C}}(k)^{\mathrm{T}} \underline{\hat{H}}(k) \\ &= \hat{c}_{00}(k) + \underline{\hat{C}}(k)^{\mathrm{T}} \left(\underline{\underline{I}} - \underline{\underline{\hat{C}}}(k)\right)^{-1} \underline{\hat{C}}(k). \end{aligned} \tag{11.13}$$

In this case one can solve for the solute–solute series function directly without iteration, since the remainder depends only upon solute–solvent functions, which have presumably already been obtained. The series function may be used to obtain the solute–solute potential of mean force within the hypernetted chain or Percus–Yevick approximation, and thence $h_{00}(r)$.

Cavities in a Hard-Sphere Solvent

Figure 11.1 shows the solute–solvent distribution function for a hard-sphere solvent in the vicinity of a cavity. The density profiles are simply this multiplied by the bulk density, $\rho_1(r) = \rho_1 g_{01}(r; R)$. It can be seen that the oscillations in the distribution function have the same period as those in the pure solvent, which corresponds to the case $D = d$. These oscillations arise from the packing of the

[1] M. J. Booth, A. G. Schlijper, L. E. Scales, and A. D. J. Haymet, Efficient solution of liquid state integral equations using the Newton-GMRES algorithm, *Comp. Phys. Comm.* **119** (1999), 122.

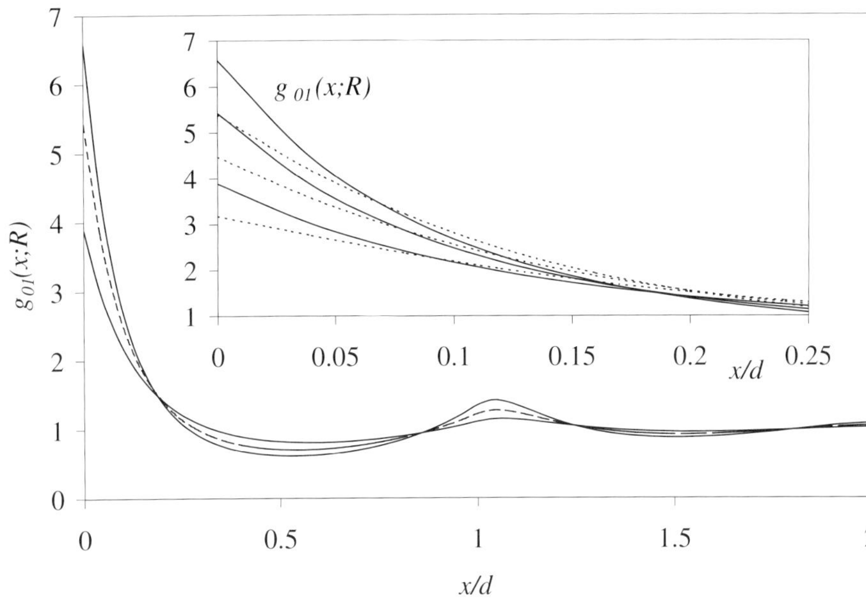

Figure 11.1: The solute–solvent radial distribution function for a hard cavity in a hard-sphere solvent. The density is $\rho d^3 = 0.7$, and the hypernetted chain approximation has been used. From bottom to top at contact the diameters of the solute are $D = 1$, 3, and $10d$, which correspond to a distance of closest approach of the solute and solvent centers of 1, 2, and $5.5d$, respectively. The inset shows on an expanded scale the effect of including the bridge function (dotted curves). The abscissa measures the distance from contact $x = r - R - d/2$.

solvent molecules; separations that correspond to an integral number of solvent layers have enhanced probability, whereas half-integral layers are unfavourable. The density at contact increases with increasing cavity diameter, as does the amplitude of the oscillations. As the curvature decreases, the propensity for solvent layering increases, which magnifies the oscillations. The inset shows the effect of including a Padé approximate to the bridge function consisting of all solvent–solvent and solute–solvent bridge diagrams of second and third order in density.[2] The bridge function tends to decrease the contact value given by the bare hypernetted chain approximation.

Figure 11.2 shows the distribution of two hard cavities in a hard-sphere solvent. Compared to the solute–solvent case, the solute–solute distribution has a much higher value at contact for the same size cavity. There is a peak in the distribution at $x = d$, which is quite marked for the largest cavity shown. This

[2] P. Attard and G.N. Patey, Hypernetted-chain closure with bridge diagrams. Asymmetric hard-sphere mixtures, *J. Chem. Phys.* **92** (1990), 4970.

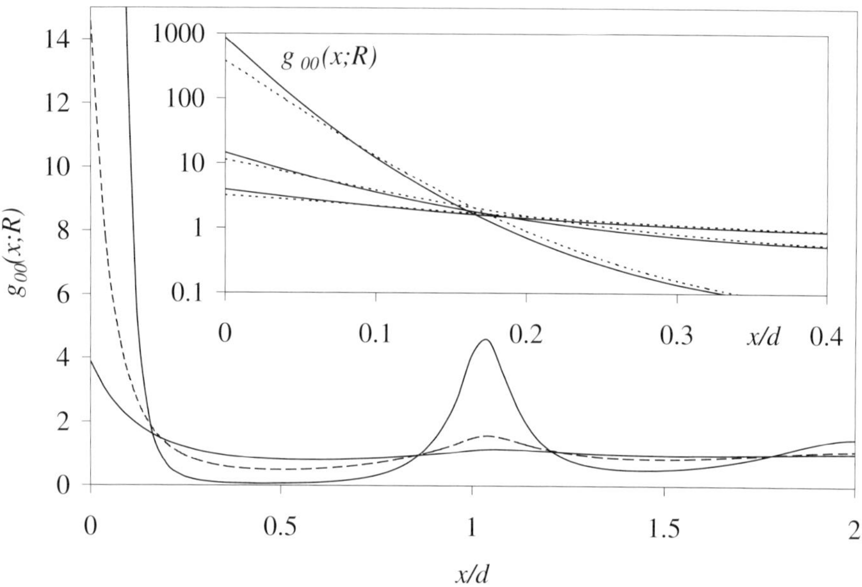

Figure 11.2: The solute–solute radial distribution function for hard cavities in a hard-sphere solvent at a density of $\rho d^3 = 0.7$, as given by the hypernetted chain approximation. The abscissa measures the surface separation. From bottom to top at contact the diameters of the solute are $D = 1$, 3, and $10d$. The inset shows on a log scale the effect of including the bridge function (dotted curves).

corresponds to being able to pack a single layer of solvent hard spheres between the surfaces of the cavities. Conversely, there is almost 0 probability of finding the two large cavities at a surface separation of $x = 0.5d$. This separation would correspond to a vacuum between the cavities for some region about the central axis, since one cannot fit a hard sphere between their surfaces in this case.

One can give an analytic approximation for solute–solute potential of mean force near contact based upon the pressure–volume work required to create such a cavity, $w_{00}(x) = -pV$. From simple geometry the volume of the exclusion region as a function of separation is $V = \pi[(D+x)^3/12 - (D+x)(D+d)^2/4 + (D+d)^3/6]$. The exclusion volume is 0 at $x = d$. For the pressure, Asakura and Oosawa,[3] after whom this depletion attraction is named, used that of the ideal gas, $p = \rho k_B T$. Although this is very inaccurate for the pressure of the present dense hard-sphere solvent, due to a cancellation of errors the consequent solute–solute potential of mean force itself is not too bad.[4] For solute diameters of $D = 1$, 3, and 10, for the potential of mean force at contact, this approxi-

[3]S. Asakura and F. Oosawa, On interaction between two bodies immersed in a solution of macromolecules, *J. Chem. Phys.* **22** (1958), 1255.

[4]P. Attard, Spherically inhomogeneous fluids. II. Hard-sphere solute in a hard-sphere solvent, *J. Chem. Phys.* **91** (1989), 3083.

mation gives $\beta w_{00}(0) =$ -0.92, -2.02, and -5.86, and the hypernetted chain with bridge function gives -1.16, -2.43, and -5.94, respectively. This is surprisingly good considering the simplicity of the approximation and demonstrates clearly the principle driving force for the adhesion of the cavities. The solute–solute distribution function is simply $g_{00}(x) = \exp -\beta w_{00}(x)$. The Asakura–Oosawa approximation hence explains the very steep increase in the contact value of the solute–solute distribution function with solute diameter, since the approximation predicts that it will scale with the exponential of the diameter cubed.

11.1.4 Asymptotics

In Section 9.6.3 the asymptotic behaviour of the total correlation functions of a solvent mixture was derived from the Ornstein–Zernike equation, using the fact that the direct correlation function decayed like the pair potential, $c_{\alpha\gamma}(r) \sim -\beta u_{\alpha\gamma}(r)$, for a power law potential, $u_{\alpha\gamma}(r) \sim a_{\alpha\gamma} r^{-\eta}$, $r \to \infty$, $\eta > d$, where d is the spatial dimension. For the case of the solute–solvent equation, similar analysis yields

$$\begin{aligned} \underline{H}(r) &\sim \underline{C}(r) + \underline{\underline{H}}(r)\underline{\hat{C}}(0) + \underline{\underline{\hat{H}}}(0)\underline{C}(r) \\ &= \left[\underline{\underline{I}} + \underline{\underline{\hat{H}}}(0)\right] \underline{C}(r) + \underline{\underline{H}}(r)\underline{\hat{C}}(0). \end{aligned} \tag{11.14}$$

Since one still has the general result that $c_{0\alpha}(r) \sim -\beta u_{0\alpha}(r)$, $r \to \infty$, which of the two terms dominates depends upon the relative rates of decay of the solute–solvent and the solvent–solvent potentials. When the latter is more long-ranged (e.g., for a cavity in a Lennard–Jones solvent), one has

$$\begin{aligned} \underline{H}(r) &\sim \underline{\underline{H}}(r)\underline{\hat{C}}(0) \\ &= \left[\underline{\underline{I}} - \underline{\underline{\hat{C}}}(0)\right]^{-1} \underline{\underline{C}}(r)\underline{\hat{H}}(0) \\ &= \left[\underline{\underline{I}} + \underline{\underline{\hat{H}}}(0)\right] \underline{\underline{C}}(r) \left[\underline{\underline{I}} + \underline{\underline{\hat{H}}}(0)\right] \underline{\hat{C}}(0). \end{aligned} \tag{11.15}$$

In this case the functional form of the asymptotic decay is determined by the solvent correlation functions, and the solute only enters through the modulation of the amplitude. Far from the solute the rate of decay of the solvent density to its bulk value is dominated by the indirect solvent–solvent correlations.

The opposite case that the solute–solvent potential is more long-ranged than the solvent–solvent potential is less realistic but nevertheless one has

$$\underline{H}(r) \sim \left[\underline{\underline{I}} + \underline{\underline{\hat{H}}}(0)\right] \underline{C}(r). \tag{11.16}$$

In this case the direct solute–solvent interactions determine the asymptotic decay of the solvent density. Often the solute interacts with the solvent in a fashion similar to that of other solvent molecules, and both terms contribute to the asymptotic decay. As for the bulk solvent case, there is a unique rate of decay shared by all of the components of the solvent density profile. This is due to the mixing of the components by the matrix multiplications, which means that the most long-ranged asymptote (out of the solute–solvent potentials $u_{0\alpha}(r)$ and the solvent–solvent potentials $u_{\alpha\gamma}(r)$) dominates.

For the case of the solute–solute interaction one has

$$\begin{aligned} h_{00}(r) - c_{00}(r) &\sim \underline{\hat{C}}(0)^{\mathrm{T}} \underline{H}(r) + \underline{C}(r)^{\mathrm{T}} \underline{\hat{H}}(0) \\ &= \underline{\hat{C}}(0)^{\mathrm{T}} \left[\underline{\underline{I}} - \underline{\underline{\hat{C}}}(0) \right]^{-1} \underline{C}(r) + \underline{\hat{C}}(0)^{\mathrm{T}} \underline{\underline{H}}(r) \underline{\hat{C}}(0) \\ &\quad + \underline{\hat{C}}(r)^{\mathrm{T}} \left[\underline{\underline{I}} - \underline{\underline{\hat{C}}}(0) \right]^{-1} \underline{C}(0). \end{aligned} \tag{11.17}$$

For the reasons discussed in Section 9.6, for integrable power law potentials the bridge function is more rapidly decaying than the potential itself, and consequently the potential of mean force is dominated by the potential and the series function. Hence the solute–solute potential of mean force goes like $w_{00}(r) \sim u_{00}(r) - k_{\mathrm{B}}T\,[h_{00}(r) - c_{00}(r)]$, $r \to \infty$. The asymptotic decay is determined by the most long-ranged of $u_{00}(r)$, $u_{0\alpha}(r)$, and $u_{\alpha\gamma}(r)$.

11.1.5 Near Criticality

The solute asymptotes just found depend upon $\underline{\underline{\hat{H}}}(0)$. As discussed in Section 9.6, the elements of the latter diverge at criticality, and hence the preceding asymptotic analysis does not hold at the critical point or along the spinodal line, although some of the general features recur. The critical analysis is simplest in the first instance for a single-component solvent. Assuming the classical Ornstein–Zernike form,

$$h_{11}(r) \sim \frac{Ae^{-\xi r}}{4\pi r}, \; r \to \infty, \tag{11.18}$$

the Fourier transform of the solvent–solvent total correlation function is

$$\hat{h}_{11}(k) \sim \frac{A}{k^2 + \xi^2}, \; k \to 0, \tag{11.19}$$

with the decay length diverging at the critical point, $\xi^{-1} \to \infty$, $\rho \to \rho_{\mathrm{c}}$, $T \to T_{\mathrm{c}}^{+}$. As mentioned in Section 9.6, for the compressibility to be positive one must have $A > 0$. The corresponding solvent–solvent direct correlation function is

$$\hat{c}_{11}(k) \sim \frac{A}{k^2 + \xi^2 + \rho_1 A}, \; k \to 0. \tag{11.20}$$

Now add to the near-critical solvent a solute particle. The solute–solvent Ornstein–Zernike equation gives

$$\begin{aligned} \hat{h}_{01}(k) &= \frac{\hat{c}_{01}(k)}{1 - \rho_1 \hat{c}_{11}(k)} \\ &\sim \hat{c}_{01}(i\xi) \frac{\rho_1 A}{k^2 + \xi^2}, \; k \to 0. \end{aligned} \tag{11.21}$$

Hence the solute–solvent total correlation function has a simple pole at $k = i\xi$, which means that the density profile about the solute decays to the bulk value at the same rate as the correlations in the bulk solvent,

$$h_{01}(r) \sim \rho_1 \hat{c}_{01}(i\xi) \frac{Ae^{-\xi r}}{4\pi r}, \quad r \to \infty. \tag{11.22}$$

Although $\hat{c}_{01}(k)$ may have a singularity closer to the origin than $k = i\xi$, in which case $h_{01}(r)$ and $c_{01}(r)$ would decay at the same rate (this is certainly the case for a Lennard–Jones or other power law solvent), one is most interested in the intermediate regime dominated by the divergence of the bulk correlation length approaching criticality and this gives the pertinent asymptotic behaviour of the solute.

The solute–solute Ornstein–Zernike equation gives

$$\begin{aligned} \hat{h}_{00}(k) &= \hat{c}_{00}(k) + \hat{c}_{01}(k)\rho_1 \hat{h}_{10}(k) \\ &\sim \frac{\rho_1^2 \hat{c}_{01}(i\xi)^2 A}{k^2 + \xi^2}, \quad k \to 0, \end{aligned} \tag{11.23}$$

again assuming that $\hat{c}_{00}(k)$ is relatively short-ranged. This shows that the interaction between solutes has the same range as the correlation length of the bulk solvent,

$$h_{00}(r) \sim [\rho_1 \hat{c}_{01}(i\xi)]^2 \frac{Ae^{-\xi r}}{4\pi r}, \quad r \to \infty, \tag{11.24}$$

and is therefore in a sense universal. Moreover, the interaction is attractive (because the numerator is positive). This last point is of interest because it says that whatever the interaction of the solutes with the solvent, whether they adsorb or desorb the solvent, in the vicinity of the solvent's critical point or line of spinodal decomposition the solutes are attracted to each other and there is a strong driving force for aggregation. Again this represents a certain universality, although the strength of the effect does depend upon the magnitude of $\hat{c}_{01}(i\xi) \approx \hat{c}_{01}(0)$.

Mixture

Now consider the multicomponent solvent. The asymptotics of a near-critical bulk mixture were analysed in Section 9.6.3, where it was shown that the solvent–solvent total correlation function decayed as

$$h_{\alpha\gamma}(r) \sim x_\alpha x_\gamma \frac{Ae^{-\xi r}}{4\pi r}, \quad r \to \infty, \tag{11.25}$$

with $A > 0$. The vector of the dyad that appears here, $\underline{x}$, is the eigenvector of the solvent–solvent direct correlation matrix at $k = i\xi$ with eigenvalue

$$\lambda(k) \sim \frac{A}{A + k^2 + \xi^2}, \quad k \to i\xi. \tag{11.26}$$

Denoting the projection on $\underline{x}$ of the vector of the solute–solvent total and direct correlation functions by $\hat{h}_x(k)$, and $\hat{c}_x(k)$, respectively, and using the fact that $\underline{\underline{\hat{C}}}(k)$ is diagonal in the basis of its eigenvectors, then the $\underline{x}$ component of the solute–solvent Ornstein–Zernike equation is

$$\hat{h}_x(k) = \frac{1}{1-\lambda(k)}\hat{c}_x(k) \sim \frac{A\hat{c}_x(i\xi)}{k^2+\xi^2},\ k \to i\xi. \tag{11.27}$$

The remaining projections are analytic here, and since $k = i\xi$ is the singularity closest to the origin (assuming that $\underline{C}(r)$ is relatively short-ranged), this gives as asymptote

$$\underline{H}(r) \sim \hat{c}_x(i\xi)\frac{Ae^{-\xi r}}{4\pi r}\underline{x},\ r \to \infty. \tag{11.28}$$

The solute–solute equation yields

$$\begin{aligned}\hat{h}_{00}(k) &= \hat{c}_{00}(k) + \underline{\hat{C}}(k)^{\mathrm{T}}\left[\underline{\underline{I}} - \underline{\underline{\hat{C}}}(k)\right]^{-1}\underline{\hat{C}}(k) \\ &\sim \hat{c}_x(i\xi)\frac{A}{k^2+\xi^2}\hat{c}_x(i\xi),\ k \to i\xi.\end{aligned} \tag{11.29}$$

Hence

$$h_{00}(r) \sim \hat{c}_x(i\xi)^2\frac{Ae^{-\xi r}}{4\pi r},\ r \to \infty. \tag{11.30}$$

As expected, this is positive, indicating a long-range attraction between the solutes in the vicinity of the critical point.

11.1.6 Solvation Free Energy

The general expression for the excess chemical potential for a single-component system was given above as Eq. (9.101). One can repeat the analysis for a multicomponent system by summing all the interactions between the partially coupled particle and the solvent species,

$$\langle -\beta\mu_\alpha^{\mathrm{ex}}\rangle = \sum_\gamma \rho_\gamma \int_V \mathrm{d}\mathbf{r}\left[h_{\alpha\gamma}(r) - v_{\alpha\gamma}(r) - h_{\alpha\gamma}(r)\sum_{n=1}^{\infty}\frac{n v_{\alpha\gamma}^{(n)}(r)}{n+1}\right]. \tag{11.31}$$

Since the excess chemical potential represents the difference in the thermodynamic potential of a system with and without a particle fixed at the origin, in the present case it represents the solvation free energy of the spherical solute, $W_0^{\mathrm{ex}} = -\mu_0^{\mathrm{ex}}$. One has the exact expression

$$\beta W_0^{\mathrm{ex}} = \sum_{\gamma>0}\rho_\gamma \int_V \mathrm{d}\mathbf{r}\left[h_{0\gamma}(r) - v_{0\gamma}^{(1)}(r) - h_{0\gamma}(r)\sum_{n=1}^{\infty}\frac{n v_{0\gamma}^{(n)}(r)}{n+1}\right]. \tag{11.32}$$

For the case of the hypernetted chain closure for the solute–solvent correlation functions, $v_{0\gamma}^{(n)} = 0$, $n \geq 2$, the solvation free energy is

$$\beta W_0^{\mathrm{ex,HNC}} = \sum_{\gamma>0}\rho_\gamma \int_V \mathrm{d}\mathbf{r}\,[c_{0\gamma}(r) - h_{0\gamma}(r)\{h_{0\gamma}(r) - c_{0\gamma}(r)\}/2]. \tag{11.33}$$

The Percus–Yevick solvation free energy is

$$\beta W_0^{\mathrm{ex,PY}} = \sum_{\gamma>0} \rho_\gamma \int_V \mathrm{d}\mathbf{r} \frac{c_{0\gamma}(r)}{h_{0\gamma}(r) - c_{0\gamma}(r)} \ln\left[1 + h_{0\gamma}(r) - c_{0\gamma}(r)\right]. \quad (11.34)$$

11.2 Planar Walls

11.2.1 Isolated Wall

Planar one-dimensional inhomogeneities are relatively common, as much from ease of analysis due to the high symmetry as from actual physical applicability. The density profile in the vicinity of a planar wall may be obtained by taking the large R limit of the spherical cavity discussed in the preceding section.[5] Defining a cylindrical coordinate system with its origin on the surface of the solute, and with the z coordinate normal to the surface, the solute–solvent Ornstein–Zernike equation may be written

$$\begin{aligned} & h_{0\alpha}(r;R) \\ & \quad = c_{0\alpha}(r;R) + 2\pi \sum_{\gamma>0} \rho_\gamma \int_{-\infty}^{\infty} \mathrm{d}z \int_0^\infty \mathrm{d}s\, s \\ & \qquad \times h_{0\gamma}(\sqrt{(R+z)^2+s^2};R)\, c_{\gamma\alpha}(\sqrt{(R+z-r)^2+s^2}), \end{aligned} \quad (11.35)$$

where the dependence of the solute–solvent correlations on the size of the solute has been shown explicitly.

Provided that the solute–solvent potential remains bounded in the large radius limit, then the solute–solvent correlation functions also remain well behaved and reduce to functions only of the distance from the interface. Hence again measuring from the surface, $r = R + y$, one may define the wall–solvent correlation functions as

$$\lim_{R\to\infty} h_{0\alpha}(R+y;R) = h_{0\alpha}(y), \quad \lim_{R\to\infty} c_{0\alpha}(R+y;R) = c_{0\alpha}(y). \quad (11.36)$$

Further, because $c_{\gamma\alpha}(t) \to 0$, $t \to \infty$, the Ornstein–Zernike convolution integral is dominated by regions $s \ll R$ and $z \approx r - R$. Provided that attention is restricted to regions of the solvent in the vicinity of the surface, $r \approx R$, this last condition implies that $|z| \ll R$, which means that $\sqrt{(R+z)^2+s^2} \sim R + z + \mathcal{O}(R^{-1})$, $R \to \infty$. Hence the limiting wall–solvent Ornstein–Zernike equation is

$$\begin{aligned} h_{0\alpha}(y) &= c_{0\alpha}(y) + 2\pi \sum_{\gamma>0} \rho_\gamma \int_{-\infty}^{\infty} \mathrm{d}z \int_0^\infty \mathrm{d}s\, s \\ &\qquad \times h_{0\gamma}(z)\, c_{\gamma\alpha}(\sqrt{(z-y)^2+s^2}). \end{aligned} \quad (11.37)$$

[5] D. Henderson, F. F. Abraham, and J. A. Barker, The Ornstein–Zernike equation for a fluid in contact with a surface, *Mol. Phys.* **31** (1976), 1291.

For a hard wall one has

$$h_{0\alpha}(y) = -1, \quad y < d_\alpha/2, \tag{11.38}$$

where $d_\alpha/2$ is the distance of closest approach of the solvent molecule to the wall. Now the z integral above is dominated by regions $z \approx y$ (because $c_{\alpha\gamma}$ is short-ranged), and one has

$$\begin{aligned}\lim_{y\to-\infty} c_{0\alpha}(y) &= -1 + 2\pi \sum_{\gamma>0} \rho_\gamma \int_{-\infty}^{\infty} \mathrm{d}z \int_0^\infty \mathrm{d}s\, s c_{\gamma\alpha}(\sqrt{(z-y)^2+s^2}) \\ &= -1 + \sum_{\gamma>0} \rho_\gamma \hat{c}_{\gamma\alpha}(0).\end{aligned} \tag{11.39}$$

For a single-component system, Eq. (9.91) shows that the right-hand side is the inverse of the isothermal compressibility. More generally for a mixture, Eq. (9.92) shows that

$$\lim_{y\to-\infty} \sum_{\alpha>0} \rho_\alpha c_{0\alpha}(y) = -(k_\mathrm{B} T \chi_T)^{-1}. \tag{11.40}$$

The fact that neither $h_{0\alpha}(y)$ nor $c_{0\alpha}(y)$ decay to 0 as $y \to -\infty$ means that their Fourier transforms do not exist (at least not numerically; they can be defined in terms of generalised functions for analytic work). One would like to use the fast Fourier transform to evaluate the convolution integral above, which reduces the number of operations from N^2 to $N \log N$, where $N \sim \mathcal{O}2^{10}$ is the number of grid points. One way to proceed is to treat the core contribution separately by rewriting the wall–solvent equation as

$$h_{0\alpha}(y) = c_{0\alpha}(y) + H_\alpha(y) + C_\alpha(y). \tag{11.41}$$

Here the contribution of the sovent direct correlation function to the Ornstein–Zernike convolution integral from within the wall ($h_{0\gamma}(y) = -1$, $y < 0$) is

$$\begin{aligned}C_\alpha(y) &= -2\pi \sum_{\gamma>0} \rho_\gamma \int_{-\infty}^{0} \mathrm{d}z \int_0^\infty \mathrm{d}s\, s c_{\gamma\alpha}(\sqrt{(z-y)^2+s^2}) \\ &= -2\pi \sum_{\gamma>0} \rho_\gamma \int_y^\infty \mathrm{d}r\, r^2 c_{\gamma\alpha}(r)[1 - y/r], \quad y > 0.\end{aligned} \tag{11.42}$$

Of course since this only depends upon the bulk solvent direct correlation functions, which are presumed already known, one does not have to iterate it. The remaining convolution integral,

$$H_\alpha(y) = 2\pi \sum_{\gamma>0} \rho_\gamma \int_0^\infty \mathrm{d}z \int_0^\infty \mathrm{d}s\, s h_{0\gamma}(z) c_{\gamma\alpha}(\sqrt{(z-y)^2+s^2}), \tag{11.43}$$

may, since $H_\alpha(y) \to 0$, $y \to \pm\infty$, be Fourier transformed,

$$\tilde{H}_\alpha(k) = \sum_{\gamma>0} \rho_\gamma \tilde{h}_{0\gamma}(k) \hat{c}_{\gamma\alpha}(k). \tag{11.44}$$

The one-dimensional Fourier transform pair is

$$\tilde{f}(k) = \int_{-\infty}^{\infty} \mathrm{d}y\, f(y) e^{iky}, \text{ and } f(y) = \frac{1}{2\pi} \int_{-\infty}^{\infty} \mathrm{d}k\, \tilde{f}(k) e^{-iky}. \tag{11.45}$$

Also, the three-dimensional Fourier transform of a radial function may be written

$$\hat{f}(k) = \frac{4\pi}{k} \int_0^{\infty} \mathrm{d}r\, r \sin(kr) f(r). \tag{11.46}$$

As for spherical solutes, one completes the wall–solvent Ornstein–Zernike equation with the usual closure approximations. The formally exact closure is just

$$h_{0\alpha}(y) = -1 + \exp\left[h_{0\alpha}(y) - c_{0\alpha}(y) + d_{0\alpha}(y) - \beta u_{0\alpha}(y)\right]. \tag{11.47}$$

The wall–solvent potential is 0 for $z < 0$, and one may have in addition a soft potential that extends from the wall into the solvent. The hypernetted chain approximation, for example, consists of the neglect of the wall–solvent bridge function, $d_{0\alpha}(y) = 0$. One solves this set of equations by iteration, successively transforming between Fourier and real space.

Bridge Diagram

For a bulk solvent it was mentioned in an earlier chapter that the hypernetted chain closure approximation could be improved by adding bridge diagrams, and that it was feasible to evaluate the first bridge diagram by Legendre transform. The Legendre transform of the Mayer f-function for a spherically symmetric solvent was given as Eq. (9.53). For the case of a planar solute, the first bridge diagram, resummed in terms of total correlation bonds and Legendre transformed, is

$$\begin{aligned} d_{0\alpha}^{(2)}(y_2) \quad &= \quad 2\pi^2 \sum_{\gamma\lambda} \rho_\gamma \rho_\lambda \sum_{n=0}^{\infty} \left(\frac{2}{2n+1}\right)^2 \int_0^{\infty} \mathrm{d}r_3\, r_3^2 \int_0^{\infty} \mathrm{d}r_4\, r_4^2\, h_{\gamma\alpha}(r_3) \\ &\times h_{\lambda\alpha}(r_4) \hat{h}_{0\gamma}^{(n)}(y_2, r_3) \hat{h}_{0\lambda}^{(n)}(y_2, r_4) \hat{h}_{\gamma\lambda}^{(n)}(r_3, r_4), \end{aligned} \tag{11.48}$$

where the Legendre coefficients of the bulk solvent–solvent total correlation functions are as given above for the Mayer f-functions, Eq. (9.53), and those of the wall–solvent total correlation function are

$$\hat{h}_{0\alpha}^{(n)}(y, r) = \frac{2n+1}{2} \int_{-1}^{1} \mathrm{d}x\, P_n(x) h_{0\alpha}(y - rx). \tag{11.49}$$

11.2.2 Interacting Walls

The above expression for the wall–solvent Ornstein–Zernike equation was derived as the large radius limit of the macrosphere–solvent equation. However, it could also have been written down directly from the diagrammatic definition of the series function. Similarly well defined is the wall–solvent bridge function, and one sees that the only role specific to the wall is in determining the appropriate geometry for the various integrals that appear.

In a similar vein, the results for two interacting walls also follow directly from the diagrammatic definitions, with two subtleties. From the symmetry of the problem most quantities scale with the cross-sectional area of the walls, and hence one generally deals with the energy per unit area, etc. In particular, the wall–wall series function per unit area is

$$b_{00}(x) = \sum_{\gamma>0} \rho_\gamma \int_{-\infty}^{\infty} \mathrm{d}s\, h_{0\gamma}(s) c_{0\gamma}(x-s), \tag{11.50}$$

where x is the separation between the surfaces of the walls. One generally avoids writing this as $h_{00}(x) - c_{00}(x)$ because neither of these two functions is defined, not even per unit area. This is because both contain terms that scale like the area to the power of the number of diagrams connected in parallel at the root points. In contrast, the root points do not form an articulation pair in either the series function or the bridge function, and hence they both scale linearly with the area.

The second subtlety in the above concerns the fact that the arguments of the wall–solvent correlation functions are signed quantities, which is in contrast to all the previous analysis for spherical solutes and solvents. It matters whether the solvent lies to the right of the wall as above, $h_{0\alpha}(z) = -1$, $z < 0$, or to the left, $h_{0'\alpha}(z) = -1$, $z > 0$. Hence these are two different types of solutes related by $h_{0'\alpha}(z) = h_{0\alpha}(-z)$, and similarly $c_{0'\alpha}(z) = c_{0\alpha}(-z)$. Two interacting walls represent the interaction of the two types of solutes, and hence the series function per unit area is, strictly speaking,

$$b_{00'}(x) = \sum_{\gamma>0} \rho_\gamma \int_{-\infty}^{\infty} \mathrm{d}s\, h_{0\gamma}(s) c_{\gamma 0'}(s-x). \tag{11.51}$$

Using the symmetry rule shows that this is equivalent to the form given above. It is not necessary to restrict attention to the symmetric system, depending upon the soft part of the wall–solvent potential.

The series function may be calculated directly from the wall–solvent correlation functions. It is evident from the form of the convolution integral that the integrand goes to 0 as $s \to \pm\infty$. However, evaluating the integral by Fourier transform is fraught with danger because neither of the two functions goes to 0 inside the wall. If one wishes to proceed down that path one must subtract these constants before transformation and add their contribution separately. Note also that the wall–solvent direct correlation function was only given for positive values of its argument in the preceding section (since negative arguments were not required in the iteration procedure). One must evaluate the bulk solvent contribution $C_\alpha(y)$ for $y < 0$ in order to obtain $c_{0\gamma}(y)$ for negative arguments.

In view of the existence of all quantities, the wall–wall potential of mean force per unit area is

$$\beta w^{\text{int}}(x) = \beta u_{00}(x) - b_{00}(x) - d_{00}(x), \tag{11.52}$$

where u and d are the wall–wall potential and bridge function, per unit area, respectively. This quantity is also called the interaction free energy per unit area; it decays to 0 as the separation goes to infinity. Since the series function may be calculated directly from the wall–solvent correlation functions, and given some approximation for the bridge function (e.g., neglect it, as in the hypernetted chain), then one can obtain the wall–wall interaction free energy without further iteration. The net pressure between the walls is the negative of the derivative of this with respect to separation. This result was originally obtained by Attard *et al.*[6]

11.2.3 Derjaguin Approximation

A very useful result can be obtained by taking the large-radius limit of the macrosphere–macrosphere Ornstein–Zernike equation and comparing it with the wall–wall series function just given. Using bipolar coordinates, the macrosphere–macrosphere series function may be written

$$\begin{aligned} b_{00}(r;R) &= \sum_{\gamma>0} \rho_\gamma \int \mathrm{d}\mathbf{v}\, h_{0\gamma}(v;R) c_{\gamma\alpha}(|\mathbf{r}-\mathbf{v}|;R) \\ &= \frac{2\pi}{r} \sum_{\gamma>0} \rho_\gamma \int_0^\infty \mathrm{d}\sigma\, \sigma h_{0\gamma}(\sigma;R) \int_{|r-\sigma|}^{r+\sigma} \mathrm{d}\tau\, \tau c_{0\gamma}(\tau;R). \end{aligned} \tag{11.53}$$

The separation of the surfaces of the macrospheres, $x \equiv r - 2R$, is small compared to their radius $|x| \ll R$, at least in the planar limit, $R \to \infty$. Defining $s \equiv \sigma - R$ and $t \equiv \tau - R$, then because the solute–solvent correlation functions are short-ranged, the integral is dominated by regions $s \ll R$ and $t \ll R$. Changing the variables of integration, the lower limit of the τ-integral becomes $|2R + x - s - R| - R = x - s$, and the upper limit may be extended to infinity, which gives

$$\begin{aligned} b_{00}(r;R) &= \frac{2\pi}{2R+x} \sum_{\gamma>0} \rho_\gamma \int_{-R}^\infty \mathrm{d}s\, [s+R] h_{0\gamma}(s+R;R) \\ &\quad \times \int_{x-s}^\infty \mathrm{d}t\, [t+R] c_{0\gamma}(t+R;R). \end{aligned} \tag{11.54}$$

Now as $s \to -\infty$, $t \to \infty$, and $c_{0\gamma}(t+R;R) \approx 0$. Therefore the integrals are dominated by regions $s \gg -R$ and $|t| \ll R$, and one has

$$b_{00}(x+2R;R) = \pi R \sum_{\gamma>0} \rho_\gamma \int_{-\infty}^\infty \mathrm{d}s\, h_{0\gamma}(s) \int_{x-s}^\infty \mathrm{d}t\, c_{0\gamma}(t), \tag{11.55}$$

[6] P. Attard, D. R. Bérard, C. P. Ursenbach, and G. N. Patey, Interaction free energy between planar walls in dense fluids: An Ornstein–Zernike approach with results for hard-sphere, Lennard–Jones, and dipolar systems, *Phys. Rev. A* **44** (1991), 8224.

valid for $R \gg x$. Dividing by πR and differentiating with respect to separation yields

$$\frac{\partial b_{00}(x+2R;R)}{\pi R \partial x} = -\sum_{\gamma>0} \rho_\gamma \int_{-\infty}^{\infty} \mathrm{d}s\, h_{0\gamma}(s) c_{0\gamma}(x-s),\ R >> x. \tag{11.56}$$

The right-hand side of this equation may be immediately identified as the negative of the wall–wall series function per unit area, as given above.

A similar result holds for the bridge function,[7]

$$\frac{\partial d_{00}(x+2R;R)}{\pi R \partial x} = -d_{00}(x),\ R >> x. \tag{11.57}$$

Realistic solute–solute potentials also obey a similar rule, $-\partial u_{00}(x+2R;R) / \pi R \partial x = u_{00}(x)$; in many cases there is no direct solute–solute interaction anyway. This means that the derivative of the macrosphere-macrosphere potential of mean force equals a geometric factor times the wall–wall interaction free energy per unit area

$$\frac{-\partial w_{00}(x+2R;R)}{\partial x} = \pi R w_{00}(x),\ R >> x. \tag{11.58}$$

The left-hand side is simply the force between the solutes, $F(x;R)$. One can work through the analysis for macrospheres of radius R_1 and R_2, and in terms of an effective radius

$$\mathcal{R} \equiv \frac{R_1 R_2}{R_1 + R_2}, \tag{11.59}$$

the force of interaction is

$$F(x;\mathcal{R}) = 2\pi \mathcal{R} w_{12}(x),\ \mathcal{R} >> x. \tag{11.60}$$

This is known as the Derjaguin approximation.[8] The reason that it is such a useful result is that once the interaction free energy per unit area has been obtained for walls, then the force between *any* sized particles with the same surface properties can be obtained by scaling with a simple geometric factor. Alternatively, the force measured for a particular pair of particles can be applied to all other size particles by multiplying by the ratio of their effective radii.

Cavities in a Hard-Sphere Solvent

The Derjaguin approximation is tested in Fig. 11.3 for two cavities in a hard-sphere solvent. It may be seen that even for these comparatively small cavities the potential is almost linear in the radius, as predicted. There is an ambiguity in the definition of the radius to be used for the scaling. One can regard the solutes

[7] P. Attard and J.L. Parker, Oscillatory solvation forces: A comparison of theory and experiment, *J. Phys. Chem.* **96** (1992), 5086.

[8] B. V. Derjaguin, Untersuchungen über die reibung und adhäsion, IV, *Kolloid Z.* **69** (1934), 155. L. R. White, On the Derjaguin approximation for the interaction of macrobodies, *J. Colloid Interface Sci.* **95** (1983), 286.

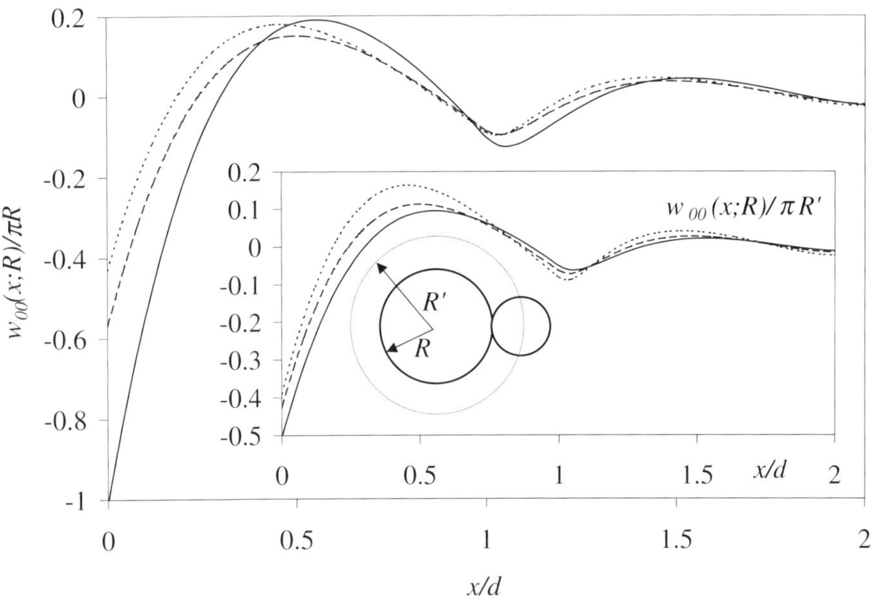

Figure 11.3: The solute–solute potential of mean force, normalised by the radius $R = D/2$, for identical hard solutes in a hard-sphere solvent at a density of $\rho d^3 = 0.7$, as given by the hypernetted chain approximation. From bottom to top at contact the diameters of the solutes are D =1, 3, and 10d. The inset uses as normalisation factor the radius of the cavity itself, $R' = (D+d)/2$. The abscissa measures the surface separation.

as the physical entities and use half their diameter as the radius. Alternatively, one can consider the cavity- or solvent-excluded region as the important quantity and use the distance of closest approach of the centres of the solvent and the solute as the scaling parameter. Comparing the inset with the main figure one can conclude that for small solutes the latter approach gives the least variant results. Obviously for large cavities the difference between the two is negligible. In general, the Derjaguin approximation improves as the radius increases.

11.2.4 Slab of Finite Thickness

In the above the walls were semi-infinite in extent, and it is of interest to consider the solvation of a wall of finite thickness L. The solvent is now excluded from the region $|z| < L/2$, and one has

$$h_{0\alpha}(z;L) = \quad 1, \ |z| < L/2. \tag{11.61}$$

(Of course in addition $h_{0\alpha}(z;L) = -1$, $|z| < (L+d_\alpha)/2$, where $d_\alpha/2$ is the distance of closest approach of the centre of the solvent to the surface.) The

Ornstein–Zernike equation is formally identical to that given above for the isolated wall, Eq. (11.37). In the present case, both $h_{0\alpha}(z; L)$ and $c_{0\alpha}(z; L)$ go to 0 as $|z| \to \infty$, so their Fourier transforms are well-defined. It should be noted that if the thickness of the slab is less than the range of the solvent–solvent potential, then the density profile at each surface of the slab is affected by the presence of solvent on the other side (i.e., it will not be identical to that for the semi-infinite half-space). Even if the wall is so thick as to preclude direct interaction of solvent across it, using an *approximate* closure can lead to an unphysical thickness dependence of the density profiles.

A quantity of interest is the solvation free energy per unit area for the slab. From the previous expression, Eq. (11.32), this is

$$\beta W_0^{\text{ex}}(L) = \sum_{\gamma>0} \rho_\gamma \int_{-\infty}^{\infty} \mathrm{d}z \times \left[h_{0\gamma}(z; L) - v_{0\gamma}^{(1)}(z; L) - h_{0\gamma}(z; L) \sum_{n=1}^{\infty} \frac{n v_{0\gamma}^{(n)}(z; L)}{n+1} \right]. \tag{11.62}$$

In the limit $L \to \infty$, the wall–solvent correlation functions become constant in the interior of the wall. This is clearly the case for the total correlation function, as given above for any thickness, and, following the derivation of Eq. (11.39), is also the case for the direct correlation function,

$$c_{0\alpha}(z; L) \to -1 + \sum_{\gamma>0} \rho_\gamma \hat{c}_{\alpha\gamma}(0), \ |z| \ll L, \ L \to \infty. \tag{11.63}$$

Recalling that the $v^{(n)}$ are convolution integrals of n solute–solvent h-bonds, which equal -1 when the solvent is inside the slab, with the $(n+1)$-body direct correlation function, which is short-ranged, then similar arguments show that inside the slab

$$v_{0\alpha}^{(n)}(z; L) \to (-1)^n \sum_{\gamma_1, \ldots, \gamma_n} \rho_{\gamma_1} \ldots \rho_{\gamma_n} \hat{c}_{\alpha\gamma^n}^{(n+1)}(0, \ldots, 0), \ L \to \infty. \tag{11.64}$$

On the right-hand side appears the n-fold Fourier transform evaluated at $\mathbf{k}^n = 0$.

The fact that these functions go to constants means that the integral that is the solvation free energy per unit area must be dominated by the interior of the slab in the limiting case. Hence to leading order it is just L times the integrand evaluated at $|z| \ll L$. However, on physical grounds, to leading order the free energy must be the pressure–volume work of inserting a cavity into the system and hence one must have

$$\begin{aligned} \lim_{L\to\infty} & \beta W_0^{\text{ex}}(L)/L \\ &= -\beta p \\ &= -\sum_{\gamma} \rho_\gamma + \sum_{\gamma_1, \gamma_2} \rho_{\gamma_1} \rho_{\gamma_2} \hat{c}_{\gamma_1\gamma_2}(0) \\ &\quad + \sum_{n=1}^{\infty} \frac{(-1)^n n}{n+1} \sum_{\gamma_1, \ldots, \gamma_{n+1}} \rho_{\gamma_1} \ldots \rho_{\gamma_{n+1}} \hat{c}_{\gamma^{n+1}}^{(n+1)}(0, \ldots, 0). \end{aligned} \tag{11.65}$$

This is an exact expression for the pressure of a solvent in terms of its many-body direct correlation functions. The first neglected term, which is of order L^{-1}, corresponds to twice the surface free energy per unit area of the wall. The term linear in density here is that of the ideal gas, and hence the terms that depend upon the direct correlation functions sum to the excess pressure. Despite the explicit appearance of powers of density, this is not a strict density expansion because of the implicit density dependence of the direct correlation functions. Given the relationship of the direct correlation functions to the density derivatives of the irreducible Mayer cluster integrals, it is not unreasonable that an expansion such as this should exist, and it would be of interest to obtain this result directly from the virial expansion.

Approximations

The hypernetted chain approximation sets $v^{(n)} = 0$, $n \geq 2$ in the expression for the solvation free energy per unit area. Using the above result, it gives for the pressure

$$\begin{aligned} \beta p^{\mathrm{HNC}} &= \sum_{\gamma} \rho_{\gamma} - \frac{1}{2} \sum_{\gamma_1,\gamma_2} \rho_{\gamma_1} \rho_{\gamma_2} \hat{c}_{\gamma_1\gamma_2}(0) \\ &= \frac{1}{2} \sum_{\gamma} \rho_{\gamma} + \frac{1}{2} \beta \chi_T^{-1}, \end{aligned} \tag{11.66}$$

where Eq. (9.92) for the compressibility has been used. For a single-component solvent this is $\beta p^{\mathrm{HNC}} = \rho[1 + \partial p/\partial \rho]/2$,[9] which is exact only to second order in density (i.e., it yields only the first correction to the ideal gas equation of state). This is in contrast to the HNC approximation for a bulk solvent, where the pressure as an integral of the virial times the radial distribution function yields the third virial coefficient exactly. This is a general rule: the approximate closures to the singlet Ornstein–Zernike equation lose a power of density in accuracy compared to the same approximation applied in the bulk.

The Percus–Yevick closure uses the approximation $v_{0\alpha}^{(n)} = -(-v_{0\alpha}^{(1)})^n/n$, and the solvation free energy follows directly from the chemical potential given above, Eq. (9.104). Hence one has

$$\beta p^{\mathrm{PY}} = \sum_{\alpha > 0} \rho_{\alpha} \frac{1 - x_{\alpha}}{x_{\alpha}} \ln[1 - x_{\alpha}], \tag{11.67}$$

where $x_{\alpha} \equiv \sum_{\gamma > 0} \rho_{\gamma} \hat{c}_{\alpha\gamma}(0)$. Expanding for low densities this is

$$\beta p^{\mathrm{PY}} \sim \sum_{\alpha} \rho_{\alpha} - \frac{1}{2} \sum_{\alpha,\gamma} \rho_{\alpha} \rho_{\gamma} \hat{c}_{\alpha\gamma}(0) + \mathcal{O}\rho^3, \tag{11.68}$$

which is the same as the HNC to the leading two orders.

[9] S. L. Carnie, D. Y. C. Chan, D. J. Mitchell, and B. W. Ninham, The structure of electrolytes at charged surfaces: The primitive model, *J. Chem. Phys.* **74** (1981), 1472.

Surface Tension

One can obtain an explicit expression for the surface energy per unit area by subtracting the pressure contribution. For a thick enough slab the solute–solvent total correlation function goes over to that of the semi-infinite wall treated above,

$$\lim_{L\to\infty} h_{0\alpha}(z;L) = h_{0\alpha}(|z| - L/2), \tag{11.69}$$

and similarly for $c_{0\alpha}$ and $v_{0\alpha}^{(n)}$. Since the pressure contribution comes entirely from within the slab, splitting the integral into two parts one obtains

$$\begin{aligned}
&\lim_{L\to\infty} [\beta W_0^{\rm ex}(L) + \beta p L] \\
&= \lim_{L\to\infty} \Bigg[2\sum_{\gamma>0} \rho_\gamma \int_0^{L/2} {\rm d}z \Bigg(- \left\{ v_{0\gamma}^{(1)}(z;L) - v_{0\gamma}^{(1)}(0;L) \right\} \\
&\quad + \sum_{n=1}^{\infty} \frac{n}{n+1} \left\{ v_{0\gamma}^{(n)}(z;L) - v_{0\gamma}^{(n)}(0;L) \right\} \Bigg) \\
&\quad + 2\sum_{\gamma>0} \rho_\gamma \int_{L/2}^{\infty} {\rm d}z \Bigg(h_{0\gamma}(z - L/2) - v_{0\gamma}^{(1)}(z - L/2) \\
&\quad - h_{0\gamma}(z - L/2) \sum_{n=1}^{\infty} \frac{n v_{0\gamma}^{(n)}(z - L/2)}{n+1} \Bigg) \Bigg] \\
&= 2\sum_{\gamma>0} \rho_\gamma \int_{-\infty}^{0} {\rm d}z \left[-\Delta v_{0\gamma}^{(1)}(z) + \sum_{n=1}^{\infty} \frac{n \Delta v_{0\gamma}^{(n)}(z)}{n+1} \right] \\
&\quad + 2\sum_{\gamma>0} \rho_\gamma \int_0^{\infty} {\rm d}z \left[h_{0\gamma}(z) - v_{0\gamma}^{(1)}(z) - h_{0\gamma}(z) \sum_{n=1}^{\infty} \frac{n v_{0\gamma}^{(n)}(z)}{n+1} \right].
\end{aligned} \tag{11.70}$$

Here

$$\Delta v_{0\gamma}^{(n)}(z) \equiv v_{0\gamma}^{(n)}(z) - (-1)^n \sum_{\gamma_1,\ldots,\gamma_n} \rho_{\gamma_1} \ldots \rho_{\gamma_n} \hat{c}_{\alpha\gamma^n}^{(n+1)}(0,\ldots,0) \tag{11.71}$$

goes to 0 as $z \to -\infty$. All the functions that appear on the final right-hand side are either bulk solvent–solvent functions, or wall–solvent functions. This then is an exact expression for twice the surface free energy per unit area of an infinitely thick wall.

11.3 Other Solute Geometries

The singlet method is very flexible in its interpretation of a solute, which is one of its great advantages. Macrospherical solutes and planar walls have already been analysed. Now three other generic types of solutes will be treated: cylinders, pores, and multimolecular species. These illustrate the range of problems that may be treated within the singlet approach.

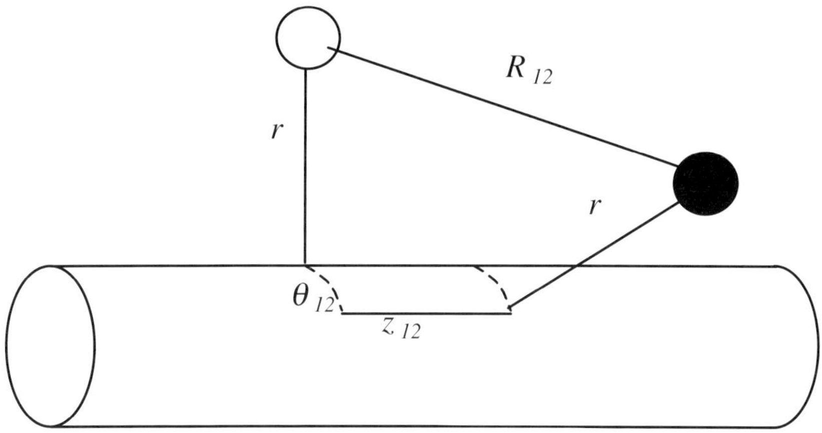

Figure 11.4: The cylindrical geometry used for the Ornstein–Zernike equation.

11.3.1 Cylinders

Whilst the singlet method was introduced by considering a multicomponent mixture, in the final analysis results could have been written down directly from the diagrammatic definitions. This in fact was the attitude taken for the planar solutes. In general then, the Ornstein–Zernike equations formally exist for any solutes. The shape and symmetry of the solute determines the most convenient coordinate system for the correlation functions, and for the evaluation of the Ornstein–Zernike convolution integral. The procedure holds for arbitrary shaped solutes, but solutes that are homogeneous in one or two of their dimensions are the most convenient. The spheres and planes treated above have sufficient symmetry, and the only remaining homogeneous objects in three dimensions are cylinders, which are now treated.

Define a cylindrical coordinate system $\mathbf{R} = (\mathbf{r}, z)$, as shown in Fig. 11.4. For an infinitely long cylindrical solute, the solute–solvent correlation functions do not depend upon z, the position of the solvent along the cylinder. They depend only upon r, the distance from the axis of the cylinder, and the cylinder–solvent Ornstein–Zernike equation is

$$\begin{aligned} \underline{H}(r_1) &= \underline{\underline{C}}(r_1) + \int \mathrm{d}\mathbf{s}_2\, \mathrm{d}z_2\, \underline{\underline{C}}(R_{12})\underline{H}(r_2) \\ &= \underline{C}(r_1) + \int_{-\infty}^{\infty} \mathrm{d}z_{12} \int_0^{2\pi} \mathrm{d}\theta_{12} \int_0^{\infty} \mathrm{d}s_2\, s_2\, \underline{\underline{C}}(R_{12})\underline{H}(r_2), \quad (11.72) \end{aligned}$$

where $R_{12} = \sqrt{z_{12}^2 + r_1^2 + r_2^2 - 2r_1r_2\cos\theta_{12}}$. The formally exact solute solvent closure is

$$h_{0\alpha}(r) = -1 + \exp\left[-\beta u_{0\alpha}(r) + h_{0\alpha}(r) - c_{0\alpha}(r) + d_{0\alpha}(r)\right]. \qquad (11.73)$$

For a hard cylinder of radius S, the cylinder–solvent potential is $u_{0\alpha}(r) = \infty$, $r < S$, and

$$h_{0\gamma}(r) = -1, \; r < S. \tag{11.74}$$

The two-dimensional Fourier transform of a circularly symmetric function is the Hankel transform of order 0. This follows since integrating the Fourier exponent over the cylindrical angle gives the Bessel function of order 0,

$$\int_0^{2\pi} \mathrm{d}\theta \, e^{-ikr\cos\theta} = 2\pi J_0(kr). \tag{11.75}$$

The Hankel transform pair is

$$\overline{f}(k) = 2\pi \int_0^\infty \mathrm{d}r \, r J_0(kr) f(r), \; f(r) = \frac{1}{2\pi} \int_0^\infty \mathrm{d}k \, k J_0(kr) \overline{f}(k), \tag{11.76}$$

which uses the orthogonality property,

$$\int \mathrm{d}\mathbf{r} \, J_0(kr) J_0(k'r) = (2\pi)^2 \delta(\mathbf{k}, \mathbf{k}'). \tag{11.77}$$

Accordingly, taking the three-dimensional Fourier transform of the Ornstein–Zernike equation yields

$$\underline{\overline{H}}(k_r)\delta(k_z) = \underline{\overline{C}}(k_r)\delta(k_z) + \underline{\underline{\hat{C}}}\left(\sqrt{k_r^2 + k_z^2}\right) \underline{\overline{H}}(k_r)\delta(k_z). \tag{11.78}$$

Here k_r and k_z are the radial and axial components of the three-dimensional Fourier vector, and the Dirac deltas appear because of the constancy of the cylinder functions in the axial direction. Integrating both sides with respect to k_z, replacing k_r by k, and rearranging, one obtains

$$\underline{\overline{H}}(k) = \left(\underline{\underline{I}} - \underline{\underline{\hat{C}}}(k)\right)^{-1} \underline{\overline{C}}(k). \tag{11.79}$$

The iterative procedure for solving the cylindrical system consists of successive Hankel transforms between real and Fourier space. Lado has given an 'almost' orthogonal discrete Hankel transform, which, for each transform, unfortunately requires $\mathcal{O}N^2$ operations, where N is the number of grid points.[10] The singlet approach to cylindrically inhomogeneous systems has been discussed by a number of authors.[11]

The interaction between two parallel cylinders can also be treated at the singlet level. As in the case of walls, for infinitely long cylinders the cylinder–cylinder total and direct correlation functions do not exist. The quantity of physical import is the interaction free energy per unit length,

$$\beta w_{00}^{\mathrm{int}}(r) = \beta u_{00}(r) - b_{00}(r) - d_{00}(r), \tag{11.80}$$

[10] F. Lado, Hypernetted-chain solutions for the two-dimensional classical electron gas, *Phys. Rev. B* **17** (1978), 2827.

[11] M. Lozada-Cassou, Fluids between walls and in pores, in *Fundamentals of Inhomogeneous Fluids* (D. Henderson, Ed.), Ch. 8, Dekker, New York, 1992. V. Vlachy and A. D. J. Haymet, Electrolytes in micropores, *Aust. J. Chem.* **43** (1990), 1961.

where u_{00}, b_{00}, and d_{00} are the cylinder–cylinder potential, series, and bridge function per unit length, respectively. In particular, the series function is

$$b_{00}(r) = \int_0^{2\pi} d\theta \int_0^{\infty} ds\, s\, \underline{H}^{\mathrm{T}}(s)\underline{C}\Big(\sqrt{r^2+s^2-2rs\cos\theta}\Big), \tag{11.81}$$

where the total and direct correlation functions pertain to an isolated cylinder. In Fourier space this is

$$\overline{b}_{00}(k) = \overline{\underline{H}}^{\mathrm{T}}(k)\overline{\underline{C}}(k). \tag{11.82}$$

As for the other solutes, it requires no further iteration beyond that required for the solute–solvent quantities to obtain the interaction between the cylinders.

For the case of nonparallel or crossed cylinders, the interaction free energy no longer scales with the length of the cylinders. Hence it exists, as do the pair correlation functions. Because of the loss of symmetry in the geometry, the result appears more complicated than for aligned cylinders, although the analysis is fundamentally no different. The result for identical cylinders, one on the z axis and the other lying in the plane $x = r$ inclined at an angle $\theta \neq 0$, is

$$\beta w_{00}^{\mathrm{int}}(r,\theta) = \beta u_{00}(r,\theta) - b_{00}(r,\theta) - d_{00}(r,\theta). \tag{11.83}$$

The series function is

$$b_{00}(r,\theta) = \int_{-\infty}^{\infty} dx\, dy\, dz\, \underline{H}^{\mathrm{T}}\Big(\sqrt{x^2+y^2}\Big)\, \underline{C}(t(x,y,z)), \tag{11.84}$$

where

$$t(x,y,z) = \big((x-r)^2 + y^2\cos^2\theta - 2yz\sin\theta\cos\theta + z^2\sin^2\theta\big)^{1/2} \tag{11.85}$$

is the perpendicular distance between a point (x, y, z) and the axis of the inclined cylinder. Differentiation of this result with respect to r gives the force between the cylinders, and differentiation with respect to θ gives the torque. For both alligned and crossed cylinders, the interaction free energy (potential of mean force) goes to 0 at large separations.

Derjaguin Approximation

It is of interest to derive the analogue of the Derjaguin approximation for two parallel cylinders. Above it was shown that in the large radius limit the force between two macrospheres, divided by π times their radius, equaled the interaction free energy between two walls. For interacting nonparallel cylinders the same result holds, with a factor of $\sin\theta$ appearing in the denominator.[12] The case of perfectly aligned cylinders is important because of its relevance to hexagonal phases of rod-like polymers and macromolecules such as DNA, and to the interaction between disc like particles adsorbed on a planar surface or linear

[12] B. V. Derjaguin, Untersuchungen über die reibung und adhäsion, IV, *Kolloid Z.* **69** (1934), 155. L. R. White, On the Derjaguin approximation for the interaction of macrobodies, *J. Colloid Interface Sci.* **95** (1983), 286.

molecules confined to a membrane. The simplest way to make the point is by an heuristic derivation. One postulates that the potential of mean force (per unit length) for solutes of radius S can, at close separations, be written as the sum of free energy densities acting between elements of area on their surface. In the large radius limit the cylinders are virtually flat, and the free energy density is taken to be that between planar walls at a local separation of $h(y) = x + y^2/S$, where x is the closest separation of their surfaces, and y is the perpendicular distance from the plane containing the two cylinders. One has

$$\begin{aligned} \beta w_{00}^{\text{int}}(x+2S;S) &= \int_{-\infty}^{\infty} w_{00}^{\text{int}}(x+y^2/S)\,dy \\ &= \sqrt{S}\int_{x}^{\infty} \frac{w_{00}^{\text{int}}(h)}{\sqrt{h-x}}\,dh. \end{aligned} \tag{11.86}$$

This shows that the potential of mean force per unit length between parallel cylinders scales with the square root of their radius, and that it is a weighted integral of the interaction free energy per unit area of planar walls. (The weight function has an integrable singularity.) As such it represents an efficient computational procedure that allows the interaction cylinders of any radius to be related to the interaction of walls (as above cylinders of different radii can be treated by replacing S by $2\mathcal{S} = 2S_1S_2/(S_1+S_2)$). However unlike the interaction between spheres, differentiation does not yield a simple expression for the force. This is rather different to the case of two spheres (or to a sphere and a plane, or to two crossed cylinders), where the Derjaguin approximation gives the total force as a geometric factor times the interaction free energy per unit area between planar walls. For parallel cylinders, or for particles adsorbed on a surface or confined to a membrane, the Derjaguin approximation only exists in the sense of this result.

11.3.2 Membranes and Pores

The spherical, planar, and cylindrical inhomogeneities that have so far been analysed illustrate the variety of solute geometries that can be utilised in the singlet approach. A conceptually different system, namely a confined system, can also be treated by the singlet method by considering the solute to be a pore. Spherical and planar geometries are the simplest. The planar slit pore will be treated in the next section, and this section will initially focus upon an isolated spherical pore of radius R, for which one has

$$\underline{H}(r) = -1,\ r > R. \tag{11.87}$$

Here $\{\underline{H}(r)\}_\alpha = \rho_\alpha^{1/2} h_{0\alpha}(r)$ is the pore–solvent total correlation function. The solvent is assumed to be in equilibrium with a bulk solution where they have number density ρ_α. It is not essential that they have a common distance of closest approach to the pore wall, so that one could have $h_{0\alpha}(r) = -1,\ r > R - d_\alpha/2$. The number of solvent molecules of a given type confined to the pore is given by

$$N_\alpha = 4\pi\rho_\alpha \int_0^R \mathrm{d}r\, r^2[1 + h_{0\alpha}(r)]. \tag{11.88}$$

In the event that one wants to model a fixed number of molecules inside the pore, one can adjust the bulk concentrations until the integrals of the individual pore–solvent distribution functions give the specified number of molecules; average quantities (but not their fluctuations) will then be the same as if pore were a closed system.

As usual the solute–solvent Ornstein–Zernike equation is

$$\begin{aligned} \underline{H}(r) &= \underline{C}(r) + \int \mathrm{d}\mathbf{s}\, \underline{\underline{C}}(|\mathbf{r}-\mathbf{s}|)\underline{H}(s) \\ &= \underline{C}(r) + 2\pi \int_0^\pi \mathrm{d}\theta \int_0^R \mathrm{d}s\, \underline{\underline{C}}(\sqrt{r^2+s^2-2rs\cos\theta})\underline{H}(s). \end{aligned} \tag{11.89}$$

Since one is generally interested in positions $r < R$, the fact that the bulk solvent–solvent direct correlation function is short-ranged means that the integrand decays to 0 for large s. Conversely, deep inside the solvent exclusion region around the pore, $r \to \infty$, one may replace $h_{0\alpha}(s)$ by -1 in the integrand because the short range of $c_{\alpha\gamma}$ keeps $s \approx r$, and one has

$$\begin{aligned} c_{0\alpha}(r) &\to -1 + \sum_\gamma \rho_\gamma \int \mathrm{d}\mathbf{s}\, c_{\alpha\gamma}(|\mathbf{r}-\mathbf{s}|) \\ &= -1 + \sum_\gamma \rho_\gamma \hat{c}_{\alpha\gamma}(0) \\ &= -(\rho_\alpha k_\mathrm{B} T \chi_T)^{-1},\ r \to \infty, \end{aligned} \tag{11.90}$$

which is the same result obtained above for planar solutes. These constant limits must be treated separately if the Fourier transform is to be used to solve the convolution integral. If one defines short-ranged functions,

$$h_{0\alpha}^{\mathrm{sr}}(r) = h_{0\alpha}(r) + 1 = g_{0\alpha}(r), \tag{11.91}$$

and

$$c_{0\alpha}^{\mathrm{sr}}(r) = c_{0\alpha}(r) + (\rho_\alpha k_\mathrm{B} T \chi_T)^{-1}, \tag{11.92}$$

then the solute–solvent Ornstein–Zernike equation becomes

$$\underline{H}^{\mathrm{sr}}(r) = \underline{C}^{\mathrm{sr}}(r) + \int \mathrm{d}\mathbf{s}\, \underline{\underline{C}}(|\mathbf{r}-\mathbf{s}|)\underline{H}^{\mathrm{sr}}(s). \tag{11.93}$$

The Fourier transforms of all these functions exist and it is straightforward to solve numerically in the usual fashion.

One of course also needs a closure relationship. The exact closure is formally identical to those given above, and the various approximations can be applied with no change.

Interacting Pores

An emulsion or aerosol consists of droplets of one immiscible liquid in a continuous phase of another fluid. At a simple level, the spherical pores described above can be used to represent the droplets, and the interactions of two such droplets give an indication of the stability of the dispersion. The potential of mean force between two such droplets is formally

$$\beta w_{00}(r) = \beta u_{00}(r) - b_{00}(r) - d_{00}(r). \tag{11.94}$$

If one focuses on the series function one has

$$b_{00}(r) = \int \mathrm{d}\mathbf{s}\, \underline{H}^{\mathrm{T}}(|\mathbf{r}-\mathbf{s}|)\underline{C}(s), \tag{11.95}$$

which integral diverges due to the fact that the functions in the integrand do not decay to 0 away from the droplet. However, if one subtracts these constants and deals with the short-ranged functions discussed above, then one has

$$b_{00}^{\mathrm{int}}(r) = \int \mathrm{d}\mathbf{s}\, \underline{H}^{\mathrm{sr;T}}(|\mathbf{r}-\mathbf{s}|)\underline{C}^{\mathrm{sr}}(s). \tag{11.96}$$

This is well defined and the interaction part of the series function goes to 0 at large separations.

Similarly the pore–pore bridge function, which consists of pore–solvent and solvent–solvent total correlation functions, goes to a constant at large separations. The constant can be eliminated by replacing two of the pore–solvent h-bonds by pore–solvent g-bonds (one connected to each pore, and not connected to the same field point), which leaves a short-ranged part that decays at the same rate as the solvent correlations of the bulk.

Semipermeable Pore

The pores treated above were either isolated or interacting through an inert phase. It is also possible to include the interactions with the surrounding medium. For example, suppose that the solvent species 1 is confined to the pore as above, $\rho_1(r) = 0$, $r > R$, and that solvent species 2 is confined to the region exterior to the pore, $\rho_2(r) = 0$, $r < R$. Hence the surface of the pore may be regarded as a semi-permeable membrane. The preceding analysis goes through largely unchanged, except that $h_{02}(r)$ is already a short-ranged function, so that

$$h_{0\alpha}^{\mathrm{sr}}(r) = h_{0\alpha}(r) + \delta_{1\alpha}, \tag{11.97}$$

and

$$c_{0\alpha}^{\mathrm{sr}}(r) = c_{0\alpha}(r) - \delta_{1\alpha} + \rho_1 \hat{c}_{1\alpha}(0). \tag{11.98}$$

Hence the solute–solvent Ornstein–Zernike equation may be written in terms of these short-ranged functions exactly as above

$$\underline{H}^{\mathrm{sr}}(r) = \underline{C}^{\mathrm{sr}}(r) + \int \mathrm{d}\mathbf{s}\, \underline{\underline{C}}(|\mathbf{r}-\mathbf{s}|)\underline{H}^{\mathrm{sr}}(s). \tag{11.99}$$

Note that the $c_{\alpha\gamma}$ that appears is that of a bulk two-component mixture with density ρ_1 of species 1 and ρ_2 of species 2. This may seem strange given the physical separation of the two components, but this is what emerges from the formalism. This indicates one limitation of the singlet procedure, wherein the physical structure of the system is not incorporated directly into the formalism.

As above one can treat the interaction of two pores, except that now the restructuring of the surrounding medium as the separation between the pores is changed is taken into account. One is interested in the potential of mean force, which means that one subtracts the large separation limit. Again this means that the interaction part of the series function may be written in terms of short-ranged functions,

$$b_{00}^{\rm int}(r) = \int {\rm d}{\bf s}\, \underline{H}^{\rm sr;T}(|{\bf r}-{\bf s}|)\underline{C}^{\rm sr}(s). \tag{11.100}$$

This is formally identical to the result given above, with the present short-ranged functions being used.

Membrane

This method of treating the interaction between two pores can also be applied to the thickness-dependent free energy of a planar membrane. The wall–wall Ornstein–Zernike equation used above gives the interaction free energy per unit area of two planar walls separated by a solvent(i.e., $z > 0$), Eqs. (11.50) and (11.52). This situation may be called a slit pore. However, as in the case of spherical pores there is no fundamental impediment to applying the result for $z < 0$, which could model solvent on either side of a bilayer membrane of thickness $|z|$. Again one must account for the fact that the wall–solvent correlation functions tend to constants deep inside the wall, and so one must work with short-range functions. Using similar arguments to the above, the interaction series function per unit area is

$$b_{00}^{\rm int}(z) = \int_0^\infty {\rm d}z'\, \underline{H}^{\rm sr;T}(z')\underline{C}^{\rm sr}(z-z'),\ z<0, \tag{11.101}$$

where the first function in the integrand is the wall–solvent distribution function, which vanishes for $z' < 0$, and the second term has been made short-ranged by subtracting the constant given by Eq. (11.39). One can also define the interaction bridge function, which has two noncontiguous wall–solvent h-bonds replaced by g-bonds, and is consequently short-ranged. Subtracting these from the membrane thickness-dependent potential gives the interaction free energy per unit area, $\beta w_{00}^{\rm int}(z) = \beta u_{00}^{\rm int}(z) - b_{00}^{\rm int}(z) - d_{00}^{\rm int}(z)$. Note that there is no bulk free energy contribution to this, which would be just the osmotic pressure of the bulk solvent times the thickness of the membrane, nor any surface free energy part. This is purely the interaction free energy per unit area, which arises from correlations between solvent molecules on the two sides of the membrane. It represents a van der Waals force, and for large thickness it should approach 0 from below, $w_{00}^{\text{int}}(z) \sim \mathcal{O}(z^{4-n})$, $z \to -\infty$, where the solvent–solvent correlation functions go like r^{-n}, $r \to -\infty$.

In some ways this method that treats the membrane as two interacting walls at negative separations is not the most obvious way of analysing the problem. What is perhaps more intuitive is to consider the membrane as a single planar wall of finite thickness t. Then all of the analysis for the isolated wall holds, with minor modifications due to the replacement of the core exclusion condition by

$$h_{0\alpha}(z) = -1, \ |z| < t/2. \tag{11.102}$$

This method and the one of the preceding paragraph treat the same problem, and each has its own advantages. The membrane free energy is obtained directly in the first approach, but not in the second, whereas the solvent profiles on both sides of the membrane are given directly by the second approach, but not the first. The two approaches illustrate another aspect of the singlet method, namely that it is possible to regard two interacting solutes as a single species, and this is the topic of the next section.

11.3.3 Dumb-bells

The flexibility of the singlet Ornstein–Zernike approach is nowhere more evident than when the solute is a cluster of particles or solvent molecules. The membrane example at the end of the preceding section showed the fundamental equivalence of treating interacting walls either as two solutes or as a single molecular species. More generally the solute can be a cluster of n solvent molecules, and then the solute–solvent density profile simply corresponds to the solvent $(n+1)$-particle distribution function. When the solute is taken to be two spherical particles, the method may be called the dumb-bell singlet approach; it is based upon the solute–solvent Ornstein–Zernike equation, as above, but it yields the properties of interacting solutes directly.[13] As such it must be distinguished from the singlet method explored in detail above that utilises the solute–solute Ornstein–Zernike equation for the interacting solute problem. Essentially the dumb-bell can be formed from any two particles at a fixed separation. A particularly interesting example is for a solute composed of two solvent molecules, because this gives the triplet distribution function of the bulk solvent. In the dumb-bell method one must take care of additional geometric factors in formulating the solute–solvent Ornstein–Zernike integral, which may then become complicated in detail though it remains conceptually straightforward. The simplest geometry is the planar case, and this section will focus upon two infinitely thick walls, which corresponds to the wall–wall analysis given above in Section 11.2.2, and which can also be considered as a slit pore.

A solute composed of two semi-infinite half-spaces separated by t has for the solute–solvent total correlation function

$$h_{0\alpha}(z;t) = -1, \ |z| > t/2, \tag{11.103}$$

[13] M. Lozada-Cassou, Fluids between walls and in pores, in *Fundamentals of Inhomogeneous Fluids* (D. Henderson, Ed.), Ch. 8., Dekker, New York, 1992.

which manifests the confinement of the solvent between the walls. That the solute actually consists of two walls will be explicitly signified by the separation appearing in the argument of the solute–solvent correlation functions. Attention will be restricted to the symmetric system, $h_{0\alpha}(-z;t) = h_{0\alpha}(z;t)$, and hence the number of confined solvent particles per unit area is

$$N_\alpha = \rho_\alpha \int_{-\infty}^{\infty} \mathrm{d}z\, [1 + h_{0\alpha}(z;t)] = \rho_\alpha t + 2\rho_\alpha \int_0^{t/2} \mathrm{d}z\, h_{0\alpha}(z;t). \tag{11.104}$$

As usual the solute–solvent Ornstein–Zernike equation is

$$\begin{aligned} \underline{H}(z;t) &= \underline{C}(z;t) + \int \mathrm{d}\mathbf{r}'\, \underline{\underline{C}}(|\mathbf{r}-\mathbf{r}'|)\underline{H}(r'_z;t) \\ &= \underline{C}(z;t) + 2\pi \int_0^{\infty} \mathrm{d}r'\, r' \int_{-\infty}^{\infty} \mathrm{d}z' \\ &\quad \times \underline{\underline{C}}\left(\sqrt{r'^2 + (z-z')^2}\right) \underline{H}(z';t). \end{aligned} \tag{11.105}$$

The closure is formally

$$h_{0\alpha}(z;t) = -1 + \exp\left[-\beta u_{0\alpha}(z;t) + b_{0\alpha}(z;t) + d_{0\alpha}(z;t)\right]. \tag{11.106}$$

In the exponent appear the solute–solvent potential, series function, and bridge function. Generally the solute–solvent potential is the sum of the potentials of the two isolated walls, and these potentials go to 0 far from the walls to allow the density to obtain its bulk value. For the confined solvent, nowhere does the density equal its bulk value except in the centre of the system in the limit $t \to \infty$.

The equations above determine the solvent profiles between two walls separated by t. For the study of slit pores these are the main quantities of interest. The wall–wall Ornstein–Zernike equation of Section 11.2.2 did not determine the density profiles between the two walls, and so for the study of slit pores the dumb-bell approach is preferable. What the wall–wall approach did give was the interaction free energy per unit area, a quantity not directly available from the formally exact dumb-bell equations given above. The separation-dependent free energy is an important quantity in the study of solute–solute interactions (it determines, for example, whether particles stick together or remain dispersed), and can be more or less directly measured. The solvation free energy, or solute chemical potential, does allow this interaction free energy per unit area to be extracted. According to Eq. (11.32), the solvation free energy per unit area is

$$\begin{aligned} \beta W_0^{\mathrm{ex}}(t) &= \sum_{\gamma>0} \rho_\gamma \int_{-\infty}^{\infty} \mathrm{d}z \\ &\quad \times \left[h_{0\gamma}(z;t) - v_{0\gamma}^{(1)}(z;t) - h_{0\gamma}(z;t) \sum_{n=1}^{\infty} \frac{n v_{0\gamma}^{(n)}(z;t)}{n+1} \right]. \end{aligned} \tag{11.107}$$

Due to the fact that the correlation functions go to constants inside the walls, as has been discussed several times above, this expression is divergent.

In the large separation limit, the correlation functions go to

$$\lim_{t\to\infty} h_{0\gamma}(z;t) = h_{0\gamma}(z+t/2) + h_{0\gamma}(-z+t/2) \tag{11.108}$$

and

$$\lim_{t\to\infty} v_{0\gamma}^{(n)}(z;t) = v_{0\gamma}^{(n)}(z+t/2) + v_{0\gamma}^{(n)}(-z+t/2), \tag{11.109}$$

where both functions on each right-hand side refer to a semi-infinite half space that confines the solvent to $z > 0$. By subtracting the limiting forms one obtains the interaction part of the solvation free energy per unit area,

$$\begin{aligned} \beta W_0^{\rm int}(t) &= \sum_{\gamma>0} \rho_\gamma \int_{-\infty}^{\infty} \mathrm{d}z \left[\Delta h_{0\gamma}(z;t) - \Delta v_{0\gamma}^{(1)}(z;t) \right. \\ &\quad \left. - \sum_{n=1}^{\infty} \frac{n}{n+1} \Delta\left\{ h_{0\gamma}(z;t) v_{0\gamma}^{(n)}(z;t) \right\} \right]. \end{aligned} \tag{11.110}$$

All the functions in the integrand are short-ranged in both directions, and since the integrand goes to 0 as $t \to \infty$, one can conclude that indeed this is a formally exact expression for the interaction free energy per unit area between the walls of the planar dumb-bell Ornstein–Zernike equation. As usual, one can implement the HNC or the PY approximations by neglecting or approximating the bridge functions. The derivative of this with respect to separation gives the net pressure.

Born–Green–Yvon Interaction

As mentioned above the density profile around a solute n-ion corresponds to the $(n+1)$-particle distribution function. The Born–Green–Yvon hierarchy also relates these two levels of distribution functions. For the dumb-bell solute the second member of the hierarchy is required, Eq. (9.74). This expresses the gradient of the pair distribution function as the average direct force plus the indirect force mediated by a third particle. Recalling that the potential of mean force is related to the radial distribution function by $g(r) = \exp -\beta w(r)$, it follows that $g'(r) = -\beta w'(r) g(r)$. Hence dividing by the pair distribution function one can rewrite the second member of the Born–Green–Yvon hierarchy as an expression for the gradient of the potential of mean force. For two spherical solutes, Eq. (9.74) becomes

$$\begin{aligned} w_{00}'(r_{12}) &= V_{00}'(r_{12}) + \sum_\alpha \rho_\alpha \int \mathrm{d}\mathbf{r}_3\, u_{0\alpha}'(r_{13}) \cos\theta \frac{g_{00\alpha}(r_{12}, r_{13}, \cos\theta)}{g_{00}(r_{12})}, \end{aligned} \tag{11.111}$$

where V_{00} is the solute–solute potential, and $u_{0\alpha}$ is the isolated solute–solvent potential. The solvent density times the solute–solute–solvent triplet distribution function divided by the solute–solute pair distribution function is the probability of finding a solute conditional upon the solutes being at $\mathbf{r}_1$ and $\mathbf{r}_2$. This is just the dumb-bell–solvent distribution function and one has

$$\begin{aligned}\frac{g_{00\alpha}(r_{12}, r_{13}, \cos\theta)}{g_{00}(r_{12})} &= g_{0\alpha}(r_{13})g_{0\alpha}(r_{23}) \exp -\beta\tau_{00\alpha}(r_{12}, r_{13}, \cos\theta) \\ &= 1 + h_{0\alpha}(\mathbf{r}_3; r_{12}). \end{aligned} \tag{11.112}$$

On the right-hand side of the first line appear the isolated solute–solvent distribution functions, $g_{0\alpha}$, whereas on the right-hand side of the second line the quantity $h_{0\alpha}$ refers to the dumb-bell–solvent total correlation function, with r_{12} being the separation between the two bells.

Strictly speaking the numerator and the denominator of the left side do not individually exist for planar walls of infinite area (cf. the discussion of the wall–wall Ornstein–Zernike equation, Eq. (11.50), above), but the right sides are well defined for both finite and infinite solutes. (The quantity τ is the solute–solute–solvent triplet potential of mean force, which is the set of connected, nonparallel diagrams with nonadjacent root points.) Inserting the dumb-bell distribution function into the Born–Green–Yvon equation gives the mean solute–solute force (and upon integration their interaction free energy). This enables the interaction of spherical and other finite solutes to be characterised.

For infinite planar solutes the analogue of the Born–Green–Yvon equation reduces to the contact theorem, as is now shown. Equation (11.111) says that the mean force between solutes is the sum of a direct part and an indirect part due to the mediation of the solvent. In the planar case one must deal with the force per unit area, but the same situation holds: the mean force per unit area is the direct pressure between the walls plus the force on one wall due to the solvent, summed over species and integrated over space, weighted with their probability. The indirect contribution clearly scales with the area of the plates, and hence it is the axial integration that gives the indirect force per unit area. That is,

$$w'_{00}(t) = V'_{00}(t) + \sum_\alpha \rho_\alpha \int_{-t/2}^{t/2} \mathrm{d}z\, u'_{0\alpha}(z)[1 + h_{0\alpha}(z;t)]. \tag{11.113}$$

Note that $\cos\theta$ has been set equal to 1 because in this planar geometry the component of force is always parallel to the line connecting the walls. It's worth mentioning that this result can be derived directly from Eq. (11.111) by taking the infinite solute radius limit, and by invoking the relationships of Section 11.2.3. Because there is no solvent on the far side of these plates, the left side represents the total pressure between the walls, $p^{\text{total}}(t) = -w'_{00}(t)$. To obtain the net pressure one subtracts the pressure due to the bulk solvent, $p^{\text{net}}(t) = p^{\text{total}}(t) - p_\infty$, and thence the interaction free energy per unit area follows by integration,

$$w^{\text{int}}(t) = \int_t^\infty \mathrm{d}t'\, p^{\text{net}}(t'), \tag{11.114}$$

with $w^{\text{int}}(t) \to 0$, $t \to \infty$.

For the solvent confined between walls, the wall–solvent potential contains a hard-wall term, which gives a Dirac δ when differentiated,[14]

$$u'_{0\alpha}(z) = \tilde{u}'_{0\alpha}(z) - k_{\rm B}T\delta(z - t/2), \tag{11.115}$$

where the first term is the soft part of the potential if any. Note that this refers to only one wall of the dumb-bell. Hence the wall–wall Born–Green–Yvon equation becomes

$$\begin{aligned} p^{\rm total}(t) &= V'_{00}(t) - \sum_\alpha \rho_\alpha \int_{-\infty}^{\infty} {\rm d}z\, [\tilde{u}'_{0\alpha}(z) - k_{\rm B}T\delta(z-t/2)]\,[1 + h_{0\alpha}(z;t)] \\ &= V'_{00}(t) + k_{\rm B}T \sum_\alpha \rho_\alpha [1 + h_{0\alpha}(t^-/2;t)] \\ &\quad - \sum_\alpha \rho_\alpha \int_{-t/2}^{t/2} {\rm d}z\, \tilde{u}'_{0\alpha}(z)[1 + h_{0\alpha}(z;t)]. \end{aligned} \tag{11.116}$$

This formally exact expression says that the total pressure (force per unit area) acting between the walls consists of the direct force per unit area between the walls, plus the average 'soft' force acting on the solvent due to one of the walls, plus the kinetic pressure at that wall due to the thermal motion of the solvent in contact with it. This is just the contact theorem, Eq.(7.93).

11.4 Inhomogeneous Ornstein–Zernike Equation

The singlet Ornstein–Zernike equation treated the density inhomogeneity as arising from the presence of a solute in an otherwise uniform system. In this approach the solvent density was a constant, the solvent–solvent pair correlations were homogeneous, and the geometry of the solute was reflected in the symmetry of the solute–solvent pair correlations. A drawback of the approach is that the solute potential, which causes the density inhomogeneity, appears on the closure equations. In practice this is the approximate part of any numerical procedure and since the solute potential can represent an extremely significant perturbation, this represents a limitation on the accuracy that can be achieved with the procedure.

An alternative approach considers the density inhomogeneity to arise from an externally applied potential. No solute appears in this approach, and the solvent density itself is nonuniform. In this method, for a single-component solvent, one invokes the inhomogeneous Ornstein–Zernike equation,

$$h(\mathbf{r}_1, \mathbf{r}_2) = c(\mathbf{r}_1, \mathbf{r}_2) + \int_V {\rm d}\mathbf{r}_3\, \rho(\mathbf{r}_3) h(\mathbf{r}_1, \mathbf{r}_3) c(\mathbf{r}_2, \mathbf{r}_3). \tag{11.117}$$

[14]More precisely, $u'(z)g(z) = u'(z)e^{-\beta u(z)}e^{\beta u(z)}g(z) = -k_{\rm B}T[{\rm d}e^{-\beta u(z)}/{\rm d}z]y(z)$. The derivative of the unit step function, the first part, gives the Dirac δ, and $y(z)$ is the cavity function, which, due to the cancellation of the $e(r)$-bond between the root points, is continuous across the wall.

For the case of a multicomponent system, the convolution integral includes a sum over the species index.

The closure equation is unchanged by the inhomogeneity,

$$h(\mathbf{r}_1,\mathbf{r}_2)+1 = \exp\left[-\beta u(\mathbf{r}_1,\mathbf{r}_2) + h(\mathbf{r}_1,\mathbf{r}_2) - c(\mathbf{r}_1,\mathbf{r}_2) + d(\mathbf{r}_1,\mathbf{r}_2)\right]. \quad (11.118)$$

As usual one generally invokes an approximation for the bridge function, either neglecting it (hypernetted chain) or expressing it as a series of series functions (Percus–Yevick). The mean spherical approximation can also be used.

Hence one has two equations for the two pair correlation functions. The density profile is also unknown, and one therefore needs a further equation to complete this inhomogeneous approach.

11.4.1 Density Equations

There are various formally exact expressions for the density profile of an inhomogeneous system. In a previous chapter the Born–Green–Yvon hierarchy was given. The first member of the hierarchy, Eq. (9.70), for a system with only singlet and pair interaction potentials, is

$$\begin{aligned}\nabla\rho^{(1)}(\mathbf{r}_1) &= -\rho^{(1)}(\mathbf{r}_1)\nabla\beta u^{(1)}(\mathbf{r}_1) \\ &\quad - \int_V \mathrm{d}\mathbf{r}_2\, \rho^{(2)}(\mathbf{r}_1,\mathbf{r}_2)\nabla\beta u^{(2)}(\mathbf{r}_1,\mathbf{r}_2). \end{aligned} \quad (11.119)$$

Hence if the inhomogeneous pair distribution function is known, this represents an integro-differential equation for the density profile.

An alternative is based upon the direct correlation function,

$$z(\mathbf{r}) = \rho(\mathbf{r}_1)\exp - c^{(1)}(\mathbf{r}_1), \quad (11.120)$$

as given in Eqs. (8.69) and (8.75). Taking the gradient of the logarithm of the density one has

$$\begin{aligned}&\nabla \ln \rho^{(1)}(\mathbf{r}_1)\Lambda^3 \\ &= -\nabla\beta u^{(1)}(\mathbf{r}_1) + \nabla c^{(1)}(\mathbf{r}_1) \\ &= -\nabla\beta u^{(1)}(\mathbf{r}_1) + \int_V \mathrm{d}\mathbf{r}_2\, \frac{\delta c^{(1)}(\mathbf{r}_1)}{\delta\rho^{(1)}(\mathbf{r}_2)}\nabla\rho^{(1)}(\mathbf{r}_2) \\ &= -\nabla\beta u^{(1)}(\mathbf{r}_1) + \int_V \mathrm{d}\mathbf{r}_2\, c^{(2)}(\mathbf{r}_1,\mathbf{r}_2)\nabla\rho^{(1)}(\mathbf{r}_2). \end{aligned} \quad (11.121)$$

This equation may be named after Triezenberg and Zwanzig (TZ).[15]

If one takes the gradient of the density itself one obtains

$$\begin{aligned}\nabla\rho^{(1)}(\mathbf{r}_1) &= \int_V \mathrm{d}\mathbf{r}_2\, \frac{\delta\rho^{(1)}(\mathbf{r}_1)}{\delta u^{(1)}(\mathbf{r}_2)}\nabla u^{(1)}(\mathbf{r}_2) \\ &= -\beta\int_V \mathrm{d}\mathbf{r}_2\, \rho_\delta^{(2)}(\mathbf{r}_1,\mathbf{r}_2)\nabla u^{(1)}(\mathbf{r}_2), \end{aligned} \quad (11.122)$$

[15] D. G. Triezenberg, and R. Zwanzig, Fluctuation theory of surface tension, *Phys. Rev. Lett.* **28** (1972), 1183.

as follows from the definition of the pair density, Eq. (9.17). Wertheim, and Lovett, Mou, and Buff (WLMB) originally derived this result.[16] Note that the external potential, $u^{(1)}(\mathbf{r})$, often includes a solvent-exclusion term (hard-wall, or hard-solute) that contributes a δ-function when the external potential is differentiated. This contact contribution from the pair correlations with a solvent molecule at the wall is often evaluated and displayed explicitly outside the integral.

The TZ and WLMB equations are related to each other by the inhomogeneous Ornstein–Zernike equation. The TZ equation may be written, using the fact that $c_\delta(\mathbf{r}_1, \mathbf{r}_2) \equiv \delta(\mathbf{r}_1, \mathbf{r}_2)/\rho^{(1)}(\mathbf{r}_1) - c^{(2)}(\mathbf{r}_1, \mathbf{r}_2)$,

$$-\nabla\beta u^{(1)}(\mathbf{r}_1) = \int_V \mathrm{d}\mathbf{r}_2\, c_\delta^{(2)}(\mathbf{r}_1, \mathbf{r}_2)\nabla\rho^{(1)}(\mathbf{r}_2). \tag{11.123}$$

Multiplying both sides of this by $\rho_\delta^{(2)}(\mathbf{r}_3, \mathbf{r}_1)$ and integrating over $\mathbf{r}_1$ one obtains

$$\begin{aligned} -\int_V \mathrm{d}\mathbf{r}_1\, \rho_\delta^{(2)}(\mathbf{r}_3, \mathbf{r}_1)\nabla\beta u^{(1)}(\mathbf{r}_1) \\ = \int_V \mathrm{d}\mathbf{r}_1\, \mathrm{d}\mathbf{r}_2\, \rho_\delta^{(2)}(\mathbf{r}_3, \mathbf{r}_1) c_\delta^{(2)}(\mathbf{r}_1, \mathbf{r}_2)\nabla\rho^{(1)}(\mathbf{r}_2) \\ = \nabla\rho^{(1)}(\mathbf{r}_3), \end{aligned} \tag{11.124}$$

where the fact that $\rho_\delta^{(2)}$ and $c_\delta^{(2)}$ are functional inverses has been used. This is just the WLMB equation. Hence if the pair correlation functions satisfy the Ornstein–Zernike equation, the WLMB and the TZ equations are equivalent for any approximate closure. In contrast, for an approximate closure the BGY equation will yield a density profile that differs from these two.

The WLMB and TZ equations are also equivalent to the second member of the BGY hierarchy, Eq. (9.74). This may be seen by invoking the latter for a homogeneous solute–solvent mixture,

$$\begin{aligned} \nabla g_{01}(r_{01}) \\ = -\beta g_{01}(r_{01})\nabla u_{01}(r_{01}) - \beta\rho_1 \int \mathrm{d}\mathbf{r}_2\, \nabla u_{01}(r_{02}) g_{011}(\mathbf{r}_{01}, \mathbf{r}_{02}). \end{aligned} \tag{11.125}$$

Replacing the solute potential by the external potential, these may be rewritten in terms of the density profile and the inhomogeneous pair total correlation function, (since the integral of the force over the unconnected diagrams vanishes),

$$\begin{aligned} \nabla\rho(\mathbf{r}_1) \\ = -\beta\rho(\mathbf{r}_1)\nabla u^{(1)}(\mathbf{r}_1) - \beta\rho(\mathbf{r}_1) \int \mathrm{d}\mathbf{r}_2\, \rho(\mathbf{r}_2)\nabla u^{(1)}(\mathbf{r}_2) h(\mathbf{r}_1, \mathbf{r}_2), \end{aligned} \tag{11.126}$$

[16] M. S. Wertheim, Correlations in the liquid–vapour interface, *J. Chem. Phys.* **65** (1976), 2377. R. Lovett, C. Y. Mou, and F. P. Buff, The structure of the liquid–vapour interface, *J. Chem. Phys.* **65** (1976), 570.

which is the WLMB equation given above.

Another equation for the density profile may be derived from the formally exact expression for the chemical potential, Eq. (9.101). This expression was derived for a uniform system, but it is evident that the analysis can be carried through for an inhomogeneous system with the partially coupled solvent molecule at $\mathbf{r}_1$, yielding

$$\begin{aligned}\langle -\beta_s \mu_s^{\text{ex}}(\mathbf{r}_1)\rangle &= -\beta V(\mathbf{r}_1) + \int \mathrm{d}\mathbf{r}_2 \rho(\mathbf{r}_2) \Big[h(\mathbf{r}_1, \mathbf{r}_2) - v^{(1)}(\mathbf{r}_1, \mathbf{r}_2) \\ &\quad - h(\mathbf{r}_1, \mathbf{r}_2) \sum_{n=1}^{\infty} \frac{n}{n+1} v^{(n)}(\mathbf{r}_1, \mathbf{r}_2) \Big] . \end{aligned} \tag{11.127}$$

Comparing with Eq. (9.96), one sees that the integral represents the local field felt by an atom at $\mathbf{r}_1$ due to the other atoms, and that the first term is the contribution from the external potential, which was not previously present. Note that the averand on the left-hand side is *the* excess chemical potential, *not* the local chemical potential $\mu - V(\mathbf{r})$ used in the previous chapter. The ideal contribution is

$$\left\langle \beta_s \mu_s^{\text{id}}(\mathbf{r}) \right\rangle = \ln \rho(\mathbf{r}) \Lambda^3, \tag{11.128}$$

and $\mu_s(\mathbf{r}) = \mu_s^{\text{id}}(\mathbf{r}) + \mu_s^{\text{ex}}(\mathbf{r})$. The chemical potential must be constant throughout the system since it is fixed by the reservoir. Hence this represents an implicit equation for the density profile,

$$\langle \mu_s(\mathbf{r}) \rangle = \mu. \tag{11.129}$$

Approximations

These exact equations for the density profile each involve an inhomogeneous pair correlation function. A common approximation is to replace this by a bulk pair function, which then bypasses the need to invoke the inhomogeneous Ornstein–Zernike equation. One needs to evaluate the bulk function at an appropriate density. One choice is the density of a uniform system at the same chemical potential as the system of interest. One can also choose the average of either the local density at the two positions in the inhomogeneous system or a nonlocal density weighted over some neighbourhood. In some cases this average density can be problematic as it may not represent a stable bulk density.

If the inhomogeneous direct correlation function is replaced by the bulk one, then one recovers the singlet hypernetted chain approximation.[17] For a solvent mixture, replacing the inhomogeneous $c_{\alpha\gamma}(\mathbf{r}_1, \mathbf{r}_2)$ by the bulk $c_{\alpha\gamma}(r_{12})$, the Triezenberg–Zwanzig equation for the density profile, Eq. (11.121), may be rewritten

$$\frac{1}{\rho_\alpha(\mathbf{r}_1)} \frac{\partial \rho_\alpha(\mathbf{r}_1)}{\partial \mathbf{r}_1}$$

[17] J. P. Badiali, M. L. Rosinberg, D. Levesque, and J. J. Weis, Surface density profile of the one-component plasma, *J. Phys. C: Solid State Phys.* **16** (1983), 2183.

$$
\begin{aligned}
&= -\beta \frac{\partial V_\alpha(\mathbf{r}_1)}{\partial \mathbf{r}_1} + \sum_\gamma \int \mathrm{d}\mathbf{r}_2 \, \frac{\partial \rho_\gamma(\mathbf{r}_2)}{\partial \mathbf{r}_2} c_{\gamma\alpha}(r_{12}) \\
&= -\beta \frac{\partial V_\alpha(\mathbf{r}_1)}{\partial \mathbf{r}_1} - \sum_\gamma \int \mathrm{d}\mathbf{r}_2 \, \rho_\gamma(\mathbf{r}_2) \frac{\partial c_{\gamma\alpha}(r_{12})}{\partial \mathbf{r}_2} \\
&= -\beta \frac{\partial V_\alpha(\mathbf{r}_1)}{\partial \mathbf{r}_1} + \sum_\gamma \int \mathrm{d}\mathbf{r}_2 \, \rho_\gamma(\mathbf{r}_2) \frac{\partial c_{\gamma\alpha}(r_{12})}{\partial \mathbf{r}_1} \\
&= -\beta \frac{\partial V_\alpha(\mathbf{r}_1)}{\partial \mathbf{r}_1} + \frac{\partial}{\partial \mathbf{r}_1} \sum_\gamma \rho_\gamma \int \mathrm{d}\mathbf{r}_2 \, h_{0\gamma}(\mathbf{r}_2) c_{\alpha\gamma}(r_{12}). \qquad (11.130)
\end{aligned}
$$

The second equality follows an integration by parts, the integrated portion vanishing, the third equality arises because $\partial f(r_{12})/\partial \mathbf{r}_2 = -\partial f(r_{12})/\partial \mathbf{r}_1$, and the final equality is due to the fact that $\rho_\gamma(\mathbf{r}_2) = \rho_\gamma[1+h_{0\gamma}(\mathbf{r}_2)]$, where the constant part gives a contribution that is independent of $\mathbf{r}_1$. Integrating this one obtains

$$
\ln[1 + h_{0\alpha}(\mathbf{r}_1)] = -\beta V_\alpha(\mathbf{r}_1) + \sum_\gamma \rho_\gamma \int \mathrm{d}\mathbf{r}_2 \, h_{0\gamma}(\mathbf{r}_2) c_{\gamma\alpha}(r_{12}), \qquad (11.131)
$$

both sides vanishing as $\mathbf{r}_1 \to \infty$. This is just the singlet hypernetted chain approximation.

For the case of the chemical potential expression for the density profile, a common practice is to invoke the same approximation as for the closure to the inhomogeneous Ornstein–Zernike equation. That is, for the hypernetted chain approximation one neglects $v^{(n)}$ for $n \geq 2$, and for the Percus–Yevick approximation one replaces the bridge diagrams by products of series diagrams. This is in contrast to the other three density equations, which are implemented exactly, and the only approximation is in the closure.

11.4.2 Planar Inhomogeneity

What makes the Ornstein–Zernike approach so popular for bulk fluids is that the convolution integral factorises upon Fourier transformation; the consequent algebraic equation is readily 'solved' for either the total or the direct correlation function. The availability of the fast Fourier transform is a significant advantage because one alternately iterates the closure in real space and the Ornstein–Zernike equation in Fourier space. For an isotropic, uniform system of spherically symmetric molecules, the pair correlation functions depend only upon the radial coordinate and are defined on a grid of size N, and the convolution integral in Fourier space takes just N evaluations. The fast Fourier transform takes $N \log_2 N$ operations. Even with $N \approx 2^{10}$, this is quite feasible, even trivial, on modern computers.

In general, inhomogeneous pair functions defined on a grid of N points in each dimension require arrays of size N^6 to store, and N^9 multiplications to evaluate the Ornstein–Zernike three-dimensional convolution integral directly. This is prohibitive for even moderate grids. Only the most symmetric inhomogeneous problems can be considered feasible, and in practice only planar and spherical inhomogeneities have been treated to date.

For a planar inhomogeneity (such as a wall, a flat membrane or interface between two phases, or a slit pore), the density is just a function of the distance from the surface, $\rho(z)$, and the pair correlations are functions of the positions of the two atoms of the solvent from the surface, z_1 and z_2, and of the lateral component of their separation, r_{12}. That is, $h(\mathbf{r}_1, \mathbf{r}_2) = h(r_{12}, z_1, z_2)$, and the actual separation between the atoms is $R_{12} = \sqrt{(z_1 - z_2)^2 + r_{12}^2}$. In this geometry the Ornstein–Zernike equation becomes

$$\begin{aligned} h(r_2, z_1, z_2) &= c(r_2, z_1, z_2) + \int_{-\infty}^{\infty} \mathrm{d}z_3 \int_0^{2\pi} \mathrm{d}\theta \int_0^{\infty} \mathrm{d}r_3\, r_3 \\ &\quad \times \rho(z_3) c(r_3, z_1, z_3) h(r_{32}, z_3, z_2), \end{aligned} \tag{11.132}$$

where the axis of the cylindrical coordinates has been set at $r_1 = 0$, and hence $r_{32} = \sqrt{r_2^2 + r_3^2 - 2r_2 r_3 \cos\theta}$.

For the case of solvent-excluding walls, the density vanishes in certain regions (e.g., $\rho(z) = 0,\ z < 0$), and the limits on the z_3 integral may be adjusted accordingly. Conversely, the density profile goes over to the bulk value far from the surface, as do the pair correlation functions. Hence for an unconfined solvent (e.g., a single wall, or membrane or interface between two phases), one typically introduces a z-cutoff, and only solves the inhomogeneous problem within the cutoff. The contribution to the convolution integral from outside the cutoff is approximated by the contribution from bulk functions.

The Fourier transform of a cylindrically symmetric function becomes a Hankel transform of order 0 (see Section 11.3.1). Hence the Hankel transform factorises the radial part of the Ornstein–Zernike equation,

$$\overline{h}(k, z_1, z_2) = \overline{c}(k, z_1, z_2) + \int_{-\infty}^{\infty} \mathrm{d}z_3\, \rho(z_3) \overline{c}(k, z_1, z_3) \overline{h}(k, z_3, z_2). \tag{11.133}$$

The one-dimensional integral that remains is readily evaluated, and hence the transformed equation is in a form suitable for solution. Typically one evaluates the convolution integral once each iteration between Hankel and real space. Alternatively, it is possible to express this as a matrix equation, with the k-dependent elements representing pairs of z-layers, and to solve explicitly the resultant equations by matrix inversion. Because one must transform back and forth from Hankel space many times, numerically a discrete orthogonal transform is desirable. Lado has given an 'almost' orthogonal transform, which appears to be quite robust in practice.[18]

Representing the density on a grid of N_z points, and with a radial grid of N_r points, it takes an array of size $N_r N_z^2$ to store the pair correlation functions, and $N_r N_z^3$ operations to evaluate the convolution integral in Hankel space. The Hankel transforms take $N_r^2 N_z^2$ operations. The latter is the limiting step and a fast version of the Hankel transform would be highly desirable.

[18] F. Lado, Hypernetted-chain solutions for the two-dimensional classical electron gas, *Phys. Rev. B* **17** (1978), 2827.

The density profile equations are also simplified by a Hankel transform. The BGY equation becomes

$$\rho'(z_1) = -\rho(z_1)\beta V'(z_1) - \frac{\rho(z_1)}{(2\pi)^2}\int_{-\infty}^{\infty} \mathrm{d}z_2 \int \mathrm{d}\mathbf{k} \times \rho(z_2)\overline{g}(k, z_1, z_2)\frac{\partial \overline{u}^{(2)}(k, z_1, z_2)}{\partial z_1}, \tag{11.134}$$

the TZ equation is

$$\rho'(z_1) = -\rho(z_1)\beta V'(z_1) + \int_{-\infty}^{\infty} \mathrm{d}z_2\, \rho'(z_2)\overline{c}(0, z_1, z_2), \tag{11.135}$$

and the WLMB equation is

$$\rho'(z_1) = -\rho(z_1)\beta V'(z_1) - \beta\rho(z_1)\int_{-\infty}^{\infty} \mathrm{d}z_2\, \rho(z_2)V'(z_2)\overline{h}(0, z_1, z_2). \tag{11.136}$$

The planar inhomogeneous Ornstein–Zernike equation has been solved for isolated[19] and for interacting[20] planar walls.

11.4.3 Spherical Inhomogeneity

For the case of a spherically inhomogeneous system, such as that around a spherical solute, that confined to a spherical pore, or that of bubbles or droplets, the density depends upon the distance from the origin, $\rho(r)$, and the pair correlations depend upon the two distances of the solvent molecules from the origin r_1 and r_2 and the cosine of the angle between them, $x_{12} = \cos\theta_{12}$. That is, $h(\mathbf{r}_1, \mathbf{r}_2) = h(r_1, r_2, x_{12})$, and the actual separations between the atoms is $R_{12} = \sqrt{r_1^2 + r_2^2 - 2r_1r_2x_{12}}$. The Ornstein–Zernike equation is

$$h(r_1, r_2, x_{12}) = c(r_1, r_2, x_{12}) + \int \mathrm{d}\mathbf{r}_3\, \rho(r_3)h(r_1, r_3, x_{13})c(r_3, r_2, x_{32}). \tag{11.137}$$

In this case a Legendre transformation factorises the angular convolution,

$$\tilde{h}^{(n)}(r_1, r_2) = \tilde{c}^{(n)}(r_1, r_2) + \frac{4\pi}{2n+1}\int_0^{\infty} \mathrm{d}r_3\, r_3^2 \times \rho(r_3)\tilde{h}^{(n)}(r_1, r_3)\tilde{c}^{(n)}(r_3, r_2). \tag{11.138}$$

This factorisation follows from the orthogonality properties of the Legendre polynomials and from the addition theorem for spherical harmonics. Using the fact that

$$\cos\theta_{32} = \cos\theta_{12}\cos\theta_{13} + \sin\theta_{12}\sin\theta_{13}\cos\phi_3, \tag{11.139}$$

[19]S. Sokolowski, On the solution of the nonuniform Percus–Yevick equation, *J. Chem. Phys.* **73** (1980), 3507. R. M. Nieminen and N. W. Ashcroft, Theory of inhomogeneous fluids, *Phys. Rev. A* **24** (1981), 560.

[20]R. Kjellander and S. Marčelja, Inhomogeneous Coulomb fluids with image interactions between planar surfaces. I, *J. Chem. Phys.* **82** (1985), 2122.

the latter is

$$P_n(x_{32}) = 2 \sum_{m=1}^{n}{}' \frac{(n-m)!}{(n+m)!} P_n^m(x_{12}) P_n^m(x_{13}) e^{im\phi_3}. \tag{11.140}$$

The orthogonality relations are

$$\begin{aligned} \int_{-1}^{1} \mathrm{d}x\, P_n(x) P_m(x) &= \frac{2}{2n+1}\delta_{nm}, \text{ and} \\ \sum_{n=0}^{\infty} \frac{2n+1}{2} P_n(x') P_n(x) &= \delta(x-x'). \end{aligned} \tag{11.141}$$

In view of this the Legendre transforms are

$$\tilde{f}^{(n)} = \frac{2n+1}{2} \int_{-1}^{1} \mathrm{d}x\, P_n(x) f(x), \text{ and } f(x) = \sum_{n=0}^{\infty} \tilde{f}^{(n)} P_n(x). \tag{11.142}$$

The algorithm for solving the spherically inhomogeneous Ornstein–Zernike equation is similar to that of the planar case, as are the considerations for making a connection with the bulk far from the inhomogeneity. The numerical complexity is identical. A discrete orthogonal (but not fast) version of the Legendre transform performs well in practice, and a fast (but not orthogonal) version is known.[21]

The equations for the density profile in spherical geometry also simplify upon Legendre transform. The WLMB equation becomes

$$\begin{aligned} \rho'(r_1) &= -\beta\rho(r_1)V'(r_1) \\ &\quad - \frac{4\pi\beta\rho(r_1)}{3} \int_0^\infty \mathrm{d}r_2\, r_2^2 \rho(r_2) \tilde{h}^{(1)}(r_1, r_2) V'(r_2), \end{aligned} \tag{11.143}$$

where $V(r) \equiv u^{(1)}(r)$ is the external potential. The TZ equation is

$$\begin{aligned} \rho'(r_1) &= -\beta\rho(r_1)V'(r_1) \\ &\quad + \frac{4\pi\rho(r_1)}{3} \int_0^\infty \mathrm{d}r_2\, r_2^2 \tilde{c}^{(1)}(r_1, r_2) \rho'(r_2). \end{aligned} \tag{11.144}$$

The BGY expression becomes

$$\begin{aligned} \rho'(r_1) &= -\beta\rho(r_1)V'(r_1) - 2\pi\beta\rho(r_1) \sum_{n=0}^{\infty} \frac{2}{2n+1} \\ &\quad \times \int_0^\infty \mathrm{d}r_2\, r_2^2 \rho(r_2) \tilde{g}^{(n)}(r_1, r_2) \frac{\partial \tilde{u}^{(n)}(r_1, r_2)}{\partial r_1}. \end{aligned} \tag{11.145}$$

[21] P. Attard, Spherically inhomogeneous fluids. I. Percus–Yevick hard-spheres: Osmotic coefficients and triplet correlations, *J. Chem. Phys.* **91** (1989), 3072. P. Attard, Lennard-Jones bridge functions and triplet correlation functions, *J. Chem. Phys.* **95** (1991), 4471. B. K. Alpert and V. Rokhlin, A fast algorithm for the evaluation of Legendre expansions, *SIAM J. Sci. Stat. Comput.* **12** (1991), 158.

The spherically inhomogeneous Ornstein–Zernike equation is particularly interesting because the inhomogeneous pair correlation function in the presence of a spherical solvent is equivalent to the triplet correlation of the uniform bulk system (see the papers by Attard in the preceding footnote). It has also been used to treat a spherical solute.[22]

The planar and spherical geometries are the only two geometries in which the inhomogeneous pair correlation functions depend upon three variables, one of them in an homogeneous fashion. In contrast, for example, for systems with cylindrical symmetry they depend upon the two distances from the axis, the mutual angle, and the mutual separation along the axis. Currently it does not appear feasible to solve the inhomogeneous Ornstein–Zernike equation in any but planar and spherical geometry (or circular or linear geometry in two dimensions).

Summary

- Inhomogeneous systems may be treated by regarding the external potential as due to a solute at infinite dilution, with the density inhomogeneity being given by the solute–solvent distribution function. The interpretation of solute is quite general, and includes spherical, planar, and cylindrical bodies and pores, and multimolecular species. The solvation free energy may also be obtained.

- The solute–solute Ornstein–Zernike equation gives the pair-wise interaction of solutes. For the case of walls, it is the interaction free energy per unit area that is the relevant quantity.

- In the asymptotic regime, the rate of decay of the density profile to its bulk value and the rate of decay of the solute–solute interaction are at least as long-ranged as the solvent–solvent correlations. A long-ranged attractive interaction is induced between solutes approaching the critical point or the spinodal line.

- An inhomogeneous Ornstein–Zernike equation can be formulated, and this this can be solved with a standard closure and one of several exact equations for the density profile in terms of the external potential and the anisotropic pair correlation functions. It is feasible to treat planar and spherical inhomogeneous systems in this fashion.

[22] P. Attard, Spherically inhomogeneous fluids. II. Hard-sphere solute in a hard-sphere solvent, *J. Chem. Phys.* **91** (1989), 3083. M. Fushiki, An anisotropic hypernetted chain approximation for the spherical cell model, *Chem. Phys. Lett.* **154** (1989), 77.

Chapter 12

Coulomb Systems

12.1 Mean Field Approximation

Coulomb systems represent an important class that includes plasmas, molten salts, and electrolytes. The latter is perhaps the most frequently occurring and it arises from the dissociation of ionic salts in polar solvents such as water. From the theoretical point of view the long-range tail of the Coulomb potential creates both opportunities and challenges for analysis. It allows a number of exact asymptotic results to be obtained, and because to a large extent it determines the properties of the electrolyte, it provides the basis for several simple, analytic approximations.[1]

The simplest model of an electrolyte is the so-called primitive model, in which the pair potential consists of a hard-sphere and a Coulomb term, $u_{\alpha\gamma}(r) = u_{\alpha\gamma}^{\rm hs}(r) + u_{\alpha\gamma}^{\rm Coul}(r)$. The hard-sphere potential is

$$u_{\alpha\gamma}^{\rm hs}(r) = \begin{cases} \infty & r < d_{\alpha\gamma} \\ 0 & r > d_{\alpha\gamma}. \end{cases} \tag{12.1}$$

where the Greek indices label the species, and $d_{\alpha\gamma}$ represents the distance of closest approach of the centres of the two ions of the given type. One usually deals with additive hard-sphere diameters, $d_{\alpha\gamma} = d_\alpha + d_\gamma$. Sometimes the finite size of the ions is instead modelled with a soft-core repulsion (e.g., r^{-12}).

Coulomb's law *in media* is

$$u_{\alpha\gamma}^{\rm Coul}(r) = \frac{q_\alpha q_\gamma}{\epsilon r}, \tag{12.2}$$

[1]There have been a number of reviews of statistical mechanical techniques and results for the electric double layer: S. L. Carnie and G. M. Torrie, The statistical mechanics of the electrical double layer, *Adv. Chem. Phys.* **56** (1984), 141. L. Blum, Structure of the electric double layer, *Adv. Chem. Phys.* **78** (1990), 171. P. Attard, Electrolytes and the electric double layer, *Adv. Chem. Phys.* **92** (1996), 1. The analysis in this chapter follows closely that of P. Attard, Asymptotic analysis of primitive model electrolytes and the electrical double layer, *Phys. Rev. E* **48** (1993), 3604.

where q_α is the charge an ion of type α. In the primitive model, the solvent is subsumed into a continuum dielectric constant: $\epsilon = 4\pi\epsilon_0\epsilon_r$ is the total permittivity, with ϵ_0 being the permittivity of free space, and ϵ_r being the relative dielectric constant of the medium. Accordingly, the Coulomb potential is the longest-ranged pair part of the interaction free energy that arises from integrating out the solvent and electronic degrees of freedom. Similarly, the ionic diameter used is generally larger than the spacing in an ionic crystal to account for the first solvation shell of the ions, which may be argued to be irremovable. The simplest version of the primitive model, a binary symmetric electrolyte with all ions being the same size and the two species having equal but opposite charge, is called the restricted primitive model.

12.1.1 Poisson–Boltzmann Approximation

The most wide-spread approach to Coulomb systems is the Poisson–Boltzmann approximation. This is a mean field approximation that replaces the local potential felt by an ion by the mean electrostatic potential at that position. The approximation can be applied to inhomogeneous systems where there is an externally applied potential (see the electric double layer below), but here it will be illustrated for a uniform system.

For a homogeneous electrolyte, the mean electrostatic potential at a distance r from an ion of type γ is

$$\psi_\gamma(r) = \frac{q_\gamma}{\epsilon r} + \sum_\lambda \rho_\lambda \int \mathrm{d}\mathbf{s}\, g_{\gamma\lambda}(s) \frac{q_\lambda}{\epsilon|\mathbf{r}-\mathbf{s}|}. \tag{12.3}$$

This comprises the direct potential due to the ion γ, plus the contribution from the other ions in the electrolyte. The first term in the integrand represents the probability of finding an ion of type λ at $\mathbf{s}$, and the second term is the contribution to the electrostatic potential at $\mathbf{r}$ from that ion.

The radial distribution function is obtained by approximating the potential of mean force by this mean electrostatic potential

$$g_{\alpha\gamma}(r) = \exp[-\beta q_\alpha \psi_\gamma(r)], \tag{12.4}$$

where $\beta = 1/k_{\mathrm{B}}T$. This is known as the Boltzmann approximation. It is an approximation because the actual potential of mean force includes the change in the local potential from the mean value due to correlations between the ions, as well as nonelectrostatic effects such as those due to the finite size of the ions (e.g., hard core). In so far as the long-range part of the Coulomb potential dominates the interaction between the ions (i.e., dilute systems), one may expect this approximation to be reliable. Considering $\rho_\alpha(r) = \rho_\alpha g_{\alpha\gamma}(r)$ as the density profile about the ion γ considered as a source of an external field, it may be seen that the Boltzmann approximation is identical to the mean field theory of Section 6.5.

These two equations may be solved simultaneously. It is traditional to recast the integral form for the mean potential into a differential equation by taking the divergence of the above. This is Poisson's equation,

$$\nabla^2 \psi_\gamma(r) = \frac{-4\pi}{\epsilon} \sum_\lambda q_\lambda \rho_\lambda g_{\gamma\lambda}(r), \tag{12.5}$$

which says that the divergence of the electrostatic potential is proportional to the charge density. The combination of the exact Poisson equation and the approximate Boltzmann form is known as the Poisson–Boltzmann approximation. In the present case it is

$$\frac{1}{r^2}\frac{\partial}{\partial r}\left(r^2 \frac{\partial \psi_\gamma(r)}{\partial r}\right) = \frac{-4\pi}{\epsilon} \sum_\lambda q_\lambda \rho_\lambda \exp[-\beta q_\lambda \psi_\gamma(r)]. \tag{12.6}$$

12.1.2 Linear Theory

In the case that the mean electrostatic potential is weak, $|\beta q_\lambda \psi_\gamma(r)| \ll 1$, which is true at high temperatures or at large separations, then one can linearise the right-hand side of this to obtain

$$\begin{aligned} \frac{1}{r^2}\frac{\partial}{\partial r}\left(r^2 \frac{\partial \psi_\gamma(r)}{\partial r}\right) &= \frac{-4\pi}{\epsilon} \sum_\lambda q_\lambda \rho_\lambda \left[1 - \beta q_\lambda \psi_\gamma(r)\right] \\ &= \kappa_{\mathrm{D}}^2 \psi_\gamma(r). \end{aligned} \tag{12.7}$$

Here the electro-neutrality condition for the bulk electrolyte has been invoked,

$$\sum_\lambda q_\lambda \rho_\lambda = 0. \tag{12.8}$$

This is an important constraint that all real systems must obey and that will recur several times below. Also, the Debye parameter has been defined,

$$\kappa_{\mathrm{D}}^2 \equiv \frac{4\pi\beta}{\epsilon} \sum_\lambda q_\lambda^2 \rho_\lambda. \tag{12.9}$$

The quantity κ_{D}^{-1} is known as the *Debye screening length*; it sets the length scale of the interactions in an electrolyte, at least in this approximation. It is in inverse proportion to the square root of the concentration, and it is shorter in multivalent electrolytes than in monovalent.

It may be confirmed by differentiation that the decaying solution to this linearised Poisson–Boltzmann equation is

$$\psi_\gamma(r) = \frac{\psi_\gamma}{r} e^{-\kappa_{\mathrm{D}} r}. \tag{12.10}$$

Accordingly, the radial distribution function is given by

$$g_{\alpha\gamma}(r) = 1 - \beta \frac{q_\alpha \psi_\gamma}{r} e^{-\kappa_{\mathrm{D}} r}. \tag{12.11}$$

From these it is clear that the mean potential due to an ion is exponentially screened, and hence so are the correlations between ions. The decay length is given by the Debye length, which as mentioned above decreases with increasing concentration and valence, and with decreasing dielectric constant. This says that the charge on each ion is more efficiently screened by the surrounding ion cloud, and that the correlations amongst the ions are increasingly short-ranged, as the coupling between the ions is increased.

Just as the electrolyte itself must be electro-neutral, so must the charge in the ion cloud surrounding an ion be equal and opposite to that on the ion, as will be proven below. This determines the prefactor,

$$\begin{aligned} -q_\gamma &= \sum_\lambda q_\lambda \rho_\lambda \int \mathrm{d}\mathbf{r}\, g_{\gamma\lambda}(r) \\ &= -4\pi\beta\psi_\gamma \sum_\lambda q_\lambda^2 \rho_\lambda \int_0^\infty \mathrm{d}r\, r e^{-\kappa_\mathrm{D} r} \\ &= -\epsilon\psi_\gamma. \end{aligned} \tag{12.12}$$

Hence the total correlation function, $h = g - 1$, is

$$h_{\alpha\gamma}(r) = \frac{-\beta q_\alpha q_\gamma}{\epsilon r} e^{-\kappa_\mathrm{D} r}. \tag{12.13}$$

This shows that the amplitude of the correlations is proportional to the product of the charges on the ions, which means that it is symmetric in the indices, as it must be.

12.1.3 Debye–Hückel Theory

The linear Poisson–Boltzmann approximation applied to electrolytes is known as Debye–Hückel theory. It may be used to obtain the thermodynamic properties of the electrolyte. In this theory the size of the ions is neglected, $d = 0$.

Using the virial expression, Eq. (7.88), Debye–Hückel theory gives for the excess pressure of the electrolyte,

$$\begin{aligned} p^{\mathrm{DH,ex}} &= \frac{-1}{6} \sum_{\alpha\gamma} \rho_\alpha \rho_\gamma \int \mathrm{d}\mathbf{r}\, r \frac{\mathrm{d}u^{\mathrm{Coul}}_{\alpha\gamma}(r)}{\mathrm{d}r} g_{\alpha\gamma}(r) \\ &= \frac{4\pi}{6} \sum_{\alpha\gamma} \rho_\alpha \rho_\gamma \int_0^\infty \mathrm{d}r\, r^3 \frac{-q_\alpha q_\gamma}{\epsilon r^2} \left[1 - \frac{\beta q_\alpha q_\gamma}{\epsilon r} e^{-\kappa_\mathrm{D} r} \right] \\ &= \frac{-\kappa_\mathrm{D}^3}{24\pi\beta}. \end{aligned} \tag{12.14}$$

Here the bulk electro-neutrality condition, $\sum_\alpha \rho_\alpha q_\alpha = 0$, has been used to convert the radial distribution function to the total correlation function, as well as the definition of the Debye length, Eq. (12.9). The ideal contribution is of course $p^{\mathrm{id}} = k_\mathrm{B} T \sum_\alpha \rho_\alpha$. The Debye–Hückel theory is valid in the limit of vanishing concentrations. This result shows that the leading correction to the ideal gas behaviour goes like $\rho^{3/2}$, not the ρ^2 predicted by the virial expansion. It can be confirmed that the individual virial coefficients diverge for the Coulomb potential, and hence this Debye–Hückel $\rho^{3/2}$ result arises from a resummation of all the virial coefficients.

The chemical potential in the zero-size mean spherical approximation is given by Eq. (9.106). Hence one has

$$\begin{aligned}
\mu_\alpha^{\rm DH,ex} &= \sum_\gamma \rho_\gamma \int {\rm d}\mathbf{r}\, u_{\alpha\gamma}^{\rm Coul}(r) \left[1 + \frac{1}{2} h_{\alpha\gamma}(r)\right] \\
&= \frac{4\pi}{2} \sum_\gamma \rho_\gamma \int_0^\infty {\rm d}r\, r^2 \frac{-\beta q_\alpha^2 q_\gamma^2}{\epsilon^2 r^2} e^{-\kappa_{\rm D} r} \\
&= \frac{-q_\alpha^2 \kappa_{\rm D}}{2\epsilon},
\end{aligned} \tag{12.15}$$

where again the bulk electro-neutrality condition has been used. In view of the fact that the grand potential density is $\Omega/V = -p$, and the fact that the Helmholtz free energy density is $F/V = \Omega/V + \sum_\alpha \rho_\alpha \mu_\alpha$, one has

$$\beta F_\alpha^{\rm DH,ex}/V = \frac{\kappa_{\rm D}^3}{24\pi} + \beta \sum_\alpha \rho_\alpha \frac{-q_\alpha^2 \kappa_{\rm D}}{2\epsilon} = \frac{-\kappa_{\rm D}^3}{12\pi}. \tag{12.16}$$

12.1.4 Modified Debye–Hückel Theory

At higher concentrations the hard core of the ions can no longer be neglected. The exponential behaviour given for the total correlation function by Debye–Hückel theory cannot hold within the hard-core region. One must have $h_{\alpha\gamma}(r) = -1$, $r < d$, assuming that the ions are all the same size. This means that the electro-neutrality condition is no longer satisfied by applying Eq. (12.13) beyond the core. Retaining the Yukawa form but scaling the prefactor to satisfy electro-neutrality, one obtains the modified linearised Debye–Hückel approximation,

$$h_{\alpha\gamma}(r) = \begin{cases} -1, & r < d \\ \dfrac{-\beta q_\alpha q_\gamma e^{\kappa_{\rm D} d}}{\epsilon[1 + \kappa_{\rm D} d]} \dfrac{e^{-\kappa_{\rm D} r}}{r}, & r > d. \end{cases} \tag{12.17}$$

A problem with this theory is that it allows the coion radial distribution function to become negative. This problem may be traced to the linearisation of the exponential, which is only valid for low potentials. Unfortunately one cannot give a closed form analytic solution to the nonlinear Poisson–Boltzmann equation in this spherical geometry. An alternative approximation is to retain the Yukawa form for the potential of mean force and to exponentiate this for the radial distribution function,

$$g_{\alpha\gamma}(r) = \exp\left[\frac{-\beta q_\alpha q_\gamma}{\epsilon r} e^{-\kappa_{\rm D} r}\right], r > d. \tag{12.18}$$

It is evident that there are certain inconsistencies in this approach (it satisfies neither electro-neutrality nor Poisson's equation). When the nonlinear terms are negligible, the linear and the nonlinear total correlation functions will be approximately equal and the theory could be said to be internally consistent. However, when the nonlinear terms make a contribution, it is arguably better to use the exponential of the potential of mean force even though this is not fully consistent with the linearisation that was assumed to solve Poisson's equation. This nonlinear version has the same asymptotic form as the modified Debye–Hückel approximation (albeit with a different prefactor), but experience shows that overall it tends to be a better approximation, in part because it always remains physical at small separations.

A more sophisticated approach is to solve the nonlinear Poisson–Boltzmann equation numerically. This approach is used widely in studies of inhomogeneous electrolytes and the electric double layer.[2] For the case of the present uniform electrolyte one should be aware that the nonlinear theory is inconsistent; for different sized ions the pair correlation functions are not symmetric in the indices, $q_\alpha \psi_\gamma(r) \neq q_\gamma \psi_\alpha(r)$.

12.1.5 Electric Double Layer

In the case of an applied electric potential $\psi^{\text{ext}}(\mathbf{r})$, the electrolyte becomes inhomogeneous and the expression for the mean electrostatic potential is

$$\psi(\mathbf{r}) = \psi^{\text{ext}}(\mathbf{r}) + \sum_\alpha \int_V \mathrm{d}\mathbf{s}\, \rho_\alpha(\mathbf{s}) \frac{q_\alpha}{\epsilon |\mathbf{r} - \mathbf{s}|}. \tag{12.19}$$

The Boltzmann equation is

$$\psi_\alpha(\mathbf{r}) = \rho_{\alpha;0} \exp[-\beta q_\alpha \psi(\mathbf{r}), \tag{12.20}$$

where $\rho_{\alpha;0}$ is the density far from the influence of the external field, $\psi(\mathbf{r}) = 0$. Rewriting the mean potential as Poisson's equation, the nonlinear Poisson–Boltzmann equation is

$$\nabla^2 \psi(\mathbf{r}) = \frac{-4\pi}{\epsilon} \sum_\alpha q_\alpha \rho_{\alpha;0} \exp[-\beta q_\alpha \psi(\mathbf{r}). \tag{12.21}$$

These equations may be compared with the mean field theory of Section 6.5. It may be seen that the expressions given there for the thermodynamic and for the fluctuation potential can be immediately applied to the Poisson–Boltzmann approximation for the electric double layer.

[2] S. L. Carnie, D. Y. C. Chan, and J. Stankovich, Computation of forces between spherical colloidal particles: Nonlinear Poisson–Boltzmann theory, *J. Colloid Interface Sci.* **165** (1994), 116. B. Honig, K. Sharp, and A.-S. Yang, Macroscopic models of aqueous solutions: Biological and chemical applications, *J. Phys. Chem.* **97** (1993), 1101. M. K. Gilson, M. E. Davis, B. A. Luty, and J. A. McCammon, Computation of electrostatic forces on solvated molecules using the Poisson–Boltzmann equation, *J. Phys. Chem.* **97** (1993), 3591. K. E. Forsten, R. E. Kozack, D. A. Lauffenburger, and S. Subramaniam, Numerical solution of the nonlinear Poisson–Boltzmann equation for a membrane-electrolyte system, *J. Phys. Chem.* **98** (1994), 5580.

12.2 Electrolytes

12.2.1 Formalism and Moment Conditions

One of the most important consequences of the long-range of the Coulomb potential is electro-neutrality. As mentioned above the total charge in an electrolyte must vanish, $\sum_\alpha q_\alpha \rho_\alpha = 0$. It is also true that the charge on each ion is balanced by an opposite charge in the ion cloud that surrounds it, as is now shown.

For a multicomponent fluid, the Ornstein–Zernike equation, Eq. (9.33), is

$$h_{\alpha\gamma}(r) = c_{\alpha\gamma}(r) + \sum_\lambda \rho_\lambda \int \mathrm{d}\mathbf{s}\, h_{\alpha\lambda}(s) c_{\lambda\gamma}(|\mathbf{r}-\mathbf{s}|). \tag{12.22}$$

Defining symmetric matrices by

$$\{\underline{\underline{H}}(r)\}_{\alpha\gamma} = \rho_\alpha^{1/2}\rho_\gamma^{1/2} h_{\alpha\gamma}(r) \tag{12.23}$$

and

$$\{\underline{\underline{C}}(r)\}_{\alpha\gamma} = \rho_\alpha^{1/2}\rho_\gamma^{1/2} c_{\alpha\gamma}(r), \tag{12.24}$$

the Ornstein–Zernike equation can be written

$$\underline{\underline{H}}(r) = \underline{\underline{C}}(r) + \int \mathrm{d}\mathbf{s}\, \underline{\underline{H}}(s)\underline{\underline{C}}(|\mathbf{r}-\mathbf{s}|), \tag{12.25}$$

with Fourier transform

$$\underline{\underline{\hat{H}}}(k) = \underline{\underline{\hat{C}}}(k) + \underline{\underline{\hat{H}}}(k)\underline{\underline{\hat{C}}}(k). \tag{12.26}$$

In view of Coulomb's law *in media*, one defines the dyadic matrix

$$\underline{\underline{Q}} = \frac{4\pi\beta}{\epsilon} \underline{q}\, \underline{q}^{\mathrm{T}}, \tag{12.27}$$

where $\beta = 1/k_{\mathrm{B}}T$ and where the components of the column vector $\underline{q}$ are

$$\{\underline{q}\}_\alpha = \rho_\alpha^{1/2} q_\alpha. \tag{12.28}$$

The trace of this matrix is

$$\mathrm{Tr}\{\underline{\underline{Q}}\} = \frac{4\pi\beta}{\epsilon} \underline{q}^{\mathrm{T}}\, \underline{q} = \frac{4\pi\beta}{\epsilon} \sum_\alpha \rho_\alpha q_\alpha^2 \equiv \kappa_{\mathrm{D}}^2, \tag{12.29}$$

where the Debye parameter was introduced above; κ_{D}^{-1} is the screening length of the electrolyte, at least at low concentrations. In view of this it is straightforward to show that

$$\underline{\underline{Q}}^{n+1} = \kappa_{\mathrm{D}}^{2n} \underline{\underline{Q}}, \tag{12.30}$$

so that although $\underline{\underline{Q}}$ itself is singular, $\mathrm{Det}\{\underline{\underline{Q}}\} = 0$, and one has

$$(\underline{\underline{I}} + \alpha \underline{\underline{Q}})^{-1} = \underline{\underline{I}} - \frac{\alpha}{1+\alpha\kappa_{\mathrm{D}}^2}\underline{\underline{Q}}. \tag{12.31}$$

Zeroth-Moment Condition

In what follows it is assumed that the total correlation function is integrable,

$$\int \mathrm{d}\mathbf{r}\, h_{\alpha\gamma}(r) \equiv \hat{h}_{\alpha\gamma}(0) < \infty, \tag{12.32}$$

which is true if the compressibility is finite. Also, the asymptotic analysis in previous chapters showed that the direct correlation function goes like $c_{\alpha\gamma}(r) \sim -\beta u_{\alpha\gamma}^{\mathrm{Coul}}(r)$, $r \to \infty$. In view of this asymptote one defines

$$\chi_{\alpha\gamma}(r) \equiv c_{\alpha\gamma}(r) + \beta u_{\alpha\gamma}^{\mathrm{Coul}}(r). \tag{12.33}$$

The working hypothesis is that $\chi_{\alpha\gamma}(r)$ is of shorter range than the Coulomb potential. (Later it will be shown to be exponentially decaying.)

The Fourier transform of the Coulomb potential is $\hat{u}_{\alpha\gamma}^{\mathrm{Coul}}(k) = 4\pi q_\alpha q_\gamma/\epsilon k^2$, and hence the Ornstein–Zernike equation may be written

$$\underline{\underline{\hat{H}}}(k) = \underline{\underline{\hat{\chi}}}(k) - \underline{\underline{Q}}k^{-2} + \underline{\underline{\hat{H}}}(k)\underline{\underline{\hat{\chi}}}(k) - \underline{\underline{\hat{H}}}(k)\underline{\underline{Q}}k^{-2}. \tag{12.34}$$

Now in the limit $k \to 0$, $k^2\underline{\underline{\hat{\chi}}}(k) \to 0$, since $\chi(r)$ decays faster than the Coulomb potential. Hence multiplying both sides by k^2 and taking the limit, in view of the integrability of the total correlation function, one must have

$$0 = -\underline{\underline{Q}} - \underline{\underline{\hat{H}}}(0)\underline{\underline{Q}}. \tag{12.35}$$

Explicitly, this is the electro-neutrality condition

$$q_\alpha = -\sum_\gamma \rho_\gamma q_\gamma \int \mathrm{d}\mathbf{r}\, h_{\alpha\gamma}(r), \tag{12.36}$$

which expresses the fact that each ion is surrounded by a cloud of ions bearing a net equal and opposite charge. The electro-neutrality condition provides constraints on the sums of the zeroth moments of the total correlation functions.

Moments

The electro-neutrality condition is also called the zeroth moment condition, and it is here appropriate to define the moments. In general the transform of an integrable function possesses a small-k Taylor series expansion; in the case of radial functions only even powers of k appear. One has

$$\underline{\underline{\hat{H}}}(k) \sim \underline{\underline{H}}^{(0)} + \underline{\underline{H}}^{(2)}k^2 + \underline{\underline{H}}^{(4)}k^4 + \ldots,\ k \to 0, \tag{12.37}$$

and similarly for $\underline{\underline{\hat{\chi}}}(k)$. This expression may be obtained by expanding the integrand of the Fourier transform, and the moments are defined as

$$\underline{\underline{H}}^{(2n)} = \frac{4\pi(-1)^n}{(2n+1)!}\int_0^\infty \mathrm{d}\mathbf{r}\, \underline{\underline{H}}(r)\, r^{2n+2}. \tag{12.38}$$

In general only a finite number of moments exist; if $h(r) \sim r^{-\eta}$, $r \to \infty$, then the moment integral is divergent for $2n + 2 - \eta > -1$. All moments exist for an exponentially short-ranged function, and the task is to show that this is indeed the case for the correlation functions of the electrolyte.

Second-Moment Condition

The electro-neutrality condition or zeroth-moment condition arises from the equality of the coefficients of k^{-2} in the small-k Taylor expansion of the Fourier transform of the Ornstein–Zernike equation. It depends upon the long-range nature of the Coulomb potential, the integrability of the total correlation function, and the assumption that the direct correlation function is faster than the Coulomb potential. The second-moment condition uses the coefficients of k^0, and it is necessary to invoke the integrability of $\chi(r)$. At this stage this will simply be assumed, pending the proof in the next section that it is actually exponentially short-ranged.

The coefficients of k^0 in the small-k Taylor expansion of the Ornstein–Zernike equation may be equated to give

$$\underline{\underline{H}}^{(0)} = \underline{\underline{\chi}}^{(0)} + \underline{\underline{H}}^{(0)}\underline{\underline{\chi}}^{(0)} - \underline{\underline{H}}^{(2)}\underline{\underline{Q}}. \tag{12.39}$$

Premultiplying by $\underline{\underline{Q}}$, this gives

$$\underline{\underline{Q}}\,\underline{\underline{H}}^{(0)} = \underline{\underline{Q}}\,\underline{\underline{\chi}}^{(0)} + \underline{\underline{Q}}\,\underline{\underline{H}}^{(0)}\underline{\underline{\chi}}^{(0)} - \underline{\underline{Q}}\,\underline{\underline{H}}^{(2)}\underline{\underline{Q}}, \tag{12.40}$$

or, using Eq. (12.35),

$$-\underline{\underline{Q}} = -\underline{\underline{Q}}\,\underline{\underline{H}}^{(2)}\underline{\underline{Q}}. \tag{12.41}$$

Explicitly this is,

$$1 = \frac{-4\pi\beta}{6\epsilon} \sum_{\gamma\lambda} q_\gamma q_\lambda \rho_\gamma \rho_\lambda \int \mathrm{d}\mathbf{r}\, h_{\gamma\lambda}(r) r^2. \tag{12.42}$$

This is known as the Stillinger–Lovett second moment condition.[3]

For an m-component electrolyte there are m zeroth-moment conditions but only one second-moment condition; the latter provides a constraint on the total sum of the second-moments of the ion pair correlation functions but does not determine the individual second-moments. Summing the second-moments with weights equal to the ionic charges results in the cancellation of the terms involving the $\hat{\chi}_{\alpha\gamma}(0)$, which accounts for the universal nature of the result; it is independent of any short-range interactions between the ions. This is a universality similar to that displayed by the electro-neutrality condition.

12.2.2 Exponential Screening

It is now shown that either the short-range part of the direct correlation function decays as the square of the total correlation function or it decays exponentially. This will complete the above proofs of the moment conditions, and it used to show that the total correlation function decays exponentially.

[3]F. H. Stillinger and R. Lovett, Ion-pair theory for concentrated electrolytes. I. Basic concepts, *J. Chem. Phys.* **48** (1968), 3858. F. H. Stillinger and R. Lovett, General restriction on the distribution of ions in electrolytes, *J. Chem. Phys.* **49** (1968), 1991.

The argument is based upon the exact closure to the Ornstein–Zernike equation, (9.43),

$$\begin{aligned} h_{\alpha\gamma}(r) &= -1+\exp\left[-\beta u_{\alpha\gamma}^{\rm Coul}(r) - c_{\alpha\gamma}(r) + h_{\alpha\gamma}(r) + d_{\alpha\gamma}(r)\right] \\ &= -1+\exp\left[h_{\alpha\gamma}(r) - \chi_{\alpha\gamma}(r) + d_{\alpha\gamma}(r)\right]. \end{aligned} \tag{12.43}$$

The exponential can be linearised since the left-hand side must decay to 0 for large r in order for $h(r)$ to be integrable. Hence

$$h_{\alpha\gamma}(r) \sim -\chi_{\alpha\gamma}(r) + h_{\alpha\gamma}(r) + d_{\alpha\gamma}(r),\ r\to\infty, \tag{12.44}$$

where the neglected terms decay as the square of the pair correlation functions.

If the total correlation function is exponentially decaying, which is to be proved, then it can be shown that both the direct correlation function and the bridge function are also exponentially decaying (see below). At this stage the contrary will be assumed, namely that the total correlation function decays as an integrable power law, $h(r) \sim r^{-\eta}$, $\eta > 3$.

The bridge diagrams are composed of h-bonds and have no nodal points between the root points. Hence at least two h-bonds must bridge between the root points, and the individual diagrams of $d(r)$ cannot decay slower than the square of $h(r)$. This result is certainly true if $h(r)$ decays as an integrable power law because then the convolution integrals are dominated by regions with the field point close to one or other of the root points (because $(r/2)^{-2\eta} \ll r^{-\eta}$). One concludes that the bridge function goes like

$$d_{\alpha\gamma}(r) \sim \mathcal{O} h_{\alpha\gamma}(r)^2,\ r\to\infty, \tag{12.45}$$

at least when $h(r) \sim r^{-\eta}$ $\eta > 3$. Strictly speaking what appears on the right-hand side should be the most long-ranged of the various $h(r)$. It will later be shown that in fact the total correlation functions between different pairs of ions all have the same range; for the present the analysis may be interpreted as applying to the longest ranged of the correlation functions.

Since in the linearisation of the closure there remains nothing to cancel the short-range part of the direct correlation function, one concludes that if $h(r)$ decays as an integrable power law, then $\chi(r)$ must decay as the square of the total correlation function,

$$\chi_{\alpha\gamma}(r) \sim \mathcal{O} h_{\alpha\gamma}(r)^2,\ r\to\infty. \tag{12.46}$$

This is sufficient to ensure that the zeroth moment of $\chi(r)$ exists, as was assumed above. It also completes the proof of the earlier assumption that

$$c_{\alpha\gamma}(r) \sim -\beta u_{\alpha\gamma}^{\rm Coul}(r),\ r\to\infty. \tag{12.47}$$

Now for the case that $h(r)$ is exponentially decaying. It will be proven by induction that in this event all of the moments of $\chi(r)$ exist and hence it is also exponentially decaying. Assume that all of the $\underline{\underline{\chi}}^{(n)}$ exist for $n \le 2m-2$, where $m \ge 1$. Since $h(r)$ is assumed exponential all of the $\underline{\underline{H}}^{(l)}$ exist. Now equate the coefficients of k^{2m} in the small-k Taylor expansion of the Ornstein–Zernike equation,

$$\begin{aligned} \underline{\underline{H}}^{(2m)} &= \underline{\underline{\chi}}^{(2m)} + \left[\underline{\underline{H}}^{(2m)} \underline{\underline{\chi}}^{(0)} + \underline{\underline{H}}^{(2m-2)} \underline{\underline{\chi}}^{(2)} + \ldots + \underline{\underline{H}}^{(0)} \underline{\underline{\chi}}^{(2m)} \right] \\ & \quad + \underline{\underline{H}}^{(2m+2)} \underline{\underline{Q}}, \end{aligned} \tag{12.48}$$

which gives $\underline{\underline{\chi}}^{(2m)}$ as the sum of a finite number of moments, which by assumption are themselves finite. Hence if all the moments $\leq 2m-2$ exist, then $|\underline{\underline{\chi}}^{(2m)}| < \infty$. The zeroth moment exists (by inspection; note that Eq. (9.129) shows that the compressibility is nonzero, $\underline{\underline{I}} + \underline{\underline{\hat{H}}}(0) > 0$), and hence by induction all moments of $\chi(r)$ exist. Thus if $h(r)$ decays exponentially, then so does $\chi(r)$.

Thus it has been shown that either all moments of $\chi(r)$ exist or if $h(r)$ decays as an integrable power law, then $\chi(r)$ has at least as many moments as $h(r)$ (since it decays as the square of the latter). This is sufficient to prove that the total correlation function decays exponentially. The proof is again by induction on the moments. Assume that all moments, $\underline{\underline{H}}^{(n)}$, exist for $n \leq 2m$, $m \geq 1$. Since $\chi(r)$ is either exponential or at least as short-ranged as $h(r)$, all of the $\underline{\underline{\chi}}^{(n)}$, $n \leq 2m$, also exist. Equation (12.48) gives $\underline{\underline{H}}^{(2m+2)}$ as the sum of a finite number of moments, which by assumption are themselves finite. Hence if all the moments $\leq 2m$ exist, then $|\underline{\underline{H}}^{(2m+2)}| < \infty$. Since the zeroth moment exists by fundamental assumption, then, by induction, all moments of the total correlation function exist. This proves that the total correlation function must be at least exponentially decaying.

From the above the short-range part of the direct correlation function is also at least exponentially decaying, and, from Eq. (12.44), so is the bridge function. For the case of a simple fluid when $h(r)$ decays as an integrable power law, both $\chi(r)$ and $d(r)$ decay as its square (see above). For the present electrolyte with an exponentially decaying $h(r)$, $\chi(r) - d(r) \sim h(r)^2/2$, $r \to \infty$, but the two functions do not necessarily individually decay as the square of the total correlation function. In what follows it will be assumed that $\chi(r)$ and $d(r)$ have a decay length strictly less than that of $h(r)$. The mathematical justification of this requires some detailed diagrammatic analysis, but the basic idea is that $\chi(r)$ is the multiply connected subset of diagrams of $h(r)$ and hence cannot be of longer range. This is perhaps clearest for the bridge function where there are always two or more h-bonds connected in parallel between the root points, and hence $d(r)$ is more short-ranged than $h(r)$, and consequently so is $\chi(r)$.

This completes the proof that the total correlation function is exponentially decaying in a bulk electrolyte, where 'exponential' covers both monotonic and damped sinusoidal behaviour. The fact that the total correlation function for an electrolyte is more short-ranged than the pair potential is in marked contrast to fluids with integrable power-law potentials; for these the total correlation function decays at the same rate as the pair potential, with a coefficient proportional to the isothermal compressibility. (Hence the total correlation function is of exactly the same range as the full direct correlation function.)

The preceding discussion considered the effect of the long-range Coulomb potential alone, since one does not expect the additional short-range interactions present in real electrolytes to qualitatively change the behaviour deduced above. One expects that the charge–charge correlations will remain exponentially screened, and that the density–density correlations will decay in proportion to any power-law potentials that are present. (In the context of the present proof, only a finite number of moments exist for these power law potentials.)

In addition to modifying the interactions between the ions, one can envisage adding solvent as a specific molecular species. The civilised model electrolyte includes a multipolar solvent, and since the multipoles can be represented as a sum of discrete charges each of which is screened, then the solvent multipolar interactions themselves are screened. Hence the ion–ion, ion–solvent, and solvent–solvent correlation functions should all decay exponentially. However if the ions themselves carry multipole moments, only the charge–charge correlations are exponentially screened; the nonspherical projections of the ion pair correlation functions (such as the dipole–dipole) decay as power laws. In summary then, irrespective of the specific short-range interactions between ions or the presence of solvent or other additives, in an electrolyte the long-range Coulomb potential can be expected to cause the charge–charge correlations to decay exponentially.

12.2.3 Debye–Hückel Theory

The zeroth- and second-moment conditions, and the exponential screening of the correlation functions, relied only upon the long-range part of the Coulomb potential and the short-range of $\chi(r)$. One of the earliest approaches to electrolytes is the Debye–Hückel theory, which was derived above using the linear Poisson–Boltzmann approximation. It can also be derived by simply neglecting the short-range part of the direct correlation function,

$$\underline{\underline{\hat{\chi}}}(k) = 0. \tag{12.49}$$

For ions of 0 size this approximation is equivalent to the mean spherical closure approximation, Eq. (9.58). The Debye–Hückel theory is expected to be valid at low ionic couplings, for example in monovalent electrolyte at low concentrations. With it the Ornstein–Zernike equation (12.26) becomes

$$\begin{aligned}\underline{\underline{\hat{H}}}(k) &= -\left(\underline{\underline{I}} + \underline{\underline{Q}}k^{-2}\right)^{-1}\underline{\underline{Q}}k^{-2} \\ &= -\left(\underline{\underline{I}} - \frac{1}{k^2+\kappa_{\rm D}^2}\underline{\underline{Q}}\right)\underline{\underline{Q}}k^{-2} \\ &= \frac{-1}{k^2+\kappa_{\rm D}^2}\underline{\underline{Q}},\end{aligned} \tag{12.50}$$

where Eqs. (12.30) and (12.31) have been used. This has inverse transform

$$\underline{\underline{H}}(r) = \frac{-e^{-\kappa_{\rm D} r}}{4\pi r}\underline{\underline{Q}}, \tag{12.51}$$

or, in component form,

$$h_{\alpha\gamma}(r) = \frac{-\beta q_\alpha q_\gamma}{\epsilon r} e^{-\kappa_{\rm D} r}. \tag{12.52}$$

This was given above as Eq. (12.13). It shows that the correlations between the ions are exponentially screened, and that their range decreases as the coupling in the electrolyte increases.

Self-consistent Screening Length

The Debye–Hückel result can be modified to take into account the finite size of the ions, Eq. (12.17). Whilst the modified form was scaled so as to satisfy the electro-neutrality condition, it may be seen that it does not obey the second-moment condition. For finite-sized ions the modified Debye–Hückel result can only satisfy the Stillinger–Lovett condition by allowing for an effective screening length to replace the Debye length. That is,

$$h_{\alpha\gamma}(r) = \begin{cases} -1, & r < d \\ \dfrac{-\beta q_\alpha q_\gamma e^{\kappa d}}{\epsilon[1+\kappa d]} \dfrac{\kappa^2}{\kappa_{\rm D}^2} \dfrac{e^{-\kappa r}}{r}, & r > d. \end{cases} \tag{12.53}$$

This satisfies the electro-neutrality condition, and κ is determined by the second-moment condition,

$$\begin{aligned} 1 &= \frac{-4\pi\beta}{6\epsilon} \sum_{\alpha\gamma} q_\alpha \rho_\alpha q_\gamma \rho_\gamma \int_d^\infty {\rm d}r \, \frac{-\beta q_\alpha q_\gamma e^{\kappa d}}{\epsilon[1+\kappa d]} \frac{\kappa^2}{\kappa_{\rm D}^2} \frac{e^{-\kappa r}}{r} 4\pi r^4 \\ &= \frac{\kappa_{\rm D}^2}{\kappa^2} \frac{1+\kappa d + (\kappa d)^2/2 + (\kappa d)^3/6}{1+\kappa d}. \end{aligned} \tag{12.54}$$

The assumption of purely exponential profiles is only expected to be valid for small κd, and one expects this result to correct the Debye–Hückel theory in at least this regime. This expression only yields sensible results for $\kappa_{\rm D} d \le \sqrt{6}$, which is the upper bound established by Stillinger and Lovett for monotonic ion correlations, $g_{++}(r) \le g_{+-}(r)$, $r > d$, in the restricted primitive model. At concentrations higher than this the correlations *must* be oscillatory; in practice the oscillations occur before the bound is reached.

One could obviously implement more sophisticated versions of the present approximation (such as a nonlinear version with exponential potential of mean force, or a version for ions with different diameters), or one can go beyond the mean spherical approximation. What is more important than the approximation itself is the underlying point: the screening length of the electrolyte is *not* the Debye length (although it is approximately equal to the Debye length at low concentrations).

It is worth mentioning that analytic solutions exist for the mean spherical approximation for the primitive model.[4] For the restricted primitive model, this

[4] E. Waisman and J. L. Lebowitz, Mean spherical model integral equation for charged hard spheres. I and II, *J. Chem. Phys.* **56** (1972), 3086 and 3093. J. S. Høye and L. Blum, The mean spherical model for asymmetric electrolytes: Thermodynamics and the pair correlation function, *Mol. Phys.* **35** (1978), 299.

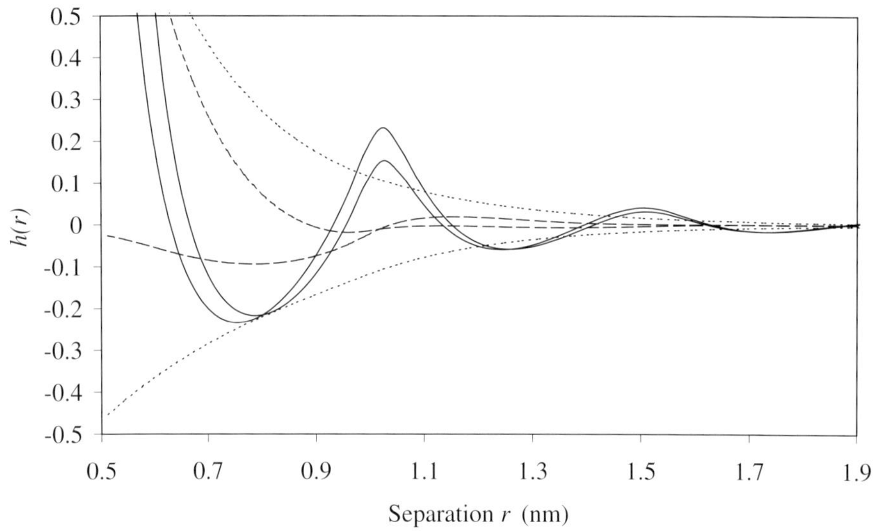

Figure 12.1: Total correlation functions in restricted primitive model electrolytes (monovalent, $\epsilon_r = 78.5$, $T = 300$ K, $d = 0.5$ nm, hypernetted chain approximation). Three regimes are shown: monotonic (0.5 M, dotted curves), oscillatory electrostatic (2 M, dashed curves), and oscillatory core (5 M, solid curves). For each pair of curves, the counterion curve is greater than the corresponding coion curve near contact.

approximation gives for the screening length $\kappa^{-1} = d/[-1+\sqrt{1+2\kappa_{\rm D} d}]$. Analytic solutions for the mean spherical approximation also exist for the 'civilised' model in which the solvent is modelled as multipolar hard spheres.[5]

12.2.4 Asymptotic Analysis

For the restricted primitive model (binary symmetric electrolyte), there are three types of asymptotic behaviour of the ion–ion correlation functions (see Fig. 12.1).[6] At low concentrations (more precisely low ion coupling) the total correlation function decays monotonically; the counterion density about an ion always exceeds the coion density. This is the exponential behaviour of Debye–Hückel theory. A transition to oscillatory behaviour occurs as the concentration is increased. First electrostatic-dominated oscillatory decay, where the oscillations are in the charge density, occurs, and the counterion and coion densities oscillate out of phase. Then at the highest couplings (core-dominated oscillatory decay), it is the number density that is oscillatory, and here the counterion

[5] L. Blum, Mean spherical model for a mixture of charged spheres and hard dipoles, *Chem. Phys. Lett.* **26** (1974), 200. J. S. Høye, J. L. Lebowitz and G. Stell, Generalised mean spherical approximations for polar and ionic fluids, *J. Chem. Phys.* **61** (1974), 3253.

[6] P. Attard, Asymptotic analysis of primitive model electrolytes and the electrical double layer, *Phys. Rev. E* **48** (1993), 3604.

and coion densities are in phase. In the core-dominated regime the period is close to the ionic diameter, whereas in the electrostatic regime it is much larger; the charge oscillations correspond to alternating shells of positive and negative charge, and hence the period in this case is at least $2d$, and can be considerably larger than this close to the monotonic–oscillatory transition. (Electrolytes with more components have a richer variety of oscillatory asymptotic phases, depending upon their symmetry and the linear combinations that may be formed.)[7]

Monotonic Asymptotic Decay

Solving the Ornstein–Zernike equation, (12.34), for the total correlation function one obtains

$$\begin{aligned}
\underline{\underline{\hat{H}}}(k) &= \left(\underline{\underline{I}} - \underline{\underline{\hat{\chi}}}(k) + \underline{\underline{Q}}k^{-2}\right)^{-1}\left(\underline{\underline{\hat{\chi}}}(k) - \underline{\underline{Q}}k^{-2}\right) \\
&= \left(\underline{\underline{I}} + \left(\underline{\underline{I}} - \underline{\underline{\hat{\chi}}}(k)\right)^{-1} \underline{q}\,\underline{q}^{\mathrm{T}} \frac{4\pi\beta}{\epsilon k^2}\right)^{-1} \\
&\quad \times \left(\underline{\underline{I}} - \underline{\underline{\hat{\chi}}}(k)\right)^{-1}\left(\underline{\underline{\hat{\chi}}}(k) - \underline{q}\,\underline{q}^{\mathrm{T}} \frac{4\pi\beta}{\epsilon k^2}\right) \\
&= \left(\underline{\underline{I}} + \underline{\tilde{q}}(k)\underline{q}^{\mathrm{T}} \frac{4\pi\beta}{\epsilon k^2}\right)^{-1}\left(\underline{\underline{I}} - \underline{\underline{\hat{\chi}}}(k)\right)^{-1} \\
&\quad \times \left(\underline{\underline{\hat{\chi}}}(k) - \underline{q}\,\underline{q}^{\mathrm{T}} \frac{4\pi\beta}{\epsilon k^2}\right). \qquad (12.55)
\end{aligned}$$

The effective charge function defined here,

$$\underline{\tilde{q}}(k) = \left(\underline{\underline{I}} - \underline{\underline{\hat{\chi}}}(k)\right)^{-1} \underline{q}, \qquad (12.56)$$

may be equivalently written

$$\underline{\tilde{q}}(k) = \underline{q} + \underline{\underline{\hat{\chi}}}(k)\underline{\tilde{q}}(k). \qquad (12.57)$$

Using Eq. (12.31) the first inverse is readily evaluated,

$$\left(\underline{\underline{I}} + \underline{\tilde{q}}(k)\underline{q}^{\mathrm{T}} \frac{4\pi\beta}{\epsilon k^2}\right)^{-1} = \underline{\underline{I}} - \frac{4\pi\beta/\epsilon}{k^2 + \Lambda(k)^2}\underline{\tilde{q}}(k)\underline{q}^{\mathrm{T}}, \qquad (12.58)$$

where a function that will become the screening length has been defined,

$$\Lambda(k)^2 = \frac{4\pi\beta}{\epsilon}\underline{\tilde{q}}^{\mathrm{T}}(k)\underline{q}. \qquad (12.59)$$

(This should be compared with the definition of the Debye length, Eq. (12.29).)

A pole in the total correlation function will occur at $k = i\kappa$, where the inverse decay length satisfies $\Lambda(i\kappa) = \kappa$. This determines its asymptotic behaviour, assuming that $\chi(r)$ is more short-ranged than $h(r)$, and hence that any singular

[7] For reviews of the various asymptotic analyses that have been performed for Coulomb systems, see the first footnote to this chapter.

behaviour of $\hat{\underline{\underline{\chi}}}(k)$ occurs further from the origin than this κ. There is a single pole corresponding to the vanishing of the scalar denominator that multiplies the remaining matrices. Hence each element of the matrix of total correlation functions has a pole located at $\Lambda(i\kappa)$, and hence each will have the same asymptotic decay. Neglecting the regular part, one has

$$
\begin{aligned}
\hat{\underline{\underline{H}}}(k) &\sim \frac{-4\pi\beta/\epsilon}{k^2+\Lambda(k)^2}\tilde{\underline{q}}(k)\underline{q}^{\mathrm{T}}\left(\underline{\underline{I}}-\hat{\underline{\underline{\chi}}}(k)\right)^{-1}\left(\hat{\underline{\underline{\chi}}}(k)-\underline{q}\,\underline{q}^{\mathrm{T}}\frac{4\pi\beta}{\epsilon k^2}\right)\\
&= \frac{-4\pi\beta/\epsilon}{k^2+\Lambda(k)^2}\tilde{\underline{q}}(k)\tilde{\underline{q}}^{\mathrm{T}}(k)\left(\hat{\underline{\underline{\chi}}}(k)-\underline{q}\,\underline{q}^{\mathrm{T}}\frac{4\pi\beta}{\epsilon k^2}\right)\\
&= \frac{-4\pi\beta/\epsilon}{k^2+\Lambda(k)^2}\tilde{\underline{q}}(k)\left(\tilde{\underline{q}}^{\mathrm{T}}(k)\hat{\underline{\underline{\chi}}}(k)-\underline{q}^{\mathrm{T}}\Lambda(k)^2/k^2\right)\\
&\sim \frac{-4\pi\beta/\epsilon}{k^2+\Lambda(k)^2}\tilde{\underline{q}}\,\tilde{\underline{q}}^{\mathrm{T}},\ k\to i\kappa,
\end{aligned}
\tag{12.60}
$$

where Eqs. (12.56) and (12.57) have been used to give the second and final equalities, respectively. Also, $\tilde{\underline{q}}\equiv\tilde{\underline{q}}(i\kappa)$. In order to exhibit the residue explicitly and to cast it in an identical form to the Debye–Hückel result, Eq. (12.50), one needs the Taylor expansion of the denominator about $k=i\kappa$,

$$
\begin{aligned}
k^2+\Lambda(k)^2 &\sim (k-i\kappa)\left(2i\kappa+\frac{4\pi\beta}{\epsilon}\underline{q}^{\mathrm{T}}\tilde{\underline{q}}'\right)+\ldots\\
&\sim (k^2+\kappa^2)\left(1+\frac{4\pi\beta}{2i\kappa\epsilon}\underline{q}^{\mathrm{T}}\tilde{\underline{q}}'\right)+\mathcal{O}(k-i\kappa)^2.
\end{aligned}
\tag{12.61}
$$

Here

$$
\begin{aligned}
\tilde{\underline{q}}' &= \left.\frac{\partial\tilde{\underline{q}}(k)}{\partial k}\right|_{k=i\kappa}\\
&= \left.\frac{\partial\hat{\underline{\underline{\chi}}}(k)}{\partial k}\right|_{k=i\kappa}\tilde{\underline{q}}+\hat{\underline{\underline{\chi}}}(i\kappa)\left.\frac{\partial\tilde{\underline{q}}(k)}{\partial k}\right|_{k=i\kappa}\\
&= \left(\underline{\underline{I}}-\hat{\underline{\underline{\chi}}}(i\kappa)\right)^{-1}\left.\frac{\partial\hat{\underline{\underline{\chi}}}(k)}{\partial k}\right|_{k=i\kappa}\tilde{\underline{q}}.
\end{aligned}
\tag{12.62}
$$

Accordingly one defines the constant

$$
\begin{aligned}
\nu &\equiv 1+\frac{4\pi\beta}{2i\kappa\epsilon}\underline{q}^{\mathrm{T}}\tilde{\underline{q}}'\\
&= 1+\frac{4\pi\beta}{2i\kappa\epsilon}\underline{q}^{\mathrm{T}}\left(\underline{\underline{I}}-\hat{\underline{\underline{\chi}}}(i\kappa)\right)^{-1}\hat{\underline{\underline{\chi}}}'(i\kappa)\tilde{\underline{q}}\\
&= 1+\frac{4\pi\beta}{2i\kappa\epsilon}\tilde{\underline{q}}^{\mathrm{T}}\hat{\underline{\underline{\chi}}}'(i\kappa)\tilde{\underline{q}}.
\end{aligned}
\tag{12.63}
$$

One now has

$$
\hat{\underline{\underline{H}}}(k)\sim\frac{-4\pi\beta}{\epsilon\nu}\frac{\tilde{\underline{q}}\,\tilde{\underline{q}}^{\mathrm{T}}}{k^2+\kappa^2},\ k\to i\kappa,
\tag{12.64}
$$

with inverse

$$h_{\alpha\gamma}(r) \sim \frac{-\beta\tilde{q}_\alpha\tilde{q}_\gamma}{\epsilon\nu}\frac{e^{-\kappa r}}{r},\ r \rightarrow \infty,\ \mathrm{Im}\{\kappa\} = 0. \tag{12.65}$$

Note that if $\underline{\underline{\chi}} = 0$, Eq. (12.56) implies that $\underline{\tilde{q}} = \underline{q}$, Eq. (12.59) yields $\kappa = \kappa_{\mathrm{D}}$, and Eq. (12.63) shows that $\nu = 1$. That is, the exact asymptote, Eq. (12.65), reduces to the Debye–Hückel result, Eq. (12.52).

Recall that $\underline{\tilde{q}}$ is given by Eq. (12.56) evaluated at $k = i\kappa$. With this definition the actual screening length, Eq. (12.59), assumes a form similar to the Debye length

$$\kappa^2 = \frac{4\pi\beta}{\epsilon}\underline{\tilde{q}}^{\mathrm{T}}\underline{q} = \frac{4\pi\beta}{\epsilon}\sum_\alpha \rho_\alpha\tilde{q}_\alpha q_\alpha. \tag{12.66}$$

The concept of an effective charge and the non-Debye screening length has been used by a number of authors.[8]

The mean electrostatic potential about an ion was given above as Eq. (12.3). Writing this as a vector in the usual fashion, the Ornstein–Zernike equation may be rewritten

$$\underline{\underline{H}}(r) = -\beta\underline{\psi}(r)\underline{q}^{\mathrm{T}} + \underline{\underline{\chi}}(r) + \int \mathrm{d}\mathbf{s}\,\underline{\underline{H}}(s)\underline{\underline{\chi}}(|\mathbf{r}-\mathbf{s}|), \tag{12.67}$$

with Fourier transform

$$\begin{aligned}\hat{\underline{\underline{H}}}(k) &= -\beta\hat{\underline{\psi}}(k)\underline{q}^{\mathrm{T}} + \hat{\underline{\underline{\chi}}}(k) + \hat{\underline{\underline{H}}}(k)\hat{\underline{\underline{\chi}}}(k) \\ &= -\beta\hat{\underline{\psi}}(k)\underline{q}^{\mathrm{T}}\left(\underline{\underline{I}} - \hat{\underline{\underline{\chi}}}(k)\right)^{-1} + \hat{\underline{\underline{\chi}}}(k)\left(\underline{\underline{I}} - \hat{\underline{\underline{\chi}}}(k)\right)^{-1} \\ &\sim -\beta\hat{\underline{\psi}}(k)\underline{\tilde{q}}^{\mathrm{T}},\ k \rightarrow i\kappa. \end{aligned} \tag{12.68}$$

The last line follows since $\chi(r)$ is more short-ranged than $h(r)$, and hence only the first term contributes to the asymptote. In other words, the mean electrostatic potential about an ion has the same range as the total correlation function, and one has

$$h_{\alpha\gamma}(r) \sim -\beta\tilde{q}_\gamma\psi_\alpha(r),\ r \rightarrow \infty. \tag{12.69}$$

Compare this with the Debye–Hückel result, Eq. (12.52), which is based upon the linear Poisson–Boltzmann approximation. One sees that it is the effective charge on the ion that gives its response to the mean electrostatic potential.

[8] G. Stell and J. L. Lebowitz, Equilibrium properties of a system of charged particles, *J. Chem. Phys.* **48** (1968), 3706. S. Alexander, P. M. Chaikin, P. Grant, G. J. Morales, P. Pincus and D. Hone, Charge renormalization, osmotic pressure, and bulk modulus of colloidal crystals: Theory, *J. Chem. Phys.* **80** (1984), 5776. P. Attard, Asymptotic analysis of primitive model electrolytes and the electrical double layer, *Phys. Rev. E* **48** (1993), 3604. R. Kjellander and D. J. Mitchell, Dressed-ion theory for electrolyte solutions: A Debye–Hückel-like reformulation of the exact theory for the primitive model, *J. Chem. Phys.* **101** (1994), 603.

Now from symmetry $h_{\alpha\gamma}(r) = h_{\gamma\alpha}(r)$, and one concludes that $\psi_\gamma(r) \propto \tilde{q}_\gamma$. The precise relationship follows from Eq. (12.65),

$$\psi_\gamma(r) \sim \tilde{q}_\gamma \frac{e^{-\kappa r}}{\epsilon \nu r},\ r \to \infty. \tag{12.70}$$

The asymptote of the potential of mean force has linear Poisson–Boltzmann form, $w_{\alpha\gamma}(r) \sim \tilde{q}_\alpha \psi_\gamma(r) = \tilde{q}_\alpha \tilde{q}_\gamma \psi(r)$, which is significant because the full symmetry of the pair correlation functions is preserved. The dyadic nature of the asymptote for a mixture with components interacting with power law potentials was noted in Section 9.6.3.

Oscillatory Asymptotic Decay

If κ is complex, then the asymptote is oscillatory. Because $h_{\alpha\gamma}(r)$ is a real, even function of r, its Fourier transform is even, $\hat{h}_{\alpha\gamma}(-k) = \hat{h}_{\alpha\gamma}(k)$, and any series expansion in k has real coefficients, $\overline{\hat{h}_{\alpha\gamma}(k)} = \hat{h}_{\alpha\gamma}(\overline{k})$, where the overline denotes the complex conjugate. Accordingly there are four poles located at $\pm i\kappa$ and $\pm i\overline{\kappa}$, and the singular part may be written

$$\underline{\underline{\hat{H}}}^{\mathrm{S}}(k) = \frac{\underline{\underline{A}}}{k^2 + \kappa^2} + \frac{\overline{\underline{\underline{A}}}}{k^2 + \overline{\kappa}^2}. \tag{12.71}$$

Hence closing the contour in the upper half plane, for the Fourier inversion (choosing κ to be in the first quadrant), one picks up the residue at $k = +i\kappa$, which is $A/(2i\kappa)$, and also its complex conjugate from the residue from the pole at $k = -i\overline{\kappa}$, which is $\overline{A}/(2i\overline{\kappa})$. The sum of these two is twice the real part of either one, and one has

$$h_{\alpha\gamma}(r) \sim 2\,\mathrm{Re}\left\{ \frac{-\beta \tilde{q}_\alpha \tilde{q}_\gamma}{\epsilon \nu} \frac{e^{-\kappa r}}{r} \right\},\ r \to \infty,\ \mathrm{Im}\{\kappa\} \neq 0. \tag{12.72}$$

The abrupt disappearance of the factor of 2 as the poles coalesce on the imaginary axis suggests nonanalyticity in the amplitude of the correlation functions at the transition from monotonic to oscillatory decay. In fact the amplitude becomes infinite, as may be seen by letting the pole just move off the imaginary k axis, $\kappa = \kappa_r + i\kappa_i$, $\kappa_i \to 0$. Expanding Eq. (12.59) one obtains

$$\begin{aligned} \kappa_r + i\kappa_i &= \left(\frac{4\pi\beta}{\epsilon} \underline{q}^{\mathrm{T}} \underline{\tilde{q}}(i\kappa_r - \kappa_i) \right)^{1/2} \\ &\sim \left(\frac{4\pi\beta}{\epsilon} \underline{q}^{\mathrm{T}} \underline{\tilde{q}}(i\kappa_r) \right)^{1/2} \left[1 - \frac{\kappa_i}{2} \frac{\underline{q}^{\mathrm{T}} \underline{\tilde{q}}'(i\kappa_r)}{\underline{q}^{\mathrm{T}} \underline{\tilde{q}}(i\kappa_r)} + \mathcal{O}(\kappa_i^2) \right] \\ &= \kappa_r - \frac{\kappa_i}{2\kappa_r} \frac{4\pi\beta}{\epsilon} \underline{q}^{\mathrm{T}} \underline{\tilde{q}}'. \end{aligned} \tag{12.73}$$

Equating the coefficients of κ_i, one obtains

$$\frac{4\pi\beta}{2\kappa_r \epsilon} \underline{q}^{\mathrm{T}} \underline{\tilde{q}}' = -i,\ \kappa_i \to 0, \tag{12.74}$$

or from the first equality of Eq. (12.63),

$$\nu \to 0, \ \kappa_i \to 0. \tag{12.75}$$

That is, the amplitude of the total correlation function becomes infinite at the oscillatory to monotonic transition.

One must be careful in interpreting this result. The vanishing of ν means that the next term in the Taylor expansion (12.61) is nonnegligible. Hence the denominator corresponds to a double pole, and the residue comes from the linear term in the numerator, which ensures that the asymptotic behaviour remains exponential. For infinitesimal but nonzero ν, the present formulae give the strict asymptote, but the regime of applicability moves to ever larger separations. For fixed r, as $\nu \to 0$ the $h(r)$ have a large contribution from the (finite) second term. For fixed nonzero ν, as $r \to \infty$ the $h(r)$ are given by the strict asymptote. So even though the amplitude diverges, so do the relevant separations, and consequently thermodynamic properties such as the internal energy remain finite at the monotonic–oscillatory transition.

Core Domination

The formally exact analysis of the previous section guarantees that the Fourier transforms of the total correlation functions have a pole at $k = i\kappa$, for κ satisfying Eqs. (12.56) and (12.59). However if there is another pole, $k = i\xi$, with $\mathrm{Re}\{\xi\} < \mathrm{Re}\{\kappa\}$, then it would be the one that determines the asymptotic behaviour of the correlations. Such a pole would correspond to the matrix $\underline{\underline{I}} - \underline{\underline{\hat{\chi}}}(i\xi)$ being singular, and this would give rise to qualitatively different behaviour. This will occur at high densities where the short-range interactions become important and give rise to the 'core-dominated' asymptote.

This regime will be treated subject to the following restriction. Locally the asymptotic charge density about an ion must vanish (because electrostatic effects are of shorter range). Hence with the amplitude of the total correlation functions in the asymptotic regime denoted by $a_{\alpha\gamma}$, one has

$$\sum_{\alpha} q_\alpha \rho_\alpha a_{\alpha\gamma} = 0. \tag{12.76}$$

This is an exact result that holds in the core-dominated asymptotic regime. One way of satisfying this equation is if

$$a_{\alpha\gamma} = a, \ \text{all species}, \tag{12.77}$$

to which case the following results are restricted. There are two cases where this equation holds exactly with no approximation. First, there is the general binary electrolyte, in which case this is the only possible solution. Second, there is a multicomponent electrolyte with the short-range interactions between the ions being identical; since it is the latter that determine the asymptote, then the total correlation functions between all the species must be asymptotically equal.

Using Eq. (12.77), and solving the Ornstein–Zernike equation for the direct correlation function near the pole, one sees that the short-range part of the direct correlation function goes like

$$\hat{\underline{\underline{\chi}}}(k) \sim \underline{\rho}\,\underline{\rho}^{\mathrm{T}} x(k),\ k \to i\xi, \tag{12.78}$$

where $\{\underline{\rho}\}_\alpha = \rho_\alpha^{1/2}$. Note that $\underline{\rho}^{\mathrm{T}}\underline{q} = 0$, and hence near the pole

$$\begin{aligned}\left(\underline{\underline{I}} - \hat{\underline{\underline{C}}}(k)\right)^{-1} &\sim \left(\underline{\underline{I}} - \underline{\rho}\,\underline{\rho}^{\mathrm{T}} x(k) + \underline{q}\,\underline{q}^{\mathrm{T}} 4\pi\beta/\epsilon k^2\right)^{-1} \\ &= \underline{\underline{I}} + \frac{\underline{\rho}\,\underline{\rho}^{\mathrm{T}} x(k)}{1 - \underline{\rho}^{\mathrm{T}}\underline{\rho}\, x(k)} - \frac{\underline{q}\,\underline{q}^{\mathrm{T}} 4\pi\beta/\epsilon k^2}{1 + \underline{q}^{\mathrm{T}}\underline{q} 4\pi\beta/\epsilon k^2}.\end{aligned} \tag{12.79}$$

Now it is the middle term which has the pole at $k = i\xi$, and neglecting the remaining regular parts one obtains

$$\begin{aligned}\hat{\underline{\underline{H}}}(k) &\sim \frac{\underline{\rho}\,\underline{\rho}^{\mathrm{T}} x(k)}{1 - \underline{\rho}^{\mathrm{T}}\underline{\rho}\, x(k)} \left(\underline{\rho}\,\underline{\rho}^{\mathrm{T}} x(k) - \underline{q}\,\underline{q}^{\mathrm{T}} 4\pi\beta/\epsilon k^2\right) \\ &= \frac{\underline{\rho}^{\mathrm{T}}\underline{\rho}\, x(k)^2}{1 - \underline{\rho}^{\mathrm{T}}\underline{\rho}\, x(k)} \underline{\rho}\,\underline{\rho}^{\mathrm{T}} \\ &\sim \frac{-2i\xi x(i\xi)}{\underline{\rho}^{\mathrm{T}}\underline{\rho}\, x'(i\xi)(k^2 + \xi^2)} \underline{\rho}\,\underline{\rho}^{\mathrm{T}},\ k \to i\xi.\end{aligned} \tag{12.80}$$

Here the denominator has been Taylor expanded, and use has been made of the fact that ξ satisfies

$$\underline{\rho}^{\mathrm{T}}\underline{\rho}\, x(i\xi) = \rho x(i\xi) = 1, \tag{12.81}$$

where the total number density is $\rho = \sum_\alpha \rho_\alpha = \underline{\rho}^{\mathrm{T}}\underline{\rho}$. It follows that

$$h_{\alpha\gamma}(r) \sim 2\mathrm{Re}\left\{ \frac{-2i\xi}{4\pi\rho^2 x'(i\xi)} \frac{e^{-\xi r}}{r} \right\},\ r \to \infty. \tag{12.82}$$

This analysis is not restricted to electrolytes; it also holds for uncharged particles, subject to the restriction (12.77). From this one sees that in the core-dominated regime, the ion–ion total correlation functions decay like damped sinusoids with a single amplitude (ξ is complex with positive real part). The surrounding counterions very rapidly neutralise the charge on an ion, and it is packing effects that determine the correlations that persist in the dense electrolyte or molten salt. The period of the oscillations corresponds approximately to the size of the ions, which contrasts with the larger period that occurs in the case of charge oscillations.

12.3 Electric Double Layer

In general solutes in a polar solvent or electrolyte acquire a charge on their surfaces, either by dissociation or dissolution of the surface species or by chemical or physical binding of ions to the surface. The counterions diffuse away from the surface and into the electrolyte so that a charge separation occurs. This is known as the electric double layer. The charge on the surface is screened by the electrolyte in the same way that the ions in the electrolyte themselves are screened. The analysis of the electric double layer proceeds in much the same way as that for inhomogeneous systems more generally, as given in Ch. 11.

12.3.1 Spherical Solutes

Double-Layer Moments

The analysis of Section 12.2.1 is here extended to the double layer about a charged solute. Following the general analysis for inhomogeneous systems, Section 11.1, one of the components of the electrolyte is taken to be a spherical solute, species $\alpha = 0$ with charge q_0. The double layer about this solute in isolation is treated by taking the infinite dilution limit, $\rho_0 = 0$.

Focussing on the solute–ion contributions to the multicomponent Ornstein–Zernike equation (12.22), one defines a vector of solute–ion total correlation functions

$$\{\underline{H}(r)\}_\gamma = \rho_\gamma^{1/2} h_{0\gamma}(r),\ \gamma > 0, \tag{12.83}$$

and similarly for the direct correlation functions. The density of ions of type γ in the double layer about the solute is $\rho_\gamma(r) = \rho_\gamma[h_{0\gamma}(r) + 1]$. As given in Eq. (11.12), the Fourier transform of the solute–ion Ornstein–Zernike equation is

$$\begin{aligned}\underline{\hat{H}}(k) &= \underline{\hat{C}}(k) + \underline{\underline{\hat{C}}}(k)\underline{\hat{H}}(k) \\ &= \underline{\hat{\chi}}(k) - \frac{4\pi\beta q_0}{\epsilon k^2}\underline{q} + \underline{\underline{\hat{\chi}}}(k)\underline{\hat{H}}(k) - \underline{\underline{Q}}\underline{\hat{H}}(k)k^{-2}.\end{aligned} \tag{12.84}$$

The ion–solute correlation functions are exponentially decaying, as may be shown using the same arguments as for the ion–ion case above. Hence all of their moments exist, and one may equate the coefficients of k^{-2} to obtain the solute electro-neutrality condition

$$-q_0 = \underline{q}^{\mathrm{T}}\underline{H}^{(0)} = \sum_\gamma q_\gamma \rho_\gamma \int \mathrm{d}\mathbf{r}\, h_{0\gamma}(r). \tag{12.85}$$

This shows that the charge on the solute is exactly cancelled by the net charge in the double layer. Either this counter-charge must have dissociated from the originally neutral surface or ions from the originally neutral bulk electrolyte must have adsorbed to the solute, giving it its charge.

Turning now to the second moment, the coefficients of k^0 in the Fourier transform of the Ornstein–Zernike equation give

$$\underline{H}^{(0)} = \underline{\chi}^{(0)} + \underline{\underline{\chi}}^{(0)}\underline{H}^{(0)} - \underline{\underline{Q}}\underline{H}^{(2)}, \tag{12.86}$$

or, using the electro-neutrality condition,

$$-q_0 = \underline{q}^{\mathrm{T}}\underline{\chi}^{(0)} + \underline{q}^{\mathrm{T}}\underline{\underline{\chi}}^{(0)}\underline{H}^{(0)} - \kappa_{\mathrm{D}}^2\underline{q}^{\mathrm{T}}\underline{H}^{(2)}. \tag{12.87}$$

In this case the second moment of the ion–solute total correlation function depends upon the zeroth moment of the short-range part of the ion–solute direct correlation function. The former is analogous to one of the individual second-moments of the ion–ion total correlation functions of the bulk electrolyte, and it will be recalled that these were not individually constrained by the second-moment condition. It was only their sum that was constrained, and because the solute is at infinite dilution, this last result can be added to the bulk second-moment condition, Eq. (12.42), without effect.

The pair-wise interaction between the solutes is given by the singlet approach when $\alpha = \gamma = 0$ in Eq. (12.22). Using Eq. (11.13), the solute–solute Ornstein–Zernike equation for identical solutes is in Fourier space

$$\begin{aligned}\hat{h}_{00}(k) &= \hat{c}_{00}(k) + \underline{\hat{C}}(k)^{\mathrm{T}}\underline{\hat{H}}(k) \\ &= \hat{\chi}_{00}(k) - \frac{4\pi\beta q_0^2}{\epsilon k^2} + \underline{\hat{\chi}}(k)^{\mathrm{T}}\underline{\hat{H}}(k) - \frac{4\pi\beta q_0}{\epsilon k^2}\underline{q}^{\mathrm{T}}\underline{\hat{H}}(k).\end{aligned} \tag{12.88}$$

Since the bulk closure remains the same, the exponential decay of the solute–solute interaction follows immediately. However, the coefficient of k^{-2} yields

$$-q_0 = \underline{q}^{\mathrm{T}}\underline{H}^{(0)}, \tag{12.89}$$

which contains nothing new, and the coefficient of k^0 yields

$$h_{00}^{(0)} = \chi_{00}^{(0)} + \underline{\chi}^{(0)\,\mathrm{T}}\underline{H}^{(0)} - \frac{4\pi\beta q_0}{\epsilon}\underline{q}^{\mathrm{T}}\underline{H}^{(2)}. \tag{12.90}$$

Using the result given above, the second-moment could be eliminated and this could be reexpressed in terms of the zeroth moments of the various correlation functions. There is no electro-neutrality condition for the solute–solute total correlation function; in this case the zeroth moment is nonuniversal and depends upon the zeroth moments of various solute correlation functions.

Asymptotic Analysis

As in the asymptotic analysis of Section 11.1.4, which was for a solute in a general solvent, a charged solute does not affect the properties of the electrolyte, and the formal asymptotic analysis for the behaviour of the double layer is virtually unchanged from that given above for the bulk electrolyte. The solute is at infinite dilution and does not contribute to the solvent correlation functions because the Ornstein–Zernike convolution integral, Eq. (12.22), is multiplied by $\rho_0 = 0$ whenever the solute correlation functions appear in the integrand. Hence the decay length κ^{-1}, the effective charge on the ions $\tilde{q}_\gamma$, $\gamma > 0$, and the scale factor ν stay the same. The Fourier transform of the solute–ion Ornstein–Zernike equation may be written

$$\underline{\hat{H}}(k) = \left(\underline{\underline{I}} - \underline{\underline{\hat{C}}}(k)\right)^{-1}\underline{\hat{C}}(k). \tag{12.91}$$

The singularity in the inverse gives the pole that determines the asymptotic behaviour of the solute–ion total correlation function. Since the inverse consists solely of the bulk ion direct correlation functions it is unchanged by the solute; the determinant vanishes at the same $k = i\kappa$ as in the bulk electrolyte. Further, the pole of the solute–solvent total correlation functions have the same residue as in the bulk electrolyte, with an effective solute charge $\tilde{q}_0$ appearing. It follows that

$$\underline{\hat{H}}(k) \sim \frac{-4\pi\beta/\epsilon}{k^2 + \Lambda(k)^2} \tilde{q}_0 \underline{\tilde{q}},\ k \to i\kappa, \tag{12.92}$$

with the effective charge on the solute being given by

$$\begin{aligned} \tilde{q}_0 &= \underline{\tilde{q}}^{\mathrm{T}} \underline{\hat{C}}(i\kappa) \\ &= q_0 + \underline{\tilde{q}}^{\mathrm{T}} \underline{\hat{\chi}}(i\kappa). \end{aligned} \tag{12.93}$$

The asymptote is

$$h_{0\gamma}(r) \sim 2\,\mathrm{Re}' \left\{ \frac{-\beta \tilde{q}_0 \tilde{q}_\gamma}{\epsilon \nu} \frac{e^{-\kappa r}}{r} \right\},\ r \to \infty. \tag{12.94}$$

As mentioned above, the solute does not affect κ, ν, or $\tilde{q}_\gamma$, $\gamma > 0$; these are all properties of the bulk electrolyte. Only $\tilde{q}_0$ depends on the nature of the solute, via the short-ranged part of the solute–solvent direct correlation functions.

The interactions between two such solutes is given by the solute–solute total correlation function. Its asymptote follows by setting $\gamma = 0$ in this result, and one sees that it depends on the square of the effective solute charge.

Because the decay length κ^{-1} is determined by the bulk electrolyte, the range of the double layer (i.e., the rate at which the ion densities decay to their bulk concentrations) is the same as the bulk screening length. This is also the range of the interactions between the solutes. Amongst other things this means that the ion profiles will become oscillatory at a point determined by the bulk electrolyte concentration, and similarly for the interaction between the solutes.

The nature of the solute contributes to the asymptote via the amplitude $\tilde{q}_0$, and the interaction between two identical solutes scales with the square of their effective charge. Hence in the monotonic regime there can only be a repulsion at large separations between two identical macroions. This follows because all quantities are real, $\tilde{q}_0$ occurs as a square, and $\nu > 0$. (At low coupling, $\nu \to 1$, the Debye–Hückel result, and at the monotonic–oscillatory transition, $\nu \to 0$, Eq. (12.75).) In the oscillatory regime it is periodically attractive, and the attractive regions can be quite large since the period of oscillations becomes infinite as one approaches the bulk oscillatory–monotonic transition.

In the core-dominated regime, the analysis proceeds as in the bulk. Assuming once more that the electrolyte ions have identical short-range interactions, the double layer about a spherical solute goes like

$$\underline{\hat{H}}(k) \sim \frac{-2i\xi x(i\xi)}{\underline{\rho}^{\mathrm{T}} \underline{\rho}\, x'(i\xi)(k^2 + \xi^2)} \underline{\rho}\, \underline{\rho}^{\mathrm{T}} \underline{\hat{\chi}}(i\xi),\ k \to i\xi, \tag{12.95}$$

with corresponding asymptote

$$h_{0\alpha}(r) \sim 2\,\mathrm{Re}\left\{ \frac{-2i\xi}{4\pi\rho^2 x'(i\xi)} \frac{e^{-\xi r}}{r} \sum_{\gamma>0} \rho_\gamma \hat{\chi}_\gamma(i\xi) \right\},\ r \to \infty. \tag{12.96}$$

Since α does not appear on the right-hand side, one concludes that the solute–ion profiles become identical at large separations whatever the direct solute–ion interaction for each species is (provided that the correlations of the ions themselves are asymptotically identical in the bulk).

The solute–solute total correlation function has a dominant pole

$$\begin{aligned} \hat{h}_{00}(k) &= \hat{c}_{00}(k) + \underline{\hat{H}}^{\mathrm{T}}(k)\underline{\hat{C}}(k) \\ &\sim \frac{-2i\xi x(i\xi)}{\underline{\rho}^{\mathrm{T}}\underline{\rho}\, x'(i\xi)(k^2+\xi^2)} \underline{\hat{\chi}}^{\mathrm{T}}(i\xi)\underline{\rho}\,\underline{\rho}^{\mathrm{T}}\underline{\hat{\chi}}(i\xi),\ k \to i\xi, \end{aligned} \tag{12.97}$$

since the solute–solute direct correlation function is more short-ranged than the total correlation function, and since $\underline{\rho}^{\mathrm{T}}\underline{\hat{C}} = \underline{\rho}^{\mathrm{T}}\underline{\hat{\chi}}$ because $\underline{\rho}^{\mathrm{T}}\underline{q} = 0$. The macroion–macroion asymptote is therefore

$$h_{00}(r) \sim 2\,\mathrm{Re}\left\{ \frac{-2i\xi}{4\pi\rho^2 x'(i\xi)} \frac{e^{-\xi r}}{r} \left(\sum_{\gamma>0} \rho_\gamma \hat{\chi}_\gamma(i\xi) \right)^2 \right\},\ r \to \infty. \tag{12.98}$$

The sum evidently represents the effective 'charge' of the solute; it gives the magnitude of the ion density profile in the double layer about the isolated solute, and its square gives the magnitude of their pair-wise interaction.

12.3.2 Isolated Planar Solutes

Wall–Ion Ornstein–Zernike Equation

The planar electric double layer is now analysed by taking the solute to be a planar wall, as was done more generally in Section 11.2. The wall–ion correlation functions depend only upon the perpendicular distance from the surface of the wall, which is assumed located to the left of $z = 0$. Assuming for simplicity a common distance of closest approach, and setting this plane at the location of the wall, one has

$$h_{0\alpha}(z) = -1,\ z < 0. \tag{12.99}$$

Of course one must also have electro-neutrality,

$$-\sigma = \int_0^\infty \mathrm{d}z\, \underline{q}^{\mathrm{T}} \underline{H}(z), \tag{12.100}$$

where σ is the surface charge per unit area on the wall.

The wall–ion Ornstein–Zernike equation may be rewritten in terms of the mean electrostatic potential,

$$\begin{aligned}\underline{H}(z) \\ &= \underline{C}(z) + 2\pi \int_0^\infty \mathrm{d}s\, s \int_{-\infty}^{\infty} \mathrm{d}z'\, \underline{\underline{C}}\left(\sqrt{(z'-z)^2+s^2}\right) \underline{H}(z'), \qquad (12.101)\\ &= -\beta\psi(z)\underline{q} + \underline{\chi}(z) \\ &\quad + 2\pi \int_0^\infty \mathrm{d}s\, s \int_{-\infty}^{\infty} \mathrm{d}z'\, \underline{\underline{\chi}}\left(\sqrt{(z'-z)^2+s^2}\right) \underline{H}(z'). \qquad (12.102)\end{aligned}$$

Here the vector of wall–ion total correlation functions has components $\{\underline{H}(z)\}_\alpha = \rho_\alpha^{1/2} h_\alpha(z)$, and similarly for the wall–ion direct correlation function $\underline{C}(z)$ and its short-ranged part $\underline{\chi}(z)$. As before one has

$$c_{0\alpha}(z) = \chi_{0\alpha}(z) - \beta q_\alpha V_0^{\mathrm{Coul}}(z), \qquad (12.103)$$

where the Coulomb part of the solute–ion potential may be written in general as

$$V_0^{\mathrm{Coul}}(\mathbf{r}) = \int \mathrm{d}\mathbf{s}\, \frac{\sigma(\mathbf{s})}{\epsilon|\mathbf{r}-\mathbf{s}|}. \qquad (12.104)$$

In the present case, the wall charge density is $\sigma(\mathbf{r}) = \sigma\delta(z)$. The mean electrostatic potential is defined by the passage from the first to the second form of the Ornstein–Zernike equation,

$$\begin{aligned}\psi(z) &= \frac{2\pi}{\epsilon} \int_{-\infty}^{\infty} \mathrm{d}z' \int_0^\infty \mathrm{d}s\, s[\sigma\delta(z') + \underline{q}^{\mathrm{T}}\underline{H}(z')] \frac{1}{\sqrt{s^2+(z-z')^2}} \\ &= \frac{-2\pi}{\epsilon} \int_{-\infty}^{\infty} \mathrm{d}z'\, [\sigma\delta(z') + \underline{q}^{\mathrm{T}}\underline{H}(z')]|z-z'| \\ &= \frac{2\pi}{\epsilon} \int_0^\infty \mathrm{d}z'\, \underline{q}^{\mathrm{T}}\underline{H}(z')z' + \frac{4\pi}{\epsilon} \int_z^\infty \mathrm{d}z'\, \underline{q}^{\mathrm{T}}\underline{H}(z')(z-z'), \qquad (12.105)\end{aligned}$$

for $z > 0$, and $\psi(z) = \psi(0)$, $z < 0$. (The electro-neutrality condition has been used to cancel the upper limit of the first integration.) The first term is independent of z, whereas the second term goes to 0 far from the wall. Hence $\psi(z) \to -\psi(0)$, $z \to \infty$. This is undesirable because in order for the density to decay to its bulk value far from the wall, the second form of the Ornstein–Zernike equation implies that $\underline{\chi}(z) \to -\beta\underline{q}\psi(0)$, $z \to \infty$ (and the closure below implies the same limiting behaviour for the bridge function). By subtracting the constant one can set the 0 of the potential to be in the bulk,

$$\psi(z) = \begin{cases} \psi(0), & z \leq 0, \\ \dfrac{4\pi}{\epsilon} \sum_\gamma q_\gamma \rho_\gamma \displaystyle\int_z^\infty \mathrm{d}z'\, h_{0\gamma}(z')(z-z'), & z \geq 0, \end{cases} \qquad (12.106)$$

and now $\psi(z)$, $\underline{\chi}(z)$, and $\underline{d}(z)$ all go to 0 in the bulk far from the wall.

Choosing the functions to vanish far from the wall is equivalent to choosing the constant of the wall–ion Coulomb potential such that

$$V_0^{\mathrm{Coul}}(z) = -\frac{2\pi\sigma}{\epsilon}|z| + \frac{2\pi\sigma}{\epsilon}S + \frac{\psi(0)}{2}, \text{ all } z, \tag{12.107}$$

where $S \to \infty$ is the lateral extent of the wall. The potential drop across the double layer is

$$\psi(0) = \frac{-4\pi}{\epsilon}\sum_\gamma q_\gamma \rho_\gamma \int_0^\infty \mathrm{d}z'\, h_{0\gamma}(z')z'. \tag{12.108}$$

Differentiating this with respect to surface charge gives the capacitance. The first moment of the total correlation function represents the dipole moment of the double layer, and $\psi(0)$ is just the drop in electrostatic potential across a dipole layer.

As in Section 11.2, the exact closure is

$$\begin{aligned} h_{0\alpha}(z) &= -1 + \exp\left[-\beta q_\alpha V_0(z) + h_{0\alpha}(z) - c_{0\alpha}(z) + d_\alpha(z)\right] && (12.109)\\ &= -1 + \exp\left[h_{0\alpha}(z) - \chi_{0\alpha}(z) + d_{0\alpha}(z)\right], && (12.110)\end{aligned}$$

where $d_{0\alpha}(z)$ is the wall–ion bridge function. Henceforth it is assumed that the ions interact with the wall only via the Coulomb potential, for $z > 0$, and that there is an infinite repulsion (hard core) for $z < 0$. Setting $\chi_{\alpha\gamma}(r)$ and $d_{0\alpha}(z)$ to 0 in the second form of the closure, Eq. (12.110), with the second form of the Ornstein–Zernike equation, Eq. (12.102), yields the nonlinear Boltzmann approximation, which is the traditional approach to the electric double layer. Note that the surface charge density does not explicitly appear in the expression for the mean electrostatic potential, Eq. (12.106), or in the alternate forms for the Ornstein–Zernike and closure equations, (12.102) and (12.110); it must be determined by the electro-neutrality condition, Eq. (12.100).

The wall–ion total correlation function that appears here may be identified with the solute–ion total correlation function of the preceding section in the limit that the radius of the macroionic solute goes to infinity,

$$\lim_{R\to\infty} h_{0\alpha}(R+z;R) = h_{0\alpha}(z), \tag{12.111}$$

where on the left side macroion–ion total correlation function appears with its dependence on the macroion radius shown explicitly, and the right side defines the wall–ion total correlation function. An identical limit holds for $\chi_{0\alpha}(z)$, but in the case of $c_{0\alpha}(z)$, $V_0(z)$, and $\psi(z)$ the limiting equality only holds up to an arbitrary constant.

As in the case of the spherical solute, the short-range part of the wall–ion direct correlation function is exponentially short-ranged, and assumed to have a decay length shorter than that of the total correlation function. This is true of the electrolyte side of the wall, but within the wall it goes to a constant, as was seen in Section 11.2. This can be seen explicitly from the Ornstein–Zernike equation (12.102), where in the limit that $z \to -\infty$ the convolution integral is dominated by regions $z' \approx z$, because $\underline{\underline{\chi}}(r)$ is exponentially short-ranged, and hence $\underline{H}(z')$ may be replaced by -1 and taken out of the integral. The result is

$$\lim_{z\to-\infty} \chi_{0\alpha}(z) = -1 + \beta q_\alpha \psi(0) + \sum_\gamma \rho_\gamma \hat{\chi}_{\alpha\gamma}(0), \tag{12.112}$$

or, in view of Eq. (9.92) and electro-neutrality,

$$\underline{\rho}^{\mathrm{T}} \underline{\chi}(z) \to -\beta \chi_T^{-1},\ z \to -\infty. \tag{12.113}$$

This result may be compared to that obtained above for non-Coulomb systems, Eq. (11.40). The fact that $\chi_{0\alpha}(z)$ goes to a constant means that its one-dimensional Fourier transform is well defined in the upper half of the complex plane.

Asymptotics of the Isolated Planar Double Layer

The asymptotic ion density profiles for the planar electric double can be obtained from the preceding analysis and the large radius limit of the solute asymptotes in Section 12.3.1. In the planar case the Fourier transforms are one-dimensional. Taking the large radius limit the relationship with the radial Fourier transform is

$$\begin{aligned}
\hat{f}(k;R) &= \frac{4\pi}{k}\int_0^\infty \mathrm{d}r\, r f(r;R)\sin kr \\
&= \frac{4\pi}{k}\int_{-R}^\infty \mathrm{d}z\, f(R+z;R)\sin\left[k(R+z)\right]\,(R+z) \\
&= \frac{4\pi}{ik}\int_{-R}^\infty \mathrm{d}z\, f(R+z;R)\sinh\left[ik(R+z)\right]\,(R+z) \\
&\sim \frac{-2\pi R e^{-ikR}}{ik}\int_{-\infty}^\infty \mathrm{d}z\, f(z)e^{-ikz},\ R\to\infty,\ \mathrm{Im}\{k\}>0, \\
&= \frac{-2\pi R e^{-ikR}}{ik}\overline{f}(k),
\end{aligned} \tag{12.114}$$

which defines the one-dimensional Fourier transform (overlined).

The Fourier transform of the Coulomb part of the wall potential turns out to be a generalised function, which may be treated by rewriting Eq. (12.107) in terms of Heaviside step functions

$$V_0^{\mathrm{Coul}}(z) = \frac{-2\pi\sigma}{\epsilon}\left[z\theta(z) - z\theta(-z)\right] + \frac{2\pi\sigma}{\epsilon}S + \frac{\psi(0)}{2}. \tag{12.115}$$

Now the Fourier transform of $z\theta(z)$ is k^{-2}, provided $\mathrm{Im}\{k\} < 0$, and the Fourier transform of $-z\theta(-z)$ is also k^{-2}, but for $\mathrm{Im}\{k\} > 0$. Using analytic continuation, the transform of the wall–ion Coulomb potential in the whole complex plane is

$$\overline{V}_0^{\mathrm{Coul}}(k) = \frac{-4\pi\sigma}{\epsilon k^2} + \left[\frac{2\pi\sigma}{\epsilon}S + \frac{\psi(0)}{2}\right] 2\pi\delta(k),\ \text{all } k. \tag{12.116}$$

With these results the effective charge on the macroion, Eq. (12.93), becomes

$$\begin{aligned}\tilde{q}_0(R) &= \underline{\tilde{q}}^{\mathrm{T}}\underline{\hat{C}}(i\kappa;R) \\ &\sim \frac{2\pi R e^{\kappa R}}{\kappa}\left[\underline{\tilde{q}}^{\mathrm{T}}\underline{\overline{\chi}}(i\kappa) - \beta\underline{\tilde{q}}^{\mathrm{T}}\underline{q}\overline{V}_0^{\mathrm{Coul}}(i\kappa)\right],\ R\to\infty \\ &= \frac{2\pi R e^{\kappa R}}{\kappa}\left[\sigma + \underline{\tilde{q}}^{\mathrm{T}}\underline{\overline{\chi}}(i\kappa)\right],\end{aligned} \tag{12.117}$$

where the fact that $\kappa^2 = (4\pi\beta/\epsilon)\underline{\tilde{q}}^{\mathrm{T}}\underline{q}$ has been used. In view of this the effective surface charge density is defined as

$$\tilde{\sigma} = \frac{1}{2}\left[\sigma + \underline{\tilde{q}}^{\mathrm{T}}\underline{\overline{\chi}}(i\kappa)\right], \tag{12.118}$$

where a factor of 1/2 is included in the definition to preserve the Poisson–Boltzmann form for the asymptote, and so that at low concentrations $\tilde{\sigma} = \sigma$. This last fact follows because in the Poisson–Boltzmann limit $\chi_{\alpha\gamma}(r) = 0$, and hence $\kappa = \kappa_{\mathrm{D}}$ and $\underline{\tilde{q}} = \underline{q}$. In this regime the linear Poisson–Boltzmann profile holds, $\underline{H}(z) = -\beta\underline{q}\psi(z)$, $z > 0$, and Eq. (12.102) shows that $\chi_{0\alpha}(z) = 0$, $z > 0$, and that $\chi_{0\alpha}(z) = -1 + \beta q_\alpha\psi(0)$, $z < 0$. In this limit $\underline{q}^{\mathrm{T}}\underline{\chi}(z) = \kappa_{\mathrm{D}}^2\epsilon\psi(0)/4\pi = \kappa_{\mathrm{D}}\sigma$, $z < 0$, and hence $\underline{q}^{\mathrm{T}}\underline{\overline{\chi}}(i\kappa_{\mathrm{D}}) = \sigma$.

The solute–ion total correlation function, Eq. (12.94), becomes

$$h_{0\gamma}(R+z;R) \sim 2\,\mathrm{Re}'\left\{\frac{-\beta\tilde{q}_0(R)\tilde{q}_\gamma}{\epsilon\nu}\frac{e^{-\kappa(R+z)}}{R+z}\right\},\ z\to\infty, \tag{12.119}$$

and one obtains in the planar limit

$$h_{0\gamma}(z) \sim 2\,\mathrm{Re}'\left\{\frac{-4\pi\beta\tilde{q}_\gamma\tilde{\sigma}}{\epsilon\nu\kappa}e^{-\kappa z}\right\},\ z\to\infty. \tag{12.120}$$

The limiting procedure is $R\to\infty$, $z\to\infty$, $z/R\to 0$. (The factor of $4\pi/\kappa$ comes from the prefactor of $\tilde{q}_0(R)$ and the factor of 1/2 in the definition of $\tilde{\sigma}$.) Here, $\mathrm{Re}'\{z\}$ means to take the real part of z if z is complex, and to take half of z if z is real, a consequence of the two mutually conjugate poles located in the upper half of the complex plane. The fact that the decay length for the density profile of the planar double layer is the same as for the spherical double layer and for the bulk electrolyte is not unexpected, and obviously the transition from monotonic to oscillatory double-layer profiles will occur at the same point as in the bulk electrolyte. Note that in planar geometry the Yukawa form of the spherical case has become a pure exponential decay, a consequence of the fact that the three-dimensional Fourier transform of the short-ranged direct correlation function has been replaced by a one-dimensional one. The asymptote for the mean electrostatic potential, Eq. (12.106), may be obtained directly with the above expression,

$$\psi(z) \sim 2\,\mathrm{Re}'\left\{\frac{4\pi\tilde{\sigma}}{\epsilon\nu\kappa}e^{-\kappa z}\right\},\ z\to\infty. \tag{12.121}$$

Again one sees that the effective charges give the response to the mean electrostatic potential, and that in the asymptotic regime the latter is proportional to the effective surface charge density.

The core-dominated regime is independent of the electrostatics, and so one can immediately take the planar limit of the core-dominated spherical solute result, Eq. (12.96), to obtain

$$h_{0\alpha}(z) \sim 2\,\mathrm{Re}'\left\{\frac{-2ie^{-\xi z}}{\rho^2 x'(i\xi)}\sum_{\gamma>0}\rho_\gamma \overline{X}_{0\gamma}(i\xi)\right\},\ \alpha>0,\ z\to\infty. \qquad (12.122)$$

Despite the charge on the wall, this shows that all of the ion profiles are asymptotically identical in the core-dominated regime.

12.3.3 Interacting Charged Walls

Wall–Wall Ornstein–Zernike Equation

The singlet approach can be applied to the problem of overlapping planar double layers using the techniques given for interacting walls in Section 11.2.2. As mentioned there, the relevant quantity is the potential of mean force per unit area, which is just the exponent of the closure equation. One has

$$\beta w_{00}^{\mathrm{int}}(z) = \beta V_{00}(z) - d_{00}(z) - \int_{-\infty}^{\infty} \mathrm{d}z'\, \underline{H}^{\mathrm{T}}(z')\underline{C}(z-z'), \qquad (12.123)$$

where $\beta = 1/k_{\mathrm{B}}T$, and where the wall–wall quantities per unit area symbolised by w, V, and d are the interaction free energy, potential, and the bridge function, respectively.

The series function per unit area, which is the convolution integral, contains the wall–ion pair correlation functions discussed above. The walls will be assumed to interact only with the Coulomb potential, in the region $z>0$, where z represents the width of the region available to the centres of the ions (cf. Eq. (12.99) above). In order to express this interaction free energy in terms of short-range functions, one needs to remove the wall–ion Coulomb potential from the integral of the wall–ion direct correlation function (cf. Eq. (12.103)). One has

$$\int_0^{\infty} \mathrm{d}z'\, \underline{H}^{\mathrm{T}}(z')\underline{q}V_0^{\mathrm{Coul}}(z-z')$$
$$= \sigma\psi(z) + \frac{2\pi\sigma^2}{\epsilon}z - \frac{2\pi\sigma^2}{\epsilon}S - \sigma\psi(0),\ z>0, \qquad (12.124)$$

where the single-wall electro-neutrality condition, Eq. (12.100), the mean electrostatic potential due to a charged wall, Eq. (12.106), the wall–ion Coulomb potential, Eq. (12.107), and the potential drop across an isolated double layer, Eq. (12.108), have all been used. By choosing the arbitrary constants in the Coulomb part of the wall–wall potential to cancel the above constants, only the mean electrostatic potential remains on the right-hand side, and hence the integral decays to 0 at large separations. Hence one chooses

$$V_{00}(z) = \frac{2\pi\sigma^2}{\epsilon} S + \sigma\psi(0) - \frac{2\pi\sigma^2}{\epsilon} z,\ z > 0, \tag{12.125}$$

and the interaction free energy per unit area becomes

$$\beta w_{00}^{\text{int}}(z) = \beta\sigma\psi(z) - d_{00}(z) - \int_{-\infty}^{\infty} \mathrm{d}z'\, \underline{H}^{\text{T}}(z')\underline{\chi}(z - z'),\ z > 0. \tag{12.126}$$

Note that $w_{00}^{\text{int}}(z) \to 0$, $z \to \infty$ (see below), and hence the surface and bulk contributions have been removed from this expression for the free energy. Differentiating this, one obtains the net pressure (force per unit area) between the walls

$$p^{\text{net}}(z) = \frac{-\partial w_{00}^{\text{int}}(z)}{\partial z}. \tag{12.127}$$

This is the total pressure of the electrolyte confined between the walls less the pressure of the bulk electrolyte imagined to be pressing in on the far sides of the walls (at infinity).

The above expressions are for two identical walls, but the generalisation to the nonsymmetric situation is immediate. If one uses the subscripts '1' and '2' to denote the two walls, then one has

$$\begin{aligned} \beta w_{12}^{\text{int}}(z) &= \beta\sigma_1\psi_2(z) - d_{12}(z) - \int_{-\infty}^{\infty} \mathrm{d}z'\, \underline{H}_1^{\text{T}}(z')\underline{\chi}_2(z - z') \\ &= \beta\sigma_2\psi_1(z) - d_{12}(z) - \int_{-\infty}^{\infty} \mathrm{d}z'\, \underline{H}_2^{\text{T}}(z')\underline{\chi}_1(z - z'). \end{aligned} \tag{12.128}$$

Asymptotics of Overlapping Planar Double Layers

The asymptotic interaction of the charged walls can be obtained from the interaction free energy, Eq. (12.126). One assumes the bridge function to be the most short-ranged function appearing in this expression, consistent with its behaviour in the bulk electrolyte and in the isolated double layer. The range of the dominant wall–ion mean electrostatic potential and the short-range series function is κ. From the transform of Eq. (12.120) one obtains

$$\underline{\overline{H}}(k) \sim \frac{-4\pi\beta\tilde{\sigma}}{\epsilon\nu\kappa} \underline{\tilde{q}} \frac{1}{\kappa + ik},\ k \to i\kappa, \tag{12.129}$$

and from Eq. (12.121), one obtains

$$\overline{\psi}(k) \sim \frac{4\pi\tilde{\sigma}}{\epsilon\nu\kappa} \frac{1}{\kappa + ik},\ k \to i\kappa, \tag{12.130}$$

assuming for the present that κ is real. Hence the wall–wall interaction free energy per unit area, Eq. (12.126), in the vicinity of the pole is

$$\begin{aligned} \beta\overline{w}_{00}^{\text{int}}(k) &\sim \left[\frac{4\pi\beta\sigma\tilde{\sigma}}{\epsilon\nu\kappa} + \frac{4\pi\beta\tilde{\sigma}}{\epsilon\nu\kappa}\underline{\tilde{q}}^{\text{T}}\underline{\overline{\chi}}(i\kappa)\right] \frac{1}{\kappa + ik},\ k \to i\kappa, \\ &= \frac{8\pi\beta\tilde{\sigma}^2}{\epsilon\nu\kappa} \frac{1}{\kappa + ik}, \end{aligned} \tag{12.131}$$

where the expression for the effective surface charge density, Eq. (12.118), has been used. It follows that the asymptote in the monotonic regime is

$$w_{00}^{\text{int}}(r) \sim \frac{8\pi\tilde{\sigma}^2}{\epsilon\nu\kappa} e^{-\kappa z},\ z \to \infty,\ \text{Im}\{\kappa\} = 0, \tag{12.132}$$

and, in the oscillatory regime, it is

$$w_{00}^{\text{int}}(r) \sim 2\,\text{Re}\left\{\frac{8\pi\tilde{\sigma}^2}{\epsilon\nu\kappa} e^{-\kappa z}\right\},\ z \to \infty,\ \text{Im}\{\kappa\} \neq 0. \tag{12.133}$$

This depends only on properties of the bulk electrolyte, κ and ν, and the effective surface charge of the isolated double layer, $\tilde{\sigma}$. This result reduces to the linear Poisson–Boltzmann theory in the Debye–Hückel limit, $\chi_{\alpha\gamma} = 0$. Since the effective surface charge occurs as a square, and because κ and ν are both positive, in the monotonic regime the force between identically charged walls is repulsive. That is, the interaction between two similar double layers is either monotonically repulsive or oscillatory, and any attractions cannot persist for all separations (although there can be a very large period of oscillation near the bulk transition).

Finally, in the core-dominated bulk regime the interaction free energy per unit area is

$$\beta w^{\text{int}}(z) \sim 2\,\text{Re}'\left\{\frac{-2ie^{-\xi z}}{\rho^2 x'(i\xi)}\left(\sum_{\gamma>0} \rho_\gamma \overline{X}_{0\gamma}(i\xi)\right)^2\right\},\ z \to \infty, \tag{12.134}$$

where again the effective surface 'charge' that described the amplitude of the decay of the density profiles at the isolated wall appears as the square in their interaction.

12.4 Algebraic Correlations along a Wall

A theme running through the preceding analysis of electrolytes and the electric double layer has been that the long-range Coulomb potential is screened and hence the correlation between ions is exponentially short-ranged, as are the density profiles in the electric double layer and interactions of two double layers. It is of interest to explore an exception to this rule. In the presence of a planar wall it is not possible to surround an ion with counterions on all sides. Hence the ions and their countercharge in the double layer have a net dipole moment, and the correlation between these dipoles decays as an inverse cubic in separation parallel to the wall.

As shown in Section 11.4.2 for the general planar inhomogeneous Ornstein–Zernike equation, the density depends only upon the distance from the wall, $\rho_\alpha(\mathbf{r}) = \rho_\alpha(z)$, and the pair correlation functions depend upon the distances of the two ions from the wall and upon their separation in the direction parallel to the wall, $h_{\alpha\gamma}(\mathbf{r}_1, \mathbf{r}_2) = h_{\alpha\gamma}(s_{12}, z_1, z_2)$. The matrix of distribution functions becomes $\{\underline{\underline{H}}(s_{12}, z_1, z_2)\}_{\alpha\gamma} = \rho_\alpha^{1/2}(z_1) h_{\alpha\gamma}(s_{12}, z_1, z_2) \rho_\gamma^{1/2}(z_2)$. In this planar geometry the inhomogeneous Ornstein–Zernike equation partially factorises upon Fourier transformation in the lateral direction, Eq. (11.133),

$$\underline{\underline{\hat{H}}}(k, z_1, z_2) = \underline{\underline{\hat{C}}}(k, z_1, z_2) + \int_0^\infty \mathrm{d}z_3\, \underline{\underline{\hat{H}}}(k, z_1, z_3)\underline{\underline{\hat{C}}}(k, z_3, z_2). \tag{12.135}$$

In this section the circumflex signifies the two-dimensional Fourier transform of a function with cylindrical symmetry, which is just the Hankel transform of order zero, Section 11.3.1.

The ion–ion potential in cylindrical coordinates is

$$\underline{\underline{u}}^{\mathrm{Coul}}(s_{12}, z_1, z_2) = \frac{1/4\pi\beta}{\sqrt{s_{12}^2 + (z_1 - z_2)^2}} \underline{\underline{Q}}(z_1, z_2), \tag{12.136}$$

with transform

$$\underline{\underline{\hat{u}}}^{\mathrm{Coul}}(k, z_1, z_2) = \frac{e^{-k|z_1 - z_2|}}{2\beta k} \underline{\underline{Q}}(z_1, z_2). \tag{12.137}$$

(The spatial dependence of the charge matrix arises from the densities that it contains.) Assuming that the inhomogeneous ion–ion direct correlation function decays like the Coulomb potential (as in the bulk and in the singlet analysis of the double layer), and hence that the function $\underline{\underline{\chi}}(s_{12}, z_1, z_2) = \underline{\underline{C}}(s_{12}, z_1, z_2) + \beta\underline{\underline{u}}^{\mathrm{Coul}}(s_{12}, z_1, z_2)$ is at least integrable, the Ornstein–Zernike equation may be alternatively written

$$\begin{aligned}
\underline{\underline{\hat{H}}}(k, z_1, z_2) &= \underline{\underline{\hat{\chi}}}(k, z_1, z_2) + \int_0^\infty \mathrm{d}z_3\, \underline{\underline{\hat{H}}}(k, z_1, z_3)\underline{\underline{\hat{\chi}}}(k, z_3, z_2) \\
&\quad - \underline{\underline{Q}}(z_1, z_2)\frac{e^{-k|z_1 - z_2|}}{2k} \\
&\quad - \int_0^\infty \mathrm{d}z_3\, \underline{\underline{\hat{H}}}(k, z_1, z_3)\underline{\underline{Q}}(z_3, z_2)\frac{e^{-k|z_3 - z_2|}}{2k}.
\end{aligned} \tag{12.138}$$

The last two terms represent the transform of the mean potential about an ion; $\phi_\alpha(s_{12}, z_1; z_2)$ is the change in the mean electrostatic potential at $\mathbf{s}_2, z_2$ due to an ion of type α being at $(\mathbf{s}_1, z_1)$,

$$\phi_\alpha(s_{12}, z_1; z_2) = \frac{q_\alpha}{\epsilon r_{12}} + \sum_\gamma \int \mathrm{d}\mathbf{r}_3\, h_{\alpha\gamma}(s_{13}, z_1, z_3)\frac{q_\gamma}{\epsilon r_{32}}\rho_\gamma(z_3). \tag{12.139}$$

This potential in the double layer plays the same role as the mean electrostatic potential about an ion in the bulk electrolyte.

As in the analysis of the bulk electrolyte for an exponentially short-ranged function all moments exist and the odd moments are 0. In this planar geometry the Hankel transform of the Coulomb potential contains odd powers of k (whereas only $1/k^2$ appeared in the bulk), and hence the correlation functions will exhibit power-law decay parallel to the wall. The moment expansion is

$$\underline{\underline{\hat{H}}}(k, z_1, z_2) = \underline{\underline{H}}^{(0)}(z_1, z_3) + |k|\underline{\underline{H}}^{(1)}(z_1, z_3) + \ldots, \tag{12.140}$$

and similarly for $\underline{\underline{\hat{\chi}}}$. Note that since $\underline{\underline{H}}(r, z_1, z_2)$ is a real even function of r the absolute value of k appears. Taking the limit that $k \to 0$ in the Ornstein–Zernike equation and one can equate the coefficients of the powers of k. (For finite $|z_1 - z_2|$ it is permissible to expand the various exponentials since z_3 is kept finite by the short-range of the total correlation function in the integrand.) Equating the coefficients of k^{-1} one has

$$0 = 0 + 0 - \frac{1}{2}\underline{\underline{Q}}(z_1, z_2) - \frac{1}{2}\int_0^\infty \mathrm{d}z_3\, \underline{\underline{H}}^{(0)}(z_1, z_3)\underline{\underline{Q}}(z_3, z_2), \tag{12.141}$$

or

$$-q_\alpha = \sum_\gamma q_\gamma \int \mathrm{d}z_2\, \mathrm{d}\mathbf{s}_2\, h_{\alpha\gamma}(s_{12}, z_1, z_2)\rho_\gamma(z_2). \tag{12.142}$$

This is just the local electro-neutrality condition in planar geometry.

The coefficient of k^0 yields

$$\begin{aligned}
&\underline{\underline{H}}^{(0)}(z_1, z_2) \\
&= \underline{\underline{\chi}}^{(0)}(z_1, z_2) + \int_0^\infty \mathrm{d}z_3\, \underline{\underline{H}}^{(0)}(z_1, z_3)\underline{\underline{\chi}}^{(0)}(z_3, z_2) + \underline{\underline{Q}}(z_1, z_2)\frac{|z_1 - z_2|}{2} \\
&\quad + \int_0^\infty \mathrm{d}z_3\, \underline{\underline{H}}^{(0)}(z_1, z_3)\underline{\underline{Q}}(z_3, z_2)\frac{|z_3 - z_2|}{2} \\
&\quad - \frac{1}{2}\int_0^\infty \mathrm{d}z_3\, \underline{\underline{H}}^{(1)}(z_1, z_3)\underline{\underline{Q}}(z_3, z_2).
\end{aligned} \tag{12.143}$$

Multiplying by $\underline{\underline{Q}}(z_0, z_1)$ and integrating over z_1 cancels most of these terms by the local electro-neutrality condition. One obtains

$$-\underline{\underline{Q}}(z_0, z_2) = \frac{-1}{2}\int_0^\infty \mathrm{d}z_1 \int_0^\infty \mathrm{d}z_3\, \underline{\underline{Q}}(z_0, z_1)\underline{\underline{H}}^{(1)}(z_1, z_3)\underline{\underline{Q}}(z_3, z_2) \tag{12.144}$$

or

$$1 = \frac{2\pi\beta}{\epsilon}\sum_{\alpha\gamma} q_\alpha q_\gamma \int_0^\infty \mathrm{d}z_1 \int_0^\infty \mathrm{d}z_2\, \rho_\alpha(z_1)\rho_\gamma(z_2)h_{\alpha\gamma}^{(1)}(z_1, z_2). \tag{12.145}$$

This shows that the first moment of the total correlation function is nonzero. Now if $\hat{f}(k) \sim \text{constant} + A|k|$, $k \to 0$, then $f(r) \sim -A/2\pi r^3$, $r \to \infty$. Hence,

$$h_{\alpha\gamma}(r, z_1, z_2) \sim \frac{-h_{\alpha\gamma}^{(1)}(z_1, z_2)}{2\pi r^3},\ r \to \infty, \tag{12.146}$$

where the numerator obeys the sum rule, Eq. (12.145). The asymptote and the sum rule for an electrolyte next to a planar wall have been discussed by Jancovici and others.[9]

[9]B. Jancovici, Classical Coulomb systems near a plane wall. I and II, *J. Stat. Phys.* **28** (1982), 43 and 263. M. Baus, Long-range correlations along an interfacial electric double layer, *Mol. Phys.* **48** (1983), 347. S. L. Carnie and D. Y. C. Chan, Correlations in inhomogeneous Coulomb systems, *Mol. Phys.* **51** (1984), 1047.

This result – that the ion–ion correlations decay as inverse cubics in the direction parallel to a planar wall – is in marked contrast to the exponential decay of the bulk electrolyte. It is a consequence of the fact that the wall prevents the electrolyte from screening the charge on an ion equally on all sides. The ion plus countercharge cloud has a net dipole moment, and the dipole–dipole interaction decays as r^{-3}. For similar reasons ions confined to a two-dimensional surface or membrane exhibit identical decay. In this case the charge–charge correlation function has a universal asymptote.[10]

Summary

- All Coulomb systems are electro-neutral and the pair correlations obey a second-moment condition. All charges are screened, and the correlations between them are exponentially decaying.

- The Poisson–Boltzmann approximation is a mean field approximation that is valid at low coupling. For a bulk electrolyte the linear version, known as Debye–Hückel theory, shows that the virial expansion is not valid for nonintegrable potentials.

- The exact asymptote for the ion correlation functions has linear Poisson–Boltzmann form, but with effective charges responding to the mean potential. The actual screening length reduces to the Debye length at low couplings. The correlations become oscillatory at high ionic couplings.

- The behaviour of the electric double layer about a charged solute is similar to that of the bulk electrolyte. The counterion charge balances the solute charge, and the ion density profiles are exponentially decaying at the same rate as in the bulk electrolyte. Asymptotically one can treat the solute as having an effective charge. The asymptotic interaction between two solutes is determined by the product of their effective charge and by the bulk decay length.

- Long-range correlations occur when the electrolyte can't screen a charge on all sides. In the case of a planar wall the pair correlations decay as inverse cubics parallel to the wall.

[10] P. Attard, R. Kjellander, and D. J. Mitchell, Interactions between electro-neutral surfaces bearing mobile charges, *Chem. Phys. Lett.* **139** (1987), 219. P. Attard, R. Kjellander, D. J. Mitchell, and Bo. Jönsson, Electrostatic fluctuation interactions between neutral surfaces adsorbed with mobile ions or dipoles, *J. Chem. Phys.* **89** (1988), 1664.

Chapter 13

Computer Simulations

Computer simulations complement the semi-analytic approaches to statistical mechanics detailed in the preceding chapters. They closely mimic the actual molecular motion of the system and provide a wealth of molecular level detail not obtainable with the other methods. In one sense simulations are the most basic approach to statistical mechanics, since one does not have to perform any diagrammatic or functional analysis, or to carry out asymptotic or other expansions. However, the fact that they represent a brute force approach also means that they yield practically exact results, and they provide useful benchmarks to test the more efficient but approximate methods. The main limitation on simulations are the size of the system that can be simulated and the length of the simulation. The fact that both of these are determined by computer power accounts for the increasing importance and use of simulations in recent years. Moreover, the development of more efficient algorithms that represent elegant, nonphysical trajectories for the system have enabled larger and more complex systems to be studied.

There are two types of simulations: Monte Carlo and molecular dynamics. The Monte Carlo method essentially evaluates the multidimensional averages of statistical mechanics by a weighted random generation of a configuration of the system. In its simplest form the weight is the Boltzmann factor for a constant temperature system, and averages are simple averages over the configurations visited. In contrast molecular dynamics is deterministic: it follows in time a single trajectory of the system on the energy hypersurface in phase space using Hamilton's equations of motion; again more sophisticated versions exist. The averages used are simple time averages. The main advantage of Monte Carlo is that it is not restricted to the physical trajectory (i.e., it is not determined rigidly by the intermolecular potential), and mathematical techniques can be used to visit important configurations so as to increase the efficiency of the simulation. The advantage of molecular dynamics is that time-dependent phenomena can be studied (e.g., diffusion, viscosity).

In what follows a brief account of the two basic methods is given. Summaries of some of the more advanced techniques are also included. Broader and more comprehensive treatments may be found in specialised monographs.[1]

13.1 Molecular Dynamics

13.1.1 Equations of Motion

In molecular dynamics one seeks the trajectory in phase space of the system. Denoting the position of the ith atom at time t by $\mathbf{q}_i(t)$ and its momentum by $\mathbf{p}_i(t)$, one proceeds from the Hamiltonian,

$$\mathcal{H}(\mathbf{q}^N, \mathbf{p}^N) = K(\mathbf{p}^N) + U(\mathbf{q}^N), \tag{13.1}$$

where the kinetic energy is

$$K(\mathbf{p}^N) = \frac{1}{2m} \sum_{i=1}^{N} p_i^2, \tag{13.2}$$

and the potential energy is

$$U(\mathbf{q}^N) = \sum_{i<j}^{N} u(q_{ij}). \tag{13.3}$$

This is obviously the simplest case, with all atoms having identical mass m, spherically symmetric pair potentials, and no external potential or many-body potentials. Hamilton's equations of motion give the velocity at a given time,

$$\dot{\mathbf{q}}_i = \frac{\partial \mathcal{H}}{\partial \mathbf{p}_i} = \mathbf{p}_i/m, \tag{13.4}$$

and the rate of change of momentum,

$$\dot{\mathbf{p}}_i = \frac{-\partial \mathcal{H}}{\partial \mathbf{q}_i} = \sum_{j=1}^{N}{}' u'(q_{ij})\hat{\mathbf{q}}_{ij}, \tag{13.5}$$

where the $j = i$ term is excluded from the summation, $\mathbf{q}_{ij} = \mathbf{q}_j - \mathbf{q}_i$, and the circumflex denotes the unit vector. The right-hand side is the total force on atom i, which can be denoted by $\mathbf{F}_i$.

In order to compute the trajectories as a function of time one must discretise these equations. Assuming that the positions and momenta are known at time t, the simplest approach is to carry out a first-order Taylor expansion,

$$\mathbf{q}_i(t + \Delta_t) = \mathbf{q}_i(t) + \Delta_t \dot{\mathbf{q}}_i(t) = \mathbf{q}_i(t) + \Delta_t \mathbf{p}_i(t)/m, \tag{13.6}$$

$$\mathbf{p}_i(t + \Delta_t) = \mathbf{p}_i(t) + \Delta_t \dot{\mathbf{p}}_i(t) = \mathbf{p}_i(t) + \Delta_t \mathbf{F}_i(t). \tag{13.7}$$

[1] M. P. Allen and D. J. Tildesley, *Computer Simulation of Liquids*, Clarendon Press, Oxford, 1987. P. Lykos (Ed.), *Computer Modelling of Matter*, ACS Symposium Series, Vol. 86, Am. Chem. Soc., Washington, DC, 1978.

This simple time stepping has an error proportional to Δ_t^2 at each step. The central difference method is a slightly more sophisticated approach based upon the third-order forward and backward differences,

$$\mathbf{q}_i(t \pm \Delta_t) = \mathbf{q}_i(t) \pm \Delta_t \dot{\mathbf{q}}_i(t) + \frac{\Delta_t^2}{2!} \ddot{\mathbf{q}}_i(t) \pm \frac{\Delta_t^3}{3!} \dddot{\mathbf{q}}_i(t) + \ldots. \tag{13.8}$$

Adding these two together, the odd terms cancel and one obtains

$$\mathbf{q}_i(t + \Delta_t) = 2\mathbf{q}_i(t) - \mathbf{q}_i(t - \Delta_t) + \frac{\Delta_t^2}{m} \mathbf{F}_i(t) + \mathcal{O}\Delta_t^4. \tag{13.9}$$

Hence each new configuration is generated from the two previous configurations. In this approach the momenta do not appear. If these are required, they can be obtained by subtracting the two differences,

$$\dot{\mathbf{q}}_i(t) = \frac{1}{2\Delta_t} \left[\mathbf{q}_i(t + \Delta_t) - \mathbf{q}_i(t - \Delta_t) \right] + \mathcal{O}\Delta_t^2. \tag{13.10}$$

A higher-order expression for the momenta can be obtained by subtracting one-eighth of the difference between the $\mathbf{q}_i(t \pm 2\Delta_t)$. These simple time-stepping algorithms have proved robust and practical. More sophisticated predictor–corrector algorithms have also been used.[2]

The smaller the time step, the more accurate will be the trajectory. However, this must be balanced against the fact that for a given number of steps, a large step allows a larger and presumably more representative region of the energy hypersurface to be visited. A good guide to the largest acceptable step can be obtained by monitoring the energy of the system. On the exact trajectory the energy is constant. In practice it fluctuates due to numerical errors arising from the finite difference approximation to the differential equation. In general a time step that conserved energy to three or four digits would be considered acceptable.

For the case of a Lennard–Jones fluid, the time scale that appears naturally is

$$\tau = \sqrt{m\sigma^2/48\epsilon}, \tag{13.11}$$

where σ is the diameter and ϵ is the well depth. For the case of argon, $\sigma = 3.41 \times 10^{-10}$ m, $\epsilon = 1.65 \times 10^{-21}$ J, and $m = 6.63 \times 10^{-26}$ kg, one has $\tau = 3.1 \times 10^{-13}$ s. Typically, for condensed materials the time step required is on the order of $\Delta_t \approx 0.1$–0.01τ.

Hard-Sphere Potential

The preceding discussion assumed a constant time step. This is appropriate for continuous potentials, where all the atoms simultaneously interact and the collision process is a continuous one. For the case of hard spheres, the collisions

[2] H. J. C. Berendson and W. F. van Gunsteren, in *Molecular Dynamics Simulation of Statistical Mechanical Systems*, Proceedings of the Enrico Fermi Summer School, Varenna, p. 43, Soc. Italiana di Fisica, Bologna, 1985.

are instantaneous and strictly binary. In such systems the trajectories of the atoms are strictly linear between collisions, and so one can readily calculate the time that the next collision will occur. Time steps correspond to successive collisions and are not equally spaced.

The discontinuous nature of the hard-sphere potential means that its differential is a generalised function, which creates problems in the direct application of Hamilton's equations. The way to proceed is to invoke energy and momentum conservation by the elastic collision. This has the consequence that for each particle the velocities perpendicular to the intermolecular vector are unchanged by the collision, whereas the two velocities along the intermolecular vector are interchanged.

13.1.2 Initiation

The above describes how the trajectory at the next instant in time is calculated, and it remains to discuss the initial starting point. The best starting point is from a previously equilibrated configuration of the same system. The second best starting point is from a previously equilibrated system at a nearby state point. The final possibility is from a regular lattice configuration with parameters chosen to yield the desired density. The worst possible choice would be a completely random arrangement of atoms. In the last two cases the velocity components of each atom are chosen independently from a Maxwell–Boltzmann distribution at a given temperature.

In most cases one must divide the simulation up into a transient equilibration phase and the equilibrium production run. That is, one must run the simulation long enough so that the effects of the starting configuration are negligible, since this by definition is an atypical configuration. In practice one monitors the potential energy, which almost always decreases steadily during the equilibration phase. Since the total energy is conserved, the kinetic energy correspondingly increases. In view of the Gaussian nature of the momentum integrals in the partition function, at equilibrium the average kinetic energy per molecule is given by

$$\langle p_i^2/2m \rangle = \frac{3k_{\mathrm{B}}T}{2}. \tag{13.12}$$

Hence the kinetic energy gives the average temperature of the system. Periodically during the equilibration stage the velocities of all the particles are rescaled to reset the kinetic temperature to the desired level.

After the equilibration phase the production part of the simulation begins. In this phase, which may not even be the bulk of the simulation, configurations for the generation of the averages of whatever quantity is of interest are accumulated. The configurations from the initial phase are discarded and are not used for the averages. The distinction between equilibration and equilibrium is not clear-cut. To some extent it is a matter of judgement as to when the changes in the kinetic temperature are no longer steady, but represent equilibrium fluctuations. One should avoid any velocity rescaling once the simulation proper has begun.

It was mentioned above that the total energy often fluctuates due to discretisation errors. A steady change in the energy is most likely due to a bug in the program. That having been said, there can occur a weak increase in the total energy due to the numerical errors introduced by the finite difference algorithm. Although in general the errors are random, the consequent energy fluctuations are more often positive rather than negative. This of course is due to the fact that entropy is a monotonic increasing function of energy: since there are more configurations of higher rather than lower energy, the random numerical errors are more likely to yield an increase rather than a decrease in total energy. This weak increase in total energy will result in a weak but steady increase in the kinetic temperature. Either one must use a small enough time step so that this increase remains acceptable over the course of the simulation or one must periodically use velocity rescaling. In the latter event, one should be cautious about carrying out averages (particularly for time correlations) across the rescaling events.

13.1.3 Averages

The trajectory followed by a molecular dynamics simulation is $\mathbf{\Gamma}(t|\mathbf{\Gamma}_0)$, where $\mathbf{\Gamma}_0$ is the point in phase space at the beginning of the equilibrium simulation. Molecular dynamics invokes a simple time average,

$$\begin{aligned} \langle g \rangle &= \frac{1}{S}\int_0^S \mathrm{d}t\, g(\mathbf{\Gamma}(t|\mathbf{\Gamma}_0)) \\ &= \frac{1}{M}\sum_{i=1}^{M} g_i, \end{aligned} \tag{13.13}$$

where $g_i \equiv g(\mathbf{\Gamma}(t_i|\mathbf{\Gamma}_0))$, t_i being one of the discrete time steps of the simulation.

In general it is not necessary to collect a value for the average at every step of the simulation. Successive steps are highly correlated and very little new information is generated. Although this redundancy does no harm, it is inefficient. Typically once every 10–100 steps is sufficient, depending upon the property being sought. Quite often the simulation itself is only used to generate configurations and a regular selection of these is periodically saved to disk. The averages can then be calculated at leisure by analysing these saved configurations.

In order to ascertain the precision of the simulated quantities, the simulation is broken into 10–100 blocks, and the averages are taken over each block independently. If the blocks are large enough, correlations between them can be ignored, and the standard deviation for the blocks can be estimated. The average value of the quantity is the average of the block averages, and the standard error is the standard deviation divided by the square root of the number of blocks. To be precise, suppose that there are m blocks and that G_j is the average value obtained from the jth block. The average value obtained from the simulation is

$$\langle g \rangle = \frac{1}{m} \sum_{j=1}^{m} G_j, \tag{13.14}$$

and the square of the standard deviation of the blocks is

$$\sigma^2 = \frac{1}{m} \sum_{j=1}^{m} [G_j - \langle g \rangle]^2 . \tag{13.15}$$

One might expect the blocks to obey a normal distribution, $\wp(G_j) = (\sigma^{-1}/\sqrt{2\pi}) \exp -[G_j - \langle g \rangle]^2/2\sigma^2$. In any case, the statistical error of the simulation is $\epsilon = \sigma/\sqrt{m}$, and one would report the result as $\langle g \rangle \pm \epsilon$. This error should be a property of the simulation and independent of the number of blocks that it is broken up into.

It should be clear that the trajectory and any averages derived from it depend upon the starting position $\mathbf{\Gamma}_0$. In so far as all points on the energy hypersurface have equal weight (i.e., $\wp(\mathbf{\Gamma}|E) \propto \delta(\mathcal{H}(\mathbf{\Gamma}) - E)$), this assumption is acceptable, which is to say that any starting point is as good as another. One also implicitly assumes that the trajectory samples all regions on the energy hypersurface, or at least a fully representative subset of them. This is difficult to ensure or to check. One can possibly run the simulation for longer times, or one could choose a variety of starting points and run several independent simulations. For simple systems problems with ergodicity are rare. For very dense or glassy systems, or for polymeric systems, or for two-phase coexistence and phase transitions, problems with metastability, entrapment, and incomplete sampling must be grappled with.

13.2 Periodic Boundary Conditions

One of the limitations of both molecular dynamics and Monte Carlo is the size of the system that can be treated. It is generally not feasible to treat more than some tens of thousands of atoms, and for more complex potentials or molecules several hundred atoms may be more realistic. Since in general one is interested in a bulk system (the thermodynamic limit), one must minimise the effects of any boundaries or walls. The way that this is commonly done is by means of periodic boundary conditions, as illustrated in Fig. 13.1.

One imagines that the cell used for the simulations is but one of an infinite number of identical cells all arranged on a simple cubic lattice. Each of the N atoms in the central cell has an image at the same relative position in all the cells. When an atom leaves the central cell, its corresponding image enters the central cell from the opposite face. In this way the number of atoms is conserved during the simulation. When an atom leaves the cell in this way one usually resets its coordinates to that of the image that has entered the cell so that the atoms in the simulation always appear to be in the central cell. This is not

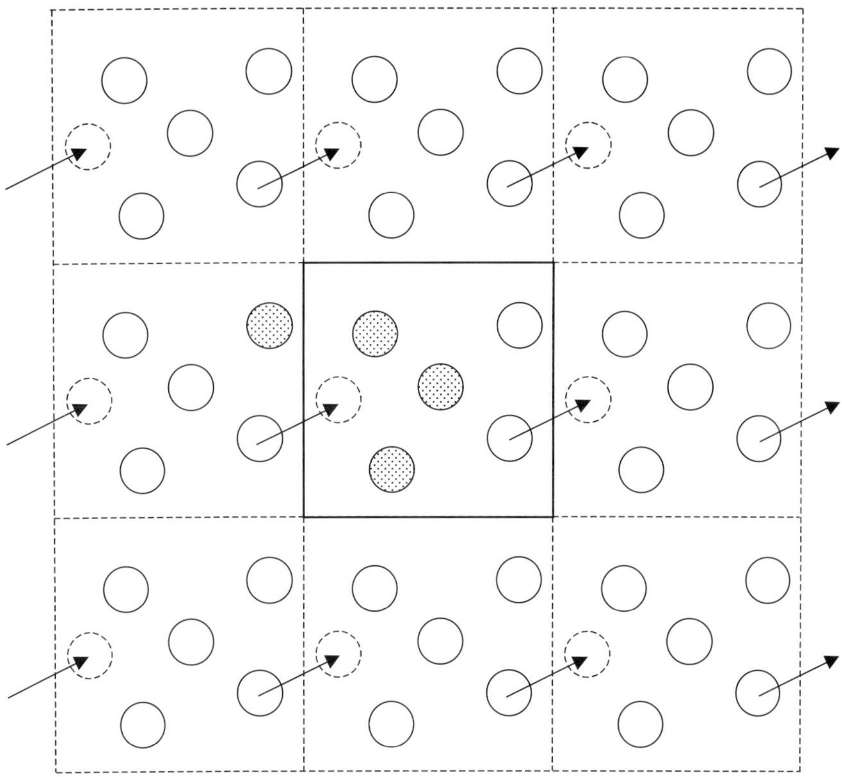

Figure 13.1: The central cell, its periodic images, and the four nearest neighbours of the moved atom.

strictly necessary, and for some properties (e.g., the self-diffusion coefficient), it is preferable to keep the real coordinates for each atom.

Each atom in the central cell interacts with the other atoms, or with one of the images of the other atoms. In choosing between an atom and its image, only the closest is used for the interaction. This is known as the minimum image convention, and is illustrated by the shaded atoms in the figure. In this way the boundaries of the cell are effectively invisible, so that each atom feels as if it is part of an infinite system. This is not entirely true because the finite size of the central cell can still have an influence on the properties being simulated. For example, if the true pair correlation function had a range that was greater than the size of the cell, then effectively each atom would feel the tail of its own interaction. To put it another way, it is not possible to simulate physical properties that range in space beyond the central cell.

Molecular dynamics requires the force on each atom, and Monte Carlo requires its interaction potential. For a pair-wise additive potential for each atom there are N such interactions to be calculated, which means $N(N-1)/2$ calculations for each time step. Hence the computational burden of the simulation grows quadratically with the number of atoms. This is ultimately what limits the system size.

This procedure is very inefficient because it takes as much effort to calculate the interaction between atoms close together, where the potential is all-important, as for atoms far apart, where the potential is weak and has little influence. One way to decrease the computational burden is to introduce a cutoff for the potential, $u(r) = 0$, $r > R_c$, and to only include atoms that are within this cutoff in the calculation of the force on each atom. Roughly speaking then, the number of calculations per atom is decreased by a factor of $4\pi R_c^3/3V$, where $V = L^3$ is the volume of the central cell. The largest that R_c can be is $L/2$, because this guarantees that each atom will interact with at most one image of every other atom. In practice R_c can be much smaller than this. For a Lennard–Jones potential a value of $R_c = 2.5\sigma$ is not uncommon, compared to system sizes on the order of $L = 10\sigma$ and larger.

The use of a cutoff potential is inappropriate for ionic or dipole systems, where it can lead to serious errors. In these cases Ewald summation[3] or reaction field[4] techniques are advised.

The use of a cutoff introduces a discontinuity in the pair potential and in the force at the cutoff. To avoid any unphysical behaviour there, in the case of molecular dynamics a constant is added to make the force go to 0 at the cutoff. That is, the force that one actually uses in the simulation is

$$\mathbf{f}_{\mathrm{MD}}(r) = \begin{cases} \mathbf{f}(r) - \mathbf{f}(R_c), & r < R_c \\ 0, & r \geq R_c. \end{cases} \tag{13.16}$$

Hopefully the cutoff is large enough that at smaller separation $\mathbf{f}(R_c)$ is relatively negligible compared to the force, $\mathbf{f}(r)$.

The average value of the simulated quantities should be corrected for the truncation of the force or potential. For example, the average potential energy per atom is

$$\begin{aligned} U/N &= \frac{1}{2N}\int_V \mathrm{d}\mathbf{r}_1\,\mathrm{d}\mathbf{r}_2\,\rho^{(2)}(r_{12})u(r_{12}) \\ &= \frac{V}{2N}\rho^2\int_V \mathrm{d}\mathbf{r}\,g(r)u(r) \\ &= 2\pi\rho\left[\int_0^{R_c}\mathrm{d}r\,r^2 g(r)u(r) + \int_{R_c}^{\infty}\mathrm{d}r\,r^2 g(r)u(r)\right]. \end{aligned} \tag{13.17}$$

[3] S. W. De Leeuw, J. W. Perram, and E. R. Smith, Simulation of electrostatic systems in periodic boundary conditions. I. Lattice sums and dielectric constant, *Proc. R. Soc. London A* **373** (1980), 27. D. M. Heyes, Electrostatic potentials and fields in infinite point charge lattices, *J. Chem. Phys.* **74** (1981), 1924.

[4] J. A. Barker, Reaction field method for polar fluids, in *The Problem of Long-Range Forces in the Computer Simulation of Condensed Matter* (D. Ceperly, Ed.), NRCC Workshop Proceedings, Vol. **9**, p. 45, 1980. H. L. Friedman, Image approximation to the reaction field, *Mol. Phys.* **29** (1975), 1533.

The first of these corresponds to the actual molecular dynamics average,

$$U_{\mathrm{MD}}/N = \frac{1}{NM} \sum_{i=1}^{M} \sum_{j=1}^{N} \sum_{k=j+1}^{N} u_{\mathrm{MD}}(r_{jk}^{(i)}), \tag{13.18}$$

where $r_{jk}^{(i)}$ is the separation between atoms j and k in the ith configuration. The tail correction that should be added to this is the second term,

$$U_{\mathrm{tail}}/N = 2\pi\rho \int_{R_c}^{\infty} \mathrm{d}r\, r^2 u(r). \tag{13.19}$$

For the tail the radial distribution function has been replaced by its asymptotic value, $g(r) \to 1$, $r \to \infty$. Often this tail integral can be evaluated analytically.

13.3 Monte Carlo

The Monte Carlo method follows a stochastic trajectory through configuration space, which is in contrast to the determinism that is molecular dynamics. However, like molecular dynamics it gives average values without evaluating the configuration integral or the partition function. The average of some function of the coordinates, $f(\mathbf{r}^N)$, is defined to be

$$\langle f \rangle = \int_V \mathrm{d}\mathbf{r}^N \, \wp(\mathbf{r}^N) f(\mathbf{r}^N). \tag{13.20}$$

For the simplest canonical case (constant temperature), the probability density is

$$\wp(\mathbf{r}^N | N, V, T) = \frac{1}{N! \Lambda^{3N} Q(N, V, T)} \exp -\beta U(\mathbf{r}^N). \tag{13.21}$$

Note that the Monte Carlo method only deals with configurations; the momentum integrals have been performed and incorporated into the thermal wavelength.

To implement this average one generates a sequence of points in configuration space according to some recipe. Let $\omega(\mathbf{r}^N)$ be the weight attached to choosing a particular point. In this case the average becomes

$$\langle f \rangle = \frac{1}{W} \sum_{\alpha=1}^{M} f(\mathbf{r}_\alpha^N) \wp(\mathbf{r}_\alpha^N | N, V, T) / \omega(\mathbf{r}_\alpha^N), \tag{13.22}$$

where $\mathbf{r}_\alpha^N$ is one of the M generated configurations, and where the total weight is $W = \sum_{\alpha=1}^{M} \wp(\mathbf{r}_\alpha^N | N, V, T) / \omega(\mathbf{r}_\alpha^N)$. One must include ω in the denominator to cancel the extra probability with which the points in the trajectory occur.

Although the weighting function is arbitrary and this formula will yield exact results for any choice, an inappropriate weight may require a prohibitively lengthy trajectory to yield reasonable results. For example, choosing the constant weight, which makes all points in the sequence equally likely, and which would correspond to choosing N points in the volume randomly, uniformly, and independently each time, fails for condensed systems because of the overwhelming probability that the configurations generated will be forbidden because some molecules overlap. One needs to bias the trajectory so that it covers the important configurations, which are those of low energy. The most common weight is one that cancels the Boltzmann factor, $\omega(\mathbf{r}^N) \propto \exp -\beta U(\mathbf{r}^N)$, so that the ratio of the probability and the weight is constant and independent of the configuration. The constant may be set equal to unity so that one has

$$\langle f \rangle = \frac{1}{M} \sum_{\alpha=1}^{M} f(\mathbf{r}_\alpha^N). \tag{13.23}$$

One sees that with this choice of weight all configurations generated contribute equally to the average.

13.3.1 Metropolis Algorithm

The crucial question is how to generate successive states along the Monte Carlo trajectory so that their weight is proportional to the Boltzmann factor. The *Metropolis algorithm* does this by forming a Markov chain in which the probability of generating the next configuration depends only upon the current configuration. The configurations of the system may be imagined to be discretised and labelled with a subscript. A trial transition matrix is defined, with elements T_{nm} that represent the probability of choosing state m for a possible move from state n. Generally one only attempts to move between neighbouring configurations, which in practice means one of the atoms of the system is displaced a small amount. For example, for some $j \in [1, N]$, $\mathbf{r}'_i = \mathbf{r}_i$, $i \neq j$, and $\mathbf{r}'_j = \mathbf{r}_j + \Delta\mathbf{a}$. Here $\mathbf{a}$ is a random vector generated for each trial move by choosing three random numbers ξ_1, ξ_2, and ξ_3, each independently and uniformly distributed on $(0, 1)$, with $a_x = (2\xi_1 - 1)$, etc. The step length Δ is typically a fraction of the atomic diameter. It is clear that this method for generating neighbouring configurations gives a symmetric trial transition matrix (i.e., $T(\mathbf{r}_i \to \mathbf{r}'_i) = T(\mathbf{r}'_i \to \mathbf{r}_i)$).

One must make a decision about whether to accept or reject each trial move, and this is where the Boltzmann weight occurs. If the energy of the trial configuration is less than or equal to the energy of the current configuration, then one accepts the move. If the energy increases, then one accepts the move with probability proportional to the Boltzmann factor of the change in energy. More generally to generate states for an arbitrary probability distribution one always accepts the trial state if it is more probable than the current state, and one accepts it with probability in proportion to the ratio of the new and old probabilities if it is less probable than the current state. This is the Metropolis algorithm. It gives for the transition probability matrix

$$\Pi_{nm} = \begin{cases} T_{nm}, & \wp_m \geq \wp_n \\ T_{nm}\wp_m/\wp_n, & \wp_m < \wp_n. \end{cases} \tag{13.24}$$

The elements Π_{nm} denote the probability of next moving to state m given that one is in state n. A move is accepted if the ratio of the probabilities is greater than a random number uniformly distributed on $(0, 1)$. If the move is rejected, then the old configuration is counted as the new configuration in the trajectory (and one chooses a new atom for a trial move). Hence the probability of remaining in the same configuration is $\Pi_{nn} = 1 - \sum_{m \neq n} \Pi_{nm}$, which, because T is normalised, clearly lies between 0 and 1. Either one can randomly choose successive atoms for a trial move or one can systematically cycle through all the atoms of the system, trying to move each one in turn. Such a cycle of one attempted move for each of the N atoms is comparable to a single molecular dynamics time step.

In general one should use a symmetric trial transition matrix, $T_{nm} = T_{mn}$, since this ensures that the transition matrix itself obeys microscopic reversibility, $\wp_n \Pi_{nm} = \wp_m \Pi_{mn}$, which is to say that the probability of observing sometime on the trajectory the transition $n \rightarrow m$ is the same as the probability of observing the reverse transition, $m \rightarrow n$. This latter condition is sufficient to ensure that $\wp$ is the eigenvector of Π with the largest eigenvalue of unity, which is to say that it is the limiting probability distribution for the Markov states generated by Π.

The Metropolis algorithm has the effect of keeping the trajectory that the system follows through configuration space close to the energy valley floor. Whilst low-energy states are favoured, states higher in energy are not totally forbidden, depending how much larger than 0 is the ratio of the probabilities of the new and old states. Obviously at lower temperatures low-energy states dominate, and at higher temperatures more energetic configurations become increasingly accessible and contribute to the average.

As in molecular dynamics, the Metropolis algorithm for Monte Carlo simulations uses a simple unweighted average over the trajectory. Both also seek to cover a large and representative region of phase space economically and efficiently. In Monte Carlo the step length plays a crucial role in determining the acceptance ratio. A small step length will give an almost 100% acceptance rate, but the trajectory will move agonisingly slowly. Whilst a large step length might appear to increase the distance travelled, it can be akin to randomly inserting an atom in the system, and because of the overwhelmingly high probability of molecular overlap in dense systems, such moves are almost certainly rejected. Again the trajectory does not move far enough to sample a representative portion of configuration space. A practical compromise is to adjust Δ by trial and error to give an acceptance rate of 30–50%.

The considerations of ergodicity and metastability are similar in Monte Carlo and in molecular dynamics. Because the trajectory followed in Monte Carlo is not a physical trajectory, one has greater flexibility in choosing the trial moves or the configuration weight in order to bypass bottlenecks or to escape trapped configurations. If ergodicity is a problem, it may be worthwhile using infrequently a trial move with a larger than normal step length in the hope that despite the low acceptance rate, the trajectory will occasionally jump to a new region of configuration space. Similarly, using a flatter weight than the Boltzmann weight (e.g., a higher temperature) means that one explores further up the sides of the energy valleys, and hence one might pass over the col between two valleys more easily. Of course for such nonphysical weights the averages are no longer simple; each configuration on the trajectory must be weighted by the ratio of the Boltzmann weight and the actual weight used in the simulation.

Other aspects of Monte Carlo are almost identical to those already discussed for molecular dynamics. One must start the simulation from an initial configuration, which ideally comes from a previously equilibrated system. If not, one must discard from the averages the initial equilibration phase. Again this may be identified by monitoring the energy or the pressure. An infinite system is mimicked by using periodic boundary conditions. The computational burden lies in evaluating the potential energy; the quadratic dependence on the number of atoms can be reduced to a linear dependence by using a potential cutoff.

13.4 Advanced Techniques

13.4.1 Neighbour Tables

Whilst it is desirable to simulate a large system, both molecular dynamics and Monte Carlo are limited by the fact that the number of pair interactions grows quadratically with the number of atoms. As mentioned above, the situation can be ameliorated somewhat by using a potential cutoff. This reduces the number of evaluations from $N^2/2$ to approximately $\rho R_c^3 N$, where ρ is the density and R_c is the cutoff. However, even though one does not have to calculate the potential for atoms beyond the cutoff, one still must calculate their distance from the central atom in order to decide whether they lie within the cutoff. Hence there are still $N^2/2$ such distances that must be calculated, and this can become the limitation on system size.

The use of neighbour tables allows one to keep track of the atoms close to a given atom and hence one need only consider such atoms to decide whether they lie within the cutoff. The two types of neighbour tables in common use are spatially based cells and atom-based tables.[5]

Spatial Neighbour Tables

Suppose that the simulation consists of N atoms in a cubic box of side length L. One subdivides the system into M^3 cubic cells of length $l = L/M$. One chooses M such that $l > R_c$, where R_c is the potential cutoff. One then creates a list of the atoms in each cell by examining the coordinates of each atom. The list is most easily maintained as a linked list in which the index of each atom in a particular cell is stored consecutively. One makes an array of size M^3 that contains the index of the first atom in the cell, and a pointer array of size N that contains at each atomic index the index of the next atom in that cell. A 0 is used in this pointer array to signify that this atom is the last atom in its cell. These two arrays allow the atoms in each cell to be identified.

[5] R. W. Hockney and J. W. Eastwood, *Computer Simulation Using Particles*, McGraw-Hill, New York, 1981.

Once the neighbour list is created, one proceeds to evaluate the interaction energy (for Monte Carlo) or force (for molecular dynamics). One loops through all the cells, and through each atom in each cell. For each atom one must examine every other atom in that cell and in each of the 26 neighbouring cells, taking into account the periodic images of the cells. The potential or force is calculated for any pair of atoms that lie within the cutoff. For the case of molecular dynamics using the fact that the force on atom i due to atom j is equal and opposite to the force on atom j due to atom i, one can halve the number of pair calculations.

Since the cost of creating the neighbour table is linear in N, one can afford to do this at every step (molecular dynamics) or cycle (Monte Carlo) of the simulation. In the case of Monte Carlo, one should have $l > R_\mathrm{c} + \Delta$, so that no atom can move within the cutoff during the cycle. (In fact, if $l \gg R_\mathrm{c}$, then one can do the update less frequently, provided that one can guarantee that no separation can change by more than $l - R_\mathrm{c}$ in the interim.) The spatial neighbour table reduces the number of calculated separations per cycle or time step from N^2 to approximately $27N\rho R_\mathrm{c}^3$, which is a substantial saving.

A variant of the spatial neighbour list uses very small cells that can contain at most one atom. One creates a large array that represents the cells, and that contains either 0 or the index of the atom in the cell. A related array gives the number of the cell associated with each atom. For calculating the potential or force for a given atom, one finds its cell and identifies the neighbouring cells that lie within the cutoff R_c (or $R_\mathrm{c} + \Delta$ in the case of Monte Carlo). This is straightforward to do and can be done at the start of the program if desired. The occupancy of these cells gives the indices of the atoms required for the separation and potential or force calculation. Because these small cells more tightly bound the cutoff sphere than the large cells, one has a saving in the number of separations that need be examined. However, the cost of setting up these larger arrays is greater, and so this method is more effective for larger systems.

Atomic Neighbour Tables

A second type of neighbour list is centred on the atoms. One defines a cutoff $R_t > R_\mathrm{c}$ and constructs for each atom a list of the indices of the atoms that fall within this neighbour cutoff. One must evaluate $N^2/2$ separations to construct the list. One neighbour array A (of size approximately $4\pi N\rho R_t^3/3$) contains all the neighbours of each atom consecutively. The pointer array B of size $N + 1$ contains the address of the the first neighbour of each atom. Hence the indices of the neighbours of atom i are stored between $A(B(i))$ and $A(B(i+1)-1)$, inclusive. If atom i has no neighbours, which could happen in a dilute gas, then $A(B(i+1)) = A(B(i))$.

The advantage of this atom based list is that it does not need to be reconstructed every time step or cycle. One must choose the interval to guarantee that no new atoms can enter the potential cutoff sphere from beyond R_t during the interval. This can be done at the start of a Monte Carlo simulation directly from the step length. It can also be done automatically during the simulation by monitoring the diffusion of every molecule during the interval and carrying out a reconstruction whenever the maximum change in separation that could occur is greater than $R_t - R_c$.[6] Typically one chooses R_t to be on the order of 10% larger than the potential cutoff, and updates the list every 5–20 cycles or time steps. Savings of a factor of 2 have been achieved with this method, and this increases with system size.

13.4.2 Isobaric Monte Carlo

The configurational probability for an isothermal system is $\wp(\mathbf{r}^N|N,V,T) \propto \exp -\beta U(\mathbf{r}^N)$, and in the Metropolis algorithm it is the change in the energy that determines the success or otherwise of a trial move. That is, if moving atom i, one calculates

$$\Delta U_i = \sum_{j=1}^{N}{}^{*} u(r'_{ij}) - u(r_{ij}), \tag{13.25}$$

where the prime signifies the trial position, and the asterisk indicates that the $j = i$ term is excluded from the sum. If ΔU_i is negative or 0 the move is accepted. Otherwise, one calculates $\exp -\beta\Delta U_i$ and accepts the move if this is greater than a random number uniformly distributed on $(0, 1)$. If the move is rejected a new atom is chosen for a trial move.

For an isobaric–isothermal system, both the energy and the volume fluctuate in connection with a temperature–pressure reservoir. The volume–configuration probability is

$$\wp(\mathbf{r}^N, V|N, p, T) \propto e^{-\beta U(\mathbf{r}^N)} e^{-\beta pV}. \tag{13.26}$$

It is straightforward to implement a Monte Carlo algorithm for this isobaric–isothermal system.[7] In this case the trial move can be either an attempt to move an atom, $\mathbf{r}'_i = \mathbf{r}_i + \Delta\mathbf{a}$, or an attempt to change the volume, $L' = L + \Delta_L(2\xi - 1)$, where L is the length of the simulation cell and ξ is a random number uniformly distributed on $(0, 1)$. Hence a trial move is now considered on the basis of the change in enthalpy, which is $\Delta U + p\Delta V$.

In practical terms, because of the periodic boundary conditions, one uses the length of the simulation cell as the unit of length, and one scales all of the coordinates such that $\mathbf{x}_i = \mathbf{r}_i/L$. Now the potential energy becomes a function of the volume of the cell, $U(\mathbf{x}^N; L)$, and changes with each change in volume. Since the partition function is

$$\frac{1}{\Lambda_V \Lambda^{3N}} \int_0^\infty \mathrm{d}V \int_V \mathrm{d}\mathbf{r}^N \, e^{-\beta U(\mathbf{r}^N)} e^{-\beta pV}$$

[6] D. Fincham and B. J. Ralston, Molecular dynamics simulation using the CRAY-1 vector processing computer, *Comput. Phys. Commun.* **23** (1981), 127.

[7] I. R. McDonald, NpT-ensemble Monte Carlo calculations for binary liquid mixtures, *Mol. Phys.* **23** (1972), 41.

$$= \frac{3L^2 L^{3N}}{\Lambda_V \Lambda^{3N}} \int_0^\infty \mathrm{d}L \int_0^1 \mathrm{d}\mathbf{x}^N \, e^{-\beta U(\mathbf{x}^N;L)} e^{-\beta p L^3}, \tag{13.27}$$

the probability distribution becomes

$$\wp(\mathbf{x}^N, L|N, p, T) \propto e^{-\beta\left[U(\mathbf{x}^N;L)+\beta p L^3 + k_\mathrm{B}T(3N+2)\ln L\right]}. \tag{13.28}$$

The quantity in brackets is sometimes called a pseudo-potential. Each trial move, whether in atomic position or in length, is assessed on the basis of the change in the pseudo-potential.

In general one does not have to recalculate all N^2 pair potential interactions following a change in volume. For example, the Lennard–Jones potential may be written

$$U(\mathbf{x}^N; L) = \frac{4\epsilon\sigma^{12}}{L^{12}} \sum_{i<j} x_{ij}^{-12} - \frac{4\epsilon\sigma^6}{L^6} \sum_{i<j} x_{ij}^{-6}. \tag{13.29}$$

Hence if one keeps track of the total energy and of the two terms separately, then one simply rescales each to find the change in energy following a trial volume move. One should add the contribution from the tail due to the change in density in assessing the change in the potential due to the change in volume. Similarly, for the case of a hard-sphere system one monitors the minimum atomic separation in the configuration. If a volume decrease rescales this closest separation to less than the atomic diameter, the volume move is rejected.

Typically, one might attempt a trial volume move once every 5–10 cycles of trial configurations. There is no point in attempting more frequent changes than this because the volume is related to the system as a whole, and the vast bulk of each configuration is unchanged by even several successive atomic moves.

13.4.3 Grand Canonical Monte Carlo

Monte Carlo algorithms have also been developed for open systems that can exchange both energy and atoms with a reservoir.[8] The probability density is

$$\wp(\mathbf{r}^N, N|\mu, V, T) \propto \frac{\Lambda^{-3N} e^{\beta\mu N}}{N!} e^{-\beta U(\mathbf{r}^N)}. \tag{13.30}$$

Hence introducing the constant $\tilde{\mu} = \mu - 3\ln\Lambda$, the pseudo-potential to be used in the Metropolis algorithm is

$$\tilde{U}(\mathbf{r}^N, N) = U(\mathbf{r}^N) - \tilde{\mu}N + k_\mathrm{B}T \ln N!. \tag{13.31}$$

Trial configurations can be the usual single atom move, in which case only the actual potential changes, or they can be an attempt to change the number of atoms in the system. Either one can attempt to delete an existing atom or one can attempt to insert a new atom somewhere in the volume. In both cases all terms in the pseudo-potential change.

[8]D. Nicholson and N. G. Parsonage, *Computer Simulation and the Statistical Mechanics of Adsorption*, Academic Press, London, 1982.

The simulation is carried out for fixed value of $\tilde{\mu}$. At the end one has the average number of atoms $\langle N \rangle$ and the most probable or equilibrium number of atoms, $\overline{N}$. For the ideal gas one has $\langle N \rangle = \Lambda^{-3} V e^{\beta \mu^{\text{id}}}$, and also $\overline{N} = \left\lfloor \Lambda^{-3} V e^{\beta \mu^{\text{id}}} \right\rfloor$. In view of the first of these, one sees that the excess chemical potential is related to the constant $\tilde{\mu}$ by

$$\mu^{\text{ex}} = \tilde{\mu} - \ln \langle N \rangle / V. \tag{13.32}$$

Insertions become increasingly difficult in dense systems because of the high probability that the new atom will overlap with an existing atom; atom-sized cavities are quite rare. Similarly, an atom tends to be located in the potential wells of its neighbours, and the large increase in energy upon its deletion makes this a rare event also. One generally does not attempt to change the number of atoms more frequently than once every 10–50 cycles. One finds that a successful insertion tends to occur in the cavity left by a previously successful deletion, and vice versa, unless one gives the system time to relax between these nonequilibrium events.

In general the decision whether an attempt to change the particle number is an insertion or a deletion move is made randomly. For a deletion the initial state is $\mathbf{r}^N$ and the trial state is $\mathbf{r}^N/\mathbf{r}_i$. For an insertion, the initial state is $\mathbf{r}^N/\mathbf{r}_i$ and the trial state is $\mathbf{r}^N$. Hence if insertions and deletions are attempted with equal probability, then the trial transition matrix is symmetric and microscopic reversibility is ensured. The label of the atom to be deleted, and the label to be given to an inserted atom, should be chosen randomly. If the trial move is successful, the remaining atoms should be relabelled to accommodate the change.

13.4.4 Inhomogeneous Systems and Preferential Sampling

The above discussion of simulation techniques assumed that the only potential present was the pair one. Three-body potentials are prohibitive because the cost of evaluating them scales with N^3. However, singlet potentials that create a density inhomogeneity are quite feasible.

In essence the two types of inhomogeneities typically simulated are due either to a solute or to confinement of the system. It is difficult to simulate the effects of an external field on a macroscopic system because of the periodic boundary conditions. In general the confined systems are either fully confined with walls on all sides (e.g., a spherical pore) or periodic in one (e.g., a cylindrical pore) or in two (e.g., a slit pore that represents a system confined between two walls) dimensions. If the walls are far enough apart one can mimic the solvation of an isolated surface (albeit two of them).

Inhomogeneities can also be caused by the presence of solutes or other inclusions in the system. Quite often one wants to carry out the simulation for a finite concentration of solutes. On other occasions the properties of the isolated solute are required. In the later case one must be aware that because of the periodic boundary conditions the lowest concentration of solute that can be simulated is $\rho_0 = 1/V$. This becomes important when the volume of the system is small, or when the solute is relatively large, because in these cases the density inhomogeneity can extend beyond the central cell and the solute begins to interact with its own image. Also, there is a limit to how large a solute can be simulated. As both the length of the system L and the solute radius R increase, the number of solvent atoms required scales with $\rho R^2(L - R)$. To avoid interactions between the solute images $L - R$ cannot be too small, and it is this that limits the solute size that can be handled.

The requirement that there be a bulk beyond the surface of the solute can mean that most of the computational effort in the simulation is directed toward moving atoms far from the solute (because there are relatively more of them). Presumably one is most interested in the solute-specific properties, which are those in the inhomogeneous region of the system, and one would like to direct most attention here. Preferential sampling implements frequent vicinal cycles, where only atoms initially in a defined neighbourhood of the solute are moved while the remaining atoms are fixed.[9] For example, one might carry out a full cycle of moving all the atoms, and then 5–10 vicinal cycles. At the start of the first vicinal cycle one identifies the vicinal atoms, which might be all atoms within a certain distance of the solute. These vicinal atoms are then moved in a number of cycles. In order to ensure microscopic reversibility one must continue to move all of the atoms initially identified as vicinal throughout the cycle, even if they move away from the neighbourhood of the solute. The vicinal atoms interact with the solute, with each other, and with the remaining atoms as normal. The sequence of a single full cycle followed by a set of vicinal cycles is repeated. In this way the coverage of configuration space in the neighbourhood of the solute is greatly improved and the statistical error for solvation properties reduced.

13.4.5 Importance and Umbrella Sampling

In generating the Markov chain for the system average, the Boltzmann weight is convenient because if focuses on the most likely configurations. However, one is not restricted to this since the method works in principle for an arbitrary weight, as shown by Eq. (13.22). NonBoltzmann sampling has been used for free energy estimation.[10] Another example might be to use an elevated temperature in the Boltzmann factor, which allows a broad region of phase space to be covered. If the configurations and their energy are saved, one can generate the average for the temperature of interest by simply removing the weight. For example, if $\beta_0 = 1/k_B T_0$ corresponds to the temperature used in the Metropolis algorithm, then the weight is $\omega(\mathbf{r}^N) \propto \exp -\beta_0 U(\mathbf{r}^N)$. Hence an average at temperature

[9] J. C. Owicki and H. A. Sceraga, Preferential sampling near solutes in Monte Carlo calculations on dilute solutions, *Chem. Phys. Lett.* **47** (1977), 600. P. K. Mehrotra, M. Mezei, and D. L. Beridge, Convergence acceleration in Monte Carlo simulation on water and aqueous solutions, *J. Chem. Phys.* **78** (1983), 3156. P. Attard, Simulation of the chemical potential and the cavity free energy of dense hard-sphere fluids, *J. Chem. Phys.* **98** (1993), 2225.

[10] G. M. Torrie and J. P. Valleau, Nonphysical sampling distributions in Monte Carlo free energy estimation, *J. Comput. Phys.* **23** (1977), 187.

$T = 1/k_{\mathrm{B}}\beta$ can be obtained from the configurations generated with this weight by removing the weight,

$$\langle f \rangle_T = \frac{1}{W} \sum_{\alpha=1}^{M} f(\mathbf{r}_\alpha^N) \exp[\beta_0 - \beta] U(\mathbf{r}_\alpha^N), \tag{13.33}$$

where $W = \sum_{\alpha=1}^{M} \exp[\beta_0 - \beta] U(\mathbf{r}_\alpha^N)$. This method works because in general the important high-temperature configurations encompass the important low-temperature ones, and more besides. The cost of evaluating an average is small compared to the cost of generating the configurations and the energy. Hence the advantage of this method is that averages at many temperatures can be obtained from a single simulation. The restriction is that one may have to generate more configurations than normal because many of them will be wasted. In practice as the temperature is decreased, the viable part of configuration space becomes relatively small, and unless the reference configurations are obtained at almost the same temperature one does not visit the important regions often enough to give a good average. In other words, there must be substantial overlap between the weight and the actual probability distribution for the method to work.

This is an example of umbrella sampling, where the actual sampling is broader than that required. One can do a similar sampling on density. If the configurations are generated on the unit cube and the different parts of the potential are stored according to how they scale with the system length, then the Boltzmann weight required to obtain a system average at a nearby density from these configurations follows by simply rescaling the parts of the potential.

Using the Boltzmann weight in the Metropolis algorithm usually works well when the quantity being averaged varies slowly compared to the exponential factor. This is not always the case, as shown by Widom's ghost particle expression for the chemical potential, Eq. (7.63),

$$e^{-\beta\mu^{\mathrm{ex}}} = \frac{1}{V} \int_V \mathrm{d}\mathbf{r}_N \left\langle e^{-\beta \Delta U(\mathbf{r}_N;\mathbf{r}^{N-1})} \right\rangle_{N-1}, \tag{13.34}$$

and the real particle version, Eq. (7.64),

$$e^{\beta\mu^{\mathrm{ex}}} = \left\langle e^{\beta \Delta U(\mathbf{r}_N;\mathbf{r}^{N-1})} \right\rangle_{N}. \tag{13.35}$$

In the first of these the configurations of the $N-1$ atoms are independent of the ghost particle at $\mathbf{r}_N$. Hence in a dense system one or more real atoms will almost certainly overlap with the ghost atom, and the averand will be 0. Conversely, in the second expression atom N is real and part of the system, which means that no overlaps will occur. However, because of the positive exponent in the averand, it is precisely these overlaps that give the greatest contribution. These are two examples where the quantity being averaged varies on a scale comparable with the Boltzmann weight, and where one should endeavour to take this into account in generating the Markov chain.

In this case, recalling that the contribution to potential from atom N is $\Delta U(\mathbf{r}_N;\mathbf{r}^{N-1}) = U_N(\mathbf{r}^N) - U_{N-1}(\mathbf{r}^{N-1})$, one may choose the weight to reflect a partial coupling of the Nth atom,

$$\omega(\mathbf{r}_N) = e^{-\beta U_{N-1}(\mathbf{r}^{N-1})} e^{-\lambda\beta\Delta U(\mathbf{r}_N;\mathbf{r}^{N-1})}. \tag{13.36}$$

(Instead of scaling the temperature by $\lambda \in (0,1)$, one could use an appropriate nonlinear coupling for the pair potential.) With this weight the second expression for the average chemical potential becomes

$$\left\langle e^{\beta\mu^{\text{ex}}} \right\rangle = \frac{\sum_{\alpha=1}^{M} e^{\lambda\beta\Delta U_\alpha}}{\sum_{\alpha=1}^{M} e^{-(1-\lambda)\beta\Delta U_\alpha}}. \tag{13.37}$$

Suitable choice of the value of the coupling constant allows more overlap between the weighted configurations and the quantities being averaged than would otherwise be the case.

The problems with chemical potential estimation can also be ameliorated by using a coupling constant integral.[11] Using Eq. (7.61) one has

$$\begin{aligned}
\langle -\beta\mu^{\text{ex}} \rangle &= \ln Q(N,V,T) - \ln Q(N-1,V,T) \\
&= \int_0^1 \mathrm{d}\lambda \, \frac{\partial}{\partial\lambda} \ln Q(N,V,T;\lambda) \\
&= \int_0^1 \mathrm{d}\lambda \, \frac{-\beta}{Q(N,V,T;\lambda)} \int \mathrm{d}\mathbf{r}^N \, e^{-\beta U_\lambda(\mathbf{r}^N)} \Delta\dot{U}_\lambda(\mathbf{r}_N;\mathbf{r}^{N-1}) \\
&= -\beta \int \mathrm{d}\lambda \, \mathrm{d}\mathbf{r}^N \, \wp(\mathbf{r}^N;\lambda|N,V,T) \Delta\dot{U}_\lambda(\mathbf{r}_N;\mathbf{r}^{N-1}) \\
&= \frac{-\beta \sum_{\alpha=1}^{M} \Delta\dot{U}_{\lambda;\alpha} \wp(\mathbf{r}_\alpha^N;\lambda)/\omega(\mathbf{r}_\alpha^N;\lambda)}{\sum_{\alpha=1}^{M} \wp(\mathbf{r}_\alpha^N;\lambda)/\omega(\mathbf{r}_\alpha^N;\lambda)}.
\end{aligned} \tag{13.38}$$

One chooses the weight to cancel the Boltzmann factor for the atoms, and in addition to provide an external potential conjugate to the coupling,

$$\omega(\mathbf{r}_\alpha^N;\lambda) = e^{\beta U_\lambda(\mathbf{r}^N)} e^{-\beta\psi(\lambda)}. \tag{13.39}$$

Hence one carries out a simulation on configuration space augmented with the extra coordinate λ. Moves in particle position and in coupling constant are attempted according to the usual Metropolis algorithm. The potential ψ is chosen to give as broad a distribution in λ as possible. A good start is to choose $\psi(\lambda) = -pV_\lambda$, where p is the pressure of the system and V_λ is the volume of the partially coupled atom. One can also update this external potential during the course of the simulation to broaden the distribution. The potential ψ has the effect of forcing the configurations important in the evaluation of the chemical potential to be sampled by the Markov chain. In the event that one cannot find

[11] K. K. Mon and R. B. Griffiths, Chemical potential by gradual insertion of a particle in Monte Carlo simulation, *Phys. Rev. A* **31** (1985), 956. P. Attard, Simulation of the chemical potential and the cavity free energy of dense hard-sphere fluids, *J. Chem. Phys.* **98** (1993), 2225.

a potential to give an adequate sample of the whole unit interval, one can break the coupling constant integral into stages with a different weight for each stage.

The probability distribution obtained from the simulation also gives the solvation free energy of the partially coupled solute atom (after the effects of the external potential ψ are subtracted). That is, if $\wp(\lambda|\psi)$ is the distribution of couplings that emerges from the Markov chain, then the solvation free energy is given by

$$W(\lambda) - W(0) = -k_{\mathrm{B}}T \ln \frac{\wp(\lambda|\psi)}{\wp(0|\psi)} - [\psi(\lambda) - \psi(0)] . \tag{13.40}$$

This expression offers an alternative route to the excess chemical potential, namely $\mu^{\mathrm{ex}} = W(1) - W(0)$.

13.4.6 Time Correlations

As the preceding sections indicate, the Monte Carlo approach has great flexibility in choosing the trajectory followed by the simulation, and this can be designed to obtain the quantity of interest in the most efficient manner. This is an advantage over the molecular dynamics approach, which is largely restricted to trajectories generated by Hamilton's equations of motion (but see below). However, the molecular dynamics method is unique in being able to study time-dependent phenomena, and these can be used to obtain physical properties, the diffusion constant, the shear viscosity, and the heat conductivity, as examples. This section illustrates the use of time correlation functions, which are discussed elsewhere in greater detail.[12]

A dynamical variable is a function of the position in phase space due to the motion along the trajectory, $A(t) = A(\mathbf{q}^N(t), \mathbf{p}^N(t))$. (This assumes that the dynamical variable does not explicitly depend on time. A similar assumption will be made for the Hamiltonian itself.) Examples include the velocity of a particular particle, $A(t) = \mathbf{p}_i(t)/m$, or the local density, $A(\mathbf{r}, t) = \sum_{i=1}^{N} \delta(\mathbf{q}_i(t) - \mathbf{r})$, or the local current, $A(\mathbf{r}, t) = \sum_{i=1}^{N} \mathbf{p}_i(t)\delta(\mathbf{q}_i(t) - \mathbf{r})/m$. The time correlation function for two such dynamical variables is

$$\begin{aligned} C^*_{AB}(t) &= \langle A(t)B(0) \rangle \\ &= \lim_{\tau \to \infty} \frac{1}{\tau} \int_0^{\tau} \mathrm{d}t' \, \langle A(t+t')B(t') \rangle \\ &= \int \mathrm{d}\boldsymbol{\Gamma}_0 \, \wp(\boldsymbol{\Gamma}_0) B(\boldsymbol{\Gamma}_0) A(\boldsymbol{\Gamma}(t|\boldsymbol{\Gamma}_0)). \end{aligned} \tag{13.41}$$

The auto-correlation function is $C^*_{AA}(t) = \langle A(t)A(0) \rangle$, and the equilibrium correlation function is $\lim_{t\to 0} C^*_{AB}(t) = \langle A(0)B(0) \rangle$. Also at large times the dynamical variables become uncorrelated, $\lim_{t\to\infty} C^*_{AB}(t) = \langle A \rangle \langle B \rangle$, where because of time homogeneity time need not be shown in the arguments of the dynamical

[12] J. P. Hansen and I. R. McDonald, *Theory of Simple Liquids*, 2nd ed., Academic Press, London, 1986. D. A. McQuarrie, *Statistical Mechanics*, Harper and Row, New York, 1976.

variables. Therefore the time correlation function is in general defined in terms of the departure from the average value,

$$C_{AB}(t) = \langle [A(t) - \langle A \rangle][B(0) - \langle B \rangle] \rangle \,. \tag{13.42}$$

It is often convenient to take the temporal Fourier transform of this,

$$\hat{C}_{AB}(\omega) = \frac{1}{2\pi} \int_0^\infty \mathrm{d}t \, e^{i\omega t} C_{AB}(t). \tag{13.43}$$

These spectral components of the time correlation functions can sometimes be measured experimentally.

Physical properties are often related to the time integral of the correlation of the derivative of the dynamical variables,

$$\gamma = \int_0^\infty \mathrm{d}t \, \left\langle \dot{A}(t)\dot{A}(0) \right\rangle . \tag{13.44}$$

The corresponding Einstein relationship is

$$2\gamma t = \left\langle [A(t) - A(0)]^2 \right\rangle . \tag{13.45}$$

As an example the velocity auto-correlation function is defined as

$$Z(t) = \frac{1}{3m^2} \langle \mathbf{p}_i(t) \cdot \mathbf{p}_i(0) \rangle \,. \tag{13.46}$$

At $t = 0$ this is essentially the average kinetic energy of the particle, $Z(0) = k_{\mathrm{B}}T/m$. In this case since there is no preferred direction of motion, the average velocity is 0 and the velocity auto-correlation function decays with time from this peak value. The diffusion coefficient is the time integral of this,

$$D = \frac{1}{3m^2} \int_0^\infty \mathrm{d}t \, \langle \mathbf{p}_i(t) \cdot \mathbf{p}_i(0) \rangle \,. \tag{13.47}$$

Alternatively

$$2Dt = \frac{1}{3} \left\langle [\mathbf{q}_i(t) - \mathbf{q}_i(0)]^2 \right\rangle . \tag{13.48}$$

This Einstein result corresponds to the diffusion probability

$$\wp(\mathbf{q}_i|\mathbf{q}_i(0), t) = \frac{1}{\sqrt{12\pi D t}} e^{-[\mathbf{q}_i - \mathbf{q}_i(0)]^2/12Dt}. \tag{13.49}$$

From the simulation the dynamical variables are obtained as a function of time. The time correlation function at time t, $C_{AB}(t)$, is obtained by multiplying $B(t_i)$ and $A(t_i + t)$ together for a set of starting times t_i along the trajectory, and averaging the result,

$$C_{AB}(t) = \frac{1}{M} \sum_{i=1}^{M} A(t_i + t) B(t_i). \tag{13.50}$$

If necessary it is actually the departure of the dynamical variable from its long time average that is used, $A - \langle A \rangle$ and $B - \langle B \rangle$. In general one tries to choose the starting times separated by more than the correlation time, which is the time taken for the time correlation function to decay to 0, $C_{AB}(t_{i+1} - t_i) \approx 0$. Little statistical advantage is gained by using more starts at closer intervals because of the redundancy in the information. After the average for the time correlation function is taken, the time integration is performed to obtain the value of the transport coefficient.

In the case of the diffusion constant for an atom of the system, the statistics can be improved by averaging over all the particles. The Einstein form of this result, which relies upon the displacement of the tagged particle, must be used with some care. This average will be invalidated whenever the particle leaves the central simulation cell and its position is reset to that of its image that enters from the other side. Often the actual position of the particle is retained for cases like this, and it is the intermolecular separation that is reduced to less than half the box length for the evaluation of the force or potential.

13.4.7 Isothermal and Isobaric Molecular Dynamics

Techniques for carrying out molecular dynamics at constant temperature and pressure have been developed. The so-called 'extended system' methods augment phase space with an extra coordinate whose purpose is to mimic an external reservoir of the desired type.[13] For the case of the isobaric system it is the volume of the system that is allowed to move by adding to the Hamiltonian a kinetic energy, $M\dot{V}^2/2 = W^2/2M$, where M is its 'mass' and W is its momentum, and a potential energy pV, where p is the reservoir pressure. In view of the periodic boundary conditions the particle coordinates are scaled to the unit cube, $\boldsymbol{\theta}_i = V^{-1/3}\mathbf{q}_i$. In order to obtain the Hamiltonian for the new coordinates one requires a canonical transformation of the second type, with generating function,[14]

$$F_2(\mathbf{q}^N, \boldsymbol{\pi}^N) = V^{-1/3} \sum_{i=1}^{N} \mathbf{q}_i \boldsymbol{\pi}_i. \tag{13.51}$$

This gives the desired result for the scaled coordinates,

$$\boldsymbol{\theta}_i = \frac{\partial F_2}{\partial \boldsymbol{\pi}_i} = V^{-1/3}\mathbf{q}_i, \tag{13.52}$$

and the conjugate momenta may be obtained from

$$\mathbf{p}_i = \frac{\partial F_2}{\partial \mathbf{q}_i} = V^{-1/3}\boldsymbol{\pi}_i. \tag{13.53}$$

---

[13] H. C. Andersen, Molecular dynamics simulation at constant pressure and/or temperature, *J. Chem. Phys.* **72** (1980), 2384.

[14] H. Goldstein, *Classical Mechanics*, Ch. 8, Addison-Wesley, Reading MA, 1950.

With these the Hamiltonian becomes

$$\begin{aligned}
&\mathcal{H}(\boldsymbol{\theta}^N, \boldsymbol{\pi}^N; V, W) \\
&= \mathcal{H}(\mathbf{q}^N, \mathbf{p}^N; V, W) + \frac{\partial F_2}{\partial t} \\
&= \frac{1}{2m} \sum_{i=1}^{N} \mathbf{p}_i^2 + U(\mathbf{q}^N) + \frac{W^2}{2M} + pV \\
&= \frac{V^{-2/3}}{2m} \sum_{i=1}^{N} \boldsymbol{\pi}_i^2 + U(\boldsymbol{\theta}^N; V^{1/3}) + \frac{W^2}{2M} + pV.
\end{aligned} \tag{13.54}$$

Hence the equations of motion are

$$\dot{\boldsymbol{\theta}}_i = V^{-2/3} \boldsymbol{\pi}_i / m, \ \dot{\boldsymbol{\pi}}_i = V^{1/3} \mathbf{F}_i \tag{13.55}$$

and

$$\dot{V} = W/M, \ \dot{W} = -p + \frac{V^{-5/3}}{3m} \sum_{i=1}^{N} \boldsymbol{\pi}_i^2 + \frac{V^{-2/3}}{3} \sum_{i=1}^{N} \mathbf{F}_i \cdot \boldsymbol{\theta}_i, \tag{13.56}$$

where $\mathbf{F}_i$ is the usual force on particle i due to the intermolecular interactions. The Newtonian form of these is instructive,

$$\begin{aligned}
\ddot{\boldsymbol{\theta}}_i &= \frac{V^{-2/3}}{m} \dot{\boldsymbol{\pi}}_i - \frac{2\boldsymbol{\pi}_i}{3m} V^{-5/3} \dot{V} \\
&= \frac{V^{-1/3}}{m} \mathbf{F}_i - \frac{2}{3} \frac{\dot{V}}{V} \boldsymbol{\theta}_i,
\end{aligned} \tag{13.57}$$

and

$$\begin{aligned}
\ddot{V} &= \frac{-p}{M} + \frac{V^{-5/3}}{3mM} \sum_{i=1}^{N} \boldsymbol{\pi}_i^2 + \frac{V^{-2/3}}{3M} \sum_{i=1}^{N} \mathbf{F}_i \cdot \boldsymbol{\theta}_i \\
&= \frac{1}{M} \left[-p + \frac{1}{3mV} \sum_{i=1}^{N} \mathbf{p}_i^2 + \frac{1}{3V} \sum_{i=1}^{N} \mathbf{F}_i \cdot \mathbf{q}_i \right] \\
&= \frac{1}{M} \left[\mathcal{P}(\mathbf{q}^N, \mathbf{p}^N) - p \right].
\end{aligned} \tag{13.58}$$

Here the instantaneous pressure has been defined, which is related to the virial, Eq. (7.29). One sees that the force acting to change the volume is proportional to the difference between the internal and the external pressures. These equations of motion conserve the augmented Hamiltonian, which is essentially the enthalpy of the system. Hence these equations of motion approximate an isenthalpic–isobaric, or (N, p, H) system.

A constant temperature system may be mimicked by augmenting phase space with an additional variable s that acts to rescale the velocities.[15] The Hamilto-

[15] S. Nosé, A unified formulation of the constant temperature molecular dynamics methods, *J. Chem. Phys.* **81** (1984), 511. H. J. C. Berendsen, J. P. M. Potsma, W. F. van Gunsteren, A. DiNola, and J. R. Haak, Molecular dynamics with coupling to an external bath, *J. Chem. Phys.* **81** (1984), 3684. W. G. Hoover, Canonical dynamics: Equilibrium phase space distributions, *Phys. Rev. A* **31** (1985), 1695.

nian may be taken to be

$$\mathcal{H}(\mathbf{q}^N, \mathbf{p}^N; s, v) = \frac{1}{2ms^2} \sum_{i=1}^{N} \mathbf{p}_i^2 + U(\mathbf{q}^N) + \frac{v^2}{2Q} + 3Nk_\mathrm{B}T \ln s, \tag{13.59}$$

where Q is the 'mass' of the new coordinate and v is its momentum. Again the Newtonian form reveals the nature of the procedure,

$$\begin{aligned} \ddot{s} &= \frac{-1}{Q}\left[\frac{-2}{2ms^3}\sum_{i=1}^{N}\mathbf{p}_i^2 + \frac{3Nk_\mathrm{B}T}{s}\right] \\ &= \frac{2}{Qs}\left[K(\mathbf{p}^N) - \frac{3Nk_\mathrm{B}T}{2}\right], \end{aligned} \tag{13.60}$$

where $K(\mathbf{p}^N)$ is the instantaneous kinetic energy of the system. If the latter is larger than the desired average value, one sees that the acceleration is positive, leading to an increase in $\dot{s}$ and eventually to s. Such an increase in s decreases the kinetic energy, as desired. The converse occurs if the kinetic energy is too low.

One problem with these extended system methods is that the 'masses' M and Q are somewhat arbitrary. Too large a value means that it takes the system too long to adjust; in the limit that these go to infinity the method reduces to the conventional (N, V, E) molecular dynamics technique. Conversely, too small a mass decouples the 'friction' term, and long-lived, undamped oscillations about the equilibrium value occur. It takes some trial and error to find an optimum value.

A second difficulty is that the equations of motion are somewhat ad hoc. There seems no compelling reason to choose these particular ones over other similar versions, nor any well-defined procedure to develop new equations for these or other systems. Also, it is unclear precisely what conventional thermodynamic system they correspond to, and how to relate the fluctuations that occur in these systems to the actual fluctuations that occur in reality.

13.4.8 Constraint Molecular Dynamics

A different approach to using molecular dynamics for an isothermal system is to alter the equations of motion. One introduces fictitious forces that constrain the system to conserve desired properties.[16] The equation for the velocity is unchanged,

$$\dot{\mathbf{q}}_i = \frac{\partial \mathcal{H}(\mathbf{q}^N, \mathbf{p}^N)}{\partial \mathbf{p}_i} = \mathbf{p}_i/m, \tag{13.61}$$

but the acceleration is given by

$$\dot{\mathbf{p}}_i = \mathbf{F}_i + \mathbf{f}_i, \tag{13.62}$$

[16] W. G. Hoover, A. J. C. Ladd, and B. Moran, High strain rate plastic flow studied by nonequilibrium molecular dynamics, *Phys. Rev. Lett.* **48** (1982), 1818. D. J. Evans and G. P. Morriss, Non-Newtonian molecular dynamics, *Comput. Phys. Rep.* **1** (1984), 297.

where $\mathbf{F}_i = -\partial\mathcal{H}/\partial\mathbf{q}_i$ is the usual force and $\mathbf{f}_i$ is the non-Hamiltonian fictitious force of constraint.

For the case of an isothermal system one wants to constrain the kinetic energy to equal the thermal energy,

$$K(\mathbf{p}^N) = \frac{1}{2m}\sum_{i=1}^{N}\mathbf{p}_i^2 = \frac{3Nk_{\mathrm{B}}T}{2}. \tag{13.63}$$

The way to do this is to insist that the time derivative vanish, $\dot{K} = 0$, or

$$0 = \frac{1}{m}\sum_{i=1}^{N}\mathbf{p}_i\cdot\dot{\mathbf{p}}_i = \frac{1}{m}\sum_{i=1}^{N}\mathbf{p}_i\cdot[\mathbf{F}_i + \mathbf{f}_i]. \tag{13.64}$$

This alone is not sufficient to determine the force of constraint. The extra information is supplied by *Gauss' principle of least constraint*, which holds that the fictional force should be as small as possible. The latter is taken in a least-squares sense, so that one should minimise

$$\epsilon = \sum_{i=1}^{N}\mathbf{f}_i^2 \tag{13.65}$$

subject to the constraint $\dot{K} = 0$. That is,

$$0 = \frac{\delta[\epsilon + \lambda\dot{K}]}{\delta\mathbf{f}_i} = 2\mathbf{f}_i + \frac{\lambda}{m}\mathbf{p}_i, \tag{13.66}$$

or $\mathbf{f}_i = \alpha\mathbf{p}_i$. The Lagrange multiplier is determined by the constraint equation,

$$\alpha = \frac{-\sum_{i=1}^{N}\mathbf{p}_i\cdot\mathbf{F}_i}{\sum_{i=1}^{N}\mathbf{p}_i\cdot\mathbf{p}_i}. \tag{13.67}$$

Accordingly, the acceleration is given by

$$m\ddot{\mathbf{q}}_i = \mathbf{F}_i - \frac{\sum_{i=1}^{N}\mathbf{p}_i\cdot\mathbf{F}_i}{\sum_{i=1}^{N}\mathbf{p}_i\cdot\mathbf{p}_i}\mathbf{p}_i. \tag{13.68}$$

These non-Hamiltonian equations of motion guarantee that the kinetic energy will remain constant during the trajectory. In this regard the fictitious force is much stronger than the friction force of the preceding section, which only keeps the kinetic energy constant on average.

Conversely, however, a system obeying these equations of motion is neither constant energy nor constant temperature. It is a constant kinetic energy system, which is not one of the systems of classical thermodynamics. Hence the fluctuations in (potential) energy are not the same as the fluctuations of a classical system. The phase space probability density that corresponds to these equations of motion is unknown.

An additional caveat is that the full consequences of altering the equations of motion are unknown. Although one has sought to minimise the influence of the fictitious force by invoking Gauss' principle, it is unclear whether the constraint itself is too strong; possibly some smaller constraint could be used. Despite these reservations, there is empirical evidence that the averages obtained with these types of non-Hamiltonian molecular dynamics are comparable to those obtained with more traditional methods. In this regard the extended system approach and the constraint methods offer flexibility in the choice of trajectory that can be fine-tuned to optimise the efficiency of the simulation. This represents an interesting convergence of molecular dynamics and Monte Carlo methods that increasingly makes them feasible for more complex systems.

13.4.9 Stochastic Molecular Dynamics

A hybrid simulation technique that combines molecular dynamics and Monte Carlo has been motivated by the underlying philosophy of this book, which seeks a physical realisation of statistical mechanics. In this method, which may be called stochastic dynamics, the external or constraint force $\mathbf{f}_i$ that appears in Eq. (13.62) is interpreted as a weak random force that arises from the reservoir. For the case of a thermal reservoir, the statistical nature of the perturbations is such that the Boltzmann distribution is ensured, $\wp(\mathbf{\Gamma}|N,V,T) \propto \exp -\beta\mathcal{H}(\mathbf{\Gamma})$.[17]

The algorithm consists of a deterministic part and a stochastic part. The trajectory may be written as a Markov chain by discretising time t in steps of length Δ_t. The 'natural' Hamiltonian evolution of the subsystem is

$$\begin{aligned} q^0_{i\alpha}(t+\Delta_t) &= q_{i\alpha}(t) + \Delta_t p_{i\alpha}(t)/m \\ p^0_{i\alpha}(t+\Delta_t) &= p_{i\alpha}(t) + \Delta_t F_{i\alpha}(t). \end{aligned} \tag{13.69}$$

(This has been written in the simplest form. In practice one often solves the natural motion to higher-order accuracy than this.) This deterministic step is augmented by a stochastic step,

$$\begin{aligned} q_{i\alpha}(t+\Delta_t) &= q^0_{i\alpha}(t+\Delta_t) \\ p_{i\alpha}(t+\Delta_t) &= p^0_{i\alpha}(t+\Delta_t) + \Delta_t f_{i\alpha}(t+\Delta_t). \end{aligned} \tag{13.70}$$

One must choose the stochastic force in such a way that the Boltzmann distribution is stationary with respect to these equations of motion. In general the evolution of a probability distribution under the action of such a transition probability is given by

$$\wp(\mathbf{\Gamma}_2; t+\Delta_t) = \int \mathrm{d}\mathbf{\Gamma}_1\, \mathcal{T}(\mathbf{\Gamma}_2;\mathbf{\Gamma}_1)\, \wp(\mathbf{\Gamma}_1;t). \tag{13.71}$$

Here $\mathcal{T}(\mathbf{\Gamma}_2;\mathbf{\Gamma}_1)$ gives the probability of being in the state $\mathbf{\Gamma}_2$ at time $t+\Delta_t$ given that the system was in the state $\mathbf{\Gamma}_1$ at time t, and it is obviously normalised,

$$\int \mathrm{d}\mathbf{\Gamma}_2\, \mathcal{T}(\mathbf{\Gamma}_2;\mathbf{\Gamma}_1) = 1. \tag{13.72}$$

[17] H. C. Andersen, Molecular dynamics simulation at constant pressure and/or temperature, *J. Chem. Phys.* **72** (1980), 2384, gives a hybrid method similar to that discussed here. In Andersen's method the momentum of a randomly chosen particle is replaced by a new value randomly drawn from the Boltzmann distribution, whereas here it is changed by a small perturbation.

For deterministic equations of motion, the transition probability is

$$\mathcal{T}(\mathbf{\Gamma}_2;\mathbf{\Gamma}_1) = \delta\Big(\mathbf{\Gamma}_1 + \Delta_t\dot{\mathbf{\Gamma}}_1 - \mathbf{\Gamma}_2\Big), \tag{13.73}$$

where $\dot{\mathbf{\Gamma}}_1 = \dot{\mathbf{\Gamma}}(\mathbf{\Gamma}_1)$. Inserting this in the evolution equation and taking the limit $\Delta_t \to 0$, it is straightforward to show that this gives Liouville's theorem for the deterministic evolution of a statistical ensemble

$$\frac{\partial \wp(\mathbf{\Gamma}, t)}{\partial t} + \nabla \cdot [\dot{\mathbf{\Gamma}}\wp(\mathbf{\Gamma}, t)] = 0. \tag{13.74}$$

Because energy is conserved by Hamilton's equations, the second term vanishes for any distribution that is a function of the Hamiltonian, and hence the Boltzmann distribution is stationary during the deterministic natural evolution.

In the stochastic case, at equilibrium, the probability of the states are proportional to their weights, which arise from the transition probabilities. If the transition from 1 to 2 is more likely than that from 2 to 1, then state 2 is more likely than state 1 by the same factor,

$$\frac{\mathcal{T}(\mathbf{\Gamma}_2;\mathbf{\Gamma}_1)}{\mathcal{T}(\mathbf{\Gamma}_1;\mathbf{\Gamma}_2)} = \frac{\wp(\mathbf{\Gamma}_2)}{\wp(\mathbf{\Gamma}_1)}. \tag{13.75}$$

Such transition probabilities are reversible,

$$\mathcal{T}(\mathbf{\Gamma}_2;\mathbf{\Gamma}_1)\wp(\mathbf{\Gamma}_1, t) = \mathcal{T}(\mathbf{\Gamma}_1;\mathbf{\Gamma}_2)\wp(\mathbf{\Gamma}_2, t), \tag{13.76}$$

which is to say that the forward transition is just as likely to occur as the reverse transition. (Note the distinction between $\mathcal{T}(1;2)$, which is the conditional probability of a particular transition, and $\mathcal{T}(1;2)\wp(2)$, which is the probability of a particular transition.) In this case the equilibrium or steady state condition is automatically satisfied

$$\begin{aligned} \wp(\mathbf{\Gamma}_2; t + \Delta_t) &= \int \mathrm{d}\mathbf{\Gamma}_1\, \mathcal{T}(\mathbf{\Gamma}_2;\mathbf{\Gamma}_1)\, \wp(\mathbf{\Gamma}_1; t) \\ &= \int \mathrm{d}\mathbf{\Gamma}_1\, \mathcal{T}(\mathbf{\Gamma}_1;\mathbf{\Gamma}_2)\, \wp(\mathbf{\Gamma}_2; t) \\ &= \wp(\mathbf{\Gamma}_2; t). \end{aligned} \tag{13.77}$$

Hence by choosing the stochastic transition probabilities $\mathcal{T}_{i\alpha}$ so that they are reversible, then the Boltzmann distribution will be stationary. This is easy to achieve since the Boltzmann distribution itself factors into the product of components of the momenta, and one only must ensure microscopic reversibility for each. A convenient choice is that of Metropolis,

$$\begin{aligned} \mathcal{T}(p_{i\alpha}; p_{i\alpha}^0) &= a\delta(\Delta_p) + \frac{1}{2p^*}\theta(p^* + \Delta_p)\theta(p^* - \Delta_p) \\ &\quad \times \left[\theta(-\Delta\mathcal{H}) + \theta(\Delta\mathcal{H})e^{-\beta\Delta\mathcal{H}}\right], \end{aligned} \tag{13.78}$$

where the rejection (same-state) factor ensures the correct normalisation and is given by

$$a = 1 - \frac{1}{2p^*} \int_{-p^*}^{p^*} \mathrm{d}\Delta_p \left[\theta(-\Delta\mathcal{H}) + \theta(\Delta\mathcal{H}) e^{-\beta\Delta\mathcal{H}} \right], \tag{13.79}$$

with clearly $0 < a < 1$. Here θ is the Heaviside step function, the change in momentum is $\Delta_p = p_{i\alpha} - p_{i\alpha}^0$, the change in energy is $\Delta\mathcal{H} = [p_{i\alpha}^2 - (p_{i\alpha}^0)^2]/2m$, and p^* is the maximum change in momentum, which is related to the maximum stochastic force from the reservoir, $f^* = p^*/\Delta_t$. One can readily confirm that with the Metropolis choice,

$$\mathcal{T}(p_2; p_1) \frac{e^{-p_1^2/2mk_\mathrm{B}T}}{\sqrt{2\pi m k_\mathrm{B} T}} = \mathcal{T}(p_1; p_2) \frac{e^{-p_2^2/2mk_\mathrm{B}T}}{\sqrt{2\pi m k_\mathrm{B} T}}, \tag{13.80}$$

and hence that the Boltzmann distribution is stationary under the total transition probability.

The implementation of the Metropolis algorithm begins with a trial move

$$p_{i\alpha}^{\mathrm{trial}} = p_{i\alpha}^0 + \Delta_p, \tag{13.81}$$

where $\Delta_p \equiv \Delta_t f_{i\alpha}$ is a random number uniformly distributed on $[-p^*, p^*]$. The change in energy of the trial move is

$$\Delta\mathcal{H}(\mathbf{\Gamma}) = \left[(p_{i\alpha}^0 + \Delta_p)^2 - (p_{i\alpha}^0)^2 \right] /2m, \tag{13.82}$$

and the move is accepted if, and only if,

$$e^{-\beta\Delta\mathcal{H}} \geq \xi, \tag{13.83}$$

where ξ is a random number uniformly distributed on $[0, 1]$. If the trial move is rejected one goes to the next component of momentum, leaving this one unchanged from its deterministic evolution.

Note that although this stochastic step only depends upon the change in kinetic energy, the phase space probability density is *not* proportional to a Boltzmann factor of the kinetic energy alone because the latter is not a constant of the natural motion and so $\wp_K$ is not an eigenvector of $\mathcal{T}$. Also, the maximum step in momentum is typically related to the root mean square velocity, $p^* = \lambda\sqrt{mk_\mathrm{B}T}$, where λ =0.1–0.01. In practice λ is decreased as equilibration proceeds. An acceptance rate of 98% is typical.

In the usual Monte Carlo simulation technique, a trial move that decreases energy is more likely to be accepted than one that gives an energy increase. However, there are many more trial moves that lead to an energy increase than there are moves that decrease energy, and in equilibrium these tendencies are in balance.[18] This equilibrium balance, and the precise way in which it is achieved,

[18] It follows that because there are a larger number of states at higher energy, random numerical errors are likely to increase the energy of the system. This explains the small but monotonic increase in energy and temperature observed in conventional constant energy molecular dynamics simulations.

is manifest by temperature equality of the subsystem and the reservoir. One can see this from the macrostate transition probability, $\mathcal{T}(E_0 + \Delta E; E_0)$, where ΔE is change in the Hamiltonian after a full step in the Markov chain. Invoking the simplest ansatz that obeys the steady state condition, namely that this is just proportional to the probability of the destination macrostate (and is 0 if $|\Delta_E| > E^*/2$), one has

$$\begin{aligned}
&\mathcal{T}(E_0 + \Delta E; E_0) \\
&= \wp(E_0 + \Delta E|T) \\
&= \frac{1}{Z'(T)} e^{S(E_0+\Delta E)/k_\mathrm{B}} e^{-(E_0+\Delta E)/k_\mathrm{B}T} \\
&= \frac{1}{Z'(T)} e^{[S(E_0)+\Delta E/T_s(E_0)]/k_\mathrm{B}} e^{-(E_0+\Delta E)/k_\mathrm{B}T} \\
&= \frac{1}{Z(E_0, E^*, T)} e^{\Delta\beta\Delta E},\ |\Delta_E| < E^*/2,
\end{aligned} \tag{13.84}$$

where the difference in the reciprocal temperature of the subsystem and the reservoir is $k_\mathrm{B}\Delta\beta \equiv T_s(E_0)^{-1} - T^{-1}$. This uses the fact that the macrostate probability is proportional to the exponential of the macrostate entropy. With this the average change in energy is

$$\langle \Delta E \rangle_{E_0} = \frac{(E^*)^2}{12k_\mathrm{B}} \left(\frac{1}{T_s(E_0)} - \frac{1}{T} \right). \tag{13.85}$$

This shows that if the subsystem temperature is too low, the average change in energy is toward higher-energy states, and vice versa. The first term is due to the greater number of states available to the subsystem at higher energies, and the second term represents the loss of entropy by the reservoir at higher subsystem energies.

It is of interest to contrast this result for the average change in an energy macrostate with the average change in momentum for a momentum microstate. Using the Metropolis algorithm of the present method, one can show that the average change in momentum for a component of momentum p at each stochastic step is

$$\langle \Delta_p \rangle_p \approx \frac{-(p^*)^2}{6mk_\mathrm{B}T} p,\ |p| > p^*/2. \tag{13.86}$$

This says that provided the speed is not too small, it is most likely to decrease from its current value. This must be the case because the Boltzmann distribution of microstates is a monotonically decreasing function of speed. Given that the system is in an unlikely state it must favour transitions to more likely states than to less likely states simply to maintain the steady state. This is in contrast to the transitions between energy macrostates found above, which can be positive or negative because the energy probability distribution is nonmonotonic due to the combination of the increase of subsystem entropy with subsystem energy and the decrease of reservoir entropy with subsystem energy. In other words, there is no internal entropy for the momentum microstate, and so the term involving the temperature of the subsystem is missing.

Summary

- Computer simulations readily yield the properties of a system of 10^2–10^4 atoms. Molecular dynamics follows the trajectory through phase space according to Hamilton's equations of motion. Thermodynamic and statistical properties for the closed system are collected as simple time averages.

- The Monte Carlo method is a stochastic approach that generates configurations of the atoms with weight proportional to the desired probability distribution, usually constant temperature. The common way of doing this is via the Metropolis algorithm. Averages are simple averages over the configurations.

- An infinite system is mimicked by using periodic boundary conditions. The pair potential is generally truncated and a tail correction added for the averages. Neighbour tables are used to facilitate the treatment of large systems. Confined and inhomogeneous systems may also be simulated.

- Monte Carlo can be used to simulate both isobaric and grand canonical systems by including volume moves and particle insertions in the configurations. One can improve the statistics of a simulation by increasing the number of moves relevant to a property of interest (preferential sampling), or by using a non-Boltzmann weight to generate more of the relevant configurations (importance or umbrella sampling).

- Molecular dynamics yields time correlation functions, which can be used to obtain the hydrodynamic transport coefficients. It can be used to simulate nonisolated systems by using extensions of phase space that include the motion of extra variables in the trajectory. Constraint methods that utilise non-Hamiltonian equations of motion, and stochastic forces that mimic the influence of a reservoir, can also be used.

Appendix A

Nature of Probability

A.1 Set Theory

Consider the finite denumerable set T of outcomes of an event. Denote the possible outcomes by the elements $\alpha = 1, 2, \ldots, m$, and the various sets of outcomes by a, b, etc. For example, the event might be rearrangements of the eggs in a carton able to hold half a dozen eggs, and the outcome might be the cell occupied by one particular egg. One could have a as the set of even numbered cells, b could be the set of cells numbered less than or equal to 3, etc.

Form an algebra consisting of the binary operators conjunction, ab (read a and b, occasionally to be written as a, b), disjunction $a + b$ (read a or b), and the unary negation operator $\overline{a}$ (read not a). The conjunction of two sets is their intersection, which is the set of elements they have in common. The disjunction of two sets is their union, which is the set of elements they have in total. The negation or complement of a set is the set of elements not in the set. In the example above, ab is an even number less than or equal to 3 (i.e., the second cell), $a + b$ is the set of numbers that is even or less than or equal to 3, which means that the egg could be in any cell but the fifth, and $\overline{a}$ is the set of uneven numbers. In contrast to ordinary algebra one has

$$aa = a, \quad a + a = a. \tag{A.1}$$

The usual rules of associativity and commutativity hold, $a(b + c) = ab + ac = (b + c)a$. In terms of the fundamental elements one has by definition $\alpha\beta = 0$, if $\alpha \neq \beta$ (0 is the empty set), and $\sum_\alpha \alpha = 1 + 2 + \ldots + m = T$.

The complement of a set is the set of all elements that are not in the set. Hence

$$a\overline{a} = 0, \quad a + \overline{a} = T. \tag{A.2}$$

Obviously $\overline{\overline{a}} = a$. For negation of composition one has the rules that

$$\overline{ab} = \overline{a} + \overline{b}, \quad \overline{a + b} = \overline{a}\overline{b}. \tag{A.3}$$

If a subset a is wholly contained within a subset b then

$$ab = a, \quad b = a + b. \tag{A.4}$$

(These two equations are the same; the right one follows by disjoining b to the left one.) If the left equation is negated, then it is converted to that on the right with b replaced by $\overline{a}$ and a replaced by $\overline{b}$,

$$\overline{a} + \overline{b} = \overline{a}, \quad \overline{b} = \overline{a}\overline{b}, \tag{A.5}$$

which is to say that not b is contained within not a. For example, the set $c = 2$ is contained within the set of even numbers, a, so that $ac = c$. Conversely the set of numbers not equal to 2, $\overline{c}$, contains the set of uneven numbers $\overline{a}$, $\overline{a}\,\overline{c} = \overline{a}$.

A.2 Laws of Probability

Attach to each of the elements of the set a positive weight ω_α, and define the probability of an element to be $\wp_\alpha = \omega_\alpha / \sum_\alpha \omega_\alpha$. The probability is a number between 0 and 1, with 0 corresponding to an impossible outcome and 1 corresponding to a certain outcome. The weight of a set a is the sum of the weights of its elements, and similarly for its probability,

$$\wp(a) = \sum_{\alpha \in a} \wp_\alpha. \tag{A.6}$$

The probability of the empty set is 0, and that of the entire set is one. For conjunction (read the probability of a and b), one has

$$\wp(ab) = \sum_{\alpha \in ab} \wp_\alpha, \tag{A.7}$$

and for disjunction (read the probability of a or b),

$$\begin{aligned} \wp(a+b) &= \sum_{\alpha \in (a+b)} \wp_\alpha \\ &= \sum_{\alpha \in a} \wp_\alpha + \sum_{\alpha \in b} \wp_\alpha - \sum_{\alpha \in ab} \wp_\alpha \\ &= \wp(a) + \wp(b) - \wp(ab). \end{aligned} \tag{A.8}$$

The final term corrects for double counting the common elements. For negation (read the probability of not a),

$$\wp(\overline{a}) = \sum_{\alpha \in \overline{a}} \wp_\alpha = \sum_{\alpha} \wp_\alpha - \sum_{\alpha \in a} \wp_\alpha, = 1 - \wp(a). \tag{A.9}$$

If the outcome of an event is known to be one of the elements in a set c, then the probabilities should be modified since c is now effectively the whole set of possible outcomes. The probability of a set a given that the outcome is in the set c is written as $\wp(a|c)$ (read the probability of a given c), and it is the weight of the common elements divided by the weight of the set, or equivalently

$$\wp(a|c) = \frac{\wp(ac)}{\wp(c)}. \tag{A.10}$$

In the case that the sets have no common elements, $\wp(a|c) = 0$, which says that given an outcome in c, then an outcome not in c is impossible. In the case that c is entirely in a, $ac = c$, then $\wp(a|c) = 1$ (i.e., if the outcome that occurred is in c, then it is certain that it is in a). If a is entirely in c ($ac = a$), then $\wp(a|c) = \wp(a)/\wp(c)$, so that the probability of a is the fraction of c that it occupies. In view of this last result one could write $\wp(a) = \wp(a|T)$. The law of disjunction also holds for conditioned probabilities, $\wp(a+b|c) = \wp(a|c) + \wp(b|c) - \wp(ab|c)$, as may be confirmed explicitly. The law of conditional probability may be rewritten $\wp(ab|c) = \wp(a|bc)\wp(b|c) = \wp(b|ac)\wp(a|c)$, and so

$$\wp(a|bc) = \frac{\wp(b|ac)}{\wp(b|c)}\wp(a|c). \tag{A.11}$$

This is known as Bayes' theorem. The denominator may be written

$$\begin{aligned} \wp(b|c) &= \wp(b(a+\overline{a})|c) \\ &= \wp(ba|c) + \wp(b\overline{a}|c) \\ &= \wp(b|ac)\wp(a|c) + \wp(b|\overline{a}c)\wp(\overline{a}|c), \end{aligned} \tag{A.12}$$

which says that the probability of an outcome may be written as the weighted sum of probabilities conditioned on exclusive outcomes.

A.3 Probability and Entropy

This section summarises the formulae for probability in terms of the weight of microstates and macrostates, and defines entropy in terms of these. Formulae for the discrete and for the continuum are given alternately.

In the discrete case the microstates of the total set are labelled by $\imath$ and the microstate weight is w_i. The total weight is

$$W = \sum_i w_i. \tag{A.13}$$

In the case of the continuum the points of the n-dimensional space are denoted by $\mathbf{x}$ and the weight density is $\omega(\mathbf{x})$. The total weight is

$$W = \int \mathrm{d}\mathbf{x}\, \omega(\mathbf{x}). \tag{A.14}$$

Note that the weight density is measured with respect to the volume element $\mathrm{d}\mathbf{x}$.

The probability that the system will be found in the microstate i is

$$\wp_i = \frac{w_i}{W}, \tag{A.15}$$

and the probability that the system will be found within $\mathrm{d}\mathbf{x}$ of $\mathbf{x}$ is

$$\wp(\mathbf{x})\,\mathrm{d}\mathbf{x} = \frac{\omega(\mathbf{x})}{W}\mathrm{d}\mathbf{x}. \tag{A.16}$$

These are evidently normalised,

$$\begin{aligned} 1 &= \sum_i \wp_i \\ &= \int \mathrm{d}\mathbf{x}\, \wp(\mathbf{x}), \end{aligned} \tag{A.17}$$

and the average of a function is

$$\begin{aligned} \langle g \rangle &= \sum_i \wp_i g_i \\ &= \int \mathrm{d}\mathbf{x}\, \wp(\mathbf{x}) g(\mathbf{x}). \end{aligned} \tag{A.18}$$

A macrostate is a set of microstates. The set of macrostates that are labelled by α are disjoint (i.e., a microstate belongs to only one such macrostate) and complete (i.e., all microstates belong to one or other of the macrostates). The characteristic function for the macrostate α is

$$\chi(i|\alpha) = \begin{cases} 1, & i \in \alpha \\ 0, & \text{otherwise.} \end{cases} \tag{A.19}$$

The continuum analogue of a macrostate is a hypersurface of the space, which in general is characterised by the constant value of $m < n$ functions, $\mathbf{F}(\mathbf{x}) = \mathbf{f}$. The constant m-dimensional vector $\mathbf{f}$ labels the hypersurface, and the set of hypersurfaces so labelled are both disjoint and complete. The characteristic function for the hypersurface $\mathbf{f}$ is

$$\chi(\mathbf{x}|\mathbf{f}) = \mathcal{D}(\mathbf{F}, \mathbf{x})\delta(\mathbf{F}(\mathbf{x}) - \mathbf{f}), \tag{A.20}$$

where the generalised Jacobean is given by

$$\mathcal{D}(\mathbf{F}, \mathbf{x})^2 = \sum_{i_1 < i_2 < \ldots < i_m}^{n} \left| \underline{\nabla}^{(m)} \underline{F} \right|^2 . \tag{A.21}$$

The dyadic matrix whose determinant appears here has elements $\partial F_i / \partial x_j$, where $i \in \{1, 2, \ldots, m\}$ and $j \in \{i_1, i_2, \ldots, i_m\}$.

The number of states in the macrostate is

$$\begin{aligned} n_\alpha &= \sum_i \chi(i|\alpha) \\ &= \sum_{i \in \alpha}, \end{aligned} \tag{A.22}$$

and the area of the hypersurface is

$$\begin{aligned} A(\mathbf{f}) &= \int \mathrm{d}\mathbf{x}\, \chi(\mathbf{x}|\mathbf{f}) \\ &= \int \mathrm{d}\mathbf{y}\mathrm{d}\hat{\mathbf{f}}\, \mathcal{D}(\mathbf{F},\mathbf{x})\delta(\mathbf{F}(\mathbf{x})-\mathbf{f}) \\ &= \oint_{\mathbf{f}} \mathrm{d}\mathbf{y}\,. \end{aligned} \tag{A.23}$$

Here $\mathrm{d}\mathbf{y}$ is the element of area of the hypersurface, which has dimension $n-m$. The relationship with the volume element is

$$\mathrm{d}\mathbf{x} = \mathrm{d}\mathbf{y}\mathrm{d}\hat{\mathbf{f}} = \mathrm{d}\mathbf{y}\mathrm{d}\mathbf{f}/\mathcal{D}(\mathbf{F},\mathbf{x}). \tag{A.24}$$

The normal to the hypersurface in the correct length scale is $\mathrm{d}\hat{\mathbf{f}}$. Evidently then, the generalised jacobean serves to convert the element $\mathrm{d}\mathbf{f}$ to the correct units. A point in space can be represented in either of the two coordinate systems $\mathbf{x} = \mathbf{x}(\mathbf{y},\mathbf{f})$, or $(\mathbf{y},\mathbf{f}) = (\mathbf{y}(\mathbf{x}),\mathbf{F}(\mathbf{x}))$. By definition, $\mathbf{F}(\mathbf{x}(\mathbf{y},\mathbf{f})) = \mathbf{f}$.

The weight of the macrostate is

$$\begin{aligned} w_\alpha &= \sum_i \chi(i|\alpha) w_i \\ &= \sum_{i\in\alpha} w_i. \end{aligned} \tag{A.25}$$

The weight density of the hypersurface measured with respect to $\mathrm{d}\mathbf{f}$ is

$$\begin{aligned} \omega(\mathbf{f}) &= \int \mathrm{d}\mathbf{x}\, \delta(\mathbf{F}(\mathbf{x})-\mathbf{f})\omega(\mathbf{x}) \\ &= \oint_{\mathbf{f}} \mathrm{d}\mathbf{y}\, \frac{\omega(\mathbf{x}(\mathbf{y},\mathbf{f}))}{\mathcal{D}(\mathbf{F},\mathbf{x}(\mathbf{y},\mathbf{f}))}. \end{aligned} \tag{A.26}$$

It is also possible to formulate a weight density measured with respect to $\mathrm{d}\hat{\mathbf{f}}$, namely

$$\begin{aligned} \hat{\omega}(\mathbf{f}) &= \int \mathrm{d}\mathbf{x}\, \chi(\mathbf{x}|\mathbf{f})\omega(\mathbf{x}) \\ &= \oint_{\mathbf{f}} \mathrm{d}\mathbf{y}\, \omega(\mathbf{x}(\mathbf{y},\mathbf{f})). \end{aligned} \tag{A.27}$$

One can use this to give a probability density $\hat{\wp}(\mathbf{f})\mathrm{d}\hat{\mathbf{f}} = \hat{\omega}(\mathbf{f})\mathrm{d}\hat{\mathbf{f}}/W$, and also the conditional probability for phase space given that the system is within $\mathrm{d}\hat{\mathbf{f}}$ of the hypersurface. However, this measure of the thickness of the hypersurface appears less relevent than the measure $\mathrm{d}\mathbf{f}$.

The probability that the system is in a microstate corresponding to the macrostate α is

$$\wp_\alpha = \frac{w_\alpha}{W}, \tag{A.28}$$

and the probability that the system is within $\mathrm{d}\mathbf{f}$ of $\mathbf{f}$ is

$$\wp(\mathbf{f})\mathrm{d}\mathbf{f} = \frac{\omega(\mathbf{f})\mathrm{d}\mathbf{f}}{W}. \tag{A.29}$$

The discrete probability is evidently normalised since

$$\sum_\alpha w_\alpha = \sum_\alpha \sum_i \chi(i|\alpha) w_i = \sum_i w_i = W, \tag{A.30}$$

which follows from the disjoint and complete nature of the macrostates,

$$\sum_\alpha \chi(i|\alpha) = 1. \tag{A.31}$$

The continuum probability density is also normalised,

$$\begin{aligned} \int \mathrm{d}\mathbf{f}\, \omega(\mathbf{f}) &= \int \mathrm{d}\mathbf{f} \int \mathrm{d}\mathbf{x}\, \delta(\mathbf{F}(\mathbf{x}) - \mathbf{f}) \omega(\mathbf{x}) \\ &= \int \mathrm{d}\mathbf{x}\, \omega(\mathbf{x}) \\ &= W. \end{aligned} \tag{A.32}$$

The disjoint and complete nature of the hypersurfaces is manifest by

$$\int \mathrm{d}\hat{\mathbf{f}}\, \mathcal{D}(\mathbf{F}, \mathbf{x}) \delta(\mathbf{F}(\mathbf{x}) - \mathbf{f}) = \int \mathrm{d}\mathbf{f}\, \delta(\mathbf{F}(\mathbf{x}) - \mathbf{f}) = 1. \tag{A.33}$$

The conditional probability for the system being in the microstate i given that it is in the macrostate α is

$$\wp(i|\alpha) = \frac{w_i}{w_\alpha} \chi(i|\alpha). \tag{A.34}$$

The conditional probability for the system being at $\mathbf{x}$ given that it is within $\mathrm{d}\mathbf{f}$ of the hypersurface $\mathbf{f}$ is

$$\wp(\mathbf{x}|\mathbf{f}) = \frac{\omega(\mathbf{x})}{\omega(\mathbf{f})} \delta(\mathbf{F}(\mathbf{x}) - \mathbf{f}). \tag{A.35}$$

It is readily verified that this conditional probability is correctly normalised.

Since the microstate belongs to one and only macrostate one has

$$\begin{aligned} \sum_\alpha \wp(i|\alpha) \wp_\alpha &= \sum_\alpha \frac{w_i}{w_\alpha} \chi(i|\alpha) \frac{w_\alpha}{W} \\ &= \frac{w_i}{W} \sum_\alpha \chi(i|\alpha) \\ &= \wp_i. \end{aligned} \tag{A.36}$$

Similarly, a point in the phase space is on one and only one hypersurface, so that one also has

$$\begin{aligned} \int \mathrm{d}\mathbf{f}\, \wp(\mathbf{x}|\mathbf{f}) \wp(\mathbf{f}) &= \int \mathrm{d}\mathbf{f}\, \frac{\omega(\mathbf{x})}{\omega(\mathbf{f})} \delta(\mathbf{F}(\mathbf{x}) - \mathbf{f}) \frac{\omega(\mathbf{f})}{W} \\ &= \wp(\mathbf{x}). \end{aligned} \tag{A.37}$$

These results are a manifestation of Eq. (A.12).

One may restrict attention to the macrostate α only and write

$$\wp^{(\alpha)}(i) = \frac{w_i}{w_\alpha}, \quad i \in \alpha. \tag{A.38}$$

In the continuum case one may restrict attention to being within $\mathrm{d}\mathbf{f}$ of the hypersurface $\mathbf{f}$ only,

$$\wp^{(\mathbf{f})}(\mathbf{y})\mathrm{d}\mathbf{y} = \frac{\omega(\mathbf{x}(\mathbf{y},\mathbf{f}))}{\omega(\mathbf{f})} \frac{\mathrm{d}\mathbf{y}}{\mathcal{D}(\mathbf{F},\mathbf{x}(\mathbf{y},\mathbf{f}))}. \tag{A.39}$$

In both cases these are the respective conditional probabilities divided by the characteristic function χ. It is evident that both restricted probabilities are normalised over the hypersurface.

Entropy

In both the discrete and the continuum cases the total entropy is

$$S = k_\mathrm{B} \ln W. \tag{A.40}$$

The entropy of the microstate may be defined as

$$S_i = k_\mathrm{B} \ln w_i, \tag{A.41}$$

and the entropy of a point in the space may be defined as

$$S(\mathbf{x}) = k_\mathrm{B} \ln \left[\omega(\mathbf{x})\Delta(\mathbf{x})\right]. \tag{A.42}$$

Consequently one may write the microstate probability as

$$\wp_i = \frac{e^{S_i/k_\mathrm{B}}}{W}, \tag{A.43}$$

and the probability density as

$$\wp(\mathbf{x}) = \frac{e^{S(\mathbf{x})/k_\mathrm{B}}}{\Delta(\mathbf{x})W}. \tag{A.44}$$

Here $\Delta(\mathbf{x})$ is a completely arbitrary volume element that is of no physical consequence.

Similarly, the entropy of a macrostate may be defined as

$$S_\alpha = k_\mathrm{B} \ln w_\alpha, \tag{A.45}$$

and consequently the macrostate probability may be written

$$\wp_\alpha = \frac{e^{S_\alpha/k_\mathrm{B}}}{W}. \tag{A.46}$$

Similarly, the entropy of the hypersurface may be defined as

$$S(\mathbf{f}) = k_\mathrm{B} \ln \left[\omega(\mathbf{f})\Delta(\mathbf{f})\right]. \tag{A.47}$$

Consequently the probability density on it may be written

$$\wp(\mathbf{f})\mathrm{d}\mathbf{f} = \frac{e^{S(\mathbf{f})/k_\mathrm{B}}}{\Delta(\mathbf{f})W}\mathrm{d}\mathbf{f}. \tag{A.48}$$

Again $\Delta(\mathbf{f})$ is arbitrary and of no consequence.

One also can write the total entropy as a sum over the macrostate entropy,

$$\begin{aligned} S &= k_\mathrm{B} \ln W \\ &= \sum_\alpha \wp_\alpha k_\mathrm{B} \ln W \\ &= \sum_\alpha \wp_\alpha \left[S_\alpha - k_\mathrm{B} \ln \wp_\alpha \right], \end{aligned} \tag{A.49}$$

and as an integral over the hypersurface,

$$\begin{aligned} S &= k_\mathrm{B} \ln W \\ &= \int \mathrm{d}\mathbf{f}\, \wp(\mathbf{f}) k_\mathrm{B} \ln W \\ &= \int \mathrm{d}\mathbf{f}\, \wp(\mathbf{f}) \left[S(\mathbf{f}) - k_\mathrm{B} \ln \{\wp(\mathbf{f})\Delta(\mathbf{f})\} \right]. \end{aligned} \tag{A.50}$$

A similar result holds for the sum over microstates,

$$S = \sum_i \wp_i \left[S_i - k_\mathrm{B} \ln \wp_i \right], \tag{A.51}$$

and for the integral over the space itself,

$$S = \int \mathrm{d}\mathbf{x}\, \wp(\mathbf{x}) \left[S(\mathbf{x}) - k_\mathrm{B} \ln \{\wp(\mathbf{x})\Delta(\mathbf{x})\} \right]. \tag{A.52}$$

One can also define the discrete restricted entropy

$$S^{(\alpha)}(i) = k_\mathrm{B} \ln w_i, \ i \in \alpha, \tag{A.53}$$

which is the same as $S(i)$ for $i \in \alpha$, and the restricted probability is proportional to its exponential. In terms of the latter the entropy of the macrostate may be written

$$\begin{aligned} S_\alpha &= k_\mathrm{B} \ln w_\alpha \\ &= \sum_{i\in\alpha} \wp^{(\alpha)}(i) \left[S^{(\alpha)}(i) - k_\mathrm{B} \ln \wp^{(\alpha)}(i) \right] \\ &= \sum_{i\in\alpha} \wp^{(\alpha)}(i) \left[S(i) - k_\mathrm{B} \ln \wp^{(\alpha)}(i) \right]. \end{aligned} \tag{A.54}$$

In the continuum the restricted entropy is

$$S^{(\mathbf{f})}(\mathbf{y}) = k_\mathrm{B} \ln \left[\frac{\omega(\mathbf{x}(\mathbf{y},\mathbf{f}))}{\mathcal{D}(\mathbf{F},\mathbf{x}(\mathbf{y},\mathbf{f}))} \Delta(\mathbf{y})\Delta(\mathbf{f}) \right], \tag{A.55}$$

and the restricted probability density is the exponential of this. In terms of the restricted probability density the entropy of the macrostate is

$$\begin{aligned} S(\mathbf{f}) &= k_{\mathrm{B}} \ln [\omega(\mathbf{f})\Delta(\mathbf{f})] \\ &= \oint_{\mathbf{f}} \mathrm{d}\mathbf{y}\, \wp^{(\mathbf{f})}(\mathbf{y}) \left[S^{(\mathbf{f})}(\mathbf{y}) - k_{\mathrm{B}} \ln \left\{ \wp^{(\mathbf{f})}(\mathbf{y})\Delta(\mathbf{y}) \right\} \right]. \end{aligned} \tag{A.56}$$

The above formalism for the continuum is invariant with respect to a change of variable. Given some $\mathbf{z}(\mathbf{x})$ one has

$$\frac{\mathrm{d}\mathbf{x}}{\mathrm{d}\mathbf{z}} = \frac{\wp(\mathbf{x})}{\wp_z(\mathbf{z})} = \frac{\omega(\mathbf{x})}{\omega_z(\mathbf{z})} = \frac{\Delta_z(\mathbf{z})}{\Delta(\mathbf{x})}. \tag{A.57}$$

A.4 Interpretation of Probability

As shown in the first chapter, there is an intimate connection between entropy and probability. Statistical mechanics uses probability constantly, as do other scientific and mathematical disciplines. Probability occurs so widely in everyday life that almost everyone has some idea of the meaning of odds, chance, and likelihood. Even more significant, although less well known, is the fact that the laws of probability are those of inductive reasoning and of rational behaviour. As such no field of human endeavour is unaffected by probability. For these reasons it is worthwhile to explore the meaning of probability in a broad context, and to be aware of various schools of thought, since these influence the specific interpretation that one makes of entropy and the other phenomena of thermodynamics and statistical mechanics.

At least three of the philosophical positions that have been adopted find some currency in statistical mechanics. The first and most common notion of probability is that of frequency, namely that the probability of the outcome of an event is proportional to the number of times that it recurs in a long sequence of trials. The second view holds that probability is a physical property. To this school belongs the idea of proportion: probabilities are the ratio of favourable possibilities to the total number of possibilities. (This was the position taken implicitly in the usage of probability in the above discussion.) A related viewpoint holds that the apparent randomness of deterministic outcomes is due to a sensitivity to the differing initial conditions between trials: the probability of an outcome is proportional to the volume of the space of initial conditions that leads to that outcome. The third and arguably the most radical position is that the probability of an outcome is the degree of reasonable belief that the outcome will occur. This subjectivist interpretation is widespread amongst Bayesian statisticians, and depends upon the connection between probability and logic, as explained below.

These three probabilities may be called frequency, proportion, and credibility, respectively. One cannot reasonably expect one of these to hold for all applications of probability, and arguably one should not speak of *the* interpretation of probability, but rather the interpretation appropriate in a particular circumstance. An interpretation of probability is viable if it obeys the laws of probability. To test this one needs to identify the outcomes and the elements of the set upon which it is based, and to show that these elements have a positive additive weight.

A.4.1 Frequency

In the frequency interpretation of probability the weight attached to the elements is just the number of times that that outcome occurred in many repeated trials of an event. For N trials the probability of the outcome α is N_α/N. The weight is obviously positive, and equally obvious is the fact that the weight of a set of outcomes is the total number of times any element in the set appeared, which is just the sum of the weights of the elements in the set.

This is enough to show that frequency satisfies the laws of probability, but it is a worthwhile exercise to treat conditional probability explicitly. The conditional probability represents counting only those outcomes of the trials that were in a particular subset, say b. The frequency with which any element in the set a occurs under this restriction, $\wp(a|b)$, is the number of times any element of a that was also in b occurred, which is N_{ab}, divided by the number of restricted trials, which is just N_b. However, this is $\wp(a|b) = N_{ab}/N_b = \wp(ab)/\wp(b)$, which is the required law of conditional probability.

Since frequency obeys the same laws as probability, it is a viable candidate for probability. As mentioned above, it is the most widespread interpretation of probability and it is a view held by commoners and professional statisticians alike. In statistical mechanics, the frequency interpretation is implicit in the time averages over a trajectory in phase space, configuration averages in computer simulations, and the ensemble averages of Gibbs. Whether or not probability can always be interpreted as a frequency, there are situations where one is genuinely dealing with a frequency, and in these cases frequencies must be manipulated according to the laws of probability.

On several points the frequency interpretation of probability is unsatisfactory. As is well known, the number of times a particular outcome occurs varies between different trials. In consequence the frequency and hence the probability that one assigns must also differ despite the fact that it is the same physical process. It is commonly recognised that the frequency approximately equals the probability, with negligible relative error in a long series of trials, but such a recognition indicates that frequency cannot be identical to probability, but is rather a consequence of it. Another problem with frequency is the fact that it can only be calculated after the trial, which means that one cannot assign a probability beforehand. Further, there are many once-only events to which one might wish to assign a probability (e.g., a horse race), and one cannot physically carry out repeated trials of these. In these case frequentists think of a fictitious trial of repetitious events and speak of probability as referring to the frequency of the outcome in this imaginary world. The ensemble averages devised by Gibbs belong to this class. Obviously to maintain the frequency interpretation of probability for once-only events requires a certain level of mental gymnastics and a vivid imagination. It is arguable that in these cases probability should more properly be classified as a belief, and one should explicitly recognize that frequency is being used as the scale of credibility.

A.4.2 Proportion

The second class of interpretations of probability take it to be a physical property of the event. In this objective view probability may be called proportion or measure. (Popper, in his objective philosophy of science, called it propensity.[1]) The most straightforward argument says that the probability of a set of outcomes is the proportion of the elementary outcomes that belong to the set, assuming that the elementary outcomes are all equivalent. This proportionality rule obviously satisfies all the laws of probability because the weights are obtained simply by counting elements. In many cases the elements are not all equivalent, and they will have attached to them a weight set by the physics of the event. Provided that this weight is positive and additive, proportion will by definition still obey the laws of probability.

Quantum mechanics takes randomness and probability as axiomatic. In contrast the existence of randomness in the deterministic world of classical mechanics is problematic. One approach that yields probability as proportion supposes that the outcome of an event is determined by the initial conditions, and that a trial of *apparently* identical events actually consists of choosing from amongst sets of slightly different initial conditions. Randomness and variability occur because the sensitivity of the system is greater than the differences between the initial conditions of each event (i.e., it is a consequence of limited resolution and control). One identifies a space of initial conditions in which each point maps to an elementary outcome, and every point has one such outcome; each element is characterised by a set of points in the space. The weight of an element is the volume of the corresponding space,

$$\omega_\alpha = \int_X dx\, \chi(x, \alpha), \tag{A.58}$$

where $\chi(x, \alpha)$ is the characteristic function (it equals 1 if the conditions x would have lead to the outcome α, and is 0 otherwise). Obviously this is a positive, additive weight, as is required to satisfy the laws of probability. Explicitly, since the conditional probability $\wp(a|b)$ is the probability of an outcome in a given that only outcomes in b are possible, one has

$$\begin{aligned} \wp(a|b) &= \frac{\int_{\mathrm{B}} dx\, \chi(x, a)}{\int_{\mathrm{B}} dx} \\ &= \frac{\int_X dx\, \chi(x, ab)}{\int_X dx\, \chi(x, b)} \\ &= \frac{\wp(ab)}{\wp(b)}, \end{aligned} \tag{A.59}$$

[1] K. R. Popper, The propensity interpretation of probability, *Brit. J. Philos. Sci.* **10** (1959), 25.

where B is the domain that maps to b, and X is the entire space. In this view, the probability of an outcome is the proportion of space that its initial conditions occupy. Bayes' theorem follows from this,

$$\wp(a|b) = \frac{\wp(b|a)}{\wp(b)}\wp(a). \tag{A.60}$$

In the interpretation of probability as measure, the probability on the left is that of a given that the outcome b occurs, which acts as a physical constraint or limitation on the possible initial conditions, and on the right $\wp(a)$ is the probability of a unconstrained by any such limitation. It is important to understand that the outcome b need not be causally related to the outcome a, even if $\wp(a|b) \neq \wp(a)$.

Probability as proportion is firmly a physical attribute that satisfies intuition that probability has an objective reality. It seems reasonable that the probability of an outcome should have an independent existence prior to an event and apart from the mind of the observer. It avoids the problems of the frequency interpretation in that it does not require repetitions, either real or imaginary, and as such can comfortably handle once-only events. The physical interpretation of probability can be criticised as being circular, since the weights of the elements are required (e.g., the measure of the space of initial conditions), and that in many cases the existence and identification of the fundamental elements may not be at all obvious. The argument in rebuttal of these criticisms is that both are characteristic of the physical process or algorithm that is the basis of the event, and whether they are known to an observer will affect only the values calculated for probability, not the objective reality of the random process itself.

A.4.3 Credibility

The third interpretation of probability is a subjective one, namely that it is the degree of rational belief that a particular outcome will occur. This approach is radical in the sense that it is the most basic (it can be traced to Bayes and Laplace), and also in the sense that it is the least conventional; professional statisticians are particularly antagonised by the subjectivist viewpoint.

In order to establish the viability of the credibility interpretation of probability, one needs to show the connection with logic. One approach is due to Cox, who cast set theory in the language of assertions.[2] For example, the element 1 could stand for the assertion 'there is an egg in the first cell', the set a for the assertion 'there are n eggs in the first carton', $b =$ 'the second set of cells is fully occupied', etc. The truth or falsity of assertions is the stuff of Aristotelean logic. Since the elements of the set are assertions about the possible outcomes of an event, then one can define *the* true element to be the one that asserts the actual outcome observed. All other elements are false. Define a set to be true if it contains the true element, and it is false otherwise. With this definition it is

[2]R. T. Cox, *The Algebra of Probable Inference*, Johns Hopkins Press, Baltimore, MD, 1961. R. T. Cox, Of inference and inquiry: An essay in inductive logic, in *The Maximum Entropy Formalism*, (R. D. Levine and M. Tribus, Eds.), MIT Press, Cambridge, MA, 1978.

easy to see that the conjunction of sets is true if all of the sets are true, and is false if any of them is false. Also the union of sets is false if all of them are false, and is true if any of them are true. These are the familiar rules of deductive logic, as enunciated by Aristotle.

Implication plays the same rôle in Aristotelean logic that proper subsets play in set theory. One says that if a is a proper subset of b, $ab = a$, then $a \Rightarrow b$ (read a implies b). It follows from set theory that if a implies b, and a is true, then b is true. Also, if $ab = a$, then $\overline{b}\overline{a} = \overline{b}$, which is to say that $\overline{b} \Rightarrow \overline{a}$. In other words, if a implies b and b is false, then a is false, which is the well-known Aristotelean syllogism.

Obviously taking assertions to be only true or false is an extreme simplification of the real world. The weight attached to statements forms a continuum, and real thinking is far richer than Aristotelean logic allows. Consider for example the case of implication, $a \Rightarrow b$. If one is told that a is true, then one deduces that b is true. However, what if one is told that b is true? On this Aristotle is silent, yet in practice most people would say that a is now more likely to be true. This is an example of inductive or inferential reasoning: the information about b has strengthened the belief in a.

This can be made clearer by way of example. Since one must break eggs to make an omelette, the assertion a = 'omelette' implies the assertion b = 'broken eggs', because $ab = a$. Hence given an omelette, one knows that eggs were broken, $a \Rightarrow b$, and if one knows that no eggs were broken then one deduces that there will be no omelette, $\overline{b} \Rightarrow \overline{a}$. If one is told that eggs were broken then one cannot be certain that an omelette is on the menu, but nevertheless one's anticipation of an omelette is the greater for knowing it; broken eggs induce salivation.

Probability theory quantifies the strength of one's beliefs when one has limited information. Suppose then in dealing with the set of assertions about the possible elemental outcomes of an event it is unknown whether an assertion about an elemental outcome is true, but rather the state of an assertion about a particular set of outcomes is known. (For example, one is told that there are n eggs in a carton.) This is clearly less information than is provided by knowing the actual elemental outcome, which would have determined the state of all the sets of outcomes. (For example, knowing the configuration gives the number of eggs in the carton, whether a particular cell is occupied, etc.) The subsets of the set of assertions about the possible outcomes are now true, false, or indeterminate, and the problem is to give a value to the indeterminate subsets that represents how likely they are to be true. (For example, told that there are eight eggs in a standard dozen-egg carton, one might conclude that it is more likely than not that the assertion that the first cell is occupied is true.)

Drawing conclusions from limited information is the problem of inference, and it is the law of conditional probability that generalizes the Aristotelean law of implication and that is in accord with the familiar practice of inductive reasoning. The formula $\wp(ab|c) = \wp(b|ac)\wp(a|c)$ does indeed behave satisfactorily. If a implies b ($ab = a$), one sees that $\wp(b|ac) = 1$, which is to say that b is true if a is true. This is of course the Aristotelean case. In the opposite case that b

implies a ($ab = b$), one has $\wp(b|ac) = \wp(ab|c)/\wp(a|c) = \wp(b|c)/\wp(a|c) \geq \wp(b|c)$, since $\wp(a|c) \leq 1$. This is what inductive reasoning suggests (if b implies a, the truth of a increases the credibility of b), but which is beyond Aristotelean logic.

The algebra of assertions and the above examples demonstrate the connection between probability and inductive or inferential reasoning. In this approach the probability of an event represents the degree of rational belief that the event will occur. Although unashamedly subjectivist, the laws of probability offer a prescription that ensures that rational individuals with the same information possess the same beliefs. Beliefs are modified in the light of new information or data, and the mechanism for this is Bayes' theorem,

$$\wp(a|bc) = \frac{\wp(b|ac)}{\wp(b|c)} \wp(a|c). \tag{A.61}$$

The difference between the probability of a that appears on each side of this equation is the data b. On the right is the degree of belief that would be assigned a prior to knowing the data, and on the left is the modified belief given the information b. In the interpretation of probability as credibility, Bayes' theorem characterises that inductive procedure known as learning from experience.

The subjectivist interpretation of probability as the degree of reasonable belief may be traced to Laplace, and is the one held most widely by Bayesian statisticians. In this interpretation, the probability of an outcome to an event is the strength with which one believes that the outcome will occur. The interpretation is nontrivial because it insists upon rational beliefs, so that two people with the same information will assign the same probability to an event. Evidently this epistemological interpretation of probability can handle the once-only events that were beyond the frequency interpretation. Indeed the imaginary trials that frequentists sometimes invent are really a mental artifice for using frequency as a gauge of credibility. There are a number of fields (eg. inference, decision theory, game theory) where it is natural and appropriate to interpret probability as a belief. The question is whether one ought always take probability to be the degree of reasonable belief, or whether there are problems in which probability is actually a physical phenomenon independent of the mind of the observer.

A.5 Subjectivity and Objectivity

A.5.1 Subjectivity in Statistical Mechanics

The principle of maximum entropy, which has some currency in statistical mechanics and which is discussed in the next section, was developed by Jaynes,[3] who strongly advocated the subjectivist view of probability and who introduced it in his formulation of statistical mechanics. The principle is very deep and general and it has had a revolutionary impact wide-spread beyond the discipline.

[3]E. T. Jaynes, Information theory and statistical mechanics, *Phys. Rev.* **106** (1957), 620; **108** (1957), 171. R. D. Rosenkrantz (Ed.), *E. T. Jaynes: Papers on Probability, Statistics, and Statistical Physics*, D. Reidel, Dordrecht, 1983.

The class of problems that it addresses are those of inference, namely given limited information or data estimate the value of some parameter. In practice one communicates the probability distribution of the likely values of the parameter, which is interpreted as indicating one's belief or confidence in those values. The principle of maximum entropy states that the probability distribution that should be used is the one that maximises the entropy constrained by the given information.

In the subjectivist view probability measures beliefs, which are modified by data or information, and entropy measures lack of information or ignorance. The more information one has, the sharper and more constrained is the probability distribution and the lower is the entropy. For this reason subjectivists often speak of the information entropy. Typically, when a subjectivist gives the number of eggs in a carton as $n \pm \delta$, they are communicating the strength of their belief about the different possible allocations in the form of a probability distribution. The known information or data (e.g., the average number observed in a trial) acts as a constraint on the possible probability distributions that they can choose. The principle of maximum entropy gives the recipe for finding the correct probability distribution given certain information. This distribution is the least biased distribution; any other distribution would either be inconsistent with the information or have a lower entropy, which implies the existence of additional constraints. Either these additional constraints are known, in which case they should be explicitly taken into account, or they are not known, in which case the lower entropy is a manifestation of an unjustified bias in ones beliefs. It is the laws of probability and the principle of maximum entropy that give the prescription for unbiased and rational beliefs.

The impact that the principle of maximum entropy has made on other fields contrasts with its influence on statistical mechanics. This is ironic since its original enunciation by Jaynes was in the context of deriving statistical mechanics, and it remains arguably a coherent and logically consistent formulation of that discipline. The reason for this resistance appears to be that most scientists regard the world as having some objective reality, which it is their happy task to uncover, and the subjectivist viewpoint is alien to this philosophy. The dependence of the outcome of an experiment upon the knowledge of the observer, as in Jaynes' interpretation of statistical mechanics, or as in the general credibility interpretation of probability, is unsettling and disconcerting to those accustomed to dealing with physical phenomena. It is arguable that this interpretation overemphasises the rôle of the observer, and that it divorces probability from the underlying physical causes. Many see it as peculiarly solipsistic to view entropy as a measure of the observer's uncertainty rather than of the disorder of the system.

A.5.2 An Example: Which Came First?

The brief description of the subjectivist position above does not do it full justice, and others would doubtless respond persuasively to the above criticisms. There are a number of reasons why many Bayesians have come to eschew objective probabilities. It is not only that the subjective interpretation is viable, and that inference is all-pervasive, but it is also because the objective interpretation can give rise to apparent paradoxes and contradictions. A simple example serves to illustrate one such paradox, namely an apparent violation of causality.

Consider once more the probability of the N-carton with n eggs gaining or losing an egg. Here the focus is on transitions that actually result in the gain or loss of an egg, so that the transition rule is to choose a cell at random in the N-carton and to take the egg out if it is occupied, or to put an egg in if it is free. Denoting the gain of an egg by '+', and a loss by '−', the probability of gaining an egg is $\wp(+|n,N) = (N-n)/N$, and that of losing an egg is $\wp(-|n,N) = n/N$. For two consecutive transitions the outcome is described by an ordered pair, with the first entry referring to the first transition, so that

$$\begin{aligned}
\wp(\{+,+\}|n,N) &= (N-n)(N-n-1)/N^2, \\
\wp(\{+,-\}|n,N) &= (N-n)(n+1)/N^2, \\
\wp(\{-,+\}|n,N) &= n(N-n+1)/N^2, \\
\wp(\{-,-\}|n,N) &= n(n-1)/N^2.
\end{aligned} \tag{A.62}$$

(For this example one assumes that all transitions are actually possible, so that $2 \le n \le N-2$.) Denote the union of both possible outcomes by a dot, so that $\wp(\{+,\cdot\}|n,N) = \wp(\{+,+\}|n,N) + \wp(\{+,-\}|n,N) = \wp(+|n,N)$. Notice that the probability of first gaining and then losing an egg, $\wp(\{+,-\}|n,N)$, is not equal to the probability of losing and then gaining an egg, $\wp(\{-,+\}|n,N)$, even though the final state consists of n eggs in both cases. In other words, the trial is not exchangeable. (Exchangeability is not central to what follows; an exchangeable trial is treated below.)

Now suppose that on the second transition an egg was gained. This should decrease the likelihood of having gained an egg on the first transition, because if first one had a gain, the probability of a loss second is increased, and that of a gain second is decreased, and vice versa. Since a gain is actually observed for the second transition, one should accordingly increase the probability that an egg was lost on the first transition. To quantify this expectation one uses Bayes' theorem in the form of conditional probability,

$$\begin{aligned}
&\wp(\{+,+\}|\{\cdot,+\},n,N) \\
&= \frac{\wp(\{+,+\}|n)}{\wp(\{\cdot,+\}|n,N)} \\
&= \frac{(N-n)(N-n-1)}{N^2}\frac{N^2}{(N-n)(N-n-1)+n(N-n+1)} \\
&= \frac{N-n-1}{N+(2n-N)/(N-n)} \\
&< \frac{(N-n)[1-1/(N-n)]}{N-N/(N-n)} \\
&= \frac{(N-n)}{N} \\
&= \wp(\{+,\cdot\}|n,N).
\end{aligned} \tag{A.63}$$

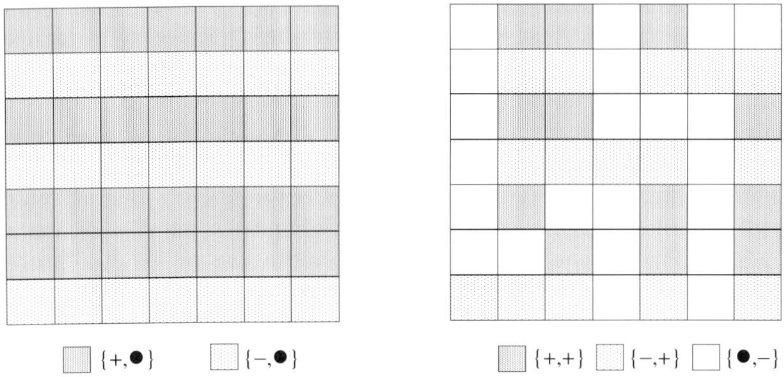

Figure A.1: The space of possibilities for two transitions of a 7-egg carton initially containing 3 eggs (in cells 1, 4, and 6). The number of each row corresponds to the number of the cell chosen first, and the number of each column corresponds to the number of the cell chosen second. In the left-hand diagram the light and dark squares represent first choosing an occupied and an unoccupied cell, respectively, and the probability of choosing an egg on the first transition irrespective of the outcome of the second may be seen to be $\wp(\{+,\cdot\}|4,7) = 4/7$. In the right-hand diagram the shaded and unshaded squares represent choosing an unoccupied and an occupied cell second. Here there are there are 15 light squares, 12 dark squares, and 22 unshaded squares, and the probability of choosing an empty cell on the first transition given that an egg was gained on the second may be seen to be $\wp(\{+,+\}|\{\cdot,+\},4,7) = 12/27$.

Since the dot represents the union of both possible outcomes, the left side could equally well be written $\wp(\{+,\cdot\}|\{\cdot,+\},n,N)$. One sees that gaining an egg on the second transition has indeed decreased the probability that an egg had already been gained on the first transition. This is illustrated in Fig. A.1.

The subjective interpretation of this result is straightforward: the strength of one's belief that an egg was gained on the first transition has been diminished by the information that an egg was gained on the second transition. There is evidently no difficulty in accepting that knowledge of the outcome of a later event can affect one's beliefs about the outcome of an earlier event.

The objective interpretation of this result is not so straightforward, and some have concluded that it violates causality.[4] That is, since the probabilities are a measure of some physical process, there is the apparent paradox that the outcome of the later process has affected the earlier process. Lest such an unphysical conclusion force the abandonment of the objective interpretation of probability, the physical transition rule needs to be examined more closely. The

[4]This example was motivated by that given in support of the pure subjectivist position by E. T. Jaynes, Clearing up mysteries –The original goals, in *Maximum Entropy and Bayesian Methods* (J. Skilling, Ed.), Kluwer, Dordrecht, 1989.

rule corresponds to choosing two cells (not necessarily different) at random in the carton, and then in turn reversing their contents. The initial space (or set of microstates) is the set of all ordered pairs of cells, each of the N^2 such pairs being equally probable. A subset of this space corresponds to gaining an egg on the second transition. Bayes' theorem quantifies by how much the proportion of ordered pairs that correspond to gaining an egg on the first transition is smaller on this subset than on the full set. The result does not say that the outcome of the second event affected the first, but rather that the set of microstates that could have led to the outcome of the second event is different to the set of all microstates. Causality rules, and the objective interpretation of probability remains viable.

It is worth noting in this example that the manipulation of the probabilities and the quantitative results were identical in both the subjective and objective interpretations. This is a general rule: the laws of probability are fixed and immutable, and it is only in their interpretation that one has any flexibility.

A.5.3 Bernoulli Trials and Prior Distributions

An example that is unavoidably subjective is as follows. From an N-egg carton containing n eggs, choose a cell at random, and denote the probability of choosing a cell occupied by an egg by $\theta \equiv n/N$. In this case do not transfer any eggs, so that the contents of the cells are unaltered. The outcome of a trial is unaffected by previous choices, and this general procedure is called an *exchangeable* or Bernoulli trial. (A familiar example of a Bernoulli trial is tossing a coin, with θ being the probability of heads.) The probability of having chosen m eggs in M trials is

$$\wp(m|M,\theta) = {}^{M}C_{m}\theta^{m}(1-\theta)^{M-m}, \tag{A.64}$$

since the order doesn't matter. The conditioning statement to the right of the bar means that the probability of choosing an egg on any one trial is θ.

The above implicitly takes an objective view of probability; θ is a fixed physical property of the system. Accordingly, in this interpretation the probability of getting l eggs in a series of L trials, given the above outcome, is just

$$\wp(l|L,m,M,\theta) = \wp(l|L,\theta) = {}^{L}C_{l}\theta^{l}(1-\theta)^{L-l}. \tag{A.65}$$

That is, the outcome of the previous M trials does not affect the outcome of the succeeding L trials because θ is a fixed and known physical attribute of the system. Likewise, $\wp(l,m|L,M,\theta) = \wp(l|L,\theta)\wp(m|M,\theta)$, since the two series of trials are independent.

To illustrate the subjective procedure a somewhat different question is posed. Suppose that one is unable to count the eggs in the carton so that the value of θ is unknown. One cannot measure θ directly, but one can estimate it from the known data, m eggs in M trials. The well-trained scientist will report the error in the measurement or estimation as $\theta \pm \Delta$, which indicates that it is a subjective belief, because obviously the actual objective physical quantity has a unique value and no error associated with it. More general and useful than the simple error estimate is the distribution of values of θ, $\wp(\theta|m, M)$, which represents the strength of one's belief that an egg will be chosen with probability between θ and $\theta + d\theta$, based on the given data.[5] Using this quantity, the probability of choosing l eggs in L trials is

$$\begin{aligned}\wp(l|L, m, M) &= \int_0^1 \wp(l, \theta|L, m, M)\, d\theta \\ &= \int_0^1 \wp(l|L, m, M, \theta)\wp(\theta|L, m, M)\, d\theta \\ &= \int_0^1 \wp(l|L, \theta)\wp(\theta|m, M)\, d\theta. \qquad \text{(A.66)}\end{aligned}$$

The first line manifests the law of disjunction, namely that the probability of an event is the sum of the probabilities of the conjunction of the event and exclusive outcomes. The second line is just the law of conditional probability. The third line results from the exchangeable nature of the trial, namely for the first factor in the integrand that the series are independent, and for the second that the same probability is to be assigned independent of the number of trials. Notice from the right-hand side that the probability of getting l eggs in L trials now depends upon the previous data (m, M).

The mechanism for modifying probabilities in the light of information or data is Bayes' theorem. In this case one has

$$\wp(\theta|m, M) = \frac{\wp(m|M, \theta)}{\wp(m|M)}\wp(\theta), \qquad \text{(A.67)}$$

where $\wp(m|M) = \int_0^1 \wp(m|M, \theta)\wp(\theta)\, d\theta$, and, as above, $\wp(\theta|M) = \wp(\theta)$. This last quantity is called the *a priori* distribution, and it represents one's initial beliefs about the probability of choosing an egg.

Such *a priori* distributions are problematic since there is no agreed recipe for their general formulation. How does one represent mathematically the state of complete ignorance? In general terms the distribution should be relatively broad and should not exclude any possibility. One finds in practice that the modified distribution becomes sharper and focussed upon the actual value of θ as more data are gathered, $\wp(\theta|m, M) \to \delta(\theta - m/M)$, $M \to \infty$. For few data however, the prediction does depend upon the initial distribution, and much discussion has been made of the optimum choice. The obvious and superficially appealing approach is to invoke a uniform distribution, the so-called Bayes–Laplace prior, which was termed by Laplace as the principle of insufficient reason. Surely complete ignorance should mean that one value of the parameter is not to be preferred over another? This principle suffers from the difficulty that ignorance

[5] One could give a somewhat artificial objective interpretation of this procedure. With $\theta = n/N$, $\wp(\theta)$ represents the probability of choosing an (n, N)-egg carton from an ensemble of cartons, and $\wp(\theta|m, M)$ represents the conditional probability that such a carton has been chosen given that the carton has yielded m eggs in M trials.

about the value of a parameter on the unit interval should imply equal ignorance about a nonlinear monotone transformation of the parameter, whereas the transformed distribution is not uniform. In addition, experience shows that the Bayes–Laplace prior gives insufficient weight to the end points.

An alternative approach is to look for symmetries in the problem. For the case of a chance parameter $\theta \in [0,1]$, the only clear symmetry for the *a priori* distribution is that between θ and $1-\theta$, $\wp(\theta) = \wp(1-\theta)$. (In the absence of any information, one should assign equal probabilities to choosing an occupied and an unoccupied cell.) What is required is a more symmetric formulation. First the domain is extended to the unit square by defining $\theta_1 = \theta$ and $\theta_2 = 1-\theta$, and one seeks the distribution $\wp(\theta_1, \theta_2) = \wp(\theta_2, \theta_1)$ on the line $\theta_1 + \theta_2 = 1$. A coordinate transformation is now made, $\alpha_i^2 = \theta_i$, $\alpha_i \in [-1, 1]$, and the constraint becomes $\alpha_1^2 + \alpha_2^2 = 1$, which is the equation of the unit circle in the (α_1, α_2)-plane. The *a priori* distribution of α_1 and α_2 should reflect the constraint by being only a function of $\alpha_1^2 + \alpha_2^2$. That is, it can be a function of the radius of the circle but not of angular position, since all positions on the circumference of a circle are equivalent. One must have that $\wp(\alpha_1, \alpha_2)\, d\alpha_1\, d\alpha_2 = \wp(\alpha_1^2 + \alpha_2^2)\delta\left(1 - \sqrt{\alpha_1^2 + \alpha_2^2}\right) d\alpha_1\, d\alpha_2$. Integrating out α_2 one obtains

$$\begin{aligned}
\wp(\alpha_1)\, d\alpha_1 &= d\alpha_1 \int \wp(\alpha_1^2 + \alpha_2^2)\delta\left(1 - \sqrt{\alpha_1^2 + \alpha_2^2}\right) d\alpha_2 \\
&= d\alpha_1\, \wp(1) \left| \frac{\sqrt{\alpha_1^2 + \alpha_2^2}}{\alpha_2} \right|_{\alpha_2 = \pm\sqrt{1-\alpha_1^2}} \\
&= \frac{\text{const.}\ d\alpha_1}{\sqrt{1-\alpha_1^2}}.
\end{aligned} \tag{A.68}$$

Using the change of variables formula for a distribution, $\wp(\alpha_1)\, d\alpha_1 = \wp(\theta)\, d\theta$, one obtains

$$\wp(\theta) = \frac{1}{\pi\sqrt{\theta(1-\theta)}}. \tag{A.69}$$

This particular *a priori* distribution of a chance parameter is sometimes called the Jeffreys prior, and there are a number of other approaches for deriving it and other *a priori* distributions.[6] The two other cases of interest are for a scale parameter, $x \in (0, \infty)$, in which case $\wp(x) \propto 1/x$, and for a location parameter, $x \in (-\infty, \infty)$, which has $\wp(x) = \text{const}$. In the case of multiple chance parameters the present argument will apply whenever it is possible to map the problem to the surface of a hypersphere.

A.5.4 Objective Beliefs

The reconciliation of the subjective and objective views of probability begins with the recognition that there exists an objective reality, but that attempts to

[6] H. Jeffreys, *Theory of Probability*, 3rd ed., Oxford, London, 1961. J. M. Bernardo and A. F. M. Smith, *Bayesian Theory*, Wiley, Chichester, 1994.

measure, to understand, and to communicate it are purely subjective. A great deal if not all of what passes as science is inference about the behaviour of the world. One acts as if the laws derived are the actual laws of the universe, and it is only when some contradiction arises that forces a change to these laws does one admit that all along they were nothing but inferences based on limited data; it is one's beliefs that have changed, not the universe. One believes in a physical universe that exists apart from one's beliefs.

The acceptance of the inferential nature of science can create confusion in a field such as statistical mechanics, because the word 'probability' is used to describe both the beliefs and the subject of the beliefs. The random nature of the physical universe is described by the probability that is a physical attribute of some stochastic process. This physical probability is distinct from one's beliefs about it. In other words, while the laws of thermodynamics and statistical mechanics are inferential and part of one's belief system, they are inferences about ontological proportion, not about epistemological credibility. One easily distinguishes between the measurement of the length of a rod, which is a belief, and the rod length itself, which has a real existence; one's treatment of the measurements ought to be in accord with the laws of probability, whereas rod lengths obey the laws of thermodynamics. It is harder to maintain the distinction between one's estimation of the probability of an event (credibility) and the actual probability of an event (proportion).

In setting out the basis for thermodynamics and statistical mechanics, it is difficult to see any alternative to assuming that the probability distributions and the consequent entropy invoked have a physical reality. As in all of science, one articulates the laws and rules as those of nature, and one proceeds as if they were absolute and independent of one's beliefs. Thermodynamics has at its basis the principle that entropy is maximised at equilibrium. This principle of maximum entropy comes not from any subjective desire to make rational and unbiased predictions based on given data (although one is occasionally so tempted), but from the objective fact that the number or weight of microstates determines both the entropy and the probability distribution. For this reason the probability distributions derived in the present work are physical quantities that arise from the total entropy of the system and the reservoir with which it can exchange a conserved quantity. In mathematical content the objective and subjective approaches are indistinguishable (as they must be, since credibility and proportion both obey the laws of probability), but perhaps the present physical interpretation of the principle of maximum entropy will make it more readily accessible and digestible than has the subjective interpretation to date.

A.6 The Principle of Maximum Entropy

The formulation of statistical mechanics pursued in this book is based upon an objective view of entropy and probability, and the use of physical exchanges of some constrained quantity with a reservoir. An alternative approach that in recent years has become increasingly widespread has been developed by Jaynes, based upon the so-called *principle of maximum entropy*.[7] The results in practice turn out to be very similar mathematically to the reservoir formalism. However, whereas the latter is physically motivated and therefore completely objective, the principle of maximum entropy is unashamedly subjective and cast in terms of the information available to the observer. Such philosophical stances have widespread ramifications, and the various interpretations of probability and entropy were discussed above. Here the 'maxent' procedure itself is outlined.

In scrambling the eggs in the first chapter it was shown that macroscopic fluxes were in the direction of increasing entropy, and that the equilibrium macrostate was the one with the greatest entropy. This is of course an explicit example of the second law of thermodynamics. Due to the intimate connection between the number of configurations and probability, it was also shown that entropy is a functional of the probability.

The principle of maximum entropy may be viewed as a generalisation of the second law of thermodynamics. In traditional thermodynamics one effectively seeks the equilibrium value of a macroscopic parameter, which is the value that maximises the entropy. In Jaynes' version of statistical mechanics one seeks the appropriate probability distribution to be assigned to the system, which is the one that maximises the entropy. This maximisation is subject to any constraints on the system. Obviously the probability distribution also gives the equilibrium value of the parameter, and much other information besides.

The principle of maximum entropy yields the least biased probability distribution. Any other choice of probability distribution would have lower entropy, and it would reflect either an additional constraint or some bias in the choice. A physical constraint, such as one prohibiting certain configurations, or one that made certain configurations more likely, narrows the probability distribution and decreases the entropy. In other words, the entropy is as large as it can be given the constraints, and the probability distribution should reflect this constrained maximisation. All known constraints should be taken into account in determining the probability distribution. Ignoring some constraints would give too broad a probability distribution and an artificially high entropy. Using a probability distribution that satisfied the known constraints but that was not optimal would by definition have a lower entropy, which either reflects an additional constraint, which should have been taken into account explicitly, or it else manifests an unjustified bias, which has no rational basis.

A typical example of assigning a probability distribution occurs in inferential problems, where one is asked to estimate some parameter given only limited information or data. In such problems there are obviously many different probability distributions that could be assigned and still satisfy the given data. The principle of maximum entropy supplies the missing rule that gives a unique rational answer to such a question. The question is whether the probability distributions of statistical mechanics are similarly inferential in nature, or whether they have an underlying physical reality.

[7] E. T. Jaynes, Information theory and statistical mechanics, *Phys. Rev.* **106** (1957), 620; **108** (1957), 171. R. D. Rosenkrantz (Ed.), *E. T. Jaynes: Papers on Probability, Statistics, and Statistical Physics*, D. Reidel, Dordrecht, 1983.

A.6.1 General Formalism

The equilibrium distribution $\wp(x)$ is required. This is obtained by considering the entropy as a functional of an arbitrary probability distribution and finding the particular distribution that maximises it. The entropy to be maximised is

$$S[\wp] = -k_B \sum_x \wp(x) \ln \wp(x). \tag{A.70}$$

This is the Gibbs–Shannon expression for the entropy, Eq. (1.12), which is the one used by Jaynes. It neglects the internal entropy of the system in the state x, and should arguably only be used when x labels equally probable microstates, but this is generally not well recognised. Note also that no reservoir or second system enters in this formalism. The relationship of the quantity on the left-hand side with the three entropies defined in the text will shortly be identified. One supposes that the probability distribution is constrained to yield the average value of the parameter,

$$\sum_x \wp(x)x = \langle x \rangle . \tag{A.71}$$

A normalisation constraint is also introduced,

$$\sum_x \wp(x) = 1. \tag{A.72}$$

The average value then is the information made available to the observer, and maximising the entropy subject to this information yields the least biased probability distribution, both of which are obviously subjective.

Introducing Lagrange multipliers, the function to be maximised without constraint is

$$S^*[\wp] = S[\wp] + \lambda' \sum_x \wp(x)x + \mu \sum_x \wp(x). \tag{A.73}$$

The functional derivative is

$$\frac{\delta S^*}{\delta \wp(x)} = -k_B \ln \wp(x) - k_B + \lambda' x + \mu, \tag{A.74}$$

the vanishing of which gives the desired probability distribution that maximises the entropy. This is

$$\overline{\wp}(x) = \frac{e^{-\lambda x}}{Z(\lambda)}, \tag{A.75}$$

where $\lambda = -\lambda'/k_B$, and the normalisation factor is

$$Z(\lambda) \equiv e^{-1+\mu/k_B} = \sum_x e^{-\lambda x}. \tag{A.76}$$

The value of the multiplier follows from the constraint equation, which may alternatively be written

$$\frac{\partial \ln Z}{\partial \lambda} = - \langle x \rangle . \tag{A.77}$$

Finally, the entropy itself may be seen to be

$$S[\overline{\wp}] = k_{\mathrm{B}} \ln Z(\lambda) + k_{\mathrm{B}} \lambda \langle x \rangle . \tag{A.78}$$

Compared to the reservoir formalism, Section 1.4.2, which gave for the probability distribution $\wp(x|\lambda) = e^{S(x)/k_{\mathrm{B}}} e^{-\lambda x}/Z'(\lambda)$, the maxent formalism gives $\overline{\wp}(x|\lambda) = e^{-\lambda x}/Z(\lambda)$. One sees that the latter is only correct if $S(x)$ is 0 or constant, which only occurs when the x states of the subsystem in isolation are equally likely. Similarly, the reservoir result for the total unconstrained entropy is $S_{\text{total}}(\lambda) = k_{\mathrm{B}} \ln Z'(\lambda)$, which contrasts with the maxent result, $S[\overline{\wp}] = k_{\mathrm{B}} \ln Z(\lambda) + k_{\mathrm{B}} \lambda \langle x \rangle$. Since $-k_{\mathrm{B}} \lambda x$ is the reservoir contribution to the total entropy in the state x, one sees that $S[\overline{\wp}]$ is the subsystem-only part of the unconstrained entropy of the total system (with $S(x) = 0$). However, the functional actually maximised, $S^*[\overline{\wp}] = S[\overline{\wp}] - k_{\mathrm{B}} \lambda \langle x \rangle + \mu \langle 1 \rangle$, is approximately equal to $S_{\text{total}}(\lambda) \approx S_{\text{total}}(\overline{x}|\lambda)$, provided that $S(x) = 0$ and neglecting the final constant term. That is, $S^*[\overline{\wp}]$ is in the thermodynamic limit the unconstrained entropy of the total system (with $S(x) = 0$). In other words, by introducing the average value as a constraint, Jaynes has effectively corrected the Gibbs–Shannon functional, Eq. (1.12), to the generalised Boltzmann expression, Eq. (1.11), at least in the case that the states x of the isolated subsystem are equally likely.

This interpretation of the maxent procedure allows a connection to be made with the reservoir formalism. It shows that the quantity $S[\overline{\wp}]$ is just the subsystem part of the total unconstrained entropy, and that the quantity $S[\overline{\wp}] - k_{\mathrm{B}} \lambda \langle x \rangle$ is the total unconstrained entropy. One concludes that in order to generalise the principle of maximum entropy to states of the isolated subsystem not equally likely, one should invoke the functional, Eq. (1.11),

$$S[\wp] = \sum_x \wp(x) \left[S(x) - k_{\mathrm{B}} \ln \wp(x) \right] , \tag{A.79}$$

together with the constraint on $\langle x \rangle$. This correction is a little artificial in the sense that if one knew $S(x)$, then one could immediately write down the probability distribution, $\wp(x|\lambda) = e^{S_{\text{total}}(x|\lambda)}/Z'(\lambda)$, which of course makes the optimisation procedure that underlies the principle of maximum entropy somewhat redundant.

A.6.2 Egg Sample

The egg distribution problem of Section 1.4.1 may be cast in a form suitable for the application of the principle of maximum entropy. What is the best possible estimate of the probability distribution given only limited information? Suppose that N, the size of the carton of interest, and $\overline{n}$, the equilibrium number of eggs in it, are known. The size of the second carton (or equivalently the total number of eggs) are not known. The principle of maximum entropy states that the appropriate probability distribution is the one that maximises the entropy subject to any constraints. The entropy to be maximised is

$$S[\wp] = \sum_{n=0}^{N} \wp_n [S_N(n) - k_B \ln \wp_n], \tag{A.80}$$

where now has been included the internal entropy of the macrostate,

$$S_N(n) = k_B \ln \frac{N!}{n!(N-n)!}. \tag{A.81}$$

The two constraints on the probability distribution are that it be normalised, $\sum_{n=0}^{N} \wp_n = 1$, and that it give the known equilibrium allocation, which is taken to equal the average allocation, $\sum_{n=0}^{N} \wp_n n = \overline{n}$. From the formalism of the principle of maximum entropy, the result for the optimum probability distribution is

$$\overline{\wp}_n = \frac{1}{Z(\alpha)} e^{S_N(n)/k_B} e^{\alpha n} = \frac{N!}{n!(N-n)!} \frac{e^{\alpha n}}{Z(\alpha)}. \tag{A.82}$$

It is the best estimate that can be made for the probability distribution given the limited information. Note that the internal entropy in this actual example is certainly not constant, and neglecting it by using the Gibbs–Shannon formula, Eq. (1.12), as in the original maxent formulation of Jaynes, will give the wrong probability distribution.

The normalising factor here is

$$Z(\alpha) = \sum_{n=0}^{N} \frac{N!}{n!(N-n)!} e^{\alpha n} = (1 + e^{\alpha})^N, \tag{A.83}$$

and the average allocation constraint gives

$$\overline{n} = \frac{\partial \ln Z(\alpha)}{\partial \alpha} = \frac{N c^{\alpha}}{1 + e^{\alpha}}, \tag{A.84}$$

or

$$e^{\alpha} = \frac{\overline{n}}{N - \overline{n}}. \tag{A.85}$$

Noting that $\overline{n}/(N-\overline{n}) = \overline{m}/(M-\overline{m})$, one sees that the Lagrange multiplier α has precisely the same value as the coefficient derived in Section 1.4.1 via the reservoir formalism. Since the probability distribution is $\wp(n|N, M, n+m) \propto \exp[S_N(n)/k_B + S_M(m)/k_B]$, one sees that the reservoir limit yields precisely the maximum entropy distribution. The parameter α is set by the properties of the reservoir, and is conjugate to the quantity that is exchangeable with the reservoir, namely the number of eggs n. This then is the physical basis of the principle of maximum entropy, and its accuracy is seen to depend upon the extent to which a finite-sized second system can be approximated as a reservoir.

The mathematical equivalence of this reservoir approach and the maximum entropy approach ought to be clear, so that in general the maximum entropy probability distribution is the exact probability distribution for a system in contact with a reservoir. This represents a physical interpretation and basis for the principle of maximum entropy, namely in terms of a reservoir that exchanges some parameter with the system. In the context of thermodynamics and statistical mechanics, the objective reservoir formulation is preferable to the subjective interpretation of the principle of maximum entropy because the systems being studied include actual physical reservoirs. Generalisations of thermodynamics and statistical mechanics may likewise be made on the basis of physical arguments in the reservoir approach, something which does not appear so straightforward in the subjective method. This is not to say that there are valid applications of the principle of maximum entropy in other fields such as inferential reasoning. However, the present book takes an objective approach in which the probability of a macrostate represents the weight of corresponding microstates, and in which entropy represents the total weight of microstates.

Appendix B

Work and Conjugate Variables

This Appendix shows that the derivatives of the entropy defined in the second chapter have the same properties as the familiar physical quantities that bear the same name. Also, pairs of conjugate variables are identified from the nature of thermodynamic work.

B.1 Energy, the First Law, and Work

Since one can consider the energy of an isolated system as dependent upon its entropy, $E(S, V, N)$, the total differential for the entropy, Eq. (2.4), can be rearranged to obtain

$$dE = TdS - pdV + \mu dN. \tag{B.1}$$

This gives the partial derivatives of the energy as

$$\left(\frac{\partial E(S, V, N)}{\partial S}\right)_{V,N} = T, \tag{B.2}$$

$$\left(\frac{\partial E(S, V, N)}{\partial V}\right)_{S,N} = -p, \tag{B.3}$$

and

$$\left(\frac{\partial E(S, V, N)}{\partial N}\right)_{V,S} = \mu. \tag{B.4}$$

Obviously one now has, for example, $T(S, V, N)$ rather than $T(E, V, N)$. The uniqueness of the thermodynamic state means that the temperature has the same value no matter what variables are used to specify the state.

Now the first law of thermodynamics will be used to confirm the physical interpretation of these derivatives. Logically, one does not need the first law. The entropy derivatives given above can be taken as definitions of temperature, pressure, and chemical potential, and the relationships between them can be deduced solely from the principle of maximum entropy (the second law). In so far as these relationships are entirely consistent with everyday experience, it is appropriate to give these quantities the same name as the familiar ones. In practice however, one generally has some mechanical notion of some or other type of work, and it is then straightforward to invoke the first law to derive the appropriate entropy derivative for the variables of interest. This is now shown in detail how this is done for the familiar pV work.

The work done by an applied pressure p_0 to change the volume of a system is $dW = -p_0 dV$.[1] This says that one must do work to decrease the volume of a system. The first law of thermodynamics states that the change in energy of a system is equal to the heat flow into the system plus the work done on the system,

$$dE = dQ - p_0 dV. \tag{B.5}$$

Compare this to the total differential given above for $dN = 0$,

$$dE(S, V, N) = T(S, V, N)dS - p(S, V, N)dV. \tag{B.6}$$

These two equations obviously must be equal, $dE \equiv dE(S, V, N)$. One cannot simply equate the coefficients of dV, because the external pressure p_0 is the change of energy of the system with volume at constant heat, whereas the system pressure $p(S, V, N) = -\partial E(S, V, N)/\partial V$ is the change of energy with volume at constant entropy, and these are not the same things. One cannot in general identify dQ with TdS, except in the case that no work is done, $dV = 0$.

In order to identify the coefficients, one needs to set $dQ = 0$ and $dS = 0$. Heat can be prevented from flowing into a system by insulating it from its surroundings, and so now such an isolated system, $dQ = 0$, will be treated. Entropy is constant in reversible processes, $dS = 0$, and increases in irreversible processes, $dS > 0$. In practice the former are distinguished by slow, quasi-static transformations, whereas irreversible changes are always characterised by sudden and rapid motion (e.g., suddenly placing a large mass on a piston). For an irreversible transformation of an isolated system one has

$$[p(S, V, N) - p_0]dV = TdS > 0. \tag{B.7}$$

If the system is compressed, $dV < 0$, as occurs if the external pressure is suddenly increased, the internal pressure is less than the external pressure, $p(S, V, N) < p_0$. For a sudden expansion, $p(S, V, N) > p_0$. This is consistent

[1] One way to see this is to imagine a piston of mass M in a gravitational field g acting on a system of cross-sectional area A and height h. Then the external pressure is the force per unit area, $p_0 = Mg/A$, and the change in volume of the system is $dV = Adh$. The change in energy of the piston, $Mgdh$, is equal and opposite to the work done by it on the system, which equals the change in energy of the system $dE = dW = -Mgdh = -p_0 dV$.

with familiar notions of pressure and volume changes. For a reversible process, $dS = 0$, and one has

$$p(S, V, N) = p_0, \tag{B.8}$$

so that the internal and external pressures are in balance. Again this is consistent with everyday experience: when the internal and external pressures are equal, the volume doesn't change. It is indeed appropriate to make the physical identification of the pressure of the system with the volume derivative of the energy at constant entropy, $p(S, V, N)$, which is equivalent to the original definition of $p(E, V, N)$ in terms of the entropy derivative at constant energy.

In general, the physical identification of thermodynamic derivatives proceeds from the mechanical notion of work, and equates the appropriate coefficients with the conjugate energy derivative at constant entropy. The procedure is valid for isolated (or insulated) systems undergoing reversible (or quasi-static) changes. The latter may be called *adiabatic* transformations,[2] but some caution should be exercised since adiabatic is often used inappropriately to denote an insulated system itself. As was seen above, isolating a system is not sufficient to ensure constant entropy. It is necessary both for the system to be insulated and for the change to be reversible to be able to identify the thermodynamic derivatives.

B.2 Conjugate Variables

Pressure and volume form a conjugate pair, as do chemical potential and number, and temperature and entropy. Conjugate pairs consist of an intensive and an extensive variable whose product has the dimensions of energy. As was defined above, extensive variables scale with the size of the system since they are linear additive quantities. Examples that have already been met are V, N, E, and S. The fundamental quantity of thermodynamics is the entropy as a function of all the extensive variables of the system. Intensive variables such as p, μ, and T do not scale with the size of the system; they may also be called field variables. In the example above the pressure is the field, and the volume may be regarded as the response of the system to an external field.

In general, one can consider an intensive field variable x and an extensive response X. One can attribute part of the total energy to external sources such that the change in the external energy with X is $dE^{\text{ext}} = x^{\text{ext}} dX$, where x^{ext} is the externally applied field. For example, changing the volume of the system by dV against an external pressure p^{ext} gives $dE^{\text{ext}} = p^{\text{ext}} dV$. This change in the external part of the energy may also be called the work done on the system. If the external field tends to decrease X (i.e., E^{ext} decreases if X

[2] In quantum mechanics, an infinitely small, infinitely slow perturbation induces what is known as an adiabatic transformation. Time-dependent perturbation theory says that the eigenstates are conserved and the system stays in the same state during such a transformation. The probability distribution of microstates is thus unchanged, as is the entropy, which is a functional of these.

decreases), then x^{ext} is positive. If the external field variable is a constant, then the external energy is $E^{\mathrm{ext}} = x^{\mathrm{ext}} X$. (In certain cases the field variable may contain contributions from the system that vary with X and the integration does not have this simple form.)

The explicit expression for the energy due to external sources serves to identify the conjugate variables. For a truly isolated system in the absence of external fields and with the extensive parameter constrained, one defines the field variable of the system as the entropy derivative at constant internal energy,

$$\left(\frac{\partial S(E^{\mathrm{int}}, X)}{\partial X} \right)_{E^{\mathrm{int}}} = \frac{x}{T}. \tag{B.9}$$

Only the relevant extensive independent variables are shown here, the remaining ones being held constant during the differentiation. Of course in the absence of the external field the internal energy is the same as the total energy.

When the external field is present, one separates the total energy into the internal part and the external part, $dE^{\mathrm{tot}} = dE^{\mathrm{int}} + dE^{\mathrm{ext}}$. Above it was shown that for a reversible volume change the internal and external pressures were in balance. This property is preserved by the above definitions. The entropy derivative at constant *total* energy is

$$\begin{aligned} \left(\frac{\partial S(X|E^{\mathrm{int}}, x^{\mathrm{ext}})}{\partial X} \right)_{E^{\mathrm{tot}}} &= \left(\frac{\partial S}{\partial E^{\mathrm{int}}} \right)_X \left(\frac{\partial E^{\mathrm{int}}}{\partial X} \right)_{E^{\mathrm{tot}}} + \left(\frac{\partial S}{\partial X} \right)_{E^{\mathrm{int}}} \\ &= \frac{-x^{\mathrm{ext}}}{T} + \frac{x}{T}. \end{aligned} \tag{B.10}$$

The first term arises because the change in entropy with internal energy is the inverse temperature, and because $dE^{\mathrm{int}} = -dE^{\mathrm{ext}}$ if the total energy is fixed. One can see that the entropy of the system is maximised at fixed total energy when the internal and the external fields are in balance, $x(\overline{X}) = x^{\mathrm{ext}}$.

This procedure may be explicitly illustrated using pressure as an example. Suppose that the system has height h and cross-sectional area A, and its upper boundary is a movable piston of mass M in a gravitational field g. In this case the external energy is $E^{\mathrm{ext}} = Mgh = p^{\mathrm{ext}} V$, where the external pressure is $p^{\mathrm{ext}} = Mg/A$. As far as the system is concerned, the origin of the external pressure is irrelevant. The total energy is $E^{\mathrm{tot}} = E^{\mathrm{int}} + E^{\mathrm{ext}} = E^{\mathrm{int}} + p^{\mathrm{ext}} V$ (this is also called the enthalpy). The external energy has a trivial effect on the entropy so that one can write $S(V|E^{\mathrm{tot}}, p^{\mathrm{tot}}) = S_0(V|E^{\mathrm{int}})$. The right-hand side is the entropy in the absence of the external pressure, and hence one defines

$$\left(\frac{\partial S_0}{\partial E^{\mathrm{int}}} \right)_V = \frac{1}{T}, \quad \left(\frac{\partial S_0}{\partial V} \right)_{E^{\mathrm{int}}} = \frac{p}{T}, \tag{B.11}$$

where p is the internal pressure of the system. One also has

$$\begin{aligned} \left(\frac{\partial S(V|E^{\mathrm{int}}, p^{\mathrm{ext}})}{\partial V} \right)_{E^{\mathrm{tot}}} &= \left(\frac{\partial S}{\partial E^{\mathrm{int}}} \right)_V \left(\frac{\partial E^{\mathrm{int}}}{\partial V} \right)_{E^{\mathrm{tot}}} + \left(\frac{\partial S}{\partial V} \right)_{E^{\mathrm{int}}} \\ &= \frac{-p^{\mathrm{ext}}}{T} + \frac{p}{T}. \end{aligned} \tag{B.12}$$

That is, the constrained entropy is maximal at the equilibrium volume, which is the one at which the internal pressure balances the external pressure. These results follow from the general formalism with $V = X$ and $p = x$.

There are a number of common examples of non-pV work. A rod or length of wire under a tension f tends to expand, so that the change in the external part of the energy due to a change in length is $dE^{\text{ext}} = -f dL$. Hence the length is equivalent to X and the tension is equivalent to $-x$, so that

$$\left(\frac{\partial S}{\partial L} \right)_{E^{\text{int}}} = \frac{-f}{T}. \tag{B.13}$$

Note how the tension is of opposite sign convention to the pressure. Note also that it has not been specified whether this particular tension is at constant volume or at constant area. Similarly one can define a surface tension that is conjugate to the area of the system, so that $dE^{\text{ext}} = -\gamma dA$, where the surface tension is

$$\left(\frac{\partial S}{\partial A} \right)_{E^{\text{int}}} = \frac{-\gamma}{T}. \tag{B.14}$$

These two examples are intended for illustration only, and more precise definitions will be required for specific cases.

Another common class of systems is those acted upon by an external field with the extensive parameter being a constrained property of the system rather than a conserved one. For example, a system of mass M in a gravitational field g changes energy with height by an amount $dE^{\text{ext}} = Mg\, dh$. Accordingly, $X \equiv h$ and $x \equiv Mg$, and hence

$$\left(\frac{\partial S}{\partial h} \right)_{E^{\text{int}}} = \frac{Mg}{T}. \tag{B.15}$$

Charge is a conserved quantity, and the conjugate variable is the electric potential. The change in the external part of the energy due to a change in the charge of the system is $dE^{\text{ext}} = \psi dQ$. Here ψ is the potential difference between the system and the point to which the charge is transferred to. If the system is at the higher potential, then transferring charge to the system increases the external part of the total energy, which is to say that a high potential tends to decrease the charge in a system. Accordingly one has $X = Q$, $x = \psi$, and hence

$$\left(\frac{\partial S}{\partial Q} \right)_{E^{\text{int}}} = \frac{\psi}{T}. \tag{B.16}$$

A more complicated class of systems is those with spatially varying external fields. Consider an external electric field that polarises dielectric material, with the total polarisation being an unconserved extensive quantity. In this example the external electric field, $\mathbf{E}^{\text{ext}}(\mathbf{r})$, is due to external charges and is the field that would be present in the absence of the dielectric material. The interaction of a given polarisation density, $\mathbf{P}(\mathbf{r})$, with this external electric field contributes to the total energy an amount

$$U^{\text{ext}} = - \int_V \mathrm{d}\mathbf{r}\, \mathbf{E}^{\text{ext}}(\mathbf{r}) \cdot \mathbf{P}(\mathbf{r}). \tag{B.17}$$

(Here the symbol U is used for the energy to avoid confusion with the electric field.) Evidently the external field tends to align the dipoles of the system, since this energy decreases as the number of dipoles pointing in the same direction as the field increases. The functional derivative of this yields

$$\delta U^{\mathrm{ext}} = -\mathbf{E}^{\mathrm{ext}}(\mathbf{r}) \cdot \delta \mathbf{P}(\mathbf{r}). \tag{B.18}$$

Hence $X = \mathbf{P}$ and $x = -\mathbf{E}^{\mathrm{ext}}$, and the entropy derivative at constant internal energy is

$$\left(\frac{\delta S}{\delta \mathbf{P}(\mathbf{r})} \right)_{U^{\mathrm{int}}} = \frac{-1}{T} \mathbf{E}(\mathbf{r}). \tag{B.19}$$

This is the field produced by an isolated system constrained to have a certain polarisation and energy. It is also the external field that would have to be applied to produce this polarisation in an equilibrated system.

When a paramagnetic sample is placed in a magnetic field $\mathbf{B}^{\mathrm{ext}}$ it becomes magnetised, with the magnetisation density $\mathbf{M}(\mathbf{r})$ pointing in the same direction as the field. The analysis is identical to that of a polarised dielectric in an electric field and one similarly obtains

$$\left(\frac{\delta S}{\delta \mathbf{M}(\mathbf{r})} \right)_{U^{\mathrm{int}}} = \frac{-1}{T} \mathbf{B}(\mathbf{r}). \tag{B.20}$$

B.2.1 Momentum

Linear additivity and conservation are the crux of the formalism for statistical thermodynamics. In classical mechanics there are just seven linear additive constants of the motion: the energy and the components of linear and angular momentum. Hence the momentum derivatives of the entropy are important quantities.

A system of mass M in uniform motion with velocity $\mathbf{v}$ has total momentum $\mathbf{P} = \sum_i \mathbf{p}_i = M\mathbf{v}$. The peculiar momentum is that measured in a reference frame moving with the system, $\tilde{\mathbf{p}}_i = \mathbf{p}_i - m_i \mathbf{v}$, so that $\tilde{\mathbf{P}} = 0$. Similarly, the peculiar kinetic energy is $\tilde{K} = \sum_i \tilde{p}_i^2/2m_i = K - Mv^2/2$, and the peculiar energy is $\tilde{E} = \tilde{K} + U$.

The temperature is the peculiar energy derivative of the entropy,

$$\frac{\partial S(\tilde{E}, N, V, \underline{P})}{\partial \tilde{E}} = \frac{1}{T}, \tag{B.21}$$

and since the number of configurations only depends upon the peculiar energy, holding the latter constant during momentum exchange fixes the entropy,

$$\frac{\partial S(\tilde{E}, N, V, \underline{P})}{\partial \underline{P}} = 0. \tag{B.22}$$

Now consider a second system with velocity $\mathbf{v}^{\mathrm{ext}}$ able to exchange momentum with the subsystem. From momentum conservation, $\mathrm{d}\underline{P}^{\mathrm{ext}} = -\mathrm{d}\underline{P}$, the change in the energy of the second system following momentum exchange is $\mathrm{d}E^{\mathrm{ext}} = \mathbf{v}^{\mathrm{ext}} \cdot \mathrm{d}\underline{P}^{\mathrm{ext}} = -\mathbf{v}^{\mathrm{ext}} \cdot \mathrm{d}\underline{P}$. Hence energy conservation gives $\mathrm{d}E = \mathbf{v}^{\mathrm{ext}} \cdot \mathrm{d}\underline{P}$. Also, the change in the peculiar energy of the subsystem is $\mathrm{d}\tilde{E} = \mathrm{d}(E - P^2/2M) = (\mathbf{v}^{\mathrm{ext}} - \mathbf{v}) \cdot \mathrm{d}\underline{P}$, whereas that of the second system is 0, $\mathrm{d}\tilde{E}^{\mathrm{ext}} = \mathrm{d}(E^{\mathrm{ext}} - (P^{\mathrm{ext}})^2/2M^{\mathrm{ext}}) = (-\mathbf{v}^{\mathrm{ext}} + \mathbf{v}^{\mathrm{ext}}) \cdot \mathrm{d}\underline{P} = 0$.

The change in the total entropy during momentum exchange at fixed total energy is

$$\begin{aligned} & \left(\frac{\partial S_{\mathrm{total}}(\tilde{E}, \underline{P}; \tilde{E}^{\mathrm{ext}}, \underline{P}^{\mathrm{ext}})}{\partial \underline{P}} \right)_{E^{\mathrm{total}}} \\ &= \left(\frac{\partial S(\tilde{E}, \underline{P})}{\partial \underline{P}} \right)_{E^{\mathrm{total}}} \\ &= \left(\frac{\partial S(\tilde{E}, \underline{P})}{\partial \tilde{E}} \right)_{\underline{P}} \left(\frac{\partial \tilde{E}}{\partial \underline{P}} \right)_{E^{\mathrm{total}}} + \left(\frac{\partial S(\tilde{E}, \underline{P})}{\partial \underline{P}} \right)_{\tilde{E}} \\ &= \frac{1}{T} \left(\mathbf{v}^{\mathrm{ext}} - \mathbf{v} \right) + 0. \end{aligned} \tag{B.23}$$

The first equality follows because the entropy of the second system doesn't change; as shown above, its peculiar energy remains fixed. This result shows that equilibrium occurs when the two systems move with the same velocity. It also follows that the momentum derivative of the entropy of the subsystem at constant energy is the negative of its centre of mass velocity,

$$\begin{aligned} \left(\frac{\partial S(\tilde{E}, \underline{P})}{\partial \underline{P}} \right)_E &= \left(\frac{\partial S(\tilde{E}, \underline{P})}{\partial \tilde{E}} \right)_{\underline{P}} \left(\frac{\partial \tilde{E}}{\partial \underline{P}} \right)_E + \left(\frac{\partial S(\tilde{E}, \underline{P})}{\partial \underline{P}} \right)_{\tilde{E}} \\ &= \frac{-\mathbf{v}}{T} + 0. \end{aligned} \tag{B.24}$$

A similar treatment of a system's macroscopic angular momentum $\underline{J}$ leads to the definition

$$\left(\frac{\partial S(\tilde{E}, \underline{J})}{\partial \underline{J}} \right)_E = \frac{-\underline{\omega}}{T}, \tag{B.25}$$

where $\underline{\omega} = \underline{J}/I$ is the rotational velocity and I is the moment of inertia of the system. One concludes that the entropy depends nontrivially on the microscopic internal energy, and only trivially on the macroscopic kinetic and rotational energies.

Appendix C

Mathematical Results

C.1 Notation and Context

It seems impossible to develop an unambiguous mathematical notation that is independent of the context in which it is used. In the text a function is generally denoted $f(x)$, where f is the function and x is the variable. Generally the argument is enclosed in parentheses, but on occasion, such as when dealing with a function of a function, brackets may be used for clarity, as in $f(g[x])$. However, brackets are also used to denote a functional, as in $f[g]$, which depends for its value not on the single point x, but on all the values of the function $g(x)$ on the appropriate interval. Parentheses (and brackets and braces) are also used to specify the hierarchy of arithmetical operations, and so, depending upon the context, the quantity $f(x+y)$ could denote either the sum of the products fx and fy or the function of the sum $x+y$.

The equality sign is used to denote both equality, as in $x = 2$, and to define a function, as in $f(x) = x^2$. The ambiguity in writing $f(x) = 2$ (does this define the constant function, or determine that $x = \sqrt{??}$) can only be resolved from the context. Occasionally the equivalence sign, $\beta \equiv 1/k_{\rm B}T$, is used to define a symbol that can be replaced by the string of symbols on the right-hand side wherever it occurs; as often as not, the equal sign is used instead.

Usually the argument of a function is merely a placeholder that assumes specific values; $f(x) = f(y)$ when $x = y$. Although it would be preferable to use different symbols to denote distinct functions, in order to prevent a proliferation of symbols, distinct functions are often in the text denoted by the same symbol; the arguments of the functions serve as well to distinguish them in these cases. For example, $S(N,V,E)$ is the entropy of an isolated system, whereas $S(N,V,T)$ is the total entropy of a subsystem in contact with a heat reservoir; these are not equal to each other even if E is numerically equal to T. It is arguable that it would be better to distinguish these by using subscripts, as in $S_{\rm T}(N,V,T)$, and this has been done when ambiguity is not precluded by the context. Because arguments are used as both variables and distinguishing marks, mixing up the order of the arguments in a function should not cause great confusion.

C.2 Partial Differentiation

In the case of partial differentiation, usually all arguments of the function are held fixed except that used for the derivative. Often this is made clear by showing the function with all its arguments in the numerator, or by placing the fixed variables as subscripts,

$$\frac{\partial S(E,N,V)}{\partial E}, \text{ or } \left(\frac{\partial S}{\partial E}\right)_{N,V}. \tag{C.1}$$

The latter procedure is always followed when it is not the arguments of the function themselves that are fixed, but rather some combination thereof.

Much of thermodynamics is an exercise in partial differentiation, and here is summarised some of the main relationships. The partial derivative of a function of two variables, $f(x,y)$, with respect to x with y held fixed is

$$\left(\frac{\partial f}{\partial x}\right)_y = \lim_{x_1 \to x_2} \frac{f(x_1,y_1) - f(x_2,y_1)}{x_1 - x_2}. \tag{C.2}$$

Sometimes one may write $\partial f(x,y)/\partial x$, or $f_x(x,y)$, or $f'(x)$, when no ambiguity exists. Because the derivative is simply the ratio of two differences, one has

$$\left(\frac{\partial f}{\partial x}\right)_y = \left(\frac{\partial x}{\partial f}\right)_y^{-1}. \tag{C.3}$$

This assumes $f(x,y) \Leftrightarrow x(f,y)$. The total differential is

$$df = \left(\frac{\partial f}{\partial x}\right)_y dx + \left(\frac{\partial f}{\partial y}\right)_x dy. \tag{C.4}$$

Setting $df = 0$ and rearranging yields

$$\left(\frac{\partial x}{\partial y}\right)_f = -\left(\frac{\partial f}{\partial y}\right)_x \left(\frac{\partial x}{\partial f}\right)_y. \tag{C.5}$$

The derivative holding some function $z(x,y)$ constant is

$$\left(\frac{\partial f}{\partial x}\right)_z = \left(\frac{\partial f}{\partial x}\right)_y + \left(\frac{\partial f}{\partial y}\right)_x \left(\frac{\partial y}{\partial x}\right)_z. \tag{C.6}$$

For the proof one simply chooses two nearby points such that $z(x_1,y_1) = z(x_2,y_2)$, and takes the limit of

$$\begin{aligned} &\frac{f(x_1,y_1) - f(x_2,y_2)}{x_1 - x_2} \\ &= \frac{f(x_1,y_1) - f(x_2,y_1)}{x_1 - x_2} + \frac{f(x_2,y_1) - f(x_2,y_2)}{y_1 - y_2}\frac{y_1 - y_2}{x_1 - x_2}. \end{aligned} \tag{C.7}$$

Finally,

$$\left(\frac{\partial f}{\partial z}\right)_y = \left(\frac{\partial f}{\partial x}\right)_y \left(\frac{\partial x}{\partial z}\right)_y, \tag{C.8}$$

as follows from the limit of

$$\frac{f(x_1, y_1) - f(x_2, y_1)}{z(x_1, y_1) - z(x_2, y_1)} = \frac{f(x_1, y_1) - f(x_2, y_1)}{x_1 - x_2} \frac{x_1 - x_2}{z(x_1, y_1) - z(x_2, y_1)}. \tag{C.9}$$

C.3 Asymptotic Analysis

For asymptotic analysis three notations are used. The tilde symbol is used for the most precise characterisation of the asymptote,

$$f(x) \sim ax^{-n},\ x \to \infty,\ \text{means} \lim_{x\to\infty} f(x) = ax^{-n}. \tag{C.10}$$

This says that for any small positive number ϵ there exists a large number $X(\epsilon)$ such that $|f(x) - ax^{-n}| < \epsilon$ for all $x > X(\epsilon)$. The so-called 'big O' notation is used to signify the functional form of the asymptote,

$$f(x) = \mathcal{O}(x^{-n}),\ x \to \infty,\ \text{means} \lim_{x\to\infty} \frac{f(x)}{x^{-n}} = M, \tag{C.11}$$

where M is a bounded constant whose (unknown) value is normally nonzero. Finally, the 'little o' notation is used to set a bound on the asymptotic behaviour,

$$f(x) = o(x^{-m}),\ x \to \infty,\ \text{means} \lim_{x\to\infty} \frac{f(x)}{x^{-m}} = 0, \tag{C.12}$$

which is to say that the function decays faster than x^{-m}. Quite often a series is terminated by ellipsis, ..., which is the same as saying that the neglected terms are little o of the last exhibited term, or big O of what is obviously the next term in the sequence.

At the simplest level asymptotic expansions can be derived as simple Taylor expansions,

$$f(x) \sim f(0) + xf'(0) + \frac{x^2}{2!}f''(0) + \ldots,\ x \to 0. \tag{C.13}$$

One of the more well-known asymptotic expansions is Stirling's approximation for the factorial,

$$\ln n! \sim n \ln n - n + \ln\sqrt{2\pi n} + \mathcal{O}n^{-1},\ n \to \infty. \tag{C.14}$$

Perhaps the most common asymptotic expansions are based upon the binomial expansion,

$$(1+x)^n \sim 1 + nx + \frac{n(n-1)}{2!}x^2 + \ldots,\ x \to 0. \tag{C.15}$$

This terminates for positive integral values of n, but not more generally. The so-called harmonic series follows from this,

$$\frac{1}{1-x} \sim 1 + x + x^2 + \ldots, \; x \to 0. \tag{C.16}$$

Obviously $x \to 0$ is the same as $x^{-1} \to \infty$; other limit points may be used. Asymptotic expansions are often differentiated and integrated fearlessly.

The distinction between a convergent series and an asymptotic expansion is that for a fixed value of the parameter x, convergent series are increasingly accurate as the number of terms in the series is increased, whereas the series representing an asymptotic expansion often diverges. An asymptotic expansion becomes more accurate for a fixed number of terms as the parameter approaches the limit point. In many cases increasing the number of terms in an asymptotic expansion means that the parameter must be taken closer to its limit point to achieve the same level of accuracy. For example, the asymptotic expansion represented by the harmonic series for fixed $|x| > 1$ is not a convergent series. However, for any $|x| < 1$ it is convergent. A more striking example is provided by the complementary error function, which has asymptotic expansion,[1]

$$\sqrt{\pi} z e^{z^2} \operatorname{erfc} z \sim 1 - \frac{1}{2z^2} + \frac{3!}{2!(2z^2)^2} - \ldots + \frac{(2m)!}{(-2)^m m!(2z^2)^m}, \; z \to \infty. \tag{C.17}$$

Clearly, for fixed z the series would diverge, but for fixed number of terms m, the tail becomes increasingly small as z increases.

For the case of integration one might have a function $f(x, y)$ that becomes sharply peaked about some x_0 in the asymptotic limit, $y \to \infty$. In this case one can perform a Taylor expansion of a more slowly varying function about the location of the peak and evaluate the integral,

$$\begin{aligned} I(y) &= \int_{-\infty}^{\infty} \mathrm{d}x \, g(x) f(x, y) \\ &\sim \int_{-\infty}^{\infty} \mathrm{d}x \, [g(x_0) + (x - x_0) g'(x_0) + \ldots] \, f(x, y), \; y \to \infty \\ &= g(x_0) I_0(y) + g'(x_0) I_1(y) + \ldots. \end{aligned} \tag{C.18}$$

One tends to find that the moment integrals are of the form $I_m = \mathcal{O} y^{-m}$, or similar, so that this procedure generates an asymptotic expansion. Successive integration by parts is another fecund source of expansions. The asymptotic evaluation of convolution integrals was discussed in §9.6. There are of course many other and more powerful asymptotic techniques.[2]

[1] M. Abramowitz and I. A. Stegun, *Handbook of Mathematical Functions*, Dover, New York, 1965.

[2] N. G. de Bruijn, *Asymptotic Methods in Analysis*, Dover, New York, 1981.

Index